42

1993

Contemporary Ergonomics 1993

**Proceedings of the Ergonomics Society's
1993 Annual Conference
Edinburgh, Scotland, 13-16 April 1993**

'ERGONOMICS AND ENERGY'

Edited by

E.J. Lovesey

Lovesey Associates
Ash Vale

Taylor & Francis
London ● *Washington DC*
1993

UK   Taylor & Francis Ltd, 4 John St, London WC1N 2ET

USA   Taylor & Francis Inc., 1900 Frost Road, Suite 101, Bristol,
            PA 19007

A catalogue record for this book is available from the
British Library

ISBN 0-7484-0070-2

Printed in Great Britain by Burgess Science Press, Basingstoke,
on paper which has a specified pH value on final paper
manufacture of not less than 7.5 and therefore 'acid free'.

Contents

MILITARY ERGONOMICS   1

Information coding in a surface naval command system   2
R.W. McLeod and A.F. Reid
Physical fitness and job requirements: Training aspects   5
M.F. Haisman and A. Duggan
Direction-of-motion sterotypes for weapon aiming   11
K.M. Wayman
A comparison of a linear and non-linear control law   17
D.J. Knowles
The assessment of a military eye protection system   22
S.J. Skinner
Reclined sitting postures   28
M. Thody and V.H. Gregg
An industrial view of MANPRINT   34
M.K. Goom
MANPRINT/Ergonomics in user friendly and biodegradable
food service packaging   40
L.E. Symington, T. Malafi, J. Kalick and M. Devine
Predictive workload analysis – RN EH 101 Helicopter   45
I.S. MacLeod, K. Biggin, J. Romans and K. Kirby

INFORMATION PRESENTATION AND RECOGNITION   51

A comparison of three map formats   52
Y. Shek, P. Gosling and P. Barber
Suspect identification by features   58
E.S. Lee, M. Densmore and T. Whalen
Efficient design of mammographic multiviewers   64
C.G. Blair-Ford, A.G. Gale, S.G. Cobb and C.M. Haslegrave

DESIGN METHODS   68

Structured notations for Human Factors specification   69
K.Y. Lim and J.B. Long
Developing user interface design tools: An analysis of
interface design practice   79
M.M. Bekker and A.P.O.S. Vermeeren
The application of rapid prototyping in the re-design of a
user interface   85
H.C.M. Hoonhout
Reliability in Ergonomics/Human Factors   91
H. Kanis
Automatic evaluation in Human Computer Interactions   97
S. Balbo and J. Coutaz
Selecting early user interface methods for use in practical
design situations   103
A.P.O.S. Vermeeren and M.M. Bekker

DESIGN METHODS – EXAMPLES 110

Development of a desk top publishing system 111
T. Wilson
User involved redesign of a clock radio 117
C.C.C. Voute, K. Kanis and A.H. Marinissen
Classification of domestic tasks to support device
procurement and design 123
V.S. Jenkins, K.Y. Lim and J.B. Long
Energy devoted to the ergonomics of product handling 129
W.P. Mossel
Feasibility and usefulness of involving users early in the
interface design process 135
A.P.O.S. Vermeeren and R. Kolli

COMPUTER ASSISTED TASKS 141

A computer based tool for accessing anthropometric
databases 142
N.I. Beagley, R.A. Haslam and K.C. Parsons
CRAMP – The development of a graphical user interface 148
M. Bontoft and J. Dillon
Using task networks to model error correction dialogues for
ASR 152
K.S. Hone and C. Baber

AUDITORY WARNINGS 158

Alarms in a coronary care unit 159
N. Stanton
Using psychophysics to predict perceived urgency 164
J. Edworthy and E.J. Hellier

HEALTH AND SAFETY 170

Human dependent failures: A schema and taxonomy of
behaviour 171
P.D. Hollywell
Risk homeostasis in a non-transportational domain 178
T.W. Hoyes and C. Baber
Passenger containment on amusement devices 184
J.A. Jackson
How and why child safety restraints in cars are misused 190
J.A. Rainford, M. Page and J.M. Porter
Safe surface temperatures of domestic products 196
K.C. Parsons
An ergonomics appraisal of the Piper Alpha disaster 202
W.H. Gibson and E.D. Megaw
Shipboard accidents in the Royal Navy 208
P.B. Jenkins

ERGONOMICS IN THE ENERGY INDUSTRY 214

An international survey of Human Factors involvement in
the nuclear industry 215
L. Herman
The development of an interactive guideline document for
control room design 221
D.L. Welch, C.C. Baker, T.M. Granda, P.J. Vingelis, J.
O'Hara and W. Brown

HUMAN ERROR AND PROCESS SAFETY 227

Human error incidents in electricity supply 228
A.I. Glendon
The development of a methodology to study maintenance
safety in chemical and nuclear power plants 234
J. Carthey and K. Rea
Ergonomics in nuclear power and process safety 240
D.L. Welch

MEMORY AND COGNITION 246

Does real time memory processing capacity decline with
age? 247
N. Morris and I. Lamb
Domestic energy management and bill design 253
D.I. Williams, J. Young and C.M. Crawshaw

VDUs IN PROCESS CONTROL 259

Screen search efficiency and spatial frequency 260
D. Scott
Evaluation of VDU sequence-based mimic displays for
process control 266
P.J. Thelwell, B. Kirwan and J. Reed
OPERA – The Office performance evaluation and rating aid 272
N. Coleman
The use of colour in CRT displays for the process industry 278
H.J.G. Zwaga
Workload in Air Traffic Control communication 284
H.B. Nijhuis

MANUAL HANDLING AND POSTURE 290

A comparative evaluation of arm rests 291
J.M. Porter, A.C. Smith and D.E. Gyi
Musculoskeleton disorders in operators of weaving
machinery 297
N.J. Wilkinson and C.M. Haslegrave
Weight gain and lifting during pregnancy 303
S. Sinnerton, K. Birch, T. Reilly and I.R. McFadyen
Menstrual cycle effects on isometric and dynamic lifting 308
K. Birch, I.R. McFadyen, T. Reilly and S. Sinnerton

SPORT ERGONOMICS 314

A physiological evaluation of ski simulators 315
T. Reilly, E. Kirton, E. McGrath and S. Coulthard
Angle specific isokinetic talocrural torque ratios 321
C.M. Graham and G. Garbutt
Assessment of muscle strength assymetry in soccer players 327
N.E. Fowler and T. Reilly

PHYSIOLOGICAL STRESS 333

The effect on working posture of a re-designed industrial
sewing machine 334
L. Guangyan and C.M. Haslegrave
Physiological measurement of stress in computer based
work 340
M. Toms and A. Masih

A postural stress analysis system for industry   346
A. Genaidy, L. Guo and R. Eckhart
A manually driven flywheel motor operates wood turning
process   352
J.P. Modak and A.R. Bapat
Oxygen cost of treadmill and road running   358
A. Duggan and J.F. Patton

SITTING, STANDING AND WORKSPACE   364

Development of a 3D approach to investigate the
ergonomics of standing   365
R.S. Whistance, L.P. Adams and R.S. Bridger
A functional sit-stand seat   370
H. Gregg and E.N. Corlett
Fishing vessel wheelhouse design – a case study   374
S. Mills

DRIVERS AND DRIVING   380

Drivers at traffic signals: A qualitative analysis   381
S.A. Robertson
A study of co-operation extended to trapped merging drivers 387
T. Wilson and M. Godin
Driving research and the Iowa driving simulator   392
J. Buck, J. Stoner, J. Bloomfield and T. Plocher
The benefits of 'Pre-information' in route guidance systems
design for vehicles   397
G.E. Burnett and A.M. Parkes
Carphone use and motorway driving   403
A.M. Parkes, S.H. Fairclough and M.C. Ashby

TELECOMMUNICATIONS   409

Human Factors guidelines for IBC designers   410
A.M. Clarke and G. Allison
Usability testing in the 'Real World'. Evaluating a multi-
function phone system   416
P.W. Jordan and K.C. Kerr
The role of experience in performance on different types of
telephone memory retrieval tasks   422
G.J. Gelderblom and A. Bremner

POSTER PRESENTATIONS   428

Practical experiences with consumer products, users and
prototyping   429
G.I. Johnson and E.P.G. van Vianen
Spatial operational sequence diagrams in usability
investigation   435
G.I. Johnson
Some ergonomic consequences of playing field hockey   441
T. Reilly and J. Temple
An ergonomic evaluation of boardsailing harnesses   445
T. Reilly, E. Brymer and M.S. Townend
Methods for user interface performance measurement   451
P.W. Jordan
A checklist for the assessment of display screen equipment 457
M. Anderson and E.D. Megaw

User stress in automatic speech recognition 463
R. Graham and C. Baber
Driver status monitoring: Can In-vehicle I.T. detect the
impaired driver? 469
S.H. Fairclough and S.J. Hirst
Human Factors checklist for service evaluation 475
A.M. Clarke, G. Allison and T. Hewson
Successful videoconferencing 478
A.M. Clarke and S.M. Pomfrett
Developing a data collation tool for MANPRINT in the U.K. 481
W.I. Hamilton and H. Walters

KEYNOTE ADDRESS 489

Applying Ergonomics in Industry: Some lessons from the
mining industry 490
G.C. Simpson

AUTHOR INDEX 504

SUBJECT INDEX 507

Military ergonomics

INFORMATION CODING IN A SURFACE NAVAL COMMAND SYSTEM

Ronald W. McLeod Ph.D.

Andrew F. Reid

Cognitech Ltd

92, Commerce Street
GLASGOW
G5 8DG

BAeSEMA Ltd
CMD Division
40-44 Coombe Road
New Malden
Surrey KT3 4QF

INTRODUCTION

The Surface Ship Command System (SSCS) being developed by BAeSEMA Ltd for the Type 23 Frigate is believed to be the first system procured by the Royal Navy, and may be the first procured by the MoD in the UK, which incorporates a Human Factors programme leading to acceptance testing of usability as an integral part of the system development.

SSCS provides its' operators with a number of different types of display each providing different views on the world and on the data in the system. Some of these displays are graphics based while some are mainly textual. This paper is concerned with the use of coding in the graphical Plan Position Indicator (PPI) display which for most of the Command team users provides their principal view on the outside world. The PPI display is a bird's eye representation of the outside world showing the objects (such as surface, sub-surface and air vehicles, land, operational areas, and reference points such as oil rigs and mines, etc) which are known to the system to exist in the world. The paper identifies the requirements for the use of coding in the Human Computer Interface (HCI), and describes the approach to developing and testing the design solutions.

Use of Coding

In SSCS information coding is used in 2 ways; i) as a means of helping the user to find information quickly and ii) as a means of conveying additional information about an object without using additional screen space. Umbers and Collier (1990) describe the use of coding techniques in process control applications. The principles discussed by Umbers and Collier are to a very large extent generalisable across many applications. However, the detailed task and usability requirements, and particularly conventions and user expectations are somewhat different in the Naval context from those which apply in the process control industry.

REQUIREMENTS
Task Requirements

In the Command System context, information coding is expected to support search and identification tasks as well as maintenance of situation awareness by supporting perceptual organisation of displayed information. The actual way coding supports tasks will depend on the precise tasks being performed and the role of each Command Team operator. For example, coding will be of little use to the picture compilation team in classifying a new object. Once the object has been classified its' colour and shape should help other operators to identify the new vehicle quickly and incorporate it into their current model of the external world.

SSCS operators are expected to be able to extract five main items of information directly from the representation of an object on the PPI display ; i) position, ii) type of object (if known), and therefore iii) its' environment, iv) whether the object is currently held by Own ship sensors, and v) whether the object is currently engaged by any weapon. It should also be readily apparent from its' appearance whether the object has any special significance for the Command team.

Coding Requirements
    The principal coding techniques used in SSCS, which were originally specified by the MoD, are; Shape, Brightness, Colour, Blink. (Position coding on a PPI display is a direct mapping onto the outside world and will not be considered further here).

    Shape: The shape of a symbol is used to indicate the type of object (e.g. a ship) and therefore, the environment in which it exists (Air, Surface or Sub-surface). There are approximately 70 different objects to be identified by shape coding.

    Brightness: The brightness of an object indicates whether the object is currently held by the Own Ship's sensors, or whether the symbol represents the position of the object the last time it was held by the sensors (known as "Dead Reckoned"). Brightness is used to code only two states (Normal or Dead Reckoned). In Royal Naval usage, a Dead Reckoned track is conventionally associated with a dim brightness state.

    Colour: Colour is used to code the identity of the object. There are four standard identities (Hostile, Unknown, Neutral, Friendly). The system also requires four background colours to be available (one standard background and three different backgrounds for operational areas).   As well as its' use in coding, colour is used as a means of structuring and organising the display.

    Blink: On the PPI display, Blink coding is used to identify two special categories of tracks, one having higher attention getting properties than the other.

Usability Requirements
    The Human Factors programme for SSCS which was a mandatory development activity, includes a programme of formal testing to demonstrate that the system can be readily used by its' intended operators.   Usability Requirements and the formal test programme are in line with MoD guidance provided during the Project Definition phase.   Performance requirements for information coding techniques are based on Discrimination, Identification, Meaning and Legibility. Operators are required to be able to accurately discriminate and identify around 70 symbols in four colours at two levels of brightness against four background colours. Operators must also be able to learn and retain the codes without undue training. Command team members will typically perform a watch of around 6 hours, although on occasion watches could be considerable longer. The means of presenting information should therefore not be irritating or cause undue visual fatigue or discomfort with periods of viewing of up to 6 hours.

DESIGN PROCESS
    The detailed design work to develop the shape, brightness, colour and blink coding was conducted in interactive sessions with User representatives based around a

commercially available, workstation-based rapid prototyping facility.   The workstation included a display which was representative of the production display. A prototyping room was established in which the light sources and the luminance on the workstation were representative of those measured at realistic positions in a Type 23 frigate.

Symbology was developed based on both the Royal Navy's previous use of symbols in Command systems, and proposed NATO symbology (NATO, 1989). The starting point for developing the use of colour was based on guidelines produced by the Captain Naval Operational Command Systems (CNOCS, 1988).   Colours were specified in UIE co-ordinates and reviewed to ensure that the production display could meet the colour standards required.

At various stages during the prototyping sessions, informal tests of discriminability and legibility were conducted using subjects who had no previous exposure to the colours. These informal tests provided more objective input to the design process than simply the opinions of the design team.

USABILITY TESTING

A series of formal trials were conducted as the basis for proving the usability of the information coding techniques developed. Trials were carried out very early in the development process and form the first step in the overall usability test programme for SSCS.   In later stages of development, trials will be conducted using production equipment and trained crews performing simulated operations in real conditions at sea. The approach adopted balanced the need to address usability issues in a formal way early in development as a basis for agreeing design details, against limitations imposed by the unrealistic conditions of the trials.

Trials were carried out on the prototyping workstation and focussed on the requirements for Discrimination, Identification, Meaning and Legibility of information. Usability criteria were established in terms of both objective measures of operator performance and subjective expressions of opinion. Trials were designed to balance the need to address all combinations of codes in a manner which made the most efficient use of resources within the constraints of a limited budget.

The test results demonstrated that trained SSCS operators should be able to discriminate 68 symbols in four colours at two levels of brightness against four background colours with an error rate of less than 1%.

CONCLUSIONS

Basing the HCI design process for BAeSEMA's Type 23 frigate Command System around a rapid prototyping facility is providing a significant capability to address Human Factors issues early in the development process. This approach has supported a very high level of user involvement in designing and testing information coding techniques.

REFERENCES
Umbers, I. G. and Collier G. D., 1990, Coding techniques for process plant VDU
        formats. Applied Ergonomics 21(3), 187 - 198
NATO (1989) Display Symbology and Colour for NATO Maritime Units. Draft
        STANAG 4xxx (Edition 1)
CNOCS (1988) User Interface Guidelines CS4441/50/88

PHYSICAL FITNESS AND JOB REQUIREMENTS: TRAINING ASPECTS

M F HAISMAN* and A DUGGAN

Army Personnel Research Establishment
Farnborough, Hants
GU14 6TD
*Pinons Lodge, Dene Close
Lower Bourne, Farnham, Surrey
GU10 3PP

One approach to the need to harmonise job demands with the
worker's capabilities is by the use of physical fitness
training. The potential benefits of various types of physical
training have been examined, benefits which may lead to reduced
musculoskeletal injuries, lower fatigue levels, improved job
performance and greater job satisfaction. Conversely the
potential disadvantages have also been considered,for example,
over-use injuries sustained in training. On balance there may
be benefits from physical training which is tailored to
specific job demands especially if used in combination with
other approaches such as placement and job modification, but
the results of any training programme should be evaluated.

INTRODUCTION

Large organizations such as the armed forces, emergency services, and
public utilities have a continuing requirement to ensure a match between
individuals in the work force and the physical demands of their jobs; this
matching has been associated with benefits including: a reduction in work-
related injuries and illnesses, and improvements in job performance and job
satisfaction(Snook,1987). Such an ideal situation is not easy to achieve
and Pedersen et al.(1989) found up to 38% of a sample of National Guard
mechanics mismatched. New requirements such as EC Directives may place
additional emphasis on this matching process. Furthermore the need to
provide Equal Opportunities in employment, and to recognise Human Rights
Codes (Fraser,1992), means that employers must be aware of the actual
requirements of each job. Physical fitness and strength were two factors,
cited by NIOSH(1981) as relevant in the context of defining safe loads, and
weak or unfit workers were recommended not to be exposed to the demands of
lifting heavy loads. Genaidy et al.(1992) reviewed the role of physical
training in increasing work tolerance limits for those engaged in Manual
Materials Handling(MMH) and they proposed a staged intervention process of
physical training to control musculoskeletal disorders.

The two EC directives which will have most impact on the physical
aspects of work are:
a. The Manual Handling Operations Regulations(HSE,1992) places duties on
employers to take steps to reduce the risks of injury. Factors to be
considered include: the task, the load, the working environment and
individual capabilities; it is recognised that an individual's state of

health, fitness and strength can significantly affect the ability to perform the task safely.

b. The Personal Protective Equipment(PPE) Proposals(HSC,1991). The use of PPE may well increase the physical demands of a job because of factors such as: the weight of the PPE, the encumbrance and stiffness, disturbance of heat dissipation, and breathing resistance of respiratory equipment. Improving physical fitness by training may alleviate the effects of wearing PPE.

Military organizations have a long experience with physical training. Activities such as marching with loads and running have traditionally featured prominently in training programmes, but such practices have been questioned with reports that excessive or inappropriate training can lead to medical problems especially with respect to the lower limbs(Jones 1983, Knapik et al.,1992). The lessons learned from the military experience may be relevant to civilian training applications.

Kroemer(1992) in reviewing many studies concerned with personnel training for safer MMH,considered that the issue of training for prevention of back injuries in MMH was confused, at best. The scope of this paper is wider, and includes the effects of physical training on the performance of all job-related physical activity, and excludes training programmes aimed to educate the worker in MMH techniques.

POTENTIAL BENEFITS OF TRAINING
Physiological and psychological changes

Physiological changes in maximum aerobic power, strength and muscle endurance, and onset of blood lactate accumulation (OBLA) with training have been well researched; endurance training also leads to partial acclimatisation to hot environments by improving heat dissipation (Wells et al. 1980). In contrast much less information is available on psychological changes. Persons who have adopted regular exercise into their life-style testify to the psychological benefits in terms of feelings of well-being, positive attitude, alleviation of stress and anxiety. It is not clear however that such benefits are found in those required to take exercise as part of a prescribed programme. From the changes which are known to take place as a result of training an improvement in job performance could be anticipated.

Training and job performance

The maximal acceptable weight lifted increased after training for muscular strength and endurance, flexibility, and cardiorespiratory endurance(Asfour et al.1984). Marcinik et al.(1987) also found an increase in performance in one of three shipboard tasks, and increased performance on the majority of fitness measures following a circuit weight training programme in naval personnel. Bernacki and Baun(1984) found a positive relationship, but probably not causal, between participation in corporate fitness programmes and above average job performance. Cox et al.(1981) showed that employees adhering to fitness programmes demonstrated lower turnover and reduced absenteeism.

Reduction of work-related injuries

Keyserling et al.(1980) found that the medical incidence rate of employees selected using isometric strength tests was approximately one-third of those selected using traditional medical criteria, but Troup et al.(1987) found that no tests of manual working capacity, or combination of tests, served reliably to identify new cases of low back pain. Chaffin et al.(1978) reported that musculoskeletal injuries were up to three times greater in workers who were over-stressed by the physical demands of the job compared with those who were under-stressed; Cady et al.(1979) showed that the fittest firefighters had fewer back injuries. Liles(1986) observed that above a critical value for Job Severity Index(JSI), based on the ratio

of the weight required to be lifted to worker capacity for lifting, the
risk of back injuries increases. Training of the abdominal muscles had no
effect on the Intra-Abdominal Pressure(IAP) developed during a sub-maximal
lifting test(Legg 1981), hence if such training does reduce the risk of
back injury, as claimed by Astrand and Rodahl(1986) it is not because of a
reduction in IAP. Genaidy et al.(1992) in their review were not able to
conclude that physical training intervention can be used to reduce
significantly the incidence of musculoskeletal disorders resulting from MMH
activities. Harwood et al.(1990) found that recruits who suffered lower
limb injuries in training had a lower knee extensor strength compared with
those with no injuries.

Reduction in fatigue

Static muscular endurance is partly a function of the proportion of
maximum voluntary contraction(MVC)utilised (Monod and Scherrer,1965), so
that improvements in MVC after training should improve endurance at a given
load. Similarly, training to enhance maximum aerobic power should allow
higher levels of work to be undertaken before fatigue effects are
encountered. Astrand(1967) showed that building workers spontaneously
chose a work load corresponding to about 40% of the individual's maximum
aerobic power,whereas fit young soldiers self-paced at 30-40%(Myles et
al.,1979). Legg and Myles(1981) found that a workload equivalent to only
21% aerobic power was subjectively acceptable for a repetitive lifting task
of 8 hours duration,possibly because of the static work component of the
task. Harma et al.,(1988) reported that moderate physical training
increased the physical fitness of female shift workers, and reduced work
dependent fatigue and musculoskeletal symptoms. Fatigued workers may be
more prone to accidents;Pheasant (1991) cites general unfitness and fatigue
as possible causes for vulnerability of the back. Long term health effects,
such as coronary heart disease risk factors, have been shown to be improved
in association with higher levels of aerobic fitness(Patton et al.,1986).

DETRIMENTAL EFFECTS OF TRAINING

Physical training is associated with risks of a variety of medical
problems depending upon the age groups of the trainees and the screening
and supervisory procedures employed (NATO,1986),and for example, strenuous
exercise even in temperate climates, can lead to heat exhaustion. A variety
of over-use injuries to the lower limbs have been encountered as a result
of excessive or inappropriate training, and training in boots has been
associated with a high injury rate(Harwood et al.,1990). Military personnel
undergoing prolonged and arduous training suffer from infections and lower
limb injuries (Riddell,1990) and recruits show an increased risk of stress
induced immunosuppression(Lewis et al.,1991). Over-training by athletes can
be associated with under-performance, disturbed sleeping and eating
patterns, depression, and lack of concentration.

EFFECTIVENESS OF TRAINING PROGRAMMES

Practical considerations include: the extent of the benefit derived,
how any benefit may be maintained, how much it can bridge any gap in
physical capability between the genders and counteract the effects of
aging, and finally the degree of specificity of the mode of training
exercise that is necessary.

Intensity, frequency and duration of training.

It is recognised that training must be demanding in order to make a
worthwhile improvement in physical fitness; for example to improve maximal
aerobic power subjects should exercise at 60 - 80 % maximum heart rate, on
3-5 occasions per week, and each session to be of 20 to 60 minutes duration
over a two month programme, furthermore the increment of aerobic power
obtained is generally greater with low initial levels. Increases in

strength can be greater and more rapid than aerobic power. Military
training of initially fit individuals can be disappointing. Patton et
al.(1980) found a drop in maximum aerobic power for those above 55
ml/kg.min and no change for those in the range 49 to 55 ml/kg.min;
similarly Haisman(1971) reported that only Physical Training Instructors
initially below the mean improved. Marcinik et al.(1985) concluded that the
overall fitness level of the average navy recruit was not significantly
altered by the standard physical training programme. Duggan and
Wansbrough(1992) found that recruits who were initially fit did not improve
in predicted maximum aerobic power, although the 2.4km run time improved,
and strength decreased. The actual changes induced are linked to the
problem of obtaining the training stimulus required by an individual in a
group programme. A maintenance programme may be required to maintain a
particular level of fitness depending on the actual physical demands of the
job.

Gender effects

A number of studies have examined gender differences in physical
performance, differences associated with the lower strength, stature and
maximum aerobic power and increased body fat of females (NATO 1986). It
appears that training can narrow this gender handicap from 22% difference
on maximum aerobic power to 18% (Daniels et al,1979), but Daniels et al,
(1982) found that females lost most of the initial improvement in aerobic
fitness over two years training and the gender gap was not significantly
narrowed.

Specificity of training

A balance must be sought on the desired mix of aerobic and strength
components of the training programme. Hickson(1980) showed that strength
training alone does not produce an increase in maximum aerobic power
whereas a combination of strength and endurance training is effective,
conversely endurance training did not improve muscle strength. Even though
combined endurance and strength programmes will improve tolerance to
loading on the whole body over more than a few minutes, high loads of brief
duration may still overload specific muscle groups. Pedersen et al.(1989)
proposed work hardening programmes which selectively strengthen critical
muscles used in particular tasks.

Training to meet a required standard

McCaig (1981) showed that trainees at a physical development course
failed fitness assessment tests because of excess body fat and upper body
muscular weakness; he recommended that candidates of weights within 20%
above the maximum and 10% below the minimum desirable weight for height
were more likely to complete the training.

TRAINING, PLACEMENT AND JOB MODIFICATION

Physical training may have beneficial effects on job performance but
training programmes should be formulated only after an analysis of physical
job demands. The results should be evaluated and appropriate adjustments
made either to the training and/or to task design, until the trainees can
meet all physical demands required of them. The use of physical training to
improve or maintain a person's suitability for a job can appear
unattractive because of potential costs and the uncertainty of benefits,
nevertheless it may offer an effective approach in some circumstances,
especially if used in combination with other approaches such as placement
and job modification.

REFERENCES
Asfour,S.S.,Ayoub,M.M. and Mital,A.,1984,Effects of an endurance and
    strength training programme on lifting capability of males,
    Ergonomics,27,4,435-442.

Astrand,I.,1967,Degree of strain during building work as related to individual aerobic work capacity,Ergonomics,10,3,293-303.

Astrand,P-O. and Rodahl,K.,1986,Textbook of Work Physiology, 3rd edn (New York: McGraw-Hill p.291).

Bernacki,E.J.and Baun,W.B.,1984,The relationship of job performance to exercise adherence in a corporate fitness program, Journal of Occupational Medicine,26,7,529-531.

Cady,L.D.,Bischoff,D.,O'Connell,E.,Thomas,P.C. and Allan,J.,1979,Strength and fitness and subsequent back injuries in fire fighters,Journal of Occupational Medicine,21,269-272.

Chaffin,D.B,Herrin,G.D. and Keyserling,W.M.,1978,Pre-employment strength testing:an updated position,Journal of Occupational Medicine,20,403-408.

Cox,M.,Shephard,R.J. and Corey,P.,1981,Influence of an employee fitness programme upon fitness,productivity and absenteeism, Ergonomics, 24,10,795-806.

Daniels,W.L.,Kowal,D.M.,Vogel,J.A. and Stauffer,R.M.,1979,Physiological effects of a military training program on male and female cadets. Aviation,Space and Environmental Medicine 50(6):562-566.

Daniels,W.L.,Wright,J.E.,Sharp,D.S.,Kowal,D.M.,Mello,R.P. and Stauffer, R.S.,1982, The effect of two years training on aerobic power and muscle strength in male and female cadets. Aviation,Space and Environmental Medicine,53,(2):117-121.

Duggan,A. and Wansbrough,P.R.I.,1992, Physical fitness tests for selecting and monitoring the training of recruits to the Parachute Regiment,Army Personnel Research Establishment Report 92R029.

Fraser,T.M., 1992, Fitness for Work.The Role of Physical Demands Analysis and Physical Capacity assessment (London: Taylor and Francis).

Genaidy,A.M.,Karwowski,W.,Guo,L.,Hidalgo,J. and Garbutt,G.,1992, Physical training:a tool for increasing work tolerance limits of employees engaged in manual handling tasks,Ergonomics,35,9,1081-1102.

Haisman,M.F.,1971, Assessment of the exercise capacity of young men, Ergonomics,14,4,449-456.

Harma,M.I.,Ilmarinen,J.,Knauth,P.,Rutenfranz,J. and Hanninen,O., 1988,Physical training intervention in female shift workers:I.The effects of intervention on fitness,fatigue,sleep, and psychosomatic symptoms, Ergonomics, 31,1,39-50.

Harwood,A.G.,Box,C.J. and Freeland,W.A.,1990,A survey of overuse lower limb injuries in British Army recruits. In Proceedings of the International Conference on Environmental Ergonomics-IV, pp.60-61.

HSE,1992, Manual Handling, Manual Handling Operations Regulations 1992 Health and Safety Executive(London: HMSO).

HSC,1991,Personal Protective Equipment at Work.Proposals for Regulations and Guidance, Health and Safety Executive Consultative Document.

Hickson,R.C.,1980, Interference of strength development by simultaneously training for strength and endurance. European Journal of Applied Physiology,45,255-263.

Jones,B.H., 1983, Overuse injuries of the lower extremities associated with marching, jogging and running: a review. Military Medicine, 148, 783-787.

Keyserling,W.M.,Herrin,G.D. and Chaffin,D.B.,1980,Isometric strength testing as a means of controlling medical incidents on strenuous jobs, Journal of Occupational Medicine,22,5,332-336.

Knapik,J., Vogel,J.A., Reynolds,K., Jones,B. and Staab,J.,1992, Injuries associated with strenuous road marching, Military Medicine 157,2,64-67.

Kroemer,K.H.E.,1992, Personnel training for safer material handling, Ergonomics, 35,9,1119-1134.

Legg,S.J.,1981, The effect of abdominal muscle fatigue and training on the intra-abdominal pressure developed during lifting, Ergonomics,24, 3,

      191-195.
Legg,S.J. and Myles,W.S.,1981, Maximum acceptable repetitive lifting
      workloads for an 8 hour workday using psychophysical and subjective
      rating methods, Ergonomics,24, 12, 907-916.
Lewis,D.,Simmons,M.D., Pethybridge,R.J.,Wright,J.M. and Chandler,H.C.,1991,
      The effect of training stress on the immune system of Royal Marine
      recruits, Institute of Naval Medicine Report 19/91.
Liles,D.H., 1986, The application of the job severity index to job design
      for the control of manual materials-handling injury, Ergonomics,
      29,1,65-76.
McCaig,R.H.,1981,Physiological assessment of Physical Development Centre
      trainees,Army Personnel Research Establishment Report 81R002.
Marcinik,E.J.,Hodgdon,J.A. and Vickers,R.R.Jr.,1985, The effects of an
      augmented and the standard recruit physical training program on fitness
      parameters,Aviation,Space and Environmental Medicine,56, 204 -207.
Marcinik,E.J., Hodgdon,J.A.,Englund,C.E. and O'Brien,J.J.,1987,Changes in
      fitness and shipboard task performance following circuit weight training
      programs featuring continuous or interval running,European Journal of
      Applied Physiology, 56,132-137
Monod,H. and Scherrer,J.,1965, The work capacity of a synergic muscle
      group,Ergonomics, 8,3,329-338
Myles,W.S., Eclache,J.P. and Beaury,J.,1979, Self-pacing during
      sustained,repetitive exercise, Aviation,Space, and Enironmental
      Medicine,50,9, 921-924.
NATO,1986, Physical fitness in Armed Forces.Final Report of RSG4.
      AC/243(Panel VIII)D/125 (Brussels:NATO).
NIOSH,1981,Work practices guide for manual lifting, DHHS, National
      Institute for Occupational Safety and Health Publication 81-
      122,Cincinnati,Ohio.
Patton,J.F.,Daniels,W.L. and Vogel,J.A.,1980, Aerobic power and body fat of
      men and women during Army basic training, Aviation, Space and
      Environmental Medicine,51,5,492-496.
Patton,J.F.,Vogel,J.A.,Bedynek,J.L.,Alexander,D. and Albright,R., 1986,
      Aerobic capacity and coronary risk factors in a middle-aged Army
      population,Journal of Cardiopulmonary Rehabilitation, 6,12,491-498.
Pedersen, D.M.,Clark, J.A.,Johns,Jr,R.E.,White,G.L. and Hoffman,S., 1989,
      Quantitative muscle strength testing: a comparison of job strength
      requirements and actual worker strength among military technicians,
      Military Medicine, 154,1,14-18.
Pheasant,S.,1991,Ergonomics,Work and Health (Basingstoke:Macmillan)pp309.
Riddell,D.I.,1990, Changes in the incidence of medical conditions at the
      Commando Training Centre,Royal Marines,Journal of the Royal Naval
      Medical Services,76,105-108.
Snook,S.H.,1987, Approaches to preplacement testing and selection of
      workers, Ergonomics,30,2,241-247.
Troup, J.D.G.,Foreman, T.K., Baxter,C.E. and Brown,D.,1987, Tests of manual
      working capacity and the prediction of low back pain. In Musculoskeletal
      Disorders at Work, ed. P.Buckle (London:Taylor & Francis), pp. 165-170.
Wells,C.L., Constable,S.H. and Haan,A.L.,1980, Training and
      acclimatization: effects on responses to exercise in a desert
      environment,Aviation,Space and Environmental Medicine 51,2,105-112.

DIRECTION-OF-MOTION STEREOTYPES FOR WEAPON AIMING

K M WAYMAN

Army Personnel Research Establishment
Ministry of Defence
Farnborough, Hants, GU14 6TD

The advent of low-cost computer technology has led to more people being exposed to computer systems and the means of controlling these systems than ever before. It is hypothesised that this learning and experience may have enhanced a stereotype control-display relationship that is currently not fully recognised by designers. The results presented here show that when a vertical joystick was used the majority of subjects expected a forward movement to move the aiming mark up on the display. The data also show that operators can be trained to use a non-preferred control display relationship but with increased training time and although no statistical difference was found between the two joystick configurations when performing a simultaneous task, all the data show that performance was slightly worse when using the non-preferred control-display relationship.

INTRODUCTION

Is the control-display relationship in the vertical plane more important today than previously recognised?

The relationship between a control movement and its effect that is expected by most of the population is known as a 'population stereotype' (Fitts 1951). Control-display relationships that conform to a stereotype are said to be 'compatible' (Fitts and Seegar 1953). Designers are apt to argue that even where stereotypes are clearly reflected in behaviour, they are of little practical importance, since the operator will quickly adapt himself to the unexpected relationship and that intensive training can reverse the effects of learning. Harlow (1951) showed that given sufficient training animals can learn to switch between conflicting patterns of response. Simon (1954), however, used a secondary task as a distracting condition and found that both mild and severe degrees of stress resulted in significantly more reversal errors being made by the subjects performing on the originally non-dominant arrangement than by those performing on the originally dominant arrangement, although both groups had apparently practised to equal performance.

In a number of military weapon systems the visual display or optical sight lies in a near vertical plane, but for various reasons the joystick is mounted on a horizontal plane, standing up vertically. Spragg, Finck and Smith (1959) found that for a horizontal joystick mounted on a vertical plane, the 'up-up' relationship (control movement up associated with display movement up) was superior to the 'up-down' relationship. When a joystick was vertical (i.e. mounted on a horizontal plane) there was less difference between the 'forward-up' and 'forward-down' relationships, although the 'forward-up' relationship was slightly superior (but not significantly so).

The literature therefore provides some information about control-display relationships in the vertical plane, but there seems to be no clear-cut expectation in the general population for one relationship to hold rather than another.

The advent of low-cost computer technology, video games systems, educational based learning systems in schools, home computer systems, etc., has led to more people being exposed to computer systems and the means of controlling these systems than ever before. It is hypothesised that this learning and experience may have enhanced a stereotype control-display relationship that is currently not fully recognised by designers and consequently not fully considered during the design process.

To evaluate this hypothesis, a series of experiments has been conducted to investigate the initial control movements of a group of subjects, without prior knowledge of the joystick display directional response in the Y-axis, and the effects of a simultaneous reaction time task on tracking performance when using a non-preferred control-display relationship.

AIMS

The aims were:

a. To determine the expected/preferred control-display relationship for a vertical joystick (i.e. mounted on a horizontal surface) and a vertical visual display.

b. To determine the effects of training when a non-preferred control-display relationship is used.

c. To determine the effects of a simultaneously performed cognitive task on tracking skills when a non-preferred control-display relationship is used.

METHOD

Experimental Design:

This experiment was conducted in three separate phases using an IBM compatible 386 PC.

Phase 1. Determination Of The Expected/Preferred Control-Display Relationship

The expected/preferred control-display relationship was determined by recording the initial reactions of a sample of the population who were required to perform a simple tracking task using a vertical display and vertical joystick (i.e. mounted on a horizontal surface). The joystick response was configured in either of two directions in the Y- axis to achieve movement of the aiming mark on the display screen (joystick forward/aiming mark up, referred to as Joystick Configuration 1 and joystick back/aiming mark up, referred to as Joystick Configuration 2). Each subject was asked to place the aiming mark on individual stationary targets that appeared on the display screen. These targets were diagonally opposite each other in the lower left, upper left, lower right, and upper right quartiles of the screen. To acquire the targets a diagonal movement of the joystick was required, involving a combined movement in azimuth as well as elevation of approximately 45 degrees. Using the joystick the operator was required to place the aiming mark inside the target square. The presentation of the four targets was randomised and the order of the two joystick configurations was alternated between subjects, to minimise any order effects.

Phase 2. Determination Of The Effects Of Training

For this part of the experiment two groups of subjects were selected from those who participated in Phase 1. Only subjects who exhibited a marked preference for the joystick forward, aiming mark up relationship were used. One group was trained using the "forward for up" configuration (Joystick Configuration 1) and the second using the "back for up" (Joystick Configuration 2) control-display relationship. During each training session each subject was required to track sixteen moving targets randomly presented individually on the display screen using a whole hand joystick. It was expected that those who were trained using the non-preferred control-display relationship would require a greater training period than those trained in the preferred control-display relationship. As a separate activity, in preparation for phase 3, all subjects were trained using the Four Choice Reaction Time test. The stimulus was randomly presented in one of the four corners of the display screen at a rate of one per second. In response to each stimulus the subjects were required to press an appropriate key on the computer keyboard. During each training session each subject was required to practise the task for two minutes.

Phase 3. The Effects Of A Simultaneously Performed Cognitive Task On Tracking Skills

In this phase of the experiment it was expected that those operators, trained to perform in a non-preferred way for a specific situation would, under cognitive overload, show degradation in their tracking task performance. Conversely, those who were trained with compatible arrangements, would show less degradation in tracking performance during the overload of their cognitive processes. The simultaneous cognitive task was a Four Choice Reaction Time Test. Each subject tracked the targets using a whole hand joystick as during the training sessions in Phase 2 and responded to the reaction time task by pressing the appropriate keys on the computer keyboard (stimulus presented at a rate of 1 per second). These two tasks were presented to the subject on the display screen simultaneously therefore potentially overloading the operator. The condition order was randomised for each subject.

Measurements

Mean Radial Error tracking scores were stored for each subject, for each target. The "hit" or "miss" score for the Four Choice Test and the reaction time for the subjects' response to each stimulus was also recorded.

RESULTS

In general, mean values have been graphically plotted and where appropriate analysed using an Analysis of Variance (ANOVA) procedure or Mann-Whitney U test. Significant differences have been separated using the Scheffe multiple range test with a 0.05 significance level. Separate analyses have been conducted for each phase of the experiment and for the Four Choice Reaction Time task and tracking performance during Phase 3.

Phase 1. Determination Of The Expected/Preferred Control-Display Relationship

Analysis of the tracking performance data shows that there was a statistically significant difference between the two joystick configurations (d.f.=1, F=15.5511, p<0.001). Figure 1 shows that the subjects' performed better when using joystick configurations 1 (forward for up) as opposed to joystick configuration 2 (back for up).

Figure 1. Phase 1 - Mean Radial Error

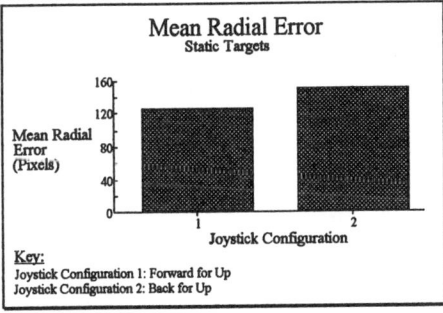

Besides the measured Mean Radial tracking error performance of each subject, observations of mistakes made during target acquisition were recorded. Whenever a subject initially moved the aiming mark in the wrong direction (relative to the position of the target) a mistake was recorded. Far fewer mistakes were recorded when using joystick configuration 1 (25) than when joystick configuration 2 (168) was used.

At the end of phase 1 each subject completed a simple questionnaire, The majority of subjects (37) preferred joystick configuration 1, of those subjects (5) who preferred joystick configuration 2 the majority had used flight simulators or had real flying experience. Two subjects expressed no preference for either of the two joystick configurations.

Phase 2. Determination Of The Effects Of Training

Analysis of the data for the first training session showed that tracking performance using joystick configuration 1 was statistically better than when using joystick configuration 2 (d.f.=1, F=163.8319, p<0.001). No statistically significant difference in tracking performance was found for training session 4 (d.f.=1, F=3.2985, p=0.0698). Figure 2 shows the Mean Radial Error for the two joystick configurations for each training session. The changes in performance illustrated in Figure 2 shows that those subjects using joystick configuration 1 exhibited little improvement in performance. By comparison the data for those subjects using joystick configuration 2 indicate a large improvement in performance during the training sessions.

Figure 2. Phase 2 - Mean Radial Error

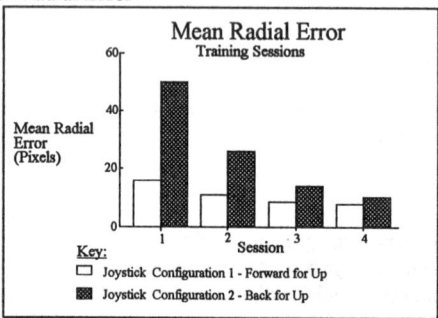

Further analysis of the training data shows that there was a statistically significant difference across training sessions for both joystick configurations (Joystick Configuration 1 - d.f.=3, F=15.3411, p<0.001, Joystick Configuration 2 - d.f.=3, F=122.1741, p<0.001). Post-hoc analysis using the Scheffe multiple range test with significance level .05 showed that session 1 was significantly worse than sessions 2, 3 and 4 for joystick configuration 1 and that for joystick configuration 2, session 1 was significantly worse than sessions 2, 3 and 4, and session 2 was significantly worse than sessions 3 and 4.

Phase 3. The Effects Of A Simultaneously Performed Cognitive Task On Tracking Performance

Comparison of the Mean Radial tracking error data using an ANOVA procedure showed that there was no significant difference between the two joystick configurations (d.f.=1, F=1.9260, p=0.1662) when the Four Choice Test was not being simultaneously performed. A statistically significant difference was not found when the Four Choice Test was being simultaneously performed (d.f.=1, F=0.0943, p=0.7589). The mean data for Radial Error are presented in Figure 3 below.

Figure 3. Phase 3 - Mean Radial Error

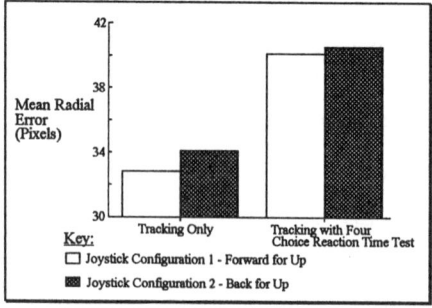

Analysis of the mean number of correct responses for the Four Choice test showed that there was no statistically significant difference between the two joystick configurations (d.f.=1, F=2.5006, p=0.1148) when the tracking task was being simultaneously performed. Figure 4 shows the mean number of correct responses.

Figure 4. Mean Number of Correct Responses

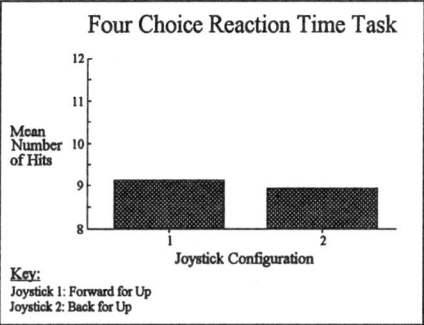

The differences in performance for the mean number of correct responses with and without the simultaneous tracking task were calculated and compared using a Mann-Whitney U test. No statistically significant difference was shown between the two joystick configurations (Z=-1.7570, 2-tailed p=0.0789).

Analysis of the mean reaction time data for the Four Choice test showed that there was no statistically significant difference between the two joystick configurations (d.f.=1, F=2.2666, p=0.1332) when the tracking task was being simultaneously performed. The reaction time mean and standard deviation (s.d.) for joystick configuration 1 was 0.5237secs (s.d.=0.0884), and for joystick configuration 2 was 0.5786secs (s.d.=0.4516). This illustrates that although there was little difference between the mean reaction times for the two joystick configurations, the standard deviation data shows that the subjects were more consistent when using joystick configuration 1 than joystick configuration 2.

The differences in performance for the mean reaction time with and without the simultaneous tracking task were calculated and compared using a Mann-Whitney U test. No statistically significant difference was shown between the two joystick configurations (Z=-1.1431, 2-tailed p=0.2530).

DISCUSSION

In the introduction to this work it was hypothesised that the learning and experience gained when using, video games, educational based learning systems in schools, home computer systems, etc., may have enhanced a stereotype control-display relationship that is currently not fully recognised by designers and consequently not fully considered during the design process.

The results presented here clearly show that most people, when using a vertical joystick (i.e. mounted on a horizontal surface), expected a forward movement on the joystick to produce an upward movement of the aiming mark on the display. The majority of the subjects who expected to pull the joystick back to produce an upward movement of the aiming mark on the display had some experience of flight simulators or real flying experience.

Although it is impossible to attribute any direct link between any learning and experience when using computer technologies and this expected control-display relationship, any experience of these systems will have influenced each individual and therefore their response to this control-display relationship.

Phase 2 of this experiment addressed the problems of training when using a non-preferred relationship. In the introduction to this work it was highlighted that designers are apt to argue that even where stereotypes are clearly reflected in behaviour, they are of little practical importance as the operator will quickly adapt himself to the unexpected relationship and that intensive training can reverse the effects of learning.

The evidence from this experiment does suggest that operators can quickly adapt themselves to the non-preferred control-display relationship when given intensive training. It is also evident that the amount of training required is considerably more than that when using a preferred relationship. This, of course, has huge implications for training costs, which today have considerable importance when calculating the total life costs for complex systems. This work has not addressed the problem of retention of skills following training. A further experiment is to be conducted on the same samples of subjects to investigate the decrease in performance following a period without any further training. It is expected that those operators using the non-preferred relationship will show a greater decrease in performance than those using the preferred relationship. If this is the case, then those operators using a non-preferred control-display relationship will require re-training more frequently and therefore increase whole life costs even further.

Phase 3 of this study addressed the effects of a simultaneously performed cognitive task on tracking skills when a non-preferred control-display relationship was used. Simon (1954), found that both mild and severe degrees of stress resulted in significantly more reversal errors being made by the subjects performing on the originally non-dominant arrangement than by those performing on the originally dominant arrangement, although both groups had apparently practised to equal performance. Although the data from this study do not show a statistically significant difference between the two joystick configurations when the simultaneous task was being performed all the data do show that performance was slightly worse when using the non-preferred control-display relationship. Further experiments are planned where the Four Choice Test stimulus will be presented at a higher rate to increase the cognitive load and therefore perhaps show greater differences in tracking performance.

At the beginning of this paper the question, 'Is the control-display relationship in the vertical plane more important today than previously recognised?' was asked. The data presented here provide strong evidence that this control-display relationship is more important today than previously recognised. The results from phase 1 show that the majority of subjects expected a forward movement of the joystick to move the aiming mark up on the display. Phase 2 showed that operators can be trained to use a non-preferred control display relationship but with increased training time. Although no statistically significant difference was found between the two joystick configurations during Phase 3, all the data do show that performance was slightly worse when using the non-preferred control-display relationship.

REFERENCES

Fitts, P. M., 1951, Engineering psychology and equipment design. Handbook of Experimental Psychology (Edited by S S Stevens, New York: Wiley and Sons); 1958, Engineering psychology. Annual Review of Psychology, 9, 267 - 294

Fitts, P. M. and Seegar, C. M., 1953, S-R compatibility: spacial characteristics of stimulus and response codes. J. Exp. Psychol., 46, 199-210

Harlow, H. F., 1951, Primate Learning. Comparative Psychology (Edited by C P Stone, New York: Prentice Hall)

Simon, C. W., 1954, The effects of stress in a dominant and non-dominant task. USAF WADC Tech. Rep. No 54-285

Spragg, S. D. S., Finck, A. and Smith, S., 1959, Performance on a two-dimensional following tracking task with miniature stick control, as a function of control-display movement relationship. Journal of Psychology, 48, 247-254

A COMPARISON OF A LINEAR AND NON-LINEAR CONTROL LAW

By D.J. KNOWLES

Army Personnel Research Establishment,
Ministry of Defence,
Farnborough, GU14 6TD.

Future long range anti-tank guided weapon systems will be required to engage targets from 500m to 5km. The primary long range tank targets have low angular velocities, whilst the secondary helicopter targets have high angular velocities. Two control laws, (for the relationship between joystick output and sight velocity), were compared using a simulator and 13 military subjects. A multi-linear control law, with the maximum output linked to the track box used to mark the target was compared with a shaped control law, and performance when engaging a variety of targets at different ranges was measured. Results showed that the multi-linear control law method produced significantly lower tracking errors when engaging both the primary long range tank targets and the secondary close range helicopter targets.

INTRODUCTION

Future long range anti-tank guided weapon (LR ATGW) systems are becoming increasingly sophisticated and versatile. They need the ability to engage targets from 500m to 5km. The primary targets are ground based, long range, stationary or slow moving vehicles, but the secondary requirement includes the engagement of close range rotary wing airborne targets, with high angular velocities.

Typically, a good quality force tracking controller has a linear output, with joystick force directly proportional to sight velocity, but many systems shape the joystick voltage output, to achieve fine control and greater tracking accuracy as the joystick output approaches zero, whilst retaining the same maximum output capability.

A pilot study was conducted at APRE using a simulated LR ATGW system and a shaped control law, designed to cope with the high angular velocities associated with the secondary targets. Results indicated that this method would not provide sufficient fine tracking control to engage the primary long range tank targets at 5km.

The problems associated with fine tracking control when engaging stationary long range targets are well known. Waygood (1972, 1974) examined the effect of viscous damping, and filtered control, to reduce hand tremor and improve tracking performance. Jensen and Wilson (1990) assessed input devices and control laws in a vibrating environment, and their results indicated that linear control laws allow significant advantages in speed and accuracy for high resolution positioning tasks, compared with non-linear control laws. These methods

17

were designed primarily to make the engagement of stationary targets easier, but were not suited to the engagement of moving targets with high angular velocities.

The future LR ATGW system has a primary function which is to engage long range tank targets. It seemed logical therefore that the control law should use a linear relationship between joystick output and sight velocity. A method of engaging targets with high angular velocities, without compromising the engagement of the primary targets, needed to be devised and investigated.

Modern weapon systems have the control laws (for the relationship between joystick output and sight velocity) configured in software rather than fixed in hardware. One additional task in the fire control sequence of the future LR ATGW is the requirement to change the size of the graticule track box to enable the system to achieve lock-on to a pre-selected target, prior to a missile being fired. The operator has to fit the track box around the target, and request lock-on, whilst he continues to track the target as accurately as possible.

In the simulated system, nine box sizes related to target size and range were used. This meant that close range targets (with high angular velocities) required a large box and long range targets needed a small box. Targets travelling broadside with high angular velocities, presented a larger profile (and needed a larger box) than those targets approaching head-on which had a smaller profile and much lower angular velocities. Crossing helicopter targets had higher angular velocities than crossing tanks, and were also physically larger in size.

Figure 1:

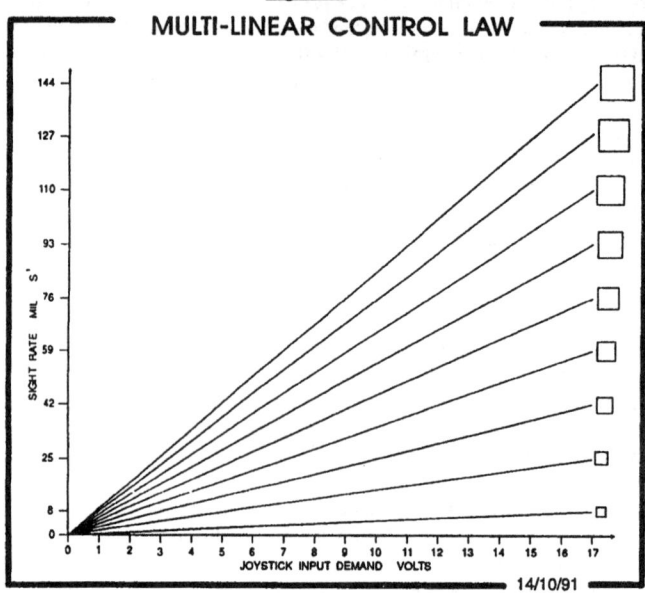

A multi-linear control law, with the maximum output linked to the size of the box used the mark the target, was devised at APRE. The linear control law with the highest output was used with the largest box, and the control law with the lowest output was used with the smallest box. At the start of an engagement sequence the largest box, (with the highest output) was displayed. It was hoped that this control law method would improve the fine tracking control needed for engaging distant tank targets without compromising the engagement of the close range helicopter targets.

A simulator experiment was designed to compare the multi-linear method with a shaped control law with the same maximum output capability.

METHOD
Man machine interface
The man machine interface consists of a force joystick (operated with the right thumb), a visual display, and two additional fire control switches, both operated with the left hand, a rocker switch (operated with the left thumb) and a trigger switch.
Fire Control Procedures
The procedures were as follows:-
        a. Track onto target with the right hand thumb controller
        b. Fit track box around target with left hand thumb rocker switch
        c. Request lock on with left hand rear trigger
        d. Continue tracking until system lock-on is achieved
        e. Press central fire button with either hand.
Control Laws
Two control laws were compared.
        Option 1 The multi-linear control law.
        Option 2 A shaped control law with a predominantly cubic function.

$$y = ax + (1-a) \, x^3$$

where:
        y = the sight rate (velocity) in mils/ sec
        x = the joystick input in volts
        a = constant

Figure 2:

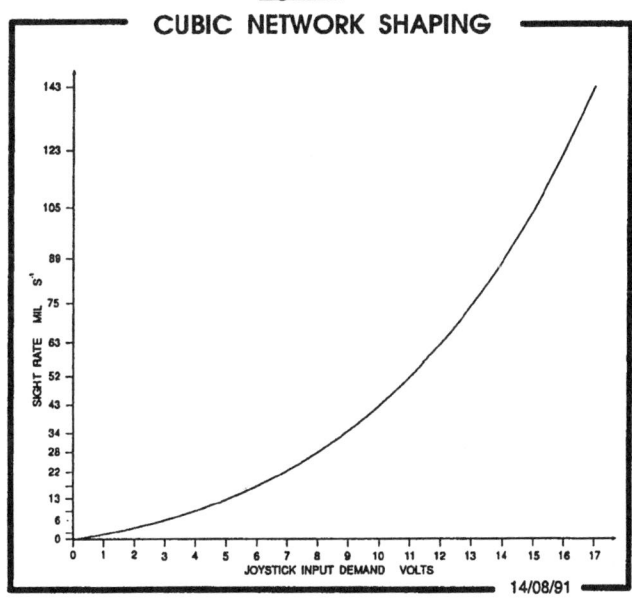

CUBIC NETWORK SHAPING

14/08/91

<u>Subjects</u>

Thirteen military subjects took part from the trials section at APRE. The soldiers were infantrymen with little or no previous weapon operator experience.

<u>Targets</u>

The simulated targets were either tanks or helicopters at ranges from 500m to 5km. They were either stationary or travelling at moderate or fast speeds and travelled in three directions, head-on, crossing, or moving diagonally. Their motion was either linear or sine wave generated. The maximum helicopter target angular velocity was 120 mils/sec.

<u>Training</u>

Each soldier was trained first on one of the control law versions and then on the other on two separate days. In each training session they were presented with 12 tank targets, and 12 helicopter targets. The control law version was alternated between subjects to minimise order effects. The training targets were presented in a fixed order starting with the stationary targets and working up to the targets with the high angular velocities.

<u>Trial design</u>

The set of primary tank targets was given first, followed by the set of secondary helicopters. The trial targets were mirror images of those used in the training sessions, to prevent familiarisation, and the order of target presentation within each set was randomised.

The control law version was alternated between subjects and each subject used each control law version on two separate days, at the same time of day. This design helped reduce order, learning and diurnal effects.

<u>Measurements</u>

During the trial, the time taken and tracking error data were collected on line at 25 Hz and stored for analysis purposes.

RESULTS

An Analysis of Variance (ANOVA) was carried out on the time taken to request lock-on and achieve system lock-on and showed no significant difference between the two control law versions, see Table 1.

Table 1: Cumulative Time in Seconds (all targets)

Procedure	Control law				
	Cubic		Multi-linear		ANOVA
	Mean	SD	Mean	SD	
Request lock-on	7.54	3.87	7.51	3.29	NS $p>0.05$
System lock-on	7.91	3.72	7.89	3.24	NS $p>0.05$

Radial tracking error (mils) over all targets was analysed using an ANOVA procedure. This showed no significant difference between the two control laws, see Table 2.

Table 2: Average Radial Tracking Error in mils (all targets)

Procedure	Control law				
	Cubic		Multi-linear		ANOVA
	Mean	SD	Mean	SD	
Request lock-on to System lock-on	0.79	1.51	0.76	1.23	NS $p>0.05$

Further analysis of the tracking error data for the two extremes of target was conducted using an ANOVA procedure, this showed that the multi-linear control law produced significantly lower tracking errors for the long range tank target (p< 0.001) and the close range helicopter target (p<0.05), than the cubic control law, see Table 3.

Table 3: Average Radial Tracking Error in mils (specific targets)

Target Type	Control law				
	Cubic		Multi-linear		ANOVA
	Mean	SD	Mean	SD	
Long Range Tank	0.20	0.07	0.12	0.05	Sig p<0.001
Close Range Helicopter	5.31	2.47	4.13	1.87	Sig p<0.05

DISCUSSION

In this experiment the multi-linear control law provided improved fine tracking control for engaging the priority long range tank targets, than the alternative shaped control law with its predominantly cubic function. In addition, the multi-linear control law also improved the weapons system's secondary helicopter target engagement capability.

The advantage of the multi-linear method was that it provided the optimum control law at the extremes the target engagement envelope, without the operator having to perform any additional procedures.

The operator's performance was improved when using the multi-linear control law if track box sizing was conducted as the target was acquired, because this automatically related the control law to the target's range and angular velocity. In the case of long range targets this resulted in a slowing and steadying of the joystick response which also helped to reduce the possibility of any overshoot.

The only identified disadvantage was the hasty selection of a box size that was too small. This resulted in the joystick feeling sluggish, and delayed target acquisition. The soldiers soon learnt, however, that increasing the box size increased the speed of response.

It is believed that the multi-linear control law concept could be implemented in many types of weaponry where there is a requirement for a long range target engagement capability, to help improve operator performance. There are also a number of other types of aiming system where this type of control law could be used; for example the control of a remote camera might be improved if the control law were directly related to the magnification factor.

REFERENCES

Jensen S E and Wilson D M, 1990: Assessment of Input Devices and their Control Laws for use in a Vibrating Environment. Unpublished MOD Report.

Waygood M, 1972: The Effect of Viscous Damping on Hand Tremor. Unpublished MOD Report.

Waygood M, 1974: The Effect of a Filtered Control on Operators' Hand Tremor. Unpublished MOD Report.

THE ASSESSMENT OF A MILITARY EYE PROTECTION SYSTEM

By S.J.SKINNER

Army Personnel Research Establishment,
Ministry of Defence,
Farnborough, GU14 6TD.

This paper describes the assessment methodology for the evaluation of a number of Human Factors issues inherent in the design of a military eye protection system The effect on a soldier's performance was studied using a battery of tests. Five different goggles were tested. The performance of the soldiers was measured for each test, and the goggles were ranked from best to worst. The results illustrate the difficulty of applying this type of methodology where each system has good and bad attributes.

INTRODUCTION

Protective eye wear is required on the modern battlefield to protect the soldier's vision from directed energy systems, and low velocity ballistic threats, such as shrapnel, and dust. There are several Human Factors implications related to bringing an eye protection system into service with the British Army. For instance, the wearer's field of view will be reduced and perception of colour will be degraded. These are of prime concern as they will both affect operator performance on a range of military tasks. The ability to discern colour is important when undertaking a target detection task and when reading a map. The comfort of the goggles is another key issue as they may have to be worn for long periods, in very restrictive conditions, when the wearer may be sleep deprived and physically fatigued.

To protect the commercial interests of each manufacturer the goggles that were tested will be referred to as goggles A to E. These are briefly described below:

Goggle A: A soft rubber frame with yellow/green lenses.
Goggle B: A hard rubber mount with a yellow/green lenses
Goggle C: A soft rubber frame with orange lenses.
Goggle D: A spectacle frame with grey lenses.
Goggle E: A spectacle frame with green lenses.

Five tests were undertaken to quantify the effects of five different types of goggles on operator performance.

The tests were:
1. Target Detection
2. Comfort
3. Display Use
4. Colour Discrimination
5. Field of View

METHOD

In this experiment the target detection experiment was followed by the comfort rankings. The subjects were then required to undergo the display usability and colour discrimination tests. The field of view test was carried out last as the subjects sometimes complained of eye strain after this test.

For each test, the performance of 15 military subjects was measured. These military subjects were all regular army infantrymen who had been screened for visual defects.

Target Detection.

During this test the subjects were required to detect a series of stationary head-on tank targets on a projected panorama, using binoculars. The targets were presented one at a time and subtended the correct visual angle to reflect their size and range. The subject was allowed 2 minutes to detect the target, if he did not detect the target within the time limit, a miss was recorded. Data were collected on the time to detect each single target.

Each subject used all five goggles and the naked eye on every visit to the target detection facility. The detection experiment was structured so that each subject visited the facility six times. The subjects used the goggles in a balanced order, so that at the end of the trial all targets had been presented in each goggle condition. All subjects had a different order of presentation of the goggles.

Comfort

The comfort of a pair of goggles is a subjective measure. After each phase of the target detection task (where the subjects used all of the goggles) the subjects were asked to rank the goggles in order of comfort. The subject was allowed to handle and re-try any of the goggles. This rank order was then transformed to scores, a score of 1 was given to the most comfortable, through to a score of 5 for the least comfortable. The naked eye condition was not included in this ranking procedure.

Display Usability.

There are several different display types already used by the Army in everyday operations. Liquid Crystal Displays (LCD)and Cathode Ray Tubes (CRTs) are used for indirect viewing of thermal imagery.

A standard acuity test, the Snellen Illiterate E test was adapted for use on a computer. The original Snellen test uses angles subtended at the eye as the benchmark of acuity. The PC version uses standard point sizes as the model. The point sizes are defined in Table 1.

Table 1: Definition of the Letter Sizes for the Snellen Illiterate E Test.

Number	Point Size	Height (mm)
1	10	2.5
2	18	4.5
3	26	6.5
4	34	8.5
5	42	10.5
6	52	13.0
7	58	14.5

The subjects were required to indicate the orientation of the E using the directional cursor keys. The available orientations were up, down, left and right. After 5 seconds, if the subject had not responded, that target was recorded as a miss. The presentation of different point sizes and orientations were randomised, but the same number of each were presented in each run. The measurements recorded were Correct/Incorrect and Response Time.

Colour Discrimination.

The use of coloured lenses has been shown to disrupt the ability of the individual to discriminate between colours (Boynton 1979). This ability is fundamental to being able to read a map effectively. Each different colour of goggle affects the colour vision characteristics of the individual's sight. The Farnsworth-Munsell 100 Hue Test was used to determine the extent of this disruption.

The Farnsworth-Munsell 100 hue test was devised by D. Farnsworth in 1943. It is a sorting test where the subject is required to arrange coloured samples in order, to make a colour sequence. The samples have a number printed on the back, a perfect run ends with all of the samples in the correct numerical sequence. The sequence that the subject accepts as correct is then plotted on a graph, and the area under this graph is measured. This area is a measure of the overall error in colour discrimination. As gross aberration was expected a re-test was used as a integral part of the trial design.

Field of View.

The field of view of each pair of goggles was measured using a Goldman Perimeter 940-K7 On arrival in the test room the subject covered one eye with a patch and then positively located his head to the chin rest using a head strap. The chin rest was then moved by the experimenter, to adjust the subject's head position to line up the pupil of his eye with the centre of the perimeter. The subject was then instructed to fixate on a marker, and to sound a warning buzzer as soon as the illuminated target came into view. The target was brought in from the periphery of the dome, towards the centre, at a steady speed. When the buzzer sounded, the experimenter marked a graph attached to the perimeter. The distance from the origin of the graph to the point marked by the experimenter is measured and used in the analysis. The target used was the largest the Goldman Perimeter was capable of producing, ($64mm^{2}$), and as the goggles reduced the quantity of light reaching the eye, the brightest target was used (relative intensity 1.00).

RESULTS & DISCUSSION

Target Detection

Data were collected on the time (in seconds) to detect single targets. The subject was given 2 minutes to detect the target: if, after this time, the target was not detected, a miss was recorded. The data collected was then transformed into target detection probabilities. These data were found to be binomially distributed. A test statistic, Z_0 was calculated to determine any significant differences in the probability data.

The analysis showed that there was no significant difference between any of the goggles, or the naked eye condition.

Comfort

After each run in the target detection facility the subjects were asked to rank the goggles in order of comfort. The naked eye condition was not included in this ranking procedure.

A non-parametric Friedman ANOVA by ranks was carried out. It was found that there was a significant difference between the goggle types (see Figure 1). Individual t-tests were conducted to identify where these differences lay.

It was found that at the 99% level ($p < 0.01$): both the Goggle A and Goggle C versions were significantly more comfortable than all the other goggles, and that Goggle E was significantly more comfortable than Goggle B.

At the 95% level ($p < 0.05$) the Goggle D was significantly more comfortable than the Goggle B.

There was no a statistically significant difference in comfort between the Goggle A and Goggle C, or between the Goggle E and Goggle D.

Figure 1 Comfort Rankings: Graph Showing Mean Plots
with 99% Confidence Limits

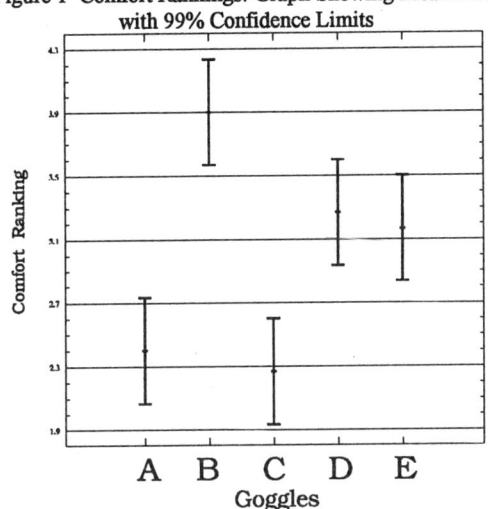

Goggle B scored badly in this test because of the hard rubber mount with a small facial contact area. This small area of facial contact left pressure marks on the subjects face after only a few minutes of use. The soft rubber of the Goggle C and Goggle A frame had a much larger area in contact with the face, this lead to the pressure of the goggle strap being placed over a much larger area. This provided a much more comfortable goggle especially when used in conjunction with binoculars. Neither of these goggles were compatible with spectacles.

Display Use

There were no statistically significant differences between the goggle types.

Using the LCD screen proved to be statistically slower than either of the two CRT displays ($p < 0.01$), but accuracy data for all three showed no significant difference.

Colour Discrimination

The subjects ability to discriminate between colours was affected by the goggles. The error data from the Farnsworth-Munsell 100 Hue Test (Table 2) was checked for skewness and was found to approximate to a normal distribution (Standardised Skewness value of 1.63).

Table 2: Table showing results for the Colour Discrimination Test

	Rank	Goggle	Error Area	Error Peak
			Arb Units	(nm)
		Naked Eye	234	
Best	1	Goggle D	290.5	
	2	Goggle As	338.5	579, 483
	3	Goggle E	369	593, 480
	4	Goggle B	381	578, 490
Worst	5	Goggle C	547.5	600. 494

An ANOVA was then applied to the data and it showed that there was a significant difference between the goggles. Goggle D was significantly better than, Goggle E, Goggle B ($p < 0.05$), and Goggle C ($p < 0.01$).

Goggle A, Goggle E, and Goggle B were all significantly better than Goggle C (p<0.01) see Figure 2.

The naked eye was significantly better than all the goggles except Goggle D (p<0.05).

Figure 2: Colour Discrimination: Graph Showing Mean Plots
with 99% Confidence Limits

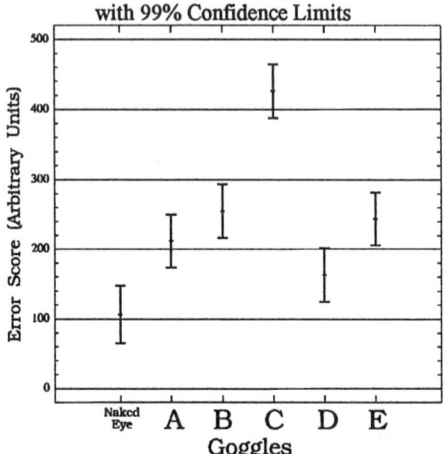

The test graphs produced by the various goggles were characteristic of different types of colour blindness. The naked eye condition was very important in the exclusion of any subjects that were colour blind. One subject was found to be Tritanomalous ('blue' colour blind), and his results were discounted. The remainder of the group performed as expected, showing the characteristics of a low discrimination group (a small error across the whole spectrum).

The Goggle A, disrupted colour perception in two areas. This disruption occurred between 590nm to 572nm, and between 497nm and 475nm. The 590-572nm range was in the "yellow" area of the spectrum, the 497-475nm in the "blue/green". The effect of this was that the goggle lens made the surroundings appear a red/purple colour. The yellows, along with most of the blues and greens are absorbed by the lens allowing only the red, and strong blues and greens to be distinguished from each other. For map reading there is very little information that is still distinguishable. Waterways, forests, and grid lines all become very difficult to distinguish. Rivers tend to be bordered by a thick blue line, which is visible when wearing the goggles, as an outline is available. Not all the nomenclature associated with waterways was visible. Swamps and marshes do not have the hard bordering, and hence will not be seen on the map.

The Goggle B had a very similar effect. There were peaks between 592-570nm and also between 500-474nm. This goggle, however, has a slightly more marked effect on the ability to discriminate between the colours.

Goggle C produced the largest error. There was one major peak, between 497nm and 480nm, the 'blue/green' area of the spectrum. There is however a large error over the whole of the spectrum, so that all of the colours are difficult to distinguish from each other, with the 497-480nm band being particularly poor.

The Goggle D did not greatly affect the soldiers ability to discriminate between colours. There was a small loss in colour discrimination ability, across the whole of the spectrum, affecting all of the wavelengths equally.

Goggle E produced a peak in the error score between 497nm and 478nm (blue/green), but also caused problems between 600nm and 590nm (Orange). The colouration of these goggles makes it difficult to perceive woods and contours, although it is still possible to identify them.

Field of View

All the goggles restricted the field of view, and all are significant from each other at the 99% level p<0.01 (except Goggle C and Goggle A, which both had the same frame), see Figure 3.

Figure 3: Field of View: Graph Showing Mean Plots
with 99% Confidence Limits

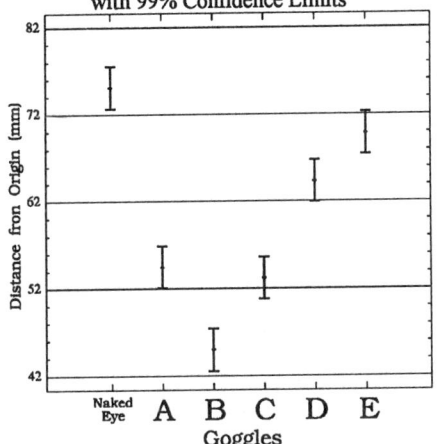

Overall Assessment

In the final recommendations we have not used any weightings to determine priorities for the tests done. For each test, the performance of the soldiers was measured and the goggles ranked. The goggle that permitted the best performance in a test was ranked 1, the worst ranked 5. Table 3 shows the overall results

Table 3: Overall Assessment

Goggles	T.Detect Rank	Comfort Rank	Display Rank	Colour Rank	View Rank	Average Rank
Goggle A	1=	2	3	2	3	2.3
Goggle E	3	3	4	3	1	2.8
Goggle C	4	1	2	5	4	3.2
Goggle B	1=	5	1	4	5	3.3
Goggle D	5	4	5	1	2	3.4

These ranks were analysed using a non-parametric ANOVA. The non-parametric ANOVA did not show a significant difference between the goggle types.

CONCLUSIONS

The overall assessment described above illustrates the difficulty of applying this type of methodology, as no one design was statistically different from any of the others.

The majority of the tests showed clear differences between each system.

Enhancements to this methodology should include some method of prioritising and or weighting each assessed attribute.

REFERENCES
Boynton R. M., 1979, <u>Human Color Vision</u>
    (New York: Holt, Rinehart and Winston)

RECLINED SITTING POSTURES: THEIR EFFECT ON HUMAN PERFORMANCE OF A VIGILANCE TASK

M. THODY*, V. H. GREGG** and R. J. EDWARDS*

* Army Personnel Research Establishment,
Farnborough, Hampshire, GU14 6TD.

** Birkbeck College, University of London,
London, WC1E 7HX.

It has been suggested that the profile of military armoured vehicles may be reduced if all the crew seats are substantially reclined. This experiment investigated the effects of operating whilst reclined over a period of 60 and 70 minutes. A computer based vigilance or monitoring task, which consisted of text messages appearing on a simulated display, was used to measure reaction times and error rates for subjects whilst operating at three angles of recline. Subjective data were also collected throughout. Body discomfort ratings showed a significant effect as a result of increased recline and a consistent decrease in performance was observed although it was not significant. The main findings and the possible implications of them are discussed.

INTRODUCTION

The task of gathering and processing information from and around the battlefield has long since been recognised as a key issue in the formula for military success. The majority of battlefield information is best gathered by stealth rather than by fighting and in order to meet the operational requirement to move about the battlefield, undetected, the vehicles must be unobtrusive and have low visual, radar, acoustic and thermal signatures. A reduction in vehicle profile could, therefore, prove an essential characteristic of the next generation of armoured reconnaissance vehicles. One suggestion put forward, which if applied would probably enable this desired vehicle characteristic to be achieved, is to recline the crewmen within the vehicle. This would have the potential for reducing the interior floor to ceiling dimensions by typically 0.1m and 0.375m for seat recline angles of 30° and 60°, respectively.

Typical tasks undertaken by the vehicle crew require periods of between 1 and 4 hours of sustained vigilance and may include observing large areas of ground through which a few enemy vehicles may pass or listening for occasional radio messages which are sent to that vehicle which occur at infrequent intervals and without warning. The aim of this project was to investigate whether or not human performance of a vigilance task is affected by operating in a reclined sitting posture.

There is a significant dearth of documented research in the area of reclined seating and its effects on human performance, particularly pertaining to ground vehicles. A number of studies have been conducted by the Royal Air Force (Glaister & Lisher, 1976; and Lisher & Glaister, 1977) to investigate the effect of reclined posture on human performance during exposure to high, sustained +G$_z$ acceleration. During these studies pilots' performance was measured for short periods, typically 4 minutes, which together with the high physical stress were not commensurate with the tasks of the vehicle mounted reconnaissance crewman. Other research within this area was related to sleep, for example Nicholson and Stone (1987) investigated aircraft passenger seats and concluded that a back rest angle of 45° was acceptable to allow a reasonable amount of sleep for the occupant.

The literature indicates that changes in levels of arousal directly affect performance, since success on a vigilance task requires the individual to have sustained attention (Monk, 1979). Yerkes and Dodson (1908) suggested that a curvilinear relationship between arousal and performance exists, so that any task is best performed at an intermediate level. Sleepiness can also have a direct effect on performance since it is a behaviour associated with lowered physiological arousal. Horne (1992) suggests that as sleepiness increases visual awareness declines, consciousness clouds and it is likely that events will be incorrectly perceived. Sleepy subjects also lower their standards of accuracy, allowing large deviations from acceptable performance to occur before corrective action is taken (Bartlett, 1943). It seems that the task of vigilance which is inherently monotonous increases drowsiness and reduces arousal which in turn reduces the performance of the vigilance task. If a state of drowsiness and lowered arousal was induced earlier in the vigilance task as the result of an additional effect, for example reclining the subject, then the decrease in performance may be accelerated.

The vigilance task chosen for the investigation was a representation of a general information display. Textual representations of radio messages were used as the stimuli. The stimuli consisted of a mixture of targets and non-targets. A number of stimuli of predetermined type and order of presentation were put together to form a scenario. Five balanced scenarios were compiled, together with a shorter one for training. The five main scenarios consisted of 286 messages which were split equally between three types. Both types consisted of three types of message:

Orders	Message giving an instruction, for example 'Move now to the barn 200 metres to your front'
Reports	Message providing information, for example 'Have moved to barn, am observing crossroads to south east'
Queries	Message requesting information, for example 'Confirm you have seen enemy vehicle at T-junction to east'

Of the 286 messages presented in each scenario, 84 were targets consisting of 28 orders, 28 reports and 28 queries. The response for a target required detection and identification of message type and the response for a non-target required only detection. All responses required a similar physical interaction with a simulated button panel.

METHOD

The general requirement for the experiment was to create a suitable environment for presenting the vigilance task to subjects in a number of reproducible seat recline angles. A Cobra Sports Car Recliner was chosen for the seat as it offered good overall postural support, including a head rest and was easily adjustable throughout the desired range of recline angles. An Apple Macintosh IIci base unit and AppleColor 13" High Resolution RGB Monitor fitted with a MicroTouch Systems Inc. touch screen were chosen as the computer hardware. Supercard version 1.5 was chosen as the software which controlled the vigilance task, logged the data and displayed the simulated interface panel.

Twelve fit, healthy male infantry soldiers from the Army Personnel Research Establishment (APRE) were used as subjects. During each experimental run they wore representative clothing for the manned armoured reconnaissance task, which consisted of a one piece military coverall, boots and a helmet.

A within subjects experimental design was used for this investigation and fully balanced to take account of the main effects. The independent variables were: seat recline angle (7°, 30°, 60°); time of day; scenario and subject. Objective measurements were made of desired response, subject's response and reaction time. Subjective measurements were made of self-reported stress and arousal, sleepiness and body discomfort. Training was completed on two days immediately prior to the experiment.

The messages appeared on a simulated colour display, one at a time, throughout the task. On presentation of a message subjects had to determine whether it was a target or non-target. The correct response for a non-target was to press the Acknowledge button. If the stimulus was a target, subjects had to determine its meaning before pressing the appropriate button, either Order, Report or Query. Once a response had been made the message disappeared and after a delay of between 1 and 20 seconds the next message was presented. The task continued until all 286 messages had been presented and responded to, lasting between 65 and 70 minutes. The simulated buttons provided audible and visual feedback to subjects when touched. No facility was provided for subjects to change their input.

The Stanford Sleepiness Scale was presented to subjects at the start, midway through and at the end of each experimental run. The stress arousal checklist (SACL) developed at Nottingham (Mackay et al, 1978; Cox and Mackay, 1985) was presented to subjects before and after each experimental run. Subjects also completed a body parts discomfort questionnaire, modified from that of Corlett and Bishop (1976) to include a specific section relating to the neck.

The procedures employed during the experiment were given full ethical approval by the APRE ethics committee prior to the investigation.

RESULTS

The statistical analyses were achieved with Statgraphics V5 based on a Dell 486P/33 PC. All results where $p<0.05$ have been accepted as statistically significant. In general the mean values with standard deviations have been tabulated (Tables 1, 2 and 3) and, where appropriate, analysed by the analysis of variance (ANOVA) technique. Separate analyses have been conducted on reaction time and error rate for non-targets, target orders, target reports, target queries and targets collectively. The subjective data have also been separately analysed and include; change in stress, change in arousal, whole body discomfort, neck discomfort, perceived effect of discomfort on performance and final sleepiness rating. These analyses considered the main effect of seat recline only.

Table 1. A summary of mean (sd) reaction times in seconds against message type and seat recline angle.

Message type	Seat recline angle		
	7°	30°	60°
Non-targets	2.4 (.77)	2.6 (.77)	3.3 (1.81)
All targets	3.9 (1.72)	4.0 (1.27)	5.4 (3.65)
Target orders	3.6 (1.48)	3.9 (1.37)	5.9 (5.43)
Target reports	3.9 (1.81)	4.2 (1.54)	5.1 (2.74)
Target queries	4.1 (2.07)	4.0 (1.18)	5.2 (3.16)

Tables 1 and 2 show a general pattern tending towards mean reaction time increasing as seated posture became more reclined. However, this pattern did not prove to be statistically significant. No definite patterns could be identified in terms of error rates when comparing seat recline angles. It is interesting to note from Table 1 that more errors were made when processing targets than non-targets, for all angles of recline but this also failed to reach significance. Error rates for queries were in the order of 5 to 6 times those for non-targets and 2 to 3 times those for target reports and orders, respectively.

Table 2. A summary of mean (sd) percentage errors against message type and seat recline angle.

| | Seat recline angle | | |
Message type	7°	30°	60°
Non-targets	11.5 (13.6)	8.3 (4.70)	12.5 (17.3)
All targets	33.5 (17.2)	32.1 (21.1)	29.9 (11.9)
Target orders	12.5 (14.5)	17.2 (15.3)	11.1 (9.88)
Target reports	26.2 (29.4)	24.4 (33.3)	28.2 (28.2)
Target queries	61.9 (21.7)	54.9 (17.6)	50.4 (12.0)

Table 3. A summary of mean (sd) subjective ratings for each seat recline angle. Variables showing significant (p<0.05) effects of recline angles are indicated *.

| | Seat recline angle | | | |
Subjective rating	7°	30°	60°	
Change in stress	-0.1 (2.71)	0.7 (1.10)	1.8 (1.69)	
Change in arousal	-2.0 (3.44)	-2.8 (3.49)	-4.5 (3.47)	
Final sleepiness	3.8 (1.60)	4.1 (2.12)	4.7 (2.21)	
Body discomfort	0.8 (1.64)	2.5 (2.88)	5.6 (3.92)	*
Neck discomfort	0.8 (1.42)	1.6 (2.46)	4.8 (4.13)	*
Perceived effect of discomfort	0.3 (0.62)	0.7 (1.27)	2.1 (1.20)	*

Table 3 shows that there was a general pattern tending toward increased seat recline angle causing an increased effect for the subjective ratings used in the investigation, although statistically significant findings only occurred for self reported whole body discomfort, neck discomfort and perceived effect of discomfort on performance.

DISCUSSION

The discomfort ratings were the only measures which showed statistically significant differences between angles of seat recline. The majority of dependent variables just failed to reach significance but an underlying pattern was observed which suggests an effect on performance as a function of seat recline angle may exist but this was not proven. Error rates provided no indication toward one angle of recline over another as they varied with message type rather than with condition. Although discomfort was experienced it caused minimum increase in stress and had little effect on the state of subjects' arousal. As expected there was a fall in the state of arousal as seat recline increased and a reduction in performance. The task was a relatively simple one and should, arguably, have been affected to a greater extent than a more complex task (Yerkes & Dodson, 1908). The full complement of tasks within the role of armoured reconnaissance are not all simple, nor do they all require prolonged periods of vigilance. It is reasonable to suggest that the effects of reduced arousal will be experienced to a greater or lesser extent depending on the task. The main task, however, is one of vigilance and visual search with possibly long periods of complete inactivity. It is possible that performance of such a task would be severely hindered by the reclined working posture. It may be necessary to design and incorporate an automatic target detection and alerting system to reduce the probability of targets being missed due to attention state or even sleep.

Although failing to reach statistical significance, the results suggest that mean reaction times differed between message types. As expected, mean reaction times for non-targets were somewhat faster than those for targets. One reason for this is that, the discrimination between targets and non-targets was detectable within the first four characters of each message presentation. These initial characters contained the proposed recipient's identifier. The subjects' identifier was J22. Any message not beginning with J22 could be immediately recognised as a non-target, for example L21A, T30, 23A. The response for non-target could then be made without giving consideration to the remainder of the message. In the case of targets more of the message was required in order to identify its type. It is interesting to note that mean reaction times for each type of target message were similar despite the differences in message length. It is possible that subjects were able to determine the type of message without reading it all. The way the messages were constructed enabled possible identification to be made at the fifth word, for example in the message 'J22 this is T30 move.....' the word 'move' is an instruction so this message is identified as an order. The proposal that this message discrimination strategy was adopted is given some supporting evidence in terms of the distribution of error rates for different message types. The error rates which occurred when responding to target queries were higher than for any other message type. The fifth word of each query presentation was either 'send' or 'confirm'. The subject could have recognised them both as instructions and, therefore, responded 'Order'. Least errors occurred for target orders as their meaning was definitely one of instruction. Another possibility for the increased number of errors which occurred with target queries is that the receipt of such a message in a real situation would require a report to be sent in response. On presentation of a query subjects may have recognised it as a request for information and responded 'Report'. Similar confusion may have occurred for all target messages as the real life response is likely to be different from the categorisation of the message, for example the response to an order may require the crewman to move the vehicle, the response to a report may be to add the information to a map and the response to a query would require preparing and sending information to another person.

With hindsight it may have been pertinent to include appropriate responses for the different types of message, for example noting down the grid he had been ordered to move to, where the enemy vehicle was reported to have been seen or that of his own location as a response to a query. It is likely that this method would have increased reaction times and may have elicited a more representative set of performance data in terms of errors rates. If further research is undertaken this should be considered.

This investigation has provided some evidence to suggest that a seated posture of 60° recline does not afford optimum vigilance performance. It is likely, however, that operational requirements will dictate external vehicle dimensions and, hence, the crewmen will have to operate in the reduced vertical space provided. Other working postures should be investigated in order to find the best compromise between task quality and user cost. A similar saving in floor to ceiling height could be achieved by sitting with the legs outstretched in front of an upright or slightly reclined torso as already used in other vehicle work stations, for example the cockpit of a Formula 1 racing car. It may be found, following further research, that none of the postures which allow adequate reduction of vertical space are ideal but some compromise may be possible to minimise the trade off between task quality and user cost.

The decrements in performance which occurred as a function of seat recline angle during this investigation were not statistically significant. The human cost, however, was significantly increased in terms of whole body discomfort, neck discomfort and perceived effect of discomfort as a function of seat recline angle. Other seated postures may exhibit similar performance with less discomfort. The implications of the human cost should not be underestimated, particularly as armoured reconnaissance crewmen may occupy their work stations and, therefore, working posture for continuous periods of 72 hours. During these extended periods crewmen may be required to conduct a number of tasks which vary in difficulty, duration and frequency.

One important aspect of the reclined seat is that it offers better quality of sleep than an upright sitting posture (Nicholson & Stone, 1987). If crewmen could get some good quality sleep, even for short periods (typically 2 to 4 hours), they may recover sufficiently to allow a more stable work rate throughout a typical battlefield mission. The general decrement in performance which may be caused by forcing crewmen to operate in a reduced height posture would have to be given due consideration. This may necessitate redesigning equipment and work procedures and possibly introducing new technology specific to the role of armoured reconnaissance. The introduction of flat screen technology to the vehicle work station may prove beneficial. For example if a flat screen was used instead of the cathode ray tube (CRT) in this investigation it would have been possible to position it above the reclined subject on the ceiling of the work station. This would enable a more relaxed posture, particularly in the neck region, potentially reducing user cost. Without further investigation this could not be assumed the ideal solution because of the possibility that the more comfortable, relaxed posture would result in a further reduction in the state of arousal and increase sleepiness which may reduce task performance.

It can be concluded, on the basis of this investigation, that a seated posture of 60° recline does not offer optimum performance. As discussed above, however, it may be necessary for soldiers to operate for long periods whilst in a reduced height posture. Further investigation is required before any recommendations, as to the best reduced height posture, can be made.

REFERENCES
Bartlett, D. E. 1943. in Stress and fatigue in human performance. ed. Hockey. (Wiley and Sons).
Corlett, E. N. and Bishop, R. P. 1976. A technique for assessing postural discomfort. Ergonomics. 19, 175-182.
Cox, T. and Mackay, C. J. 1985. The measurement of self reported stress and arousal. British journal of psychology. 76, 183-186.
Glaister, D. H. and Lisher, B. J. 1976. The effect of a reclined sitting position on psychomotor performance during exposure to high, sustained +G$_z$ acceleration. IAM Report No.561.
Horne, J. 1992. Stay awake, stay alive. New scientist. 4 January 1992, 20-24.
Lisher, B. J. and Glaister, D. H. 1977. The effect of acceleration and seat back angle on performance of a reaction time task. IAM Report No. 563.
Mackay, C. J., Cox, T., Burrows, G. C. and Lazzerini, A. J. 1978. An inventory for the measurement of self reported stress and arousal. British journal of social and clinical psychology. 17, 283-284.
Monk, T. H. 1979, 1984. Search. Chapter 9 in Sustained attention and human performance. Warm, J. S. (Wiley).
Nicholson, A. N. and Stone, B. M. 1987. Influence of back angle on the quality of sleep in seats. Ergonomics, 1987. Vol.30. 7, 1033-1041.
Yerkes, R. M. and Dodson, J. D. 1908. The relation of strength of stimulus to rapidity of habit formation. Journal of comparative neurological psychology. 18, 459-482.

This paper is a summary of a project undertaken at APRE (MOD) in part fulfilment of the requirements for the MSc Ergonomics degree, University of London. The views expressed are those of the authors and do not necessarily represent those of APRE, MOD or Her Majesty's Government.

AN INDUSTRIAL VIEW OF MANPRINT

Michael K. Goom

System Concepts Department (FPC500)
British Aerospace Defence Ltd.
Dynamics Division
BRISTOL
BS12 7QW

This Paper is concerned with the practical implementation of the MANPRINT initiative. It provides a very brief overview of what MANPRINT is and why it is important. It is based on a four year study of how to implement MANPRINT in one particular firm. It outlines the tasks that a contractor must undertake and some of the difficulties that this entails. In particular the need to express tasks and people in a manner that is both compatible and comprehensible to development engineers is established. The paper also proposes a model for the investigation of the trade offs in terms of user numbers, ability levels, and training that could assist system developers.

WHAT IS MANPRINT AND WHY IS IT IMPORTANT?

Definition

The definition of the US MANPRINT programme is as follows:-

"The Manpower and Personnel Integration (MANPRINT) programme is a comprehensive management and technical initiative to enhance human performance and reliability during weapon system and equipment design, development, and production."

The focus of MANPRINT is to integrate technology, people, and force structure to meet mission objectives under all environmental conditions at the lowest possible life-cycle cost.

The Problems that lead to MANPRINT

There are two principal problems that caused MANPRINT to be brought into being. The first concerns population and attitude changes, and the second concerns the increasing complexity of modern systems.

a) The Demographic Trough

There have been dramatic changes in the population distribution of most western countries. The numbers of people in the 16–20 year old range that are available to be recruited has fallen. Figure 1 shows the American demographic trough. There has been approximately a 28% fall from a peak during the late 70's and early 80's to the level available now. In Britain we do not have a demographic trough we have a demographic slide. Figure 1 also shows a 35% fall in numbers available for UK recruitment over a similar period and there is no significant recovery in the first decade of the next century.

Changes in peoples attitude to the geographical disruption that characterise the military life style is also having an effect on the aptitude levels of people putting themselves forward for recruitment. Despite the current unemployment figures fewer of the more able candidates are coming forward.

Figure 1. Demographic changes in the United States of America and the United Kingdom (1980—2005)

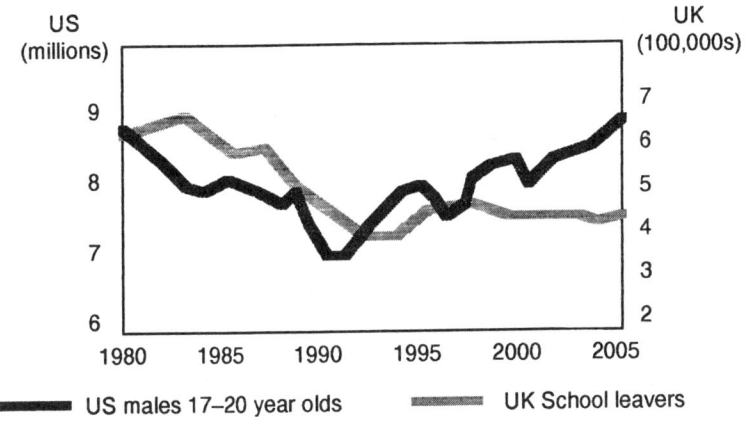

b) Skill Creep

The second main problem is the increase in the level of complexity of current systems. Until very recently there was a requirement to utilise the latest technology. Giving the user the 'best' equipment was seen as the way to provide the best military systems. There was the implicit assumption that people could be recruited and trained to operate whatever was produced. Unfortunately the more complex equipment required more able users and longer training periods. This need for ever increasing ability levels has been dubbed 'skill creep'. On a number of systems the difference between skills required and skills available was such that the systems failed and the MANPRINT programme was initiated.

Five MANPRINT Rules of Thumb

Miles (1991) uses 5 rules of thumb to show how the decisions a designer makes in relation to the equipment design effectively determine the tasks the user must perform and the overall system performance. The 5 rules are:–

• User Performance Affects System Performance,

  (The user is an integral part of the system)

• Skill is a Function of Aptitudes and Training,

• Measure User performance by Time and Accuracy

• Equipment Design Determines User Tasks

• Make the Designer Responsible for User Performance

The MANPRINT Domains

For the implementation of MANPRINT a number of specific areas have been identified. However, it must be stressed that MANPRINT is succeeding where many previous initiatives failed because it concentrates on the integration and trade offs between these areas (or domains). Some of the MANPRINT type programmes use variations on these areas but essentially they are the same. The six domains used by the Army both in the United States and United Kingdom are manpower, personnel, training, human factors engineering, health hazard assessment, and system safety.

Manpower refers to the total number of people (operators, maintainers, trainers, support staff, etc.) associated with the system.

Personnel refers to the types of people required. This includes their physical , mental characteristics together with their knowledge and experience.

Training considers how the required extra skills and knowledge can best be delivered to the personnel who will have to use the equipment.

Human Factors Engineering is concerned with optimum allocation of functions to man and machine and fitting the tasks and jobs to the user characteristics.

Health Hazard Analysis considers those aspects of the system such as fumes, vibrations shocks, etc. that could cause injury, illness or reduce job performance of personnel.

System Safety refers to inherent ability of the system to be used and maintained without accidental injury to personnel.

THE CONTRACTOR'S MANPRINT TASK

Identifying potential solutions

The contractor's task is to analyse the mission and generate a number of options that could possibly satisfy the need. This is shown in Figure 2. Then to determine which of the options produces the best performance. In the MANPRINT context this means breaking each of the options down into constituent tasks and determining which produces the best mapping onto the user population.

Figure 2. The Contractor's MANPRINT Task

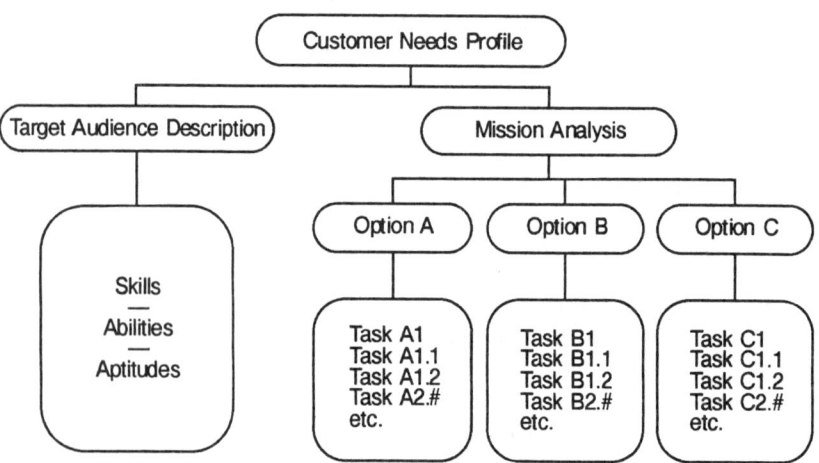

The task analysis is one of the key elements in achieving a regime in which this mapping can be approached in a systematic manner. It should also be pointed out that a large number of other processes in system development rely on an analysis of the tasks to be performed. Unfortunately there can be a wide gulf between a task analysis produced by a design engineer and a human factors specialist. Great efforts are now being directed to understanding the process of analysing tasks to produce task lists and descriptions that are suitable for all the subsequent steps, (Kirwan & Ainsworth, 1992). There are many people within industry who are looking for methods that allow task analyses to be undertaken in a consistent manner, quickly and efficiently. It should be emphasised the 'task analysis' in the above context refers to the three stages; data collection, description and analysis (Stammers et al, 1990).

Matching tasks and people

To achieve the match between tasks and people, the description of the potential users needs to be in terms of skills, abilities and aptitudes that can be equated to the task description parameters derived for each option. The Target Audience Description (TAD) is the second key element of an effective MANPRINT implementation and its optimum structure and content is vital to success. This is discussed in the next section.

The equipment designer determines the user tasks

It is important to remember that the development of the hardware and software portions of the system effectively defines the tasks that the users must undertake. It is therefore important that the systems designers are working with task and people descriptions that they understand. It is also important that they realise that they are also designing the users jobs.

Engineers and designers in industry are used to developing individual pieces of equipment. As ergonomists who are used to taking a multi-disciplinary approach, it is too easy to overlook this particular aspect. Demonstrating to the system developers that their equipment design can have a profound effect on the users tasks and hence the overall system performance is often the MANPRINT challenge in industry.

DESCRIPTIONS OF USERS

Contents of a Target Audience Description (TAD)

A joint MoD/Industry Working Group has spent a considerable time investigating the information that could be made available in the TAD, and what information designers could use effectively. The co-operation between the Ministry and industry has been extremely useful in establishing what user information is available and establishing additional information industry may require. It should be emphasised that this issue is far from resolved and any constructive ideas would be most welcome.

Physical attributes

The physical attributes are relatively straight forward. Size and strength data are readily available and understood (and miss-used) by many engineers. Anthropometic information is available from DEF STAN 00-25 and MIL SPEC 1472. In a similar way vision, hearing ..... information is available. However, unless the information and guidelines are in a form that are easy to locate and apply they will be ignored. System developers are usually working against very strict time scales and have many more (better defined) problems to address than the time allows.

'Mental' attributes

The specification of the perceptual, cognitive, psychomotor performance that the users are likely to exhibit is far less well defined. How do you specify the potential tracking ability of the proposed user population in such a way that the system developer can assess options? Ergonomists can achieve this, but, it has usually been on a case by case basis. The object of MANPRINT is to design the human contribution into ALL future systems and in the case of the author's division of British Aerospace this represents some 20–30 projects every year. For this reason the case by case approach is no longer a viable option.

Knowledge

The knowledge that members of the potential user population already posses could be a major driver in the design of the system under consideration. Training is one of the most costly items in the whole life cycle of a military system, and an option that builds on existing knowledge may be preferable to s simpler, cheaper (initially) one that does not. The question is – how can this information be communicated to the development engineers?

Organisation

When describing the users it is often overlooked that they exist within a framework that imposes considerable influence on the ways in which they can operate. All too often engineers and designers (and ergonomists) look at users as individuals only. The constraints of the military command structure are frequently missed by designers until it is too late and a significant portion of the allocation of function has been frozen.

The target audience description must provide information on which users have the authority and knowledge to undertake which tasks in addition to which have the ability to undertake different types of tasks.

THE TRADE OFF TASKS

The development of the task and target audience descriptions will be an iterative process, however, if we assume that they can been specified adequately the question of trade off arises.

MANPRINT differs from many of the earlier attempts to formally integrate human factors into military and civilian programmes in that it stresses the interdependency of the areas (or domains). This laudable aim is easy to talk about, but at a practical level in an industrial setting it provides an interesting challenge. To make MANPRINT work we need to be able to assess the implications of trading factors such as numbers of people against the ability level of those people; the performance implications of increasing the training time against recruiting more able people.

Three levels of trade off have been identified these are:-

a)   Manpower, Personnel & Training

b)   All six MANPRINT Domains

c)   MANPRINT and the other project parameters

Only the first and theoretically simplest is considered at present.

Manpower, Personnel and Training

One of the tasks facing the application of MANPRINT is the MPT trade off. This considers the optimum mix of numbers, ability levels and training. There are a number of models that can analyse a potential system solution in terms of the probable cost that it implies. They can also check that any constraints on the maximum numbers have not been breached. However most are designed for the analysis of systems after they have been specified. The MPT trade off model that the system developer requires includes the tasks.

The model that we are currently developing is shown in Figure 3. It consists of linkages between the three domains, but it has at its centre the tasks that the system option requires. Using the task descriptions that contain information on the abilities required for the tasks and estimates of the training load, both basic and continuation, the system developer can begin to investigate the implications of task combinations. The aim is to allow the identification of the critical tasks and make an assessment of the effects of their elimination or automation using sensitivity analysis.

At this stage the model contains a number of rules to link the domains. In some cases are highly speculative. However it is the authors contention that there exists,

Figure 3. A Model for Manpower, Personnel, Training trade off against tasks

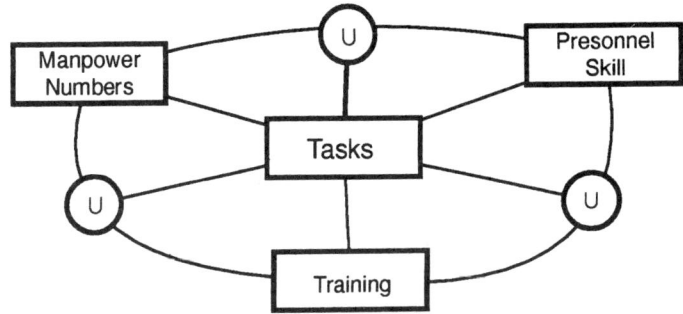

in the literature and in the bottom of researchers cabinets, sufficient information to produce a viable model. It was only by trying to build the model that the area where information was needed were identified. It is also proving of value in testing whether knowledge of a particular piece of information would indeed help the design process.

The types of information required

The information required is of the type that would commonly be used by training specialists, those involved with recruitment, system safety officers and military planners.

What is the effect of aptitude on training times?

How do you trade numbers against aptitude?

Selection of training strategy and methods.

How many privates can a corporal be expected to look after?

As can be seen the questions are simple enough. Finding suitable information and fusing it into a model that is suitable for system designers is one of our current goals.

CONCLUSIONS

MANPRINT offers an ideal opportunity to integrate human factors into military system design. The formal requirement is there from the customer, and the financial incentive is there for industry. There is an enthusiasm for the gains that MANPRINT is offering. The urgent need is to apply the research findings from many areas of ergonomics within a framework that is practical for industry. What are required are methods that can be simply applied and that produce results that can be used to assist system design. Most importantly the methods must provide useable results and they must be achievable now or in the very near future if the enthusiasm is to be sustained. Finally it must be stressed that for a successful implementation of MANPRINT within the UK the close co-operation between the MoD and industry must continue.

REFERENCES

Kirwan, B. and Ainsworth, L. K., (Eds) 1992 A Guide to Task Analysis, (London: Taylor & Francis)

Miles, J. L., 1991, Handbook for Conducting Analysis of the Manpower, Personnel, and training Elements for a MANPRINT Assessment, US Army Research Institute Research Note No 91-43

Stammers, R. B., Carey, M. S., Astley, J. A., 1990, Task Analysis. In Evaluation of Human Work. eds. J. R. Wilson & E. N. Corlett. (London : Taylor & Francis)

MANPRINT/ERGONOMICS IN USER FRIENDLY AND BIODEGRADABLE MILITARY FOOD SERVICE PACKAGING

L. E. SYMINGTON, T. MALAFI, J. KALICK & M. DEVINE

Behavioral Sciences Division
Soldier Science Directorate
U.S. Army Natick Research, Development & Engineering Center
Natick, Massachusetts 01760-5020 USA

The U.S. Army Natick RDE Center addressed several
food packaging ergonomic issues including establishing a
profile of Navy users relevant to design, marketing and
use of packaging and other items made from biodegradable
polymers. New packaging with appealing graphics and
user information was created for the U.S. individual
military combat ration. It also provided functional
innovations such as a ziplock closure while retaining
camouflage characteristics. Field and laboratory
evaluation data not only show preference and improved
ergonomics, but the identical entree was rated more
acceptable in the new packaging.

INTRODUCTION

One of the five essential capabilities of the U.S. Army
Soldier System is the sustainability of the soldier. Food and
water are obviously the major contributors to this capability and
are critical for all military services. The U.S. Army Natick
Research, Development, & Engineering Center (Natick) is
responsible for this area for the U.S. Department of Defense.

The packaging of food, and to a lesser extent, water has some
particularly unique military requirements. Any field ration
packaging, for example, must be camouflaged, have enough strength
to resist damage from being carried into field combat situations
by a combatant and, in the future, will likely be required to be
totally biodegradable. The latter constraint becomes
particularly critical for the U. S. Navy when, effective January
1994, Annex V of the Marine Pollution Treaty (MARPOL) will
prohibit the dumping of plastic waste into the world's oceans.

This paper will examine some of the MANPRINT/ergonomics issues
Natick is pursuing in food packaging and containers.

Three of the critical issues in the marine biodegradable area
are communicating new product information and benefits to Navy
personnel, maximizing usage through the creation of user friendly
products and waste disposal systems and insuring compliance with

changes in shipboard procedures brought about by the new products. Natick MANPRINT work in these areas includes the establishment of a profile of U.S. Navy users relevant to the design, marketing and use of packaging and other items made from biodegradable polymers. In addition, user evaluations of specific products are being conducted and data will be presented from an evaluation of an all paper cup developed as an alternative to the plastic lined cup currently used.

Human factors personnel at Natick have designed new packaging for the Meal-Ready-to-Eat (MRE), the U.S. individual military combat ration, by creating appealing graphics, including user information and providing functional innovations, while retaining camouflage characteristics. The prime functional innovation was ziplock external packaging for alternative later uses, such as protecting a pair of dry socks from the elements -- a critical capability for preventing trenchfoot and other debilitating foot conditions.

U.S. NAVY USER PROFILE

As a necessary first step in a marketing program concerning the use of biodegradable materials at sea, Natick is developing a psychographic profile of the Navy consumer with respect to marine environmental issues. Since the purpose of such a profile is to provide information about the eventual user of a product in terms of his/her attitudes, activities and interests, this psychographic analysis explores Navy consumers' views on issues related to the plastics problem in the marine environment. The profile includes sailors' knowledge and opinions about the problem, their perceptions of its seriousness, and their attitudes toward possible solutions (e.g., biodegradable materials). This type of information will aid in the development of communication strategies to educate Navy consumers and to market any products.

Method

A study was conducted in late 1991 on two Navy ships in San Diego, CA and and two in Pearl Harbor, HI. Three hundred sixty nine sailors, 90% being enlisted personnel, completed a survey with scaled opinion, multiple choice factual and open-ended opinion questions.

Results and discussion

The sailors surveyed were concerned about the plastics problem at sea and viewed the dumping of plastics overboard as a serious threat to the marine environment. Many also agreed that making waste disposal at sea less environmentally harmful should be a priority. Using plastics that decompose naturally in the environment when discharged (biodegradable) was mentioned by more sailors than any other method as being the best way for the Navy to solve the plastics disposal problem.

Over half the sailors questioned saw compliance with current waste disposal procedures as "easy" or "very easy." In spite of

this, only 10% of them reported having had no problems with the
system. Most frequently mentioned concerns were finding space on
board ship (61%), odors emanating from waste (55%), separating
plastic from other trash (55%) and lack of sailor compliance
(50%). Of all difficulties, finding storage space and lack of
sailor compliance were rated as most serious by most sailors.

The majority of sailors (87%) could identify a simple
definition of "biodegradable." Less consensus was demonstrated
concerning the impact of using materials made from biodegradable
polymers, i.e., the effect of the dumping and disintegration of
biodegradable plastic on the marine environment. A relatively
large percentage of sailors responded to questions concerning
beliefs about biodegradable materials with "don't know." For
example, when asked how long it would take a biodegradable
plastic to dissolve at sea, 44% of the sailors admit that they
don't know.

For the most part, all communication strategies were perceived
as relatively effective means for communicating information about
biodegradable materials. Direct orders by superiors was seen as
most effective, followed by information at Quarters, information
in the Plan-of-the-Day, presentations by other sailors and
videos. By comparison, mailing brochures was seen as the least
effective method.

Conclusion

In summary, the results of this study indicate that sailors
are open to information dealing with marine environmental issues
and biodegradable materials. The sailors see the plastics
problem as real and think it worth doing something about.
Furthermore, although sailors appear to have some idea, albeit
relatively simplistic, about the concept of "biodegradability,"
they are less clear as a group about the concept of using such
materials. They do not, however, seem negative about relying on
biodegradable materials to solve the plastics problem at sea.
In fact, many believe that using products made from these
materials is the best way for the Navy to deal with the problem.
Therefore, rather than having to change an already existing
negative attitude, the opportunity exists to create positive
beliefs about biodegradable materials. However, some of the
problems cited, such as lack of sailor compliance with
procedures, will need to be addressed since high levels of
compliance are essential to the success of any waste disposal
program.

USER EVALUATION OF AN ALL PAPER CUP

As a first step in user evaluation of specific products,
Natick human factors/MANPRINT personnel tested an all paper cup
that was developed as an alternative to the standard plastic
lined paper beverage cups currently used by the Navy.

The evaluation was conducted on board two ships at the U.S.
Naval Station, Newport, RI. At one noontime meal, each ship
received only one type of beverage cup with no other beverage

glasses available. The beverages that were consumed were not controlled in any way. One hundred fifty-seven enlisted sailors, all male, used the cups and completed questionnaires evaluating them in terms of sturdiness, ease of use, shape, insulation properties, flavor transfer, desire to use again and overall acceptability.

On a 7-point scale, the all paper cup was rated as significantly sturdier than the plastic lined cup. In addition, sailors professed significantly greater overall liking for the all paper cup. Sailors´ estimates of the ease of drinking from the cup did not differ significantly between the two.

The insulation properties of the cups were also evaluated. Ninety percent of the sailors reported that the cups maintained the temperature of the beverage. Those who used the plastic lined cup were more likely to report that it did not preserve the beverage temperature (16% plastic lined vs. 3% all paper). It should be noted, however, that since 80% of the sailors drank a cold beverage, no definite conclusions can be drawn about either cup´s insulation abilities for hot liquids.

In conclusion, the all paper cup was very well received by the sailors, and all results indicate that the all paper cup was acceptable to users in every evaluation category; and it was recommended that it be adopted as an alternative to the plastic lined cup.

CONSUMER ORIENTED FIELD RATION PACKAGING

U.S. Army field rations are often not consumed in sufficient quantity to maintain optimal performance. Traditional efforts to deal with this situation have focused primarily on food product formulation, processing and preservation technologies. Recently, Natick human factors personnel have been expanding the strategy to maximize ration acceptability and consumption to include social and situational influences. The food service industry has repeatedly shown that packaging can also be a major factor encouraging product usage.

Human factors personnel at Natick have designed new packaging for the Meal Ready-to-Eat (MRE), the U.S. individual military combat ration. Seventy two soldiers/consumers participated in six focus groups (12 soldiers in each) in order to design layouts for MRE packaging prototypes which would promote a positive product image. A focus group involves a trained moderator who facilitates and focuses group interaction. Qualitative ideas generated by these groups, which included colors, names, information needs, shapes and sizes for foods and their packaging were used to develop final graphic layouts and several mock-up designs for the packages. Three final packaging prototypes were designed from these data, and in a subsequent field evaluation at Fort Ord, CA, one packaging type was chosen for further comparison to the current MRE package.

Method

A total of 192 soldiers participated in a Fort Cambbell, KY
lunch study of three packaging alternatives in three different
field locations. The meal was identical for each group.

Two of the packaging systems were based on the design
developed in the earlier focus groups and field test. The first
of these, the FIELD package, was a medium olive-green color, had
a tear strip and ziplock closure and the name FIELD BREAK printed
in yellow. The food contents were listed below the menu number
in yellow. The second, the DESERT package, was a beige, tear
strip, ziplock pouch with the FIELD BREAK logo printed in light
tan from the bottom to the top of the bag. A bullet on the right
side stated, "A well-balanced and complete meal." The entree
name, chicken stew, was printed in green on the right side along
with the other food contents. The third, the MRE package, was
designed to be identical in appearance to the current MRE package
except for the FIELD BREAK name. It was dark brown with black
lettering and had the words FIELD BREAK, MENU NO. 3, CHICKEN STEW
printed in a repeat pattern from top to bottom. Soldiers at each
site were provided with a meal pouch, given a brief description
of the tear strip and ziplock closure of the bags and an
acceptance/consumption questionnaire.

Results and Discussion

Soldiers rated a series of attributes focusing on the
appearance, functionality and design of the packages. The
following attributes received significantly higher ratings from
those evaluating the FIELD and DESERT packages: More Attractive,
Interesting, Unique, Colorful, Informative, Easier to Store,
Easier to Read, Communicating Contents Well, and Having Labeling
that Makes the Food More Appealing. All three groups were
enthusiastic about the tear strip and the ziplock. Many reported
several alternative uses for the ziplock outer package -- the
most common being to keep socks dry. Despite the fact that the
chicken stew entree was identical in all three packages, FIELD
and DESERT package users rated the chicken stew as significantly
more acceptable than did the MRE package users.

Finally, questions concerning content labeling indicated
soldiers preference for a system that proposed using colored bar
graphs showing the nutritional values in the ration next to
recommended daily guidelines. They also wanted to know the
caloric values.

REFERENCE

Malafi, T. and Devine, M., 1992, Navy Consumers' Beliefs About
Marine Environmental Issues and Biodegradable Materials. U.S.
Army Natick RDE Center Technical Report NATICK/TR-92/037.

The opinions expressed are the authors' and do not necessarily
represent those of the U.S. Army Natick R, D, and E Center,
Department of the Army or Department of Defense.

Predictive Workload Analysis - RN EH101 Helicopter

*I. S. MacLeod, K. Biggin, J. Romans, K. Kirby.

Westland Helicopters Limited (WHL)
Yeovil
Somerset, BA20 2YB

*Aerosystems International (Ael)
Yeovil
Somerset, BA20 2AL

The RN EH101 maritime patrol helicopter is to enter service with the Royal Navy near the end of the decade. The prime contractorship for this weapon system development was awarded to an IBM led IBM ASIC/WHL consortium in early 1992. A Human Engineering Programme Plan (HEPP) was mandated by the MoD contract and is being managed by WHL. A major HEPP requirement is for a predictive workload analysis of aircrew tasks. This paper covers aspects of the predictive workload analysis conducted by WHL in consultation with Ael. The predictive analysis includes two complementary approaches to the study of workload. One uses a time line based 'snap-shot' approach which considers mission task time requirements versus the task time available and task associated attentional demand characteristics. The second approach uses a novel technique examining task related decision processes and their associated errors. This paper concentrates on the second approach. To set the paper's context, the 'snap-shot' approach to the analysis is briefly discussed. The results from the predictive workload analysis will be validated under full simulation conditions in mid 1994.

INTRODUCTION

The overall process of RN EH101 workload analysis was composed of a series of analyses that covered mission analyses, task analyses, empirical observation and the use of knowledge elicited from Subject Matter Experts (SMEs). The overall analysis process will be briefly discussed as an introduction to a more detailed discussion on the decision and error analysis component of the study. A necessary aspect of all the analyses was an emphasis throughout on the importance of customer participation in the conduct of the study.

Purpose of predictive workload analysis

The purpose of the this workload analysis was to allow an early prediction of EH 101 aircrew workload as an aid to the assessment of the efficacy of the weapon platform design. Workload can have many definitions depending on the context and the audience. For the purpose of this paper the definition by Hart (1982) will be used, namely:

"Workload is a subjective experience caused by external and internal factors such as motivation, ability, expectations, training, timing, stress, fatigue and circumstances in addition to the number, type and difficulty of tasks performed, effort expended, and success in meeting requirements."

On the assumption that excessive peaks and troughs in workload are generally not conducive to the best performance of a mission, the specific purposes of the analysis were considered to be:

- The location of peaks and troughs in aircrew workload;
- The determination of the causes of those peaks and troughs;
- To suggest solutions to the amelioration of unwanted workload (through design, operating procedures or training).

Mission analysis process

The mission analysis process was based on high level mission descriptions supplied by MoD. A questionnaire was designed to allow the MoD and SMEs to aid in the careful selection of the most pertinent segments of the missions for subsequent workload analysis. Such a careful selection is important as:

- It is sensible to focus the analysis on the areas of the mission most pertinent to the study of workload;
- Given the finite resources allocated to any study, it is necessary to bound the scope of the analysis process;
- Careful selection of the area of study gives important face validity to the basis of the analysis process.

The selected mission segments spanned nearly two hours of mission performance over three different RN EH101 missions. As no one method of studying workload possesses a high construct validity (not surprising considering the disagreements on what workload is), the intention of the study was to analyse workload in detail using several different approaches as a means of promoting the study's criterion validity. To aid the replicability and face validity of the study, the subsequent workload analysis plan was designed to promote the traceability of the analysis process. Such traceability was considered essential to the study to ensure that:

- The logic and sequences of the analysis process were rational and understandable.
- Each step of the overall analysis process could be defined and related to the other steps.
- Analysis could be performed on any other mission segments using the same method and with the minimum performance of any new analysis.

Task analyses

The analysis of operators work tasks is an essential prerequisite to any predictive study of workload. Through analysis, tasks must be defined with respect to their purpose, nature, complexity, time to complete and interrelationships with other tasks.

Therefore, the details on the essential tasks for the performance of any of the chosen missions were elicited from SMEs. Using these essential tasks, a 'story-line' was created to cover the chosen segments for each missions. This 'story-line' was amended, amplified and agreed with SMEs in order to achieve a realistic mission description suitably for the support of following and more detailed task analyses.

The 'story-lines' were then amplified to create Operational Sequence Diagrams (OSDs) at the level of aircrew subtask/activity. The particular timings and specific detail of equipment related items in the OSDs were determined through detailed empirical observations on aircrew operation of equipments intended for the aircraft. The OSDs were then presented to the MoD and approved. Static time-line and attentional demand studies of workload were then conducted at the lowest level of detail possible but in line with the OSD subtask depiction and subtask sequences. Figure One depicts the sequence of the workload analyses described above and illustrates its relationship with the subsequent and dynamic decision and error analysis process.

Figure 1. Summary Diagram of Analyses

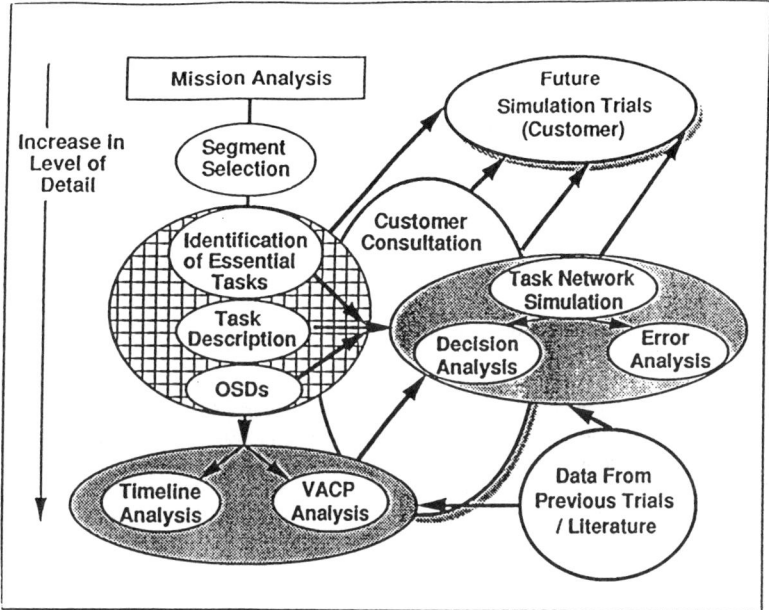

DISCUSSION ON HUMAN DECISION PROCESSES AND ERROR
General
   Prior to discussion on the decision and error analysis, it is necessary to discuss the underlying tenets of the analysis. Firstly, the human ability to make decisions directly contributes to the quality of their work performance. Decision processes are deemed to be required by an operator in any situation where there is more than one obvious course of action. Decision processes include the scope for making errors. However, error processes may or may not involve decision processes.

   The ability to make decisions can be partly developed by appropriate training. Indeed, one of the benefits to be gained from the results of an analysis of operator decision processes should be the tailoring of operating procedures and training to develop and support the operator's decision processes.

   However, regardless of the quality of training, decision processes are also affected by forces outside the conscious control of the operator, including influences from external events that may not have been forseen by training. In addition, operator decision processes are highly affected by the nature and demands of the work task, the working environment, organisational influences, the experience of the operator and the individual characteristics of the operator.

   To lessen the effects of confounding influences on the decision process, procedures are necessary to make decision forms as predictable as possible. Thus, one of the purposes of operating procedures is to aid the operator's ability to make decisions. In support, the presentation of task-related information must be unambiguous. The presentation of information must also be appropriate in form and time to the operator's requirements for evidence to support decision processes.

   If possible all decision processes should be decomposed into a series of simpler forms (such as decision on a choice of two alternatives), few of which are in themselves critical to the successful completion of a mission. Decisions are based on judgement which is a cognitive operation that may or may not lead to an action or choice. However, choice does commit the decision maker to action. Decision processes will deteriorate if confounding influences affect operator judgement and force decisions processes from a simple to a multi-choice selection.

Decision and error effect human performance and hence human reliability. The analysis of human reliability is concerned with the accuracy of human work performance and its converse of human error. Human error can be taken to encompass any action adverse to the completion of a task or an overall mission. Human error occurs in many forms that appear to be common to all human operators.

Error Forms

To allow the determination of a Human Error Probability (HEP) associated with a task, or a task based decision, the possible human errors must be first identified, their causes determined and the errors then quantified in terms of frequence of occurence and severity of effect. To define the initial probabilities of an error occuring, task information must be initially matched to either probability data obtained from the literature or from the judgements obtained from SMEs.

The main emphasis of this error analysis was on critical operator errors associated with critical decision processes. Here, critical is defined as pertaining to any action or decision that places the achievement of mission goals in jeopardy. To allow SMEs to understand the concept of error, it was necessary to devise and present categories of error forms and error severity. Basically errors were seen as mainly attributable to the performance requirements of a particular task or to the influence of operator ideosyncrasies. Table One gives the five forms of error considered by the study and the error severity classifications used.

Table 1. Classification of Error Forms and Severity

ERROR FORM	EXAMPLE
1. Deliberate Breach of Rules.	I decided not to follow standard practice in this case because I thought of a better method.(Trial & Error).
2. Incorrect Application of Rules.	I believe that I am following correct procedures but your telling me that my method is not correct in the light of the evidence (Tversky & Kahneman (1974), Reason (1990)).
3. Misperception of the Evidence.	Not perceiving or mistakenly perceiving information (Hancock (1987), Reason (1990)).
4. Mistake.	Mistakenly thinks he knows the correct process thus makes a mistake in the execution of that process (Rasmussen (1986)).
5. Slips and Lapses.	Have high skill and following the correct procedure but attention wanders or focuses on something extraneous (Rasmussen (1986)).
SEVERITY CLASS/MARK	DESCRIPTION
1.	Certainly won't complete the task successfully.
2.	High probability of failure.
3.	Fair probability of failure.
4.	Low probability of failure.
5.	Effects the time taken to complete and not success.

DISCUSSION ON DECISION AND ERROR STUDY AND ANALYSIS

Purpose

The purpose of the decision and error analysis was to support the overall study of predictive workload for the RN EH101 but particularly to investigate:

- The effects of operator error on subsequent decision processes;

- Operator task performance errors and associated decision processes using agreed mission scenarios;
- The areas of operator task performance most critical to mission success;
- The extent to which the results corroborate or clarify the results obtained from previous analyses;
- The areas of operators' mission performance that may require support through one or more of the following:
  * Changes in design;
  * Formulation of operating procedures;
  * Formulation of tactics;
  * A specific emphasis during personnel selection and training.

Assumptions used

The decision and error analysis was performed under the following assumptions:
- That errors made in the performance of a task influence subsequence decisions;
- That the critical decisions to be examined reside at the tactical level of operator performance;
- That error influence is a function of error probability and error severity;
- That all error is mapped against time and considering:
  * A basic environmental model (e.g. time in contact with a target);
  * Operator task execution times.

Tool

The study used the PC based task analytic simulation language encapsulated in MicroSAINT for Windows (MSW). The use of MSW allowed the study of critical decisions and errors through the dynamic simulation of various possible avenues to operator task completion using monte carlo rules. The tool supports a pragmatic but stochastic approach to the problem of studying human performance. Empirically collected performance data can be utilised and models can be behavioural orientated to depict crew interrelationships and the effects of mission events on human performance. Moreover, MSW allows the presentation of a map of the model in a form easily understandable to SMEs.

Inputs and their relevance to this use of MSW models

The inputs to the MSW models were several and were chosen to complement each other. Thus task error assessments obtained from the SMEs were compared with actual results obtained from operations and trials; the designed performance of the aircraft and its planned equipments was compared with the mission performance requirements inherent in the mission 'story-lines'. Finally, the reliability of the operator in the performance of his tasks was considered alongside the planned equipment performance, this to determine their joint influence on the probability of achieving mission success as defined by the 'story-lines' depicted in MSW.

Form of results

Broadly, the form of the study results allowed the following areas to be investigated:
- The efficacy of tactical decision and the associated probability of mission success;
- Queries from SMEs concerning usage of the aircraft equipments;
- The extent to which established design supported operator performance, especially the reliable and timely completion of assigned tasks;
- The causes of unacceptable workload as determined by previous analyses.
- New avenues on the use and combination of aircraft equipments and crew.

DISCUSSION ON METHOD
The decision and error analysis was conducted by two Human Factors Engineers over a 7 month period. Over 10 SMEs aided the study. MSW was hosted on a 50Mhz 486 PC using DOS 5, Windows 3.1, VGA and over 4Mb of both RAM and space on the hard disk.

Data from the SMEs on task error probabilities and severities was input into MSW under the values of the error forms and severities obtained from completed questionnaires. The SMEs were given detailed instructions prior to each session.

The seed data for the MSW models was obtained from:
- The mission 'story-lines';
- A series of questionnaires completed by the SMEs to determine the task error probabilities and severities for each model;
- Crew performance data obtained from in-service simulators and aircraft trials;
- Error data obtained from the literature;
- Task timing data obtained from trial simulations on the planned RN EH101 aircraft mission and management systems.
- Flight performance data obtained from MoD and WHL.
- Data on the functionality and performance requirements for the proposed aircraft equipments obtained from various sources.

This paper is too short to allow detailed coverage of MSW characteristics and the detailed model build. Basically, error calculation algorithms were devised to utilise the various types of seed data. The results obtained through the algorithms were used to influence the MSW task timings, the decision/choice logic of the model and the values of certain model variables.

Each stage of the study was documented and agreed with MoD. The penultimate models were 'tuned', post a final appraisal by the SMEs, and then run several hundred times to obtain the final results.

IN CONCLUSION
This form of investigation using MSW is novel in the UK but has precedent in North America (e.g. Griffith and Stewart (1991)). Considering the current analysis performed with the method, it is believed that it may be used to sensibly investigated human decision and error processes. In addition, it has shown its worth as a means of assessing and investigating areas of operator performance critical to the success of RN EH101 missions. The method has also demonstrated complementarity to the other forms of workload analyses used in the study.

REFERENCES
Griffith, W.E. and Stewart, J.E., 1991, Computer Simulation Model of Cockpit Crew Coordination: A Crew-Level Error Model for the U.S. Army's Blackhawk Helicopter, Research Report 1601, U.S. Army Research Institute for the Behavioural and Social Sciences, Fort Rucker.
Hancock, P.A., 1987, Human Factors Psychology, (Amsterdam: Elsevier).
Hart, S.G., 1982, Theoretical basis for workload assessment research at NASA-Ames research center, in Frazier, M.L. and Crombie, R.B. (Eds), Proceedings of the workshop on flight testing to identify pilot workload and pilot dynamics, Edwards Air Force Base, CA, AFFTC-TR-82-5.
Rasmussen, J., 1986, Information Processing and Human Machine Interaction, (New York: Elsevier).
Reason, J., 1990, Human Error, (Cambridge: Cambridge University Press).
Tversky, A. and Kahneman, D. 1974, Judgement under uncertainty: Heuristics and biases, Science, 185, 1124 - 1131.

Information presentation and recognition

A COMPARISON OF THREE MAP FORMATS

Y. SHEK[1], P. GOSLING[2] and P. BARBER[3]

[1]Defence and Civil Institute of Environmental Medicine,
1133 Sheppard Ave. West, P.O. Box 2000,
North York, Ontario, M3M 3B9, CANADA

[2]Army Personnel Research Establishment,
Farnborough, Hampshire, GU14 6TD, U.K.

[3]Psychology Department, Birkbeck College,
Malet Street, London, WC1E 7HX, U.K.

In this study, operator performance on three types of map systems were compared: conventional map and radio, digital map with symbols and text, and digital map with symbols only. Measures included: reaction time, error rates, NASA-TLX subjective workload scale, and time estimation. Reaction time results show highly significant differences between the three experimental conditions as predicted. Error rates were not statistically significant. However, the trend was in favour of a digital-symbols only system over digital-symbols-text and conventional systems. NASA-TLX results show significance between the conventional system and the two digital map systems but not between the two digital map systems. Time estimation results show significance when not balanced for time-on-task but show no significance when time-on-task was taken into consideration.

INTRODUCTION

The present study focuses on the attentional aspect of map using. The assumption for human factors input and task design is that maps should be tailored to the context-task-user configuration (Barber et al., 1984). The main goal of this study is to compare map systems for field use by measuring human performance. Also important are the different measures of human performance. Included are objective and subjective measures.

Presently, there are at least three ways to achieve a map-reading and enemy engaging task by the vehicle commander of a tank or armoured vehicle: (1) conventional map and verbal commands; (2) digital map on a visual display unit (VDU) that has NATO symbology and includes text; (3) digital map on a VDU that only includes the presentation of symbols without text. The objective of this study is to compare the latter two types of representation by comparing operator performance (reaction times and error rates). The conventional method will be included so as to see the magnitude of performance change between it and the new digital systems. It is believed that the magnitude of the differences between system 1 (conventional) and systems 2 or 3 (digital) would allow an overall view of how the present system compares with the future systems, and how the two future systems compare to each other. Such comparisons would hopefully yield results of human performance limitations associated with systems and multiple resource theory (Wickens, 1984).

Objective measures

Primary task measures such as reaction time are simple to use and sensitive to workload changes even for conditions involving different sensory modalities. Reaction time measure assumes that total reaction time equals the sum of the duration of a number of component processing stages (e.g. perceptual encoding, response selection, etc.) (Wickens, 1984). Thus, the more mental steps are involved in the decision making, the longer the latency of the response should be expected. In the present study, one is not concerned about each of the components of the decision process, but more so in the whole of the decision itself, since it is the final decision that is of practical interest.

Error rates were recorded also as a measure of performance. Error rates are often analysed in conjunction to reaction times as a primary measure. Any time-accuracy trade-offs may be noted and analysed when the two are used as measures together. The present goal in using error rates as a measure is to predict human reliability with respect to each of the three systems/conditions (Adams, 1982; Sheridan, 1981).

Subjective measures

The NASA-TLX scale is made up of six subscales: mental demand, physical demand, temporal demand, performance, effort, and frustration. These six factors are thought to be independent factors, each contributing to the overall perception of workload (Hart & Staveland, 1988). This is particularly useful since mental workload is thought to be multidimensional in nature (Liu & Wickens, 1992 unpublished). The second part of the TLX test involves 15 pairwise and exhaustive comparisons of the six factors mentioned above. The TLX was used in this study since it is capable of solving workload problems in many applied settings, including systems development and workload prediction (Nygren, 1991). The TLX not only takes into account the multidimensional nature of workload, but also individual differences. Also, the final score is a percentage and can be treated as continuous data in the analysis. Unlike many subjective scales that involve non-parametric analysis, TLX data can be analysed using more robust parametric tests.

Each session/experimental condition was followed by asking the subject how long he thought he took to complete the task, excluding trial runs. This time estimate (TE) in minutes was compared to the real time in minutes. The difference in absolute value was then analysed statistically (see TE in Results). The absolute value was also divided by the total time used (as percentage) and analysed statistically (see 'Time' in Results). These measures were included to test the applicability of Liu & Wickens' results (1992 unpublished) to this study. Liu & Wickens' (1992 unpublished) study showed that time estimation intervals produced while performing two simultaneous tasks had significantly greater length and larger variability than those produced while performing one task alone or performing a well practiced task.

EXPERIMENTAL METHODOLOGY

The three conditions (see Introduction above) were compared using a repeated measures within-subjects design. Performance was judged by comparing reaction times, error rates, NASA-TLX results, and time estimation results, by ANOVA and protected t-tests. The three tasks for the three systems were designed to be balanced for compatibility and difficulty. To control for learning and fatigue effects, the three conditions were presented in a balanced schedule over the three days of experiment. Also, there were three different trial scenarios (similar in difficulty) - one for each of the conditions - so to avoid learning effects. The scenarios, as well, were presented in a balanced schedule over the three days (Latin square design).

<u>Apparatus</u>

The Macintosh Quadra 700 with the Apple color High Resolution RGB monitor were used to generate the three experimental conditions. Condition 1 did not require the monitor to be used however, since the stimuli were in the form of an auditory message generated by the Quadra 700 unit with voice input. For condition 1, subjects used paper maps (with plastic covers) that were printed out from scanned copies of local maps. These paper maps were the same as the digital versions for conditions 2 and 3, which were presented on the VDU. Non-permanent markers were made available so that subjects could mark grid coordinates on their plastic covered paper maps. Pens and papers were available to the subjects for recording information given "over the radio". The subjects used the Mac UnMouse touchpad as the input device.

Forms used included: a standard consent form and the NASA-TLX forms (one after each condition). Also, a 'rules reminder sheet' was posted on the wall behind the monitor for the subject to refer to concerning which actions to take under certain trials.

A training manual was used to elevate the monitor such that the subject had a 15-30° viewing angle of the screen at the relaxed seated position. The average luminosity measure for all three conditions at the distance of the subject's eyes was 5800±200 LUX, using a Megatron DL4 Digital Lightmeter. The software that generated the three conditions was written by P. Gosling. The programming was done by using *Supercard*.

<u>Subjects</u>

Eighteen male subjects from the subject pool at the APRE volunteered to participate. The subjects consisted of 16 infantry soldiers of Private and Corporal ranks and two civilian staff from the APRE. Each of the soldiers had a few years of experience in the army before the experiment, but none, including the two civilians had any tank commanding experience. All were between the ages of 20 and 32. All subjects had normal visual acuity, with or without corrective lenses.

<u>Experimental procedures</u>

The experiment consisted of three sessions for each subject. Subjects returned for sessions 2 and 3 on consecutive working days. For all three sessions, the experimenter first read the instructions for the condition task to the subject. The subject was then asked to practice on a few pre-programmed trial runs. Subjects were not allowed any breaks in between trials. Thus, even though condition 1 took much longer, the subjects had no breaks because of conditions 2 and 3 (too short for a break). However, not all the subjects completed all the trials in condition 1 since it required more time per trial than conditions 2 and 3. The trials were stopped when it reached the end of the one-hour session. The computer recorded all reaction times and errors (see RT and Err in Results) for each trial session and subject. The time estimation was followed by administering a NASA-TLX subjective workload scale. The subject was debriefed after the last session and was thanked for his participation.

RESULTS

Objective data for the experiment include reaction times (RT) and error rates (Err). ANOVA and separate t-tests show that RT is significantly different for all three conditions, with $F = 31.446$, $p = .0000$. Error rates however, did not differ between the three conditions, with $F = 1.322$, $p = .2756$. Separate t-tests were not carried out for Err results since ANOVA shows no differences. Graph 1 below shows differences between RT's for the three conditions. Graph 2 shows no differences between Err's for the three conditions, and the boxes overlap. Even though there is no significant difference between the three conditions for Err, nonetheless there appears to a trend - with less errors for condition 2 and even fewer for condition 3. This trend is in line with the absence of a RT-Err interaction (parallel trend) supports the predicted results in that condition 3 is superior to condition 2 for both reaction time and error rates.

ANOVA and separate t-tests show that TLX scores are significantly different between conditions 1 and 2, and 1 and 3 only, but not between 2 and 3. ANOVA show significant differences between TE's for the three conditions. This is in line with predictions for workload, and are similar to RT results. Similar to RT and TLX results however, TE data does not seem to have homogeneity of variance (requiring Satterthwaite's solution in the t-test analysis, from Howell, D.C., 1982). 'Time' results show no significant differences for any of the conditions. This is quite contrary to predictions and with TE results ('Time' = balanced TE). 'Time' results would be more reliable than TE results, since TE does not account for temporal length of the task.

Table 1 below gives an ANOVA and t-test summary for all measures.

Graph 1. Reaction time in seconds versus Conditions 1, 2, and 3.

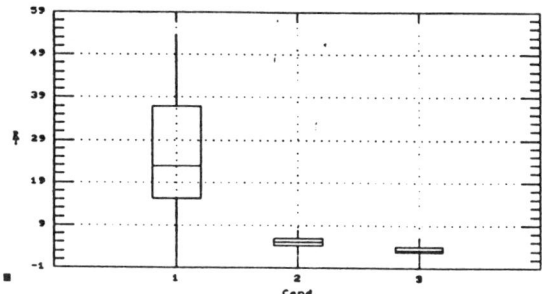

Graph 2. Error rates (percentage) versus Conditions 1, 2, and 3.

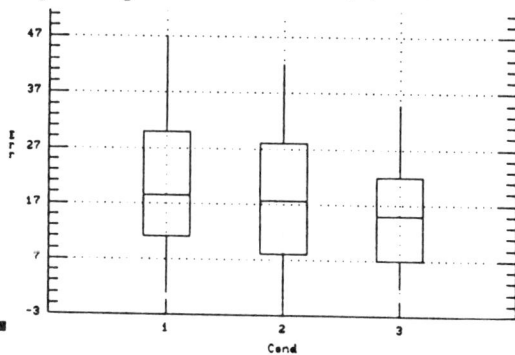

Table 1. ANOVA and t-test summary for all measures.

MEASURE	F-RATIO	SIG. LEVEL	T-TEST
RT	31.446*	.0000	* for all conditions
ERR	1.322	.2756	no differences
TLX	8.85*	.0005	* for conditions 1-2 and 1-3, but not 2-3
TE	21.160*	.0000	* for all conditions
TIME	1.479	.2375	no differences

DISCUSSION

The results obtained from the experiment in general seem to support a system of symbols-only presentation over a system with text and symbols. This is in line with Wickens' (1984) multiple resource theory which states that a manual task could be better performed if graphics were the mode of presentation. Reaction time results are highly significant and favour system 3 over system 2, which in turn is superior to system 1. Although the objective measure of errors was not significant, there seems to be a trend for lower error rates for system 3, then system 2, and lastly for system 1. This trend is in line with RT results and multiple resource theory as well.

Subjective measure using the NASA-TLX yielded results in favour of systems 2 and 3 over system 1. The TLX may not have been sensitive enough to pick out any differences between systems 2 and 3. On the other hand, this may not necessarily be an inadequacy of the TLX. Perhaps the differences between system 2 and 3 are not intrinsically that different after all. It could be that there are indeed differences (results show median lower for system 3), and that the differences are not statistically significant but are *practically* significant. This should be noted for future designing purposes.

Time estimation is an interesting measure for workload. At first, it seems that TE is sensitive to workload as RT is. When TE is corrected for (Time results), it no longer shows any significant differences between conditions. It appears that TE is correlated with RT itself ($r^2 = .36$). Since RT immediately influences total time spent on task (same number of trials for each condition not same amount of time allotted), and TE and RT are correlated, TE could then be dependent on total time spent on the task (condition). Thus, it may be assumed that the significant results found for TE is an artefact. The new measure of workload (Time) does not seem to be sensitive to the workload differences in this experiment, since the data points are percentages of the total time on the task. The finding that TE is significant could be attributed to the fact that the subject was busy for a longer period of time (and subjects were accurate at estimating lapsed time). This supports Sheridan's (1980) proposal that workload is related to time busy. For further studies, it would be interesting to control for length of time spent on each task. This was not done for this experiment, since exposure to only 10 minutes of condition 1 would yield very few data points. The results of non-significance between the three conditions in terms of time estimation (Time results) are not surprising considering the limits of this measure as discussed by Liu & Wickens (1992, unpublished). In their study, time estimation does not discriminate between the two activities to which automaticity has developed (as tasks are in this study), but does reflect the presence or absence of non-automated monitoring task. Since automaticity is assumed to have developed for the three present conditions (especially for digital systems), it is doubtful that we would see differences in time estimation. The authors insist that subjective scales such as the TLX on the other hand, is sensitive to such differences in workload, especially when the TLX is broken down into its subscales and compared on those terms.

Aside from greater memory requirement (i.e. colour red represents "contact with the enemy") and possible interference (spatial-spatial information), condition 3 seems to fare better than condition 2 in general. Perhaps the weak point about condition 2 (spatial and verbal) is that it requires more searching skills and eye movements. Could the requirement for greater eye movements alone cause better performance for condition 3 and not the effects according to multiple resource theory? This does seem like a reasonable question to ask when examining RT results. However, superior performance in terms of error rates cannot be explained by this visual search theory. Thus, this complaint about the task design of condition 2 is not a legitimate one.

Although there was a mismatch between condition 2 and condition 3 in terms of different types of symbols used, there was no requirement in the *knowledge* domain to cause any mismatch between system variable and knowledge variable. There was no requirement in condition 2 for knowledge of NATO symbols to accomplish the experimental task. Similarly, there was no requirement in condition 3 for knowledge of pictorial symbols to accomplish the same task. The only knowledge required for both tasks was to identify friendlies and enemies by colour (red or blue) recognition. Thus, it seems the argument that condition 2 and 3 were not matched may be dismissed in terms of it affecting experimental results. It would be interesting though, to test same systems (both pictorial presentations) in future studies.

Ergonomics implications, applications and recommendations

The results of this study are in line with multiple resource theory which predicts that condition 3 would be advantageous over condition 2 and 1 since a manually controlled system would best be presented in symbols-only mode, visually. In terms of theoretical implications, this study shows that multiple resource theory is applicable to this particular human-computer interaction system.

In terms of practical implications then, the use of symbols-only in a digital map system is advisable. However, the significant results are perhaps limited to a system such as condition 3. It is recommended that future systems be similar in design to that of condition 3's. It is recommended that simple pictorial representations (used in the symbols-only condition) be used instead of NATO symbols. A system similar to condition 3 would benefit from using *both* simple pictorial symbology and symbols-only presentation. It is believed that the use of NATO symbols in condition 2 did not influence experimental results however, as discussed above. One could also argue that the reason condition 3 is more advantageous to condition 2 is that condition 2 requires greater visual search time and that the advantage is not due to its suitability to manual mode response. However, the results for error rates cannot account for the better performance in condition 3 (though not statistically significant). Thus, not only is condition 3 superior in terms of fitting multiple resource theory and using simple pictorial representations, but it does not require unnecessary visual search or eye movements as does a text-symbols system.

REFERENCES

Adams, J.A., 1982, Issues in human reliability, Human Factors, 24, 1-10.

Barber, P.J., Carver, M.K., & Laws, J.V., 1984, Investigation of the human aspects of automated data processing techniques. Prepared for MA3 Division RARDE under contract no. 2012/03 RARDE.

Hart, S.G. & Staveland, L.E., 1988, Development of NASA-TLX (Task Load Index): Results of empirical and theoretical research. In P.A. Hancock and N. Meshkati (Eds.), Human mental workload. pp. 139-183.

Howell, D.C., 1982, Statistical methods for psychology. PWS Publishers, Boston, MA.

Liu, Y. & Wickens, C.D. (submitted to Ergonomics, May 1992). Mental workload and cognitive task automation: An evaluation of subjective and time estimation metrics. Unpublished.

Nygren, T.E., 1991, Psychometric properties of subjective workload measurement techniques: implications for their use in the assessment of perceived mental workload, Human Factors, 33(1), 17-33.

Sheridan, T., 1980, Mental workload: what is it? Why bother with it? Human Factors Society Bulletin, 23, 1-2.

Sheridan, T., 1981, Understanding human error and aiding human diagnostic behavior in nuclear power plants. In J. Rasmussen and W.B. Rouse (Eds.), Human detection and diagnosis of system failures. New York: Plenum Press.

Wickens, C.D., 1984, Engineering psychology and human performance, Charles E. Merrill Publishing Company.

SUSPECT IDENTIFICATION BY FEATURES

ERIC S. LEE
HELEN DENSMORE
Management Science
St. Mary's University
Halifax, N.S., Canada
elee@bootless.stmarys.ca
(902) 420-5734

THOM WHALEN
Behavioural Research Group
Communications Research Centre
3701 Carling Avenue
Ottawa, Canada
thom@rick.doc.ca
(613) 990-4683

Correct suspect identification of known offenders by
witnesses deteriorates rapidly as more are examined in
mugshot albums. Feature approaches, where mugshots are
displayed in order of similarity to witnesses'
descriptions, increase success by reducing this
number. We address two issues: the effect on system
performance of alternative measures of similarity
between police and witness feature descriptions, and
the number of police officers who must rate each
suspect. Four experiments, using a system for 640
mugshots, suggest system performance is optimized for
a Euclidean metric and two police raters.

INTRODUCTION

The identification of suspects by witnesses or victims is
paramount in solving crimes. Three methods are commonly used to
identify suspects: verbal descriptions; composite methods such as
Photofit or Identikit; and mugshot albums.

Verbal descriptions lack sufficient detail and accuracy to be
of much use (Ellis 1984). Composite procedures are relatively
unsuccessful, because they do not generate an accurate, detailed
image of the suspect (Ellis et al. 1978). The mugshot album
approach is more successful, because people are good at face
recognition (Ellis et al. 1989).

Nevertheless, the album approach suffers fundamental problems.
The task is time-consuming, and confusing as witnesses often
examine thousands of photos. The probability of selecting the
correct suspect (a hit) decreases rapidly after the first 100-200
examined. The probability of selecting the wrong person (a false
alarm) increases rapidly (e.g., Ellis et al. 1989, Lenorovitz and
Laughery 1984).

In response to such difficulties, researchers turned to the
feature approach to image retrieval (Ellis et al. 1989, Harmon
1973, Lenorovitz and Laughery 1984, Lee and Whalen 1993). A
feature describes a visual aspect of images. In suspect
identification, hair length and neck size are two aspects used to
describe facial characteristics of offenders. In our approach, a
single police rater describes each database image on system
features using 5-point scales (e.g., complexion: fair 1 2 3 4 5
dark). Witnesses describe a suspect on the same scales. Database
images are displayed to users in rank order of similarity to their
description of a target. The objective of feature systems such as

ours is to increase identification success by reducing the number of irrelevant mugshots examined by witnesses (i.e., by reducing the retrieval rank of a suspect's photo).

Harmon (1973) introduced the feature approach to suspect identification. However, Ellis' (1989) investigations established its value over the traditional album approach.

Empirical tests suggest feature systems possess valuable qualities (Ellis et al. 1989, Harmon 1973, Lee and Whalen 1993). First, system performance is good with target suspects generally retrieved within the first 10 mugshots, thereby increasing the probability witnesses will correctly identify suspects. Second, they tolerate witness errors. Third, they permit uncertainty. A user can omit features without terminating the retrieval process. Fourth, they require no training or instructions. The method therefore is well suited to non-expert users.

This success has been achieved despite marked differences among feature approaches in feature measurement, the prompting of witnesses with feature queries, the use of intermediaries, the measurement of similarity between witness and police codings, and the number of police raters coding each suspect. Little research has been directed toward investigating these differences.

The present paper addresses methodological and theoretical issues raised by differences among the three feature approaches of Harmon, Ellis, and our own. Four experiments are discussed which explore the effect on system performance of alternative algorithms for measuring similarity between witness and police codings, and the number of police officers rating each suspect.

OUR RETRIEVAL SYSTEM

The database consists of 640 official mugshot photos of known offenders. (In contrast, Harmon and Ellis used photos of non-offenders.) Colour photos were taken under standard conditions -- frontal view of face from the shoulder up (90 x 125 mm prints). The suspects are all white males, aged 18-33 (99.5% under age 27).

Each mugshot was coded on 90 facial features by one of 12 raters (males and females in their early twenties). The raters received no training or instructions. Raters coded directly from the photo which was always available for inspection, as would be the case if police officers coded the mugshots. Coding time per mugshot was approximately 5-10 minutes. Each feature is coded on a 5-point Likert scale (e.g., narrow nose 1 2 3 4 5 broad nose).

Witnesses work directly on the system without an intermediary. Feature queries appear on the screen in succession. They describe suspects using the same 90 5-point scales. Database images are displayed to users in rank order of similarity to their description of a target. (See Lee and Whalen (1993) for details).

METRIC EFFECTS: THE MEASUREMENT OF SIMILARITY

Feature systems rank mugshots in terms of similarity between witness and police descriptions. Ranking depends on the method used to measure similarity. The present analysis explores effects of alternative measures of similarity on system performance.

Similarity is measured by distance between feature

descriptions. Many methods of measuring distance which can be used in feature systems are given by the Minkowski distance function:

$$d_j = \frac{\left(\sum_i \mid F_{ij} - f_i \mid^m \right)^{1/m}}{c} \qquad [1]$$

where the sum is taken across the c features rated by both witness and police officer, F_{ij} is the database coding on feature i for mugshot j, f_i is the witness's coding of the suspect on feature i, d_j is the distance between the database's featural description of mugshot j and the witness's description of the suspect, and m is the Minkowski distance metric. The two most commonly used distance functions are Euclidean ($m = 2$) and city-block ($m = 1$) metrics. Euclidean is straight-line distance between two places (i.e., "as the crow flies"). City-block is the distance you would walk through city streets to get from one place to another.

Ellis ranks mugshots in terms of the number of feature matches between witness and police descriptions, a degenerate form of city-block metric known as a Hamming distance metric. In our system, we count a match whenever witness and police codes are within $\pm h$ units of each other. Thus, for $h = 0$, witness and police codes match only if the feature values are identical (i.e., $f_i = F_{ij}$); for $h = 2$, codes match if feature values are within 2 points of each other on our 5-point scales.

The effect of different distance metrics on system performance can be determined by synthetic experiment, that is, without conducting a separate experiment on the system for each metric. Witness feature descriptions from 4 experiments were re-entered repeatedly on our retrieval system, each time presetting the system to use a different metric. Retrieval rank for a given search was determined for four Minkowski metrics ($m = 1,2,3,4$) and four Hamming metrics ($h = 0,1,2,3$).

Witness descriptions, obtained from 3 previous experiments, were reanalyzed for present purposes (Lee and Whalen, 1993). In each experiment, subject witnesses had 10 sec to examine a target mugshot. (Ellis' testing procedure was followed to increase comparability of our systems.) For each target, subjects answered 90 feature queries using 5-point scales on our computer. Skipping features was permitted in all experiments.

In Experiment 1, 5 subject witnesses were tested. Five target mugshots, randomly selected from the database, were presented, one at a time to each subject. No feedback was provided between trials. Subjects were encouraged to code all features, guessing if necessary. Retrieval rank was determined synthetically for each metric. In Experiment 2, 5 subject witnesses were tested on 5 target mugshots each. Guessing was discouraged. In Experiment 3, 15 subjects each searched for a randomly-selected single mugshot. Guessing was not actively discouraged.

For Experiment 4, data from the multiple raters experiment were reanalyzed to investigate similarity effects. Ten subject witnesses searched for a single target suspect each. Guessing was discouraged. The database consisted of averaged feature codings by

4 database raters for 200 of the original 640 mugsghots.
Mean retrieval rank for these experiments are displayed in
Table 1. The effect of Minkowski and Hamming metrics were analyzed
separately. The effect of Minkowski metric was significant in
Experiments 1 and 2, $F(3,12)$ = 6.82 and 4.56, p < .001, but not in
Experiments 3 and 4. Newman-Keuls multiple-comparison tests
indicated city block metric (m = 1) was significantly worse than
all other Minkowski metrics in the first experiment. In the
second, m = 4 metric was significantly worse.
    The effect of Hamming metric was also significant in
Experiments 1 and 2, $F(3,12)$ = 9.69 and 18.77, p < .001, but not
in the other two. Newman-Keuls indicated h = 0, 1, and 3 are all
significantly worse than h = 2 in Experiment 1. In Experiment 2, h
= 3 was significantly worse than all others.
    The best Minkowski metric (m = 2) was compared with the best
Hamming (h = 2) for each experiment. A priori sign tests indicated
a significant performance advantage for the Euclidean metric (m =
2) over the best Hamming metric in Experiment 2, p < .001 (23
versus 0 with 2 ties) and marginally significant advantage in
Experiment 4 (6 versus 1 with 3 ties), p < .06.

Table 1. Mean retrieval rank as a function of distance metric for
4 experiments.

Exp. No.	Database Size	Distance Metric				Hamming Distance			
		m=1	m=2	m=3	m=4	h=0	h=1	h=2	h=3
1	640	38	17	13	14	124	62	22	75
2	640	3	4	10	19	13	5	27	77
3	640	18	13	13	17	65	30	34	62
Mean	640	20	11	12	17	65	36	28	72
4	200	2	2	2	3	17	6	6	20

MULTIPLE POLICE RATERS

    Feature approaches differ in the way police code database
mugshots: physical measurement or subjective judgment. In Ellis'
system, mugshots are coded by direct physical measurement of some
features from the photo itself (e.g., the distance between the
eyes). In contrast, in subjective coding systems, one or more
police coders rate suspects' mugshots on Likert subjective scales
(e.g., 5-point scale for hair length: short 1 2 3 4 5 long).
(Regardless of how police code features, all systems, physical or
subjective, require witnesses to use subjective scales.)
    Ellis asserts subjective coding is error prone if many coders
(i.e., police officers) have to rate each mugshot, they may rate
the same offender quite differently on any given feature.
Therefore, multiple coders may be required to reduce this error by
averaging their ratings. Harmon (1973) averaged the subjective
ratings of 10 judges to form the database coding for each feature.
As Ellis points out, however, such an approach becomes time-

consuming and expensive as the number of coders is increased. We
argue that the effect of variation between database and witness
codings will introduce unwanted error. The question is whether
this error variation is sufficient to impair retrieval
performance. It is not if these errors tend to be small relative
to true differences between suspects. The successful performance
of our system in Experiments 1 to 3, in which only a single police
coder was used, supports our contention that multiple judges may
not be required for subjective systems.

To assess the effect of multiple raters directly, we tested 10
subject witnesses on systems using 1, 2, 3, or 4 raters. Four
raters coded the same 200 mugshots, a subset of the original 640
(always available for inspection). We built all possible systems
using 1 to 4 raters -- a, b, c, d, ab, ac, ..., abcd -- by
averaging feature codings across raters for each mugshot. Each
witness viewed a suspect's photo for 10 sec before describing him
on our scales (the original 90 features plus 16 new ones).
Similarity between police and witness descriptions was measured by
a Euclidean metric. For each witness, retrieval rank was averaged
across all systems with the same number of raters. Thus, retrieval
rank equalled 9.4, 4.4, 2.5 and 1.0 for the first witness.

A repeated-measures analysis of variance on these averaged
scores indicated a significant effect of number of raters, $F(3,27)$
= 11.08, $p < .001$ (see Table 2). The Geisser-Greenhouse
conservative F test, used to compensate for heterogeneity among
variances and covariances, also indicated a significant effect of
raters, $F(1,9) = 11.08$, $p < .01$. Both the linear and quadratic
trends are significant, $F(1,27) = 25.55$ and 6.81, $p < .001$ and
.05. Retrieval rank did not differ significantly for 2, 3, or 4
raters, but was significantly worse for a single rater than for
more than one rater (Newman-Keuls multiple-comparison test).
Euclidean distance between police and witness codings increased
significantly as the number of raters is increased from 1 to 4,
$F(3,27) = 557.55$, $p < .001$.

Table 2. System performance as a function of number of police
raters.

No. of RATERS	Retrieval Rank	Worst Rank	Mean Absolute Deviation	Mean Euclidean Distance
1	8.0	45	.767	.142
2	3.4	29	.715	.124
3	2.6	16	.693	.119
4	2.2	10	.678	.115

DISCUSSION

Results from 4 experiments support our contention that
performance depends upon the algorithm used to measure similarity
between witness and police descriptions. The best metric is
Euclidean, $m = 2$ ($m = 3$ is almost as good). Though not the best
metric in each experiment, it was consistently good in all.

We expected Minkowski to outperform Hamming metrics, and they did. Minkowski metrics take the size of deviations between witness and police feature descriptions into account. Hammming metrics only count the number of matches. Nevertheless, Hamming $\underline{h}$ = 2 performed surprizingly well. We are currently exploring the effect of varying $\underline{h}$ from feature to feature.

The differences in effectiveness among the Minkowski metrics is explained by errors. An error is a deviation between witness and database description of some feature. A single large error results in an increase in distance of $(5-1)^1$ = 4, $(5-1)^2$ = 16, $(5-1)^3$ = 64, $(5-1)^4$ = 256 for the four metrics (e.g., codings of 5 versus 1 when a 5-point scale is used). Increasing the metric $\underline{m}$ has the effect of penalizing errors more heavily and the size of this difference in penalty increases with error severity. If subjects tend to make few large errors (and therefore deviate minimally from the database description of the target) while incorrect database mugshots tend to differ maximally from the target on at least a few features, then increasing the metric should improve subject performance. System performance may deteriorate, however, for high values of $\underline{m}$, because even a single large deviation between witness and police descriptions may incorrectly increase the rank of a suspect's mugshot. Thus, we expected a U-shaped function with optimal performance for $\underline{m}$ = 2. We found a U-shaped function, though both $\underline{m}$ = 2 and 3 performed very well.

Our experimental results provide little justification for Harmon's averaging feature judgments across 10 police raters. Two are sufficient. System performance is good even with a single rater (mean retrieval rank = 8.0). However, two raters are significantly better than one (rank = 3.4), while additional raters have little effect on system performance. The improvement in performance is undoubtedly attributable to the significant decrease in Euclidean distance between police and witness codings as the number of raters is increased from 1 to 4.

REFERENCES

Ellis, H.D., 1984, Practical aspects of facial memory. In *Eyewitness Testimony: Psychological Perspectives*, eds. G. Wells and E. Loftus (New York: Cambridge University Press).

Ellis, H.D., Davies, G.M. and Shepherd, J.W., 1978, A critical examination of the Photofit system for recalling faces. *Ergonomics*, 21, 297-307.

Ellis, H.D., Shepherd, J.W., Shepherd, J., Klin, R.H. and Davies, G.M., 1989, Identification from a computer-driven retrieval system compared with a traditional mug-shot album: A new tool for police investigations, *Ergonomics*, 32, 167-177.

Harmon, L.D., 1973, The recognition of faces. *Scientific American*, 229, 70-82.

Lee, E.S. and Whalen, T., 1993, Computerized feature retrieval of images: Suspect identification. *Ergonomics*, in press.

Lenorovitz, D.R. and Laughery, K.R., 1984, A witness-computer interactive system for searching mug files. In *Eyewitness Testimony*, eds. G.Wells and E. Loftus (New York: Cambridge University Press).

Efficient Design of Mammographic Multiviewers

C.G. Blair-Ford[1], A.G. Gale[2], S.G. Cobb[1], & C.M. Haslegrave[1]

[1]Dept. of Production Engineering and Production Design,
Nottingham University
[2]Applied Vision Research Unit, University of Derby,
Mickleover, Derby

The advent of breast cancer screening for all women in the U.K., in the 50-64 year age range, requires that radiologists visually examine large numbers of mammograms (circa 10,000 cases a year) for signs of this disease. A dedicated mammogram multiviewer is used to present the cases for scrutiny. This device is essentially a workstation which can store hundreds of cases and simultaneously display several for visual inspection, together with a means of control for the operator. The research reported here investigates the design of such multiviewers in relationship to established ergonomic principles and official government guidelines. It is suggested that existing guidelines could be improved. Finally design recommendations are made

INTRODUCTION

Screening for breast cancer was introduced throughout the U.K. following the Forrest report (1986). Essentially all women between the ages of 50 and 64 are invited for mammographic examination every three years. This means that annually there is the potential of screening some 1,000,000 women. These mammograms (at least two Xrays per woman) are visually examined by experienced radiologists around the country at 120 screening centres. It is important that their task is efficiently designed to aid their task of visually detecting any possible cancer at its earliest.

In breast screening the radiologists have to visually detect very early signs of cancer which are often, for instance in the case of microcalcifications, only a few microns in size. Consequently visual examinations are made both by naked eye and with the aid of a magnifying glass, coupled, with the need to alter viewing distance. These approaches shift the spatial frequency spectra of the image thus making any potential abnormalities more visible (Gale et al., 1979). However there is a cost involved in terms of the postures adopted in examining the mammograms.

The present work addresses this issue of the interface between the radiologists and these medical images. The mammograms are examined on large electro-mechanical dedicated viewing boxes or multiviewers. These workstations consist of an illuminated vertical surface to back-light the Xrays, some means of transporting the films over this surface, the capacity to store the mammograms of several hundreds of women, a writing surface, together with some means of operator control for these various operations.

METHOD
In the U.K. there are about four commonly employed multiviewers. Two different types are used locally and this research concentrated on these as they are good exemplars of the alternative designs currently in use.

One of them (manufactured by Philips) comprises a large illuminated vertical surface on which typically the two mammograms of 16 women (4 cases wide by 4 high) can be simultaneously displayed mounted within a clear frame. The radiologist examines these cases from a seated position and by moving horizontally across the work area on chair can bring different cases into close inspection. The frame of 16 mammograms can be moved vertically to bring either the top or bottom 8 cases to an appropriate height for inspection and further display panels of mammograms can be electrically brought to the illuminated surface.

The other multiviewer (Radx) is similar in that the radiologist sits at the workstation, however only 6 cases (3 cases wide by 2 high) are presented. The chief difference is that the radiologist can electronically move the top or bottom series of cases to bring them into an appropriate position for inspection without the need for the radiologist to physically move their lateral seated position.

A hierarchical task analysis was conducted to break the overall inspection process down into its constituent sub-tasks so as to define it in detail. Additionally a link table was generated to examine the relationship between the various multiviewer controls.

Physical measurements
Guidance notes for Health Authorities on the design of multiviewer illuminators for breast screening have been produced by the NHS (1989). Additionally the DHSS have published a consultation draft document which considers the design recommendations for Xray illuminators (1991). Both documents are concerned chiefly with issues such as safety, performance and technical requirements. Some consideration is given to ergonomic criteria and the DHSS document specifically addresses some ergonomic issues more fully. Direct measuremtns were made of the two chosen multiviewers and data were compared to these recommended values.

Videotaped study
A video study was then made of two experienced radiologists as they examined one day's caseload on both machines.

Interviews
Structured interviews were carried out with an additional 22 radiologists, randomly selected, who were involved in screening at other screening centres. The interviews addressed the following areas of; seat layout and posture, controls, usefulness of features, and machine dimensions.

RESULTS
The task analysis elucidated the sequence of steps involved in examining mammograms, on both multiviewers, and their hierarchical relationship. These encompassed delivery of the particular case for examination to the recording of diagnostic data on computer. The link table further summarised the relationship between the controls.

Physical measurements
Table 1 compares the DHSS recommendations to those of Clark and Corlett (1984) and also Pheasant (1990). It is apparent from these data that they are broadly similar. However a problem was the amount of forward leg room - in inspecting mammograms the radiologist has, at times, to lean as close to the illuminator as possible. Pheasant (1990) gives the value for this measurement as 425mm for a male (95th percentile in buttock-knee length and 5th

percentile in abdominal depth) and Grandjean (1990) arrives at a value of 600mm depth at knee height. Measured values for both machines were 225mm (Philips) and 315mm (Radx).

In terms of the height of the worktops for both multiviewers these were measured as 800 mm (Philips) and 730 mm (Radx). Whilst the latter is within BS5940 specifications for desk workstations, both values are outside the range given by Grandjean (1981) for males.

TABLE 1 : COMPARISON OF ERGONOMIC DIMENSIONS

DIMENSION (in mm.)	DHSS RECOMMENDATIONS ON X-RAY ILLUMINATORS & VIEWING WORKSTATIONS	CLARK & CORLETT (1984) EQUIVALENT	PHEASANT (1990) EQUIVALENT
1. Seat height	440	424 (5th %ile woman)	400 (Popliteal height)
2. Clearance under worktop	670	750 (95th %ile man)	670
3. Worktop height	720	760 (95th %ile man)	720
4. Vertical zone for location of controls : height from worktop	50-100	-	-
5. Vertical zone for location of single bank illuminators : height from controls	800-1400	-	-
6. Seated eye level : small woman	1110	-	1490
7. Seated eye level : large man	1290	1024 (General eye level)	1250
8. Worktop depth	600	610	650 (Forward grip reach 5th %ile woman)

Videotaped study

Analysis of the video data demonstrated differences in posture adopted by the radiologists in that each radiologist adopted a characteristic posture. Both radiologists adopted unhealthy postures for both machines. In particular with the Philips machine a sideways leaning posture was more often employed. This is partly because the Radx is physically not as wide and partly because with the latter the cases can be transported to the preferred seating position, rather than the radiologist having to physically traverse to that case in order to inspect it closely. As mentioned above both machines had somewhat limited forward leg space which affected the working posture.

In addition to posture information the videotaped recordings elicited data on the subtasks of the inspection process itself. In general these comprise identifying the case to be examined into the computer system, physical examination with or without a magnifying glass and recording of diagnostic data. All computer entry of data is via barcode (c.f. Preistland et al., 1989). The majority of the time per case was found to be spent in identifying and recording diagnostic decisions. some 24-27% of the time was spent in close-up and detailed examination of the case and 2-4% of the time spent in normal visual examination without the use of a magnifying glass.

Interviews

Of the interviewed radiologists, 18 used the Radx and three the Philips machine. Respondents generally (72%) spent an hour or less (24%) at a time using the multiviewers. This was the time taken to examine a typical screening day's worth of cases - some 100 women. The majority (81%) reported no body pain/discomfort and 63% reported not using the backrest of the seat. Some 72% considered that they had enough leg space.

DISCUSSION

Whilst the present study is limited in the number of radiologists studied it still gives insight into pertinent factors which can affect the performance of radiologists in this demanding task.

The measurement made of the two multiviewers generally demonstrated that they were acceptable and approximated available ergonomic recommendations applicable to such multiviewers. However the major factor of lack of forward leg space was seen as crucial, forcing the adoption of poor working posture, which was confirmed from the video recordings. It is intriguing that the interviews with radiologists did not elicit major complaints about this factor.

In conclusion it is argued that more consideration must be given to the adequate user-centred design of multiviewers so as to improve the overall efficiency of breast screening.

REFERENCES

Clark, T.S. and Corlett, E.N., 1984, The ergonomics of workspaces and machines - a design manual, London Taylor and Francis.

Forrest P. , 1986, Breast cancer screening- report to the health ministers of England, Wales, Scotland and Northern Ireland, London,HMSO.

Gale, A.G., Johnston, F. and Worthington, B.S., 1979, Psychology and Radiology, In Research in Psychology and Medicine, ed. D.J. Oborne, M.M. Gruneberg and J.R. Eiser (London, Academic press), pp 453-461.

Grandjean, E., 1988, Fitting the task to the man, London Taylor and Francis.

NHS Estates, 1991, Consultation draft- health building note 6 - radiology department, London, NHS.

NHS Procurement Directorate, 1989, Guidance notes for Health Authorities on illuminators for breast cancer screening mammograms, London, NHS.

Pheasant, S., 1990,Bodyspace, London, Taylor and Francis.

Preistland, R.N., Gale, A.G., Roebuck, E.J. and Worthington B.S.,1989, A new bar-coding system for use in breast cancer screening. In Contemporary Ergonomics 1989, ed. E.D. Megaw, (London, Taylor and Francis),pp 303-308.

Design methods

STRUCTURED NOTATIONS FOR HUMAN FACTORS SPECIFICATION

LIM, K. Y. AND LONG, J. B.

Ergonomics Unit,
University College London,
26 Bedford Way, London WC1H 0AP.

The paper illustrates how structured notations may be used to support the specification of organisational hierarchies, conceptual level tasks, domain semantics, human-computer interaction and user interface design. It argues for structured notations (as opposed to formal or algebraic notations) since their graphical characteristics have been reported to facilitate design communication with users. Thus, user feedback elicitation and design validation would be supported better throughout system development. In the future, it is expected that the structured notations illustrated in the paper would be used more widely since : they address the above concerns of human factors design; they have been incorporated into a structured human factors method; and off-the-shelf computer-based tools to support the notation are already available, e.g. PDF™ [1]

INTRODUCTION

In recent years, the scope of human factors (HF) involvement in interactive system[2] development has become wider. For instance, increasing emphasis is placed on earlier and continued HF input throughout the design cycle (see Moraal and Kragt, 1990; Lundell and Notess, 1991). Hence, HF contributions are targeted not only at design evaluation, but also at design specification and implementation. Although designers are increasingly aware of the need for wider HF involvement, appropriate incorporation of its inputs remains hindered by notational inadequacies. In particular, the demands associated with design specification, namely the format, granularity and specificity of design descriptions, have not been satisfied by existing HF notations. This requirement for better notations is also implicated by the following observations in recent reports:

(a) current HF specifications are insufficiently specific to support design collaboration between HF designers and software engineers. This situation is aggravated further by the development of increasingly complex and sophisticated interactive systems. For instance, in the context of safety critical system development, task specifications have to be detailed enough to support design simulation, workload assessment and probabilistic human reliability assessment. Thus, Brooks (1991) emphasised that task specifications should describe the hierarchical structure and operational control of the user's task. Consequently, it

---

[1] PDF is a trademark of Learmonth and Burchett Management Systems (LBMS) Limited.

[2] The term 'system' implies a human-machine system which performs tasks in a particular environment, work being achieved in the process. In the context of the system, HF is concerned primarily with user behaviour with respect to computers, while software engineering is concerned primarily with computer behaviour with respect to humans.

follows that more powerful notations need to be developed to support more specific design specification;

(b) the uptake of HF may be hindered by inadequate guidance on the process of design derivation. Specifically, HF has been criticised for focusing predominantly on *what* should be done and not on *how* design requirements could be met (Sutcliffe, 1989). The imbalance in focus may be an artefact of the traditionally late recruitment of HF to design. To rectify the problem, wider and more explicit conceptions of the HF design process have been defined and developed into structured methods (see Blyth and Hakiel, 1989; Damodaran, Ip and Beck, 1988; Lim, Long, and Silcock, 1992; Sutcliffe, 1988). As a result, existing notations have been enhanced to accommodate the specification of a wider range of intermediate HF design products. The notations described in this paper are a product of such developments.

In summary, an appropriate HF notation should fulfil two pre-requisites, namely it should rectify the inadequacies of existing HF notations, and accommodate additional specification demands arising from wider HF involvement in system development. Keeping these notational requirements in view and consistent with a wider conception of HF design[3] (see Lim and Long, 1992a and b; Lim, Long and Silcock, 1992), a survey of existing HF and software engineering notations was conducted. In particular, the notations of software engineering *structured* methods were selected. This selection is consistent with requirement (c) above (i.e. communication with users), since the graphical notations of these methods were considered more promising for this purpose than the algebraic notations of formal methods (see Finkelstein and Potts, 1985; Fitter and Green, 1979; Hares, 1987) and more specific than existing HF notations. For a more detailed account of notational requirements, see Lim and Long (1992d). The selected notations were then tested for their capability of supporting HF specifications. In between testing cycles, suitable notational extensions were also proposed. A detailed account of the process of notational development may be found in Lim et al (1990a and b). Presently, the notations developed to support an in-house structured HF method are reviewed to illustrate the outcome of the research. In particular, the review illustrates how *adapted and extended* software engineering structured notations[4] may be used for describing specific HF design concerns.

STRUCTURED NOTATIONS TO SUPPORT HUMAN FACTORS SPECIFICATION

Generally, HF design may involve specifications at the organisational level (or super-ordinate system level); task level (or sub-system level); and interaction level (or input/output level). An account of the notations used by the method for HF specification at these design levels follows.

At the organisational level, the following notations may be applied :

Semantic nets. This notation may be used for two types of specifications, namely to define organisational structure (e.g. organisation job chart); and to specify relations between entities, concepts, events and processes associated with the domain of the system. Semantic nets are essentially tree diagrams comprising nodes and numerically indexed inter-node relations (see Figure 1).[5] Rules of thumb that support the construction of a semantic net are as follows : its contents should adequately cover the scope of the system; its nodes should exclude device-specific information; and its inter-node relations should include taxonomic and composite relationships.

Network diagrams. This notation may be used to describe the content, direction and types of explicit information flow between system (and sub-system) entities (implicit information flows may also be included if appropriate). Figure 2 shows two types of information flow, namely

---

[3] The conception was used to identify general categories of HF information that are relevant to system design.

[4] Some of the notations originate from the Jackson System Development method. The notations chosen were shown to be suitable for task description (see Carver, 1988; Carver and Cameron, 1987; Sutcliffe, 1988).

[5] The reader is referred to Lim, Long and Silcock (1992) for a more detailed example of a semantic net specification.

obligatory communication of information (denoted by a circle, e.g. c1 and c2), and periodic information requests (denoted by a diamond, e.g. r1 and r2). Note that system entities are denoted by rectangles and the flow of information corresponds to the direction of the linking arrow. The Figure also indicates that the information exchanged is detailed in a supporting table.

Figure 1. A semantic net description of the system domain

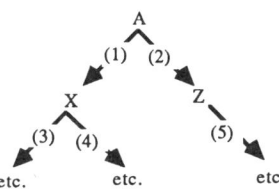

Node	Description	No.	Relation
A	Description of node A.	(1)	Description of relation (1).
		(2)	Description of relation (2).
X	etc.	etc.	etc.

Figure 2. A network diagram description of information flows

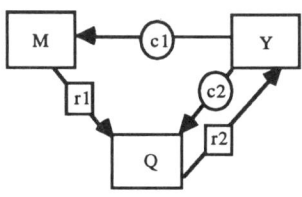

(1) Y is obliged to send reports c1 and c2 to M and Q respectively.

(2) Q may request information r1 from M, while Y may request information r2 from Q.

Info ID	Sender	Receiver	Info Status	Content
c1	Y	M	Routine	etc.
r1	etc.	etc.	etc.	etc.

Function flow diagrams. This notation may be used to describe the performance criteria specified for each sub-system. For instance, criteria such as capacity, accuracy, range, response time, etc., may be specified for a particular set of function modules (performed by one or more sub-systems) so that specific organisational goals may be achieved at acceptable costs. A simple example of such a specification involving different operators is shown in Figure 3. Function flow diagrams may be used to facilitate high level assessments of the safety and reliability of a system. For instance, alternative function flow diagrams (representing different paths for achieving the same organisational goal at comparable performance levels) may be constructed to support the computation and comparison of failure probabilities. Similar assessments at a lower level of specification may be supported by structured diagrams (see later).

Discussions on conceptual design options may thus be supported by such specifications. In this way, organisational goals and performance levels may then be specified sufficiently to constrain sub-system level design.

At the sub-system or task level, the following notations may be applied :

Structured diagrams. This notation may be used to specify the structure and control relationships between operations; sub-tasks; discrete and related tasks; and user and computer functions. For instance, structured diagrams may be used to specify when computer support functions should be presented to the user, i.e. to contextualise these presentations to the user's execution of the task.

Figure 3. A function flow diagram description of system performance

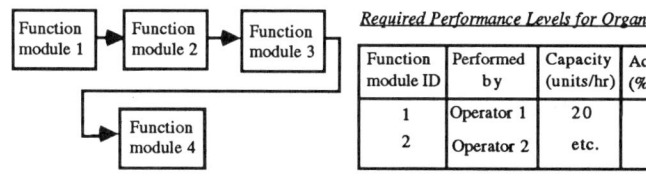

Required Performance Levels for Organisation Goal 1

Function module ID	Performed by	Capacity (units/hr)	Accuracy (% error)	Range	Response time (hr)
1	Operator 1	20	10	+10	1
2	Operator 2	etc.	etc.	etc.	etc.

Notational constructs that may be applied in structured diagram specifications comprise the following set: sequence, selection, iteration, hierarchy, posit and quit. With the exception of the sequence construct (no symbol), each of these constructs is represented by a box with a symbol at either the top right- or left-hand corner (see Figure 4). Condition statements are indexed numerically at the bottom of the boxes (see Figure 4c for an example of conditions for particular selections and termination of iterations). Generally, a structured diagram description is read from top to bottom and from left to right (in that order). When creating a structured diagram description, two simple rules should be followed, namely :

(a) the leaves of the diagram (i.e. boxes at the bottom of the diagram) should comprise actions (except 'quit' boxes which indicate the sub-node box to which control is to be returned);
(b) boxes at the *same* horizontal level in the diagram should have the *same* construct, e.g. all 'selection' boxes.

Figure 4. Notational constructs for structured diagram specification

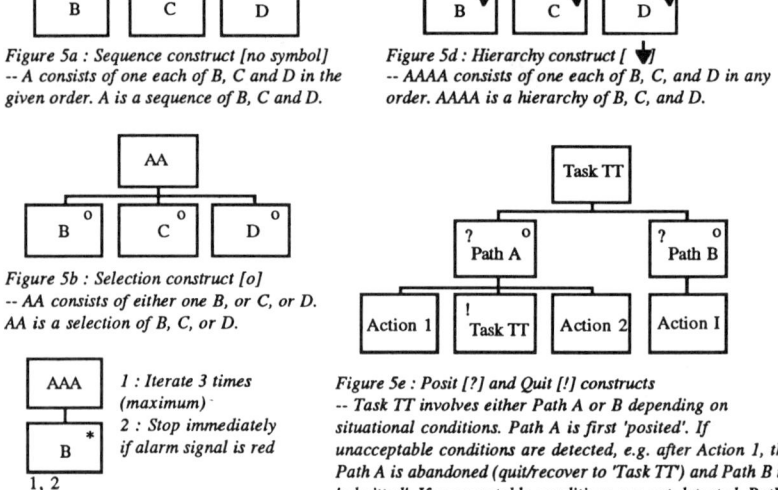

Figure 5a : Sequence construct [no symbol]
-- A consists of one each of B, C and D in the given order. A is a sequence of B, C and D.

Figure 5d : Hierarchy construct [▼]
-- AAAA consists of one each of B, C, and D in any order. AAAA is a hierarchy of B, C, and D.

Figure 5b : Selection construct [o]
-- AA consists of either one B, or C, or D. AA is a selection of B, C, or D.

1 : Iterate 3 times (maximum)
2 : Stop immediately if alarm signal is red

Figure 5c : Iteration construct [*]
-- AAA consists of zero or more Bs. AAA is an iteration of B.

Figure 5e : Posit [?] and Quit [!] constructs
-- Task TT involves either Path A or B depending on situational conditions. Path A is first 'posited'. If unacceptable conditions are detected, e.g. after Action 1, then Path A is abandoned (quit/recover to 'Task TT') and Path B is 'admitted'. If unacceptable conditions are not detected, Path A is completed. In other words, the posit and quit constructs are used together to describe uncertain events, i.e. events which may involve recovery steps to rectify erroneous assumptions.

Individual boxes in a structured diagram description may also be differentiated to indicate functions allocated to the user and the computer (see Figure 5).

Figure 5. Structured diagram specification of functions allocated to the user and computer

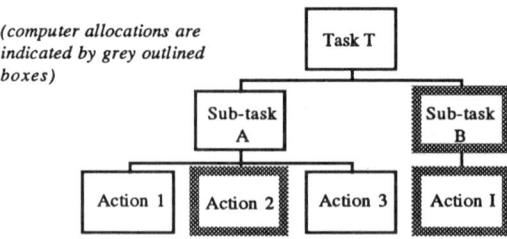

(computer allocations are indicated by grey outlined boxes)

Apart from indicating function allocation, additional representation rules of the structured diagram notation are included to support the specification of task inter-leaving, concurrency and multiplicity (see Figure 6a to c). An account of these task concepts and notational rules follows:

(a) **inter-leaved tasks** are discrete tasks whose operations are inter-woven. Thus, the user is required to execute a *current* task and monitor a *background* task. At a pre-specified point of task execution, the status of the tasks is reversed. To describe such tasks, separate structured diagrams with common actions are constructed and aligned vertically (see Figure 6a). Task inter-links are established by specifying common actions *across* the structured diagrams. Note that common actions are denoted by boxes assigned the same identifier, e.g. actions {2, 4, 6} of Tasks X and Y in Figure 6a;

(b) **concurrent tasks** are discrete tasks that are performed at the same time, e.g. data input while monitoring displays. To describe such tasks, separate structured diagrams are constructed and aligned horizontally (see Figure 6b). In addition, the root node (top box) of each diagram is assigned a common root identifier, e.g. Operator 1(data input) and Operator 1(display monitoring);

(c) **multiple tasks** describe many units of the same task being performed at the same time. To describe such tasks, the root node (top box) is layered and the number of units is indicated, e.g. n=4+(action 'C')/hour. The latter indicates that more than 4 units of the same task may be performed in the specified time period, and the last unit is terminated at action 'C' (see Figure 6c).

Semantic net and function flow diagrams. These notations may be used for HF specifications at the sub-system level in the same manner as illustrated for organisational level specifications.

Performance tables. These tables may be constructed to describe tests and metrics to be applied on proposed design solutions. Generally, the tabular formats suggested by Whiteside et al (1985), and Blyth and Hakiel (1989) would be suitable for the following specifications :

(1) *who* should be tested, e.g. a particular user group;
(2) *what* tasks test subjects are required to perform, e.g. particular benchmark tasks;
(3) *how* tests should be conducted, e.g. laboratory simulation of the work environment;
(4) *what* test metrics should be used, e.g. task speed and accuracy (objective assessment); and attitude questionnaire (subjective assessment);
(5) *what* performance levels are expected, e.g. worst and best levels for a particular prototype.

These performance specifications establish the basis for HF design at the interaction level, and define the prescribed tests to which proposed solutions should be subjected.
At the interaction level, the notations described earlier may be applied in the same manner as for organisational and sub-system levels. Thus, these notations are not discussed further. Instead, a case-study illustration is used to show how the notations supplemented by screen diagrams (pictures), may be used to describe interactive behaviours between the user and computer; e.g. how computer functions and messages are specified with respect to the task execution context of the user. A simplified illustration involving a digital network security management system follows.
Figures 7 to 11 show a partial set of HF specifications for a user interface design. The design specifications address the following HF concerns:

Interactive task and potential user errors. Figure 7 shows part of the network manager's inputs required to achieve on-line task goals of the (case-study) system. The structured diagram specification, termed an interaction task model, also indicates the relationship between specific display screens and the on-line task of the network manager, e.g. Screen 4B (S4B) and the selection of a 'show user list' button (see Figure 7 -- the circle and arrow indicate the part of the diagram that has been enlarged). Note that grey bubbles in the Figure indicate the point at which specified screens are removed or 'consumed', followed by an immediate presentation of the next screen. For instance, following input of the 'show user list' selection, Screen 4B (S4B) is 'consumed' and either Screen 5B-1 (S5B-1) or Screen E1-em3 (SE1-em3) is triggered. The screen that is actually triggered depends on whether the required inputs have been made correctly (see later).

Figure 6. Structured diagram specification of inter-leaved, concurrent and multiple tasks

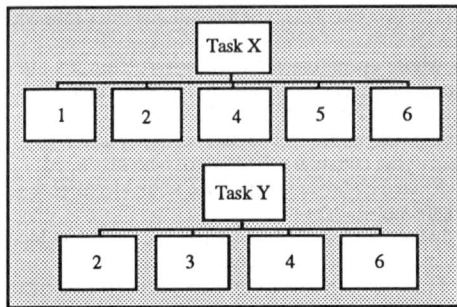

Figure 7a : Structured diagram specification of inter-leaved tasks X and Y. The common actions are {2, 4, 6}.

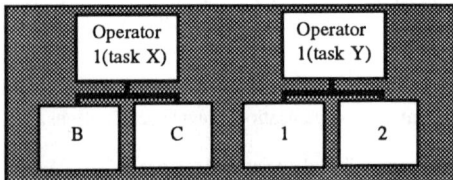

Figure 7b : Structured diagram specification of Operator 1's ability to perform task X and Y concurrently.

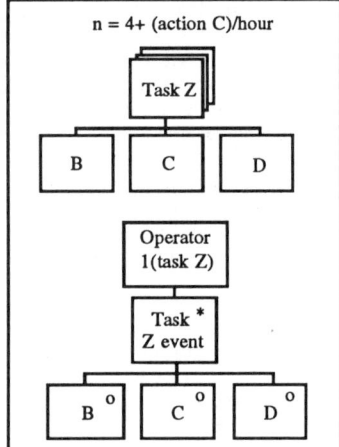

Figure 7c : Structured diagram specification of Operator 1's performance of multiple units of task Z without first having to complete each task unit. More than four units may be performed per hour, and the task terminates on action C at the fifth unit.

<u>Context and timing for presenting computer functions to support the user's task (including error handling).</u> Figure 8 is a structured diagram specification that augments the specifications described by Figure 7 (the circle and arrow indicate the part of the diagram that has been enlarged). In particular, it shows how some of the screens (e.g. Screen E1-em3, Screen 4B) are mapped onto and actuated with respect to the actions of the network manager (described in Figure 7). For instance, the streams of network manager actions and screen actuations are inter-linked as follows:

Manager actions: 'show user list' ---> 'select user name'
Screen actuations 1: Screen 4B ---> Screen 5B-1
Screen actuations 2: Screen 4B ----> Screen E1-em3 ----> Screen 4B ----> Screen 5B-1

Together, Figures 7 and 8 specify that <u>Screen Actuations 1</u> is applicable if all inputs required to select the 'show user list' has been made correctly, while <u>Screen Actuations 2</u> is applicable if an input error has occurred, i.e. if the 'show user list' button has not been clicked prior to the 'select' button, an additional error screen is triggered (namely, Screen E1-em3 or SE1-em3).

<u>Error, feedback and help messages.</u> Figure 9 augments the specifications in Figures 7 and 8 with a tabular index of screen messages, e.g. error messages. As an illustration, consider Screen E1-em3 that was referred to earlier. The screen annotation indicates the display screen of pictorial format E1 (not shown) with an error message identified as number 3 (em3). The content of this message is specified in Figure 9.

<u>Screen design and behaviour of screen objects.</u> Figure 10 also augments the specifications in Figures 7 and 8 with pictorial specifications of computer screens, e.g. Screens 4B (S4B). The pictures may either be drawn on paper (to-scale or dimensioned), or prototyped using a computer-based tool. Additional information on the screen contents is provided in an accompanying table, and the behaviours of individual screen objects are described using structured diagrams (see Figure 11).

Figure 7. Part of an interaction task model for network security management

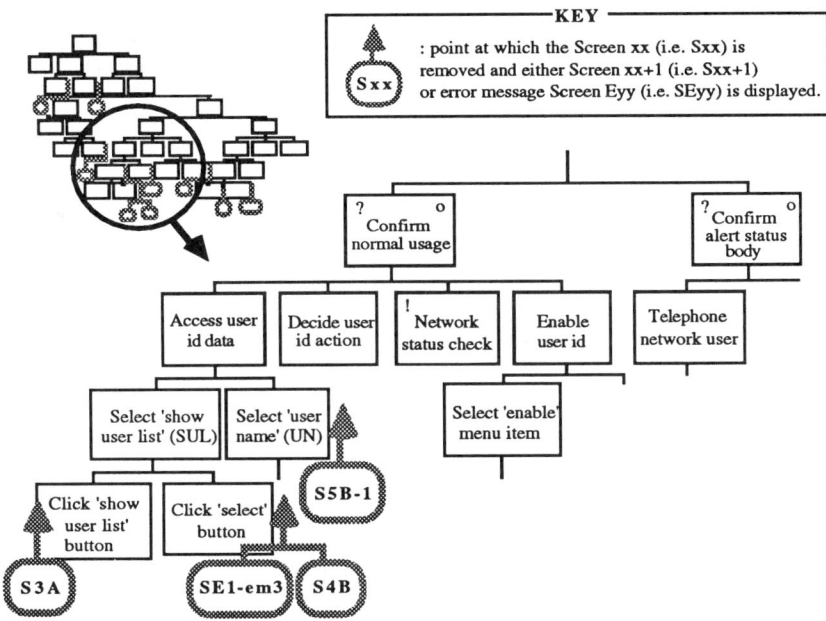

Figure 8. Part of a screen actuation specification for network security management

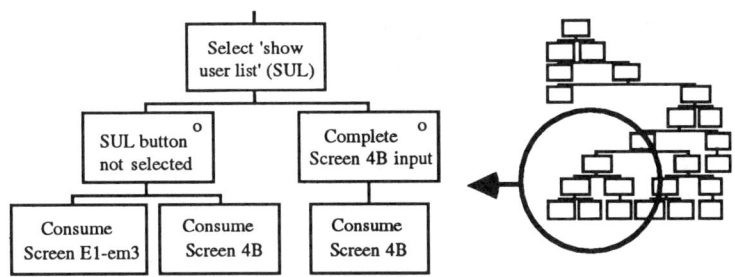

Figure 9. Part of an error message index for network security management

Message Number	Message Content
em1	Sorry, your log-on inputs are incorrect. Your session will be terminated.
em2	Please indicate a host and/or user report action by selecting either the 'Delete' or 'Pending' radio button.
em3	Please indicate the required security action by selecting a radio button from the 'Security Action Selection Menu'. Do this BEFORE clicking the 'Select' button.
em4	Please select a user name from the user name display window..........etc.

Interactive task scenarios are thus designed and documented using the above specifications. As an illustration, the scenario for this particular case-study is as follows. Having been alerted to a security violation, the network manager is required to access the computer database to gather information on the network user involved. To this end, the manager double-clicks the security icon in Screen 3A to activate the network security application (see Figure 9a). The input 'consumes' Screen 3A and triggers Screen 4B (see Figure 7). Thus, a menu offering a selection of three actions is presented to the network manager, namely 'Search Connection', 'Show User List' and 'Show Access Points' (see Figure 9b). The manager then clicks one of the three radio buttons to indicate the desired selection and confirms the input by clicking the 'Select' button (see Figure 7). If a radio button was not selected prior to clicking the 'Select' button, an error message screen, namely Screen E1-em3 is activated (see Figures 7, 8 and 11). However, if the inputs were made correctly, Screen 4B would be 'consumed' and Screen 5B-1 presented (see Figures 7 and 8). Using the latter screen (not shown pictorially), the network manager may then specify what network information should be extracted from the database for the network user concerned. On the basis of the information, the manager would then decide whether the user should be contacted to establish possible causes of the security violation (see Figure 7). Thus, appropriate responses to the violation are determined, e.g. if the event is due to a password mis-key, the manager may restore the user's account using other computer functions (not shown).

The above account completes an illustration of how structured notations may be used to support comprehensive HF specification of a system design.

Figure 10. Pictorial screen layout specification of Screen 4B for network security management

é	**File**		
.....as for Macintosh User Interface Environment	Open Quit	**Security Action Selection Menu** ◉ Search Connections ◉ Show User List ◉ Show Access Points (**Select**)	

Screen Object	Description	Design Attributes
File (menu bar)	Offers 'Open' and 'Quit' menu items. 'Open' allows the network manager to open host and user reports. 'Quit' allows the manager to quit the security application.	Behaviour as per standard Macintosh menu items.
etc.	etc.	etc.

Figure 11. Part of an interface model specification of the SecurityPending folder for network security management

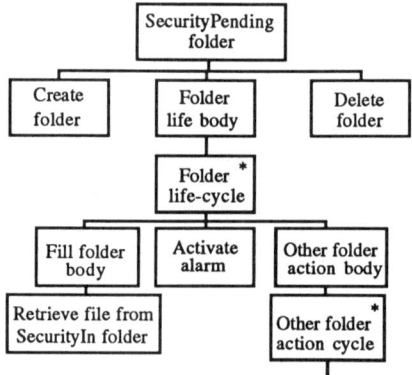

CONCLUSION
The paper illustrates how structured notations can be used to support wider HF involvement in system design. In particular, the notations may improve the effectiveness of HF involvement in three ways, namely :

(a) by supporting more complete and detailed HF specifications at various stages of system design. Since the specifications have been set explicitly within the system development context, HF contributions may be incorporated more efficiently by software engineers;
(b) by supporting the elicitation of user feedback. In particular, the improvements in HF specifications outlined in (a) above, would support more comprehensive user feedback throughout system design. Feedback elicitation is also facilitated by the graphical nature of structured notations since they would be understood better by users. Thus, invalid design assumptions may be identified earlier and eliminated more effectively;
(c) by exploiting the benefits of a common notation. Thus, by recruiting structured notations from existing software engineering methods, HF design specifications would be understood better by software engineers.

For the above reasons, the uptake of HF contributions would improve. Additional improvements would also accrue if an explicit HF design process were established, e.g. via a structured HF method (Lim and Long, 1992a and b; Lim, Long and Silcock, 1992). Such a method could also be integrated with similarly structured software engineering methods. Thus, the scope, format, granularity and timing of HF contributions may be configured appropriately to support a more effective uptake.

ACKNOWLEDGEMENTS
Part of the research was carried out for the Procurement Executive, Ministry of Defence (RARDE, UK). Views expressed in the paper are those of the authors and should not be attributed to the Ministry of Defence.

REFERENCES
Blyth, R. C. and Hakiel, S. R., 1989, A user interface design methodology and the implication for structured system design methods. In Proc. IEE Third Int. Conf. on Command, Control, Comms. and Management Information System, (Bournemouth, May 2-4, UK).
Brooks, R., 1991, Comparative task analysis: an alternative direction for HCI science. In J. Carroll (ed.), Designing Interaction : Psychology at the Human Computer Interface, Cambridge Univ. Press.
Carver, M. K., 1988, Practical experience of specifying the human computer interface using JSD. In Proc. ES 1988 Conf., (Manchester, April 11-15, UK), Taylor&Francis, 177-182.
Carver, M. K. and Cameron, J., 1987, The JSD method: a framework for the specification of the human computer interface. Internal Report of LBMS Ltd.
Damodaran, L., Ip, K. and Beck, M., 1988, Integrating HF principles into structured design methodology: a case- study in the UK civil service. In : H. J. Bullinger et al (eds.), Information Technology for Organisational Systems, 235-241, Elsevier Science.
Finkelstein, A. and Potts, C., 1985, Evaluation of existing requirements extraction strategies. FOREST Project Report R1.
Fitter, M. and Green, T. R. G., 1979, When do diagrams make good computer languages? Int. J. Man-Machine Studies, 11, 235-261.
Hares, J., 1987, Methods for a longer life. Computer News/Databases, August 6, pg. 18.
Lim, K. Y. and Long, J. B., 1992a, A method for (recruiting) methods: facilitating HF input to system design. In Proc. ACM CHI'92 Conf., (Monterey, May 3-7, USA), 549-556.
Lim, K. Y. and Long, J. B., 1992b, Rapid prototyping, structured methods and the incorporation of HF in system design. In Proc. East-West Int. Conf. on HCI (EWHCI'92), (St. Petersburg, August 4-8, CIS), 407-417.
Lim, K. Y. and Long, J. B., 1992c, Computer-based tools for a structured HF method. In Proc. Int. Conf. on Computer-Aided Ergonomics and Safety, (Tampere, May 18-20, Finland), Elsevier Science, 71-79.
Lim, K. Y. and Long, J. B., 1992d, Task analysis for system design: current problems, requirements and solutions for improving the applicability of task analysis to system design. To appear in Proc. EACE Task Analysis in HCI Workshop, (Schaerding, June 9-

11, Austria).

Lim, K. Y., Long, J. B. and Silcock, N., 1990a, Requirements., research and strategy for integrating HF with structured analysis and design methods: the case of the JSD method. In Proc. ES 1990 Conf., (Leeds, April 3-6, UK), Taylor & Francis, 32-38.

Lim, K. Y., Long, J. B. and Silcock, N., 1990b, Motivation, research management and a conception for structured integration of HF with system development methods: an illustration using the JSD method. In Proc. 5th European Conf. on Cog. Ergonomics,(Urbino, September 3-6, Italy), Golem Press, 359-374.

Lim, K. Y., Long, J. B. and Silcock, N., 1992, Integrating HF with the JSD method : an illustrated overview. Ergonomics, 35, 10, 1135-1161.

Lundell, J. and Notess, M., 1991, HF in software development : models, techniques, and outcomes, In : P. Robertson, G. M. Olson, and J. S. Olson (eds.), Proc. CHI'91 Conf., (New Orleans, May, USA), 145-152.

Moraal, J. and Kragt, H., 1990, Macro-Ergonomic design : the need for empirical research evidence, Ergonomics, 1990, 33(5), London : Taylor Francis, 605-612.

Sutcliffe, A., 1988, Some experiences in integrating specification of HCI within a structured system development method. In Proc. BCS HCI'88 Conf., (Manchester, September 5-9, UK), Cambridge Univ. Press, 145-160.

Sutcliffe, A., 1989, Task analysis, systems analysis and design: symbiosis or synthesis? Interacting with Computers, 1, 1, 6-12.

Whiteside, J., Jones, S., Levy, P. S., and Wixon, D., 1985, User performance with command, menu and iconic interfaces. In Borman and Curtis (eds.), HF in Computing Systems II, Elsevier Science, 185-191.

DEVELOPING USER INTERFACE DESIGN TOOLS: AN ANALYSIS OF INTERFACE DESIGN PRACTICE

Mathilde M. Bekker & Arnold P.O.S. Vermeeren

Delft University of Technology,
Faculty of Industrial Design Engineering
Jaffalaan 9
2628 BX Delft
The Netherlands

To determine what user interface design tools are needed by interface designers we interviewed sixteen designers about their experiences with design practice. The results showed that tools are needed to support communication during the design projects. A further analysis resulted in a list of interesting issues related to communication problems: differences between disciplines, recipient design, resolution of the representations used, working interactively, capturing history and evolution of design in representations. A major part of the list is related to the representations used during communication. We conclude that new tools are needed to enhance communication between the members of a design team and between the design team and other people involved in the design process.

INTRODUCTION

Research by Gould et al. (1985) and Bellotti (1990) shows that design methods do not seem to be used as often as expected by the developers of these methods. The well-known design principle, early focus on users and tasks (Gould et al., 1985), is also applicable to the development of design methods itself. Therefore, we decided to perform an analysis of user interface design in practice in order to determine what kind of methods or tools would be most useful to designers and what requirements have to be met by these tools in order to be useful in practice.

Rosson et al. (1988) and Curtis et al. (1988) have studied design practice to determine what kind of tools are needed by user interface designers. Rosson et al. (1988) focussed on design projects with a major user interface part. They concluded that there are two types of design processes: incremental and phased development processes respectively, and that each requires its own kind of tools. They mentioned a number of tools that designers need ranging from comprehensive, easy-to-use prototyping tools to design-record keeping tools. Curtis et al. (1988) focussed on design processes of large systems and concluded that the following three activities should be supported in software development; knowledge sharing and integration, communication and coordination, and change facilitation. Both these studies were based on design practice in the USA.

In the present study we focussed on user interface design practice in the Netherlands. We focussed on small teams and the emphasis of the design projects was on interface design instead of on systems design.

An interview was developed containing questions about various aspects of design projects. We gathered information about designers' needs based on their descriptions of design projects. Furthermore, we asked designers what kind of tools they would need to improve the design process.

In this paper we discuss the set-up of the experiment, the results and the consequences of these results for the development of design methods.

METHODS

Since we wanted to gather extensive and thorough descriptions of a large number of design projects we decided to use interviews. The use of longitudinal studies would have been too time consuming and hard to realize.

Interview design

Based on earlier research on design practice (Bekker & Vermeeren, 1992) four important activities of design projects were selected to be included in the interview: gathering information about users, decision making and task division between team members, gathering information about the application domain and asking for advice from parties outside of the design team.

During the semi-structured interviews one interviewer asked the main questions, while an assistant monitored the interview and sometimes indicated interesting topics that should be discussed in more detail. We videotaped each interview for later analysis.

Subjects were asked to briefly describe one design project on which they would focus while answering the questions. In the first part of the interview the five selected aspects of design processes were discussed. Subjects were asked to describe what their experiences were, whether they were satisfied during the design project and whether they had had different experiences in other projects.

In the second part of the interview subjects were asked to describe with what kind of tools they thought the design process could be improved.

Subjects

Most of the subjects worked for a software or computer company (5) or worked as a freelance designer (5). Three subjects worked for a non-computer company and three worked in a research environment. Their positions ranged from programmers (2), graphic designers (2) and interface designers (10) to project managers (2).

During two group discussions, with a total of 8 designers, and eight individual interviews information was gathered about their experience with interface design.

Analysis

Transcripts were made of the videotapes of the interviews and of the group discussions. The transcripts were scanned for problems mentioned by the designers that referred to designers' needs for design methods or tools.

To get an impression of the behavioural level at which the problems had occurred, we distinguished (according to Curtis et al., 1988) between problems at the individual level (designer himself), at team level (the designer and his team members), and at project and company level (the designer and project managers or higher management). The problems were also categorized according to activities during the design project: information gathering, designing, prototyping, evaluating or implementing. The number of designers that mentioned a particular problem, and the number of activities during which such a problem was experienced, gave an indication of its importance.

During the last part of the interview the designers described with what kind of tools the design process could be improved. These tools were categorized according to the aims of the tools. While describing the kind of tools needed to improve the design process, designers often mentioned specific requirements that would have to be met by these tools.

We determined what kind of tool would be most useful to designers, based on the problems that were experienced, and the tools that were mentioned by the designers. In order to gain a more thorough understanding of the activities that have to be supported by this tool, the data were scanned for remarks that explained in more detail in what type of context the tool would have to be used.

RESULTS

To give an impression of the projects that were studied in the experiment we will first provide some general information about these design projects. Then we will give an overview of the kind of problems that were experienced by designers, the tools that, according to the designers, could improve the design process, and the requirements that would have to be met by these tools. Finally we will present a number of issues that are related to the problem that was mentioned most by the designers.

General information about the design projects

The survey included design projects that were highly variable in nature. Team size varied from 3 to 30 team members. The shortest project lasted one month, while the two longest projects lasted 3 years. A wide variety of applications and products was described, including telecommunication applications (e.g., business telephone), point of information systems (e.g., museum information system), consumer products (e.g., hifi installation) and process control information systems.

Table 1. A matrix of design problems in two dimensions: behavioural level (designer, team and organisation) and activities during the project. N : number of designers that mentioned a particular problem.

	designer	N	team	N	organisation	N
overall	*design process *representing users *usage of tools /methods	3 1 4	*design process *involvement of users *usage of tools *communication about design solutions	6 2 1 6	*design process *constraints given by clients	7 1
analy-sis	*lack of technical information *lack of information about users	2 3	*lack of technical information *lack of information about users	3 10	*information supply by clients *uncertainty of target user groups *project organisations	2 2 3
design	*representing the users *designing itself *design process	1 3 4	*involving and representing the users *communication using tools *communicating about design ideas *design process	5 5 7 9	*involving the users *communication using tools *constraints given by client *design process/methods	1 2 1 2
simu-lation	*making simulations	2	*making simulations	3	-	
evalu-ation	*what to evaluate *how to evaluate	1 4	*involve users early in the process *how to evaluate	2 1	*being allowed to do user-testing	6
imple-men-tation	-		*communicating about design solutions	2	*constraints given by clients	1

Problems mentioned by designers

During the interviews designers mentioned many problems that they had experienced during their work.

We classified the list of problems that were mentioned by the designers according to two dimensions: the activities during which the problems had occurred, and the behavioural level (individual, team, project and organisation) at which they had occurred. This way we were able to list all the problems in a matrix depicted in table 1.

Design activities:

Some problems are very typical for a specific design activity, such as the problem of making simulations. Other problems occur at many phases, especially problems concerning communication and the involvement of users occur during almost all activities.

Behavioural levels:

Problems at the individual level were mostly related to the use of tools, how to apply design methods, and how to involve or represent the users. At the team level, communication problems were mentioned most. At the organisational level, communication problems and constraints given by clients are often mentioned. Involving and representing users is an important problem at every level.

Two types of problems occurred during many activities and at many behavioural levels: communication with team members, clients and other people involved in the design project (11 designers) and involvement of users in the design projects (9 designers).

Table 2. Tools needed to improve the design process and requirements that have to be met by these tools according to the designers that participated in the interviews. N : number of designers that mentioned a particular tool, M : number of designers that mentioned a requirement.

tools	N	requirements	M
Prototyping tools	9	-easy to use -quick building of prototype -possibility to work interactively	3 2 2
Communication facilitating tools	6	-use jargon and concepts that relate to communication partners	4
User involvement tools	5	-help decide how to involve users and evaluate results of user tests	3
Project management tools	4	-keep and manage documents used during design projects	3
Decision support tools	4	-help gather information to support design decisions	3
Interface description tools	3	-describe relations between design decisions and solutions	3

Tools mentioned by designers

When designers were asked how the design process could be improved, they mentioned a number of tools. Table 2 shows an overview of the tools and requirements that were mentioned by more than one designer.

Tools needed by designers

The information about the tools that designers need, based on the tools they mentioned, and the problems they had experienced complement each other. The list of problems indicates that designers need tools to communicate with team members, clients and users, and to involve users. These problems were experienced at several behavioural levels and during a major part of the design process. When designers were asked how the design process could be improved, they mentioned mostly communication tools, and tools that help represent the design (like prototyping tools).

Both the descriptions of the design projects and the solutions mentioned by the user interface designers show that designers need tools to support their communication activities.

Figure 1. Communication problems between designer and team members, clients and other people involved in design projects.

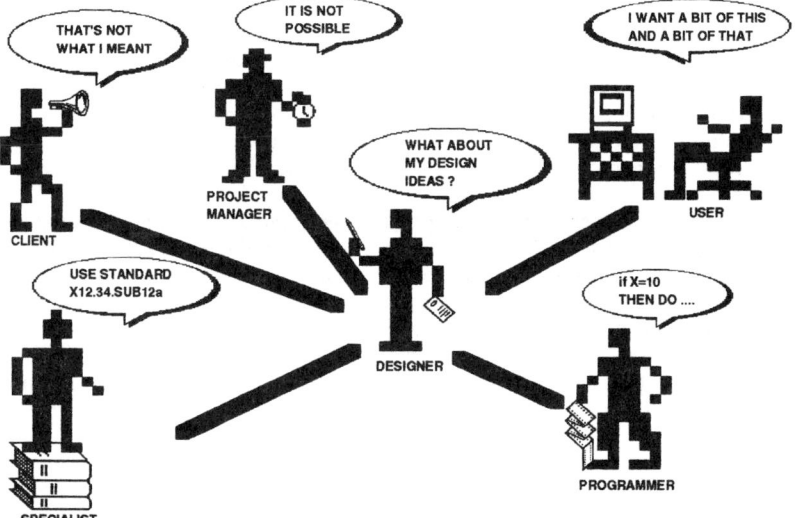

Communication problems

To gain more insights into the context in which communication problems (the most important problems) had occurred, a thorough analysis of the transcripts was performed. It revealed a number of interesting issues related to communication problems.

Two of these issues are related to the representations used during communication. First, representations can relay different levels of resolutions, e.g., representations made with pen and paper have a low resolution and representations made with a computer have a high resolution. Four designers mentioned that the resolution of design representations influences the way in which communication partners perceive them: a design solution is more definitive when a representation has a high resolution than when it has a low resolution. Second, representations can be used to capture the evolution and the history of the design process. Three designers mentioned that they had reused representations in later phases of the design project, and two other designers said that they had used the history of design representations to reconsider design decisions.

Two issues are related to the team members who are the communication partners of the designer. First, differences in communication between disciplines sometimes resulted in problems during the design process. Designers mentioned that team members belonging to different disciplines use different sorts of representations (3 subjects), and have different priorities in showing information in representations (3). Second, recipient design (i.e., tailoring of communication depending on communication partner) was used by designers that communicated with other team members. Three designers gave examples of situations in which emphasis was given to those aspects of the representation that would be most appealing to the communication partner.

Another issue is related to the interaction between the designer, the team members and representations involved in the communication. Designers mentioned that they lacked the possibility to work interactively on the making of representations. Three designers mentioned that making representations together with other team members enhanced the agreement among the team members.

DISCUSSION
Comparison with other studies
We compared the results of this study with other studies that analysed design practice to determine whether the results can be generalized towards other situations (countries and team sizes) and whether we have obtained new insights into the needs of designers.

Rosson et al. (1988) do not mention the need for communication tools as such. They mainly focus on design tools for specific activities, such as prototyping and idea generation; communication is mentioned as a part of these activities. Curtis et al. (1988) found that one of the three most salient problems during design were breakdowns in communication and coordination.

Their conclusions about communication problems are similar to some of the aspects we found: the importance of establishing common terminology and representational conventions and the importance of capturing design history and evolution to cross temporal boundaries. The way these problems manifested themselves in our study, is slightly different, because our study focussed on smaller teams and the emphasis of the design projects was on interface design instead of on systems design. However, the similarity between the conclusions shows that the situation we described is not only typical for the Dutch situation, but is typical for interface design practice in general.

CONCLUSIONS
We conclude that tools are needed to enhance communication between team members and other people involved in the design process. We found a number of aspects that play an important role during communication. Some of these are related to the representations used, while others are related to the communication partners.

Further research is required to determine how communication activities of designers can best be supported. Therefore, our future research plans will consist of a number of experiments in which we focus on the use of various representations by multi-disciplinary design teams while they communicate.

REFERENCES
Bekker, M.M. and Vermeeren, A.P.O.S. , 1992, Human computer interface design in practice: variability, constraints and information sources. In Contemporary Ergonomics 1992, (Taylor & Francis, London), pp. 352-357.
Bellotti, V.M.E., 1990, Applicability of HCI techniques to systems interface design. Doctoral thesis submitted at Queen Mary and Westfield College, London, .
Curtis, B., Krasner, H., and Iscoe, N. , 1988, A field study of the software design process for large systems., Communications of the ACM, Vol.31, 11, pp. 1268-1287.
Gould, J.D. and Lewis, C., 1985, Designing for usability: Key Principles and what designers think, Communications of the ACM, 28, 3, pp. 300-311.
Rosson, M. B., Maass, S. and Kellogg, W. A., 1988, The designer as user: building requirements for design tools from design practice. Communications of the ACM, Vol. 31, 11, pp. 1288-1298.

THE APPLICATION OF RAPID PROTOTYPING IN THE (RE)DESIGN OF A USER INTERFACE

H.C.M. HOONHOUT

Psychonomics Department, Utrecht University
2 Heidelberglaan, 3584 CS Utrecht,
The Netherlands

In this paper, two human factors tools, guidelines
and rapid prototyping, are discussed with respect to
their potential usefulness to designers. The question is
raised whether these tools, in the form as they are now
widely promoted, are useful to designers. It is proposed
that in order to make these human factors tools useful
for designers, they should be adapted to the needs of
the designers. Furthermore, it is argued that in the
communication between human factors specialists and
designers, a prototyping tool can be useful.

INTRODUCTION

User interfaces of software applications were some years ago
usually limited in their design flexibility. The manufacturer of
the systems determined the appearance of the interface. If their
customers had particular requirements, the manufacturer would, if
possible, implement these into the design. But otherwise, the
design of the systems and of the user interface had usually just
one single fixed format, determined by the manufacturer. In fact,
the manufacturer was in his design of the user interface also
limited by the technology then available (e.g. layout options, use
of colour). Implementing the system's user interface was also a
job for specialists, that required extensive training in
programming languages.

Modern software applications allow a much larger number of
interface design options and an increased flexibility of display
design. Another development is that manufacturers provide now many
software applications with facilities for the customers to design
the interface themselves, i.e. the manufacturer no longer provides
the design of the interface, but just provides the software

'tools' to make an interface. With these facilities, the system owners can make their interface the way they would like to have it.

Superficially, this shift of responsibility for the interface design from the manufacturer to their customers may seem a attractive development in response to an increasing demand for more flexibility of systems, and in particular of the interface. But whereas the manufacturers have been building their experience with interface design over many years, for most of their customers making an interface will not be an everyday job.

If Human Factors has achieved one thing in software design, than it is the increased awareness of the importance of usable and understandable systems, or, as it is often called, the userfriendliness of systems.

System owners may acknowledge how crucial the user interface is in achieving effective and efficient use of their system, and may well be able to recognize an example of a bad user interface. However, this does not imply that they know what determines a good interface design. As is the case with most software design aspects, the task of making a proper interface is more difficult than it initially seems to be.

In fact, if a system owner decides to make the interface himself, he is faced with at least three issues, that could impede the development of a satisfactory user interface. The first issue is that using the interface design tool may prove to be more difficult than was initially expected. The danger may be that the designer will be inclined to apply only those facilities of the interface design tool that appear to be the least complex to use. The design of the interface will then be driven by 'technological' constraints, rather than by system requirements.

The second issue concerns the level at which the system owner can adapt the software application to his needs; usually only the interface of a software application can be adapted by him, the underlying structure is in most cases not changeable. This may impose severe constraints on interface design that usually cannot be solved without help of the manufacturer.

The last, and most critical, issue is a lack of knowledge and experience in interface design and design procedures by the system owner. In most cases, the persons who have to make the interface will not be experienced in interface design. The danger is that a design-by-intuition approach will be adopted, i.e. the designer assumes, as a guiding principle in design decisions, that what he considers to be understandable, will also be readily understood by the enduser.

What designers need are tools and aids to assist them in the design of the interface. These tools should aim to make designers aware of human factors aspects of user interface design, and of a user-centred design approach. In this paper, two of such tools are discussed, i.e. guidelines and standards, and rapid prototyping. The question is raised whether these tools, in the form as they are now widely promoted, meet the needs of designers. It is proposed that to make these human factors tools useful for designers, they should be adapted, and more important, they should be integrated, to meet the needs of the designers.

HUMAN FACTORS GUIDELINES

Since the first introduction of computer systems, a large body of knowledge about interface design has been compiled. This knowledge is derived from several sources, e.g., basic psychological research, human factors studies, plain hands-on experience of designers and users, and experience and skill available from the field of graphic design. Apart from the literature dealing with particular design aspects or research, a variety of guidelines and standards addressing human-computer interface issues have been published over the years.

Guidelines and standards can be considered as a way of making this body of human factors knowledge potentially accessible to designers. The guidance provided may range from general discussions about interface design, to relatively specific requirements for interface design. In this respect, Smith (1986) makes the following distinction between guidelines and standards. Guidelines present a series of generally stated recommendations for user interface design, and a standard gives a set of (generally stated) requirements of user interface design. Where as guidelines describe and recommend, standards prescribe and enforce.

As the growing number of guidelines and standards documents shows (see for example Tullis (1988), and Grossman (1992), for an overview of currently available guidelines and standards), they are a widely accepted and popular medium for the transfer of human factors expertise and knowledge in the field of HCI. What, however, not always is made explicit in guidelines and standards, is who the intended users are. It is likely, however, that guidelines and standards may be appreciated by human factors specialists quite differently than by designers.

Appreciation of guidelines by designers

Mosier and Smith (1986) conducted a survey of the users (human factors specialists as well as system designers) of a guidelines report for software design. A complaint by the designers who responded was that the guidelines were difficult to fit in with their task. Their complaints were: the information provided is too general, is not relevant for the particular system, or impractical for the particular system; in many cases the set of guidelines seemed to give conflicting information, in which cases designers found it difficult to decide which part of the guidance was to be followed. Another problem reported was that the guidelines report needed more structuring that could help in finding particular information on a particular design problem more easily.

Empirical basis for guidelines and standards

A more fundamental problem with HCI guidelines and standards is that most of these documents do not indicate on what basis the information is selected. This makes it difficult to assess the value of the provided information. Parts of the guidelines and standards are well founded in the results of basic or applied research, but other parts are not (Tullis, 1988). A large part of the available software guidelines is not based on experimental data and quantitative performance measurements, but is still based on expert judgement and accumulated practical experience. Where this is the case, the guidance provided should be treated with care. Professional prudence should guide here the application of the recommendations provided. In some cases it may be necessary to verify the decisions made in e.g. user trials. It cannot be denied that with no relevant data at hand, an 'educated guess' as offered by the guidelines is to be preferred above an untutored opinion of an individual designer. However, for the designers it will not make a convincing case for guidelines usage, and it will certainly not enhance ease of use of guidelines and standards by the designers.

Usability of guidelines

The knowledge collected in guidelines is often more qualitative than quantitative of nature. Designers, however, prefer quantitative data, that they can readily apply. It interferes with their actual task if they have to interpret the generally stated guidelines to derive specific design specifications and rules. What is needed is a 'translation' of these general guidelines into system specific design specifications that can be more readily understood and applied by the designers of a system.

To make them relevant for a broad range of interfaces, guidelines have to be stated in a general, task-independent and system independent way. This implies, however, that translating general guidelines into application specific design rules cannot be done properly without first having analyzed and described the specific context in which the particular application will be used. This involves an analysis of the tasks requirements, and of the requirements of the users. Other sources of information that may have to be consulted, are staff regarding additional system constraints and requirements, or organizational requirements. When making general guidelines system specific, it is essential that the intended system's users are involved. Current system usage should be recorded, and user requirements should be determined, to ensure that the proposed design features will meet their needs.

Involvement of the designer in this process of collecting information about task and user requirements, etc., and especially user experience with current system, may help to make the designer become aware of the relevance of this type of information as an input for design decisions. Involvement could take the form of an iterative procedure in which the human factors specialist collects the information and drafts the design rules, and presents it to the designer for comment.

MAKING USABLE HUMAN FACTORS TOOLS FOR DESIGNERS

When 'translating' general guidelines into a set of system specific design rules, the human factors specialist should realize that his responsibility is two-fold. First, he must ensure that in the specification of the specific design rules the context of the application is accounted for, i.e. the task and user requirements, organizational and housestyle requirements, and possible constraints due to the hardware, or other system aspects already decided. Secondly, he must ensure that the resulting design rules document has a format that will optimally support the designer in his task. Thus, the human factors approach of user-centred design has to be adopted at two levels: that of the end-user of the system, and that of the user of the human factors design tool.

The document containing the specific design rules should have a format in which the rules are made clear to the designer with examples and explanations. From a set of general guidelines those relevant for the particular system at hand should be selected, and modified and reworded to make them application specific and understandable to non-specialists. The information acquired in interviews, questionnaires, etc. from staff and users, form a basis on which the specific design rules can be formulated. Detailed descriptions and examples could be added. If feasible, an explanation can be provided, i.e. what a specific rule tries to prevent, or reversely, tries to promote. This may also be useful in an evaluation of the design at a later stage, since in this way adequate test criteria are provided.

However, the resulting document may still be rather abstract, lacking an integration of the different design rules. Therefore, a tool is required to demonstrate to the designers the application of the specific design rules in the design of the system's interface.

PROTOTYPING AND USER TRIALS

In recent years, a design tool called (rapid) prototyping has become popular. The development of software tools that can be used to emulate user interfaces has made the building of prototypes a reasonably cost-effective proposition. Rapid prototyping seems to meet more support than criticism (Wilson and Rosenberg, 1988). It is generally considered to be a useful means of testing the appropriateness of a user interface in an early phase of the design cycle. It can be used to test relatively easy several design options concurrently. Especially in those cases that human factors knowledge does not have a definite, empirically validated answer, and a design decision is made on the basis of educated guesses and professional judgement, prototyping can help to evaluate the proposed solutions.

Without denying the advantages, rapid prototyping is nothing
more than a tool, that requires good 'craftsmanship' to get the
full benefit from it. It is not productive or efficient to use a
rapid prototyping tool just for casual inspection of interface
proposals. To use it effectively in the design process, a test
procedure is required. Adequate test methods have to be adopted,
to collect relevant data about the usability of the proposed
system's interface. This involves, among other things, a proper
selection of tests that represent the actual tasks, and a
representative sample of the intended users. The point here is
that the test data should provide information that not just tells
(in a quantitative way) how quick a task can be completed, and
with how many errors, etc., but helps to understand why the user
is making the errors, etc. Finally, the results do not provide a
direct answer to redesign of the interface, but the results will
need to be interpreted. This all means that rapid prototyping used
in this way is hardly a feasible tool for designers.

Use of the prototyping tool in the design process may prove,
however, to be highly efficient in discussions with staff and
designers about the design principles and the specific design
rules. It integrates the different rules, and can for the designer
be a great support in his task of actually designing the user
interface.

REFERENCES

Grossman, J.D., 1992, Color conventions and Application Standards.
    In Color in electronic displays, eds. H. Widdel and D.L. Post
    (New York: Plenum Publishing Corporation), pp. 209-218.
Mosier, J.N. and Smith, S.L., 1986, Application of guidelines for
    designing user interface software, Behaviour and Information
    Technology, 5(1), 39-46.
Tullis, T.S., 1988, Screen design. In Handbook of Human-Computer
    Interaction, ed. M. Helander (Amsterdam: Elsevier Science
    Publishers BV), pp. 377-411.
Wilson, J. and Rosenberg, D., 1988, Rapid Prototyping for User
    Interface Design. In Handbook of Human-Computer Interaction,
    ed. M. Helander (Amsterdam: Elsevier Science Publishers BV),
    pp. 859-875.

RELIABILITY IN ERGONOMICS/HUMAN FACTORS

H. Kanis

Delft University of Technology
School of Industrial Design Engineering,
Department of Product and Systems Ergonomics,
Jaffalaan 9, 2628 BX Delft, The Netherlands

Within constituent disciplines of Ergonomics/Human Factors criteria for outcomes of measurement such as reproducibility, reliability and validity are dealt with differently. This variety is amply present in Ergonomics literature. Especially the term reliability is the centre of ambiguity and confusion. On the basis of a literature study it is proposed to drop the term reliability as a criterion for random variation in measurement outcomes in favour of the term reproducibility. Another recommendation is to graft the analysis of measurement error on statistical techniques that are described in the literature as conclusive rather than on current approaches such as correlating test-retest outcomes which generally can be demonstrated to be insignificant.

INTRODUCTION

Ergonomics is the meeting place of research traditions from different areas. Ideally, this 'melting-pot' of various disciplines should provide favourable circumstances for the emergence of interdisciplinarity rather than multidisciplinarity. This, however, turns out to be not always the case in the field of Ergonomics. In this respect, monodisciplinary traditions can be seen to live on in the use of criteria for outcomes of measurement, such as reproducibility, reliability and validity.
Since measuring lies at the heart of science, the significance and range of measurement outcomes are a matter of obvious concern in any research area.
Actually, two phenomena can be distinguished (cf. Carmines and Zeller, 1979):
• random variation in the outcomes of repeated measurements, and
• systematic deviation of measurement outcomes if contrasted with external evidence.
The object of the study dealt with in this paper is twofold.
First, the paper will disembark upon the question how the dichotomy - random variation vs. systematic deviation - is addressed within disciplines that can be considered as important 'constituents' of Ergonomics/Human Factors, i.e. technical sciences and social sciences. The attention will primarily be focussed on criteria for random variation in outcomes of measurements; systematic deviation will mainly be dealt with as delimitation.
Given this restriction, the second question involves the way random variation in measurement outcomes is dealt with in Ergonomics/Human Factors literature.

METHOD

In order to answer the first research question, textbooks, 'historical' publications and other literature have been studied in the field of:
- technical sciences, e.g. ISO-documents, and
- social sciences, in particular in the area of psychology and educational research.

The second question of how random variation in measurement outcomes is actually treated in the literature, is answered by the consultation of three volumes (1989, 1990 and 1991) of *Applied Ergonomics, Ergonomics, Human Factors, Contemporary Ergonomics* and *Proceedings of the Human Factors Annual Meeting.* Titles and abstracts of papers/articles have been screened on the occurrence of the words 'repeatability', 'reproducibility', 'reliability' and 'validity', including compounds.

In the analysis, the description of measurement by Stevens (1951, p.1) - "the assignment of numerals to objects or events according to rules" - serves as a conceptual framework to chart the issue. It is true that this often quoted definition has been seriously critisized as being (too) liberal. It has been argued for example that 'numerals' ought to be replaced by 'numbers' in view of the availability of a measurement unit, whilst furthermore Stevens hardly specifies any 'rule' (cf. Berka (1983), p.24/p.162). However, within the scope of the present paper, the given description of measurement is satisfactory as to the concepts it offers for a global inventory. As yet, a wide interpretation is given to 'objects/events'. These terms should be conceived as characteristics attributed to objects, to events or to phenomena. 'Rules' refer to procedures or methods as empirical activities, which should warrant that the assignment of 'numbers' - or possibly of 'numerals' - is not haphazard or random, but systematic and imitable.

RESULTS
Criteria in technical sciences
Random variation in outcomes of measurement
   In technical sciences an international standard has been agreed on for the definition of random variation in measurement outcomes (ISO, 1986). Two measures for the dispersion of outcomes are introduced:
* repeatability as the least significant difference between outcomes when a measurement is repeated on "identical test material" according to the same method under the same circumstances (same operator, same apparatus, same laboratory), after a short period of time;
* reproducibility, which is defined likewise, provided that the operators should be different, just as the apparatus and the laboratory, while no time constraint is indicated.
Comparison with Stevens' description for measurement shows that the repetition of a measurement implies that both the 'object' and the 'rule' are intended to remain unchanged. This illustrates that the distinction between repeatability and reproducibility is grafted upon circumstances that can be specified to a high degree, as is not unusual in technical research. In this paper the mentioned distinction is considered to be of minor importance.
In the ISO-document, measurement results are conceived to be of interval- or ratio-level and to conform to an additive statistical model based on the decomposition of observations in a so-called 'true value' and an error-term (see Lumsden (1976) for limitations of this model and particularly for criticisms on the platonic notion of 'true value'). Differences between outcomes are generally seen as normally distributed, which facilitates the specification of a least significant difference on the basis of a confidence interval.

Systematic differences between outcomes of measurement
   In the ISO-document no separate criterion is introduced for differences in outcomes that turn out to be systematic. If this occurs, external criteria are needed to indicate which outcomes deviate and which do not. Conclusions of that type concern the criterion 'validity'. As has already been noted by Ebel (1961), this criterion is rarely discussed in the technical sciences, presumably because in this area the

operationalization of concepts has traditionally been considered to be more or less self-evident. One of the few places where the word 'validity' is mentioned explicitly, is an ISO-IEC document (1982).

Criteria in social sciences
Random variation in outcomes of measurement

What in technical sciences is termed as reproducibility, is indicated as reliability in social sciences, be it that repetition of measurement is achieved in a different way. Especially in test theory as developed in psychology and educational research, human activities cannot be tested many times in succession for people gain experience and learn. Therefore, reliability is determined on the basis of a test across different subjects. In addition, test items are sampled from a domain, for example words for a spelling test or items in an intelligence test, which makes repetition of a test feasible. Reliability is defined for a test on a group of subjects as the ratio between the 'true' variance and the total (observed) variance (given the fact that outcomes can be conceived to be at least of interval level).

If the ' 'true value'-error model' is adopted, it can easily be shown that the indicated ratio - which is called reliability coefficient - equals the correlation between the outcomes if a test is repeated (see e.g. Nunnally, 1979). Thus, in Stevens' terms, reliability is established by measuring different 'objects', unlike reproducibility in technical sciences where the 'object' should ideally be the same.

The computation of a reliability coefficient by correlating test-retest results across different subjects is a wide-spread approach but has been equally criticized in the literature. A large correlation coefficient, e.g. approaching 1, means that the variation between subjects (largely) dominates the within subjects variation, without giving a quantitative clue about the latter which indeed concerns the measurement error (cf. Burton, 1991). Similarly, a small correlation coefficient, e.g. approaching 0, does not imply that the measurement error is large if the interindividual variability is known to be small. Altman and Bland (1983) conclude that the correlation coefficient, as a measure of association rather than of agreement, cannot cope with replicated data. These authors point at alternatives based on statistics of differences between paired results of a repeated measurement. If heteroskedasticity is absent or can be eliminated, this approach resembles the establishing of reproducibility in technical sciences as described in the ISO-document (1986).

Pseudo-random variation in outcomes of measurement

Another difficulty concerns the repeatability of a measurement. Although not incorporated in Stevens' description, measuring always involves interaction between a 'rule' and an 'object'. This interaction may vary, especially between different human beings, for instance interviewees who may perceive/interpret the 'same' topic differently. Then the question arises as to what extent a conceived reality can be assumed to exist 'out there', that is: invariant for 'rules' to establish it. This is all the more so since in this type of research, generally, linguistic amenability of that 'reality' has to be taken for granted in order to make measuring possible at all. Thus, matters boil down to the criterion 'validity'.

These conceptual intricacies may be exhibited in particular by the research which employs ratings. Sometimes external criteria are available, like in the study of estimated frequencies by Hancock and Klockars (1991). If this is not the case, as a rule the (dis)agreement between the judgements by different raters of the 'same' phenomenon serves as the criterion. In the literature, variability in ratings regularly appears to be discussed only quantitatively, without regard of possible diversity between raters in the quality they assess (cf. Tinsley and Weiss, 1978). In addition, systematic differences between raters are usually indicated in terms of error. However odd that may sound, it might explain why agreement between raters is termed as reliability, such as 'interrater reliability', be it sometimes bracketed together with validity (Saal, Downey and Lahey, 1983). Beside this conceptual confusion, in Stevens' terms - as far as still applicable - there may be as many 'rules' as there are raters. However, conception of raters as 'measuring instruments' is a misnomer which completes the operational deficiences as to 'random deviation' in measurement outcomes in this field of social sciences.

Systematic deviation of measurement outcomes
   The concept of validity, which so far has been encountered a few times, has been largely developed within the social sciences as a criterion for systematic deviation of measurement outcomes or of generalizations in time/place derived from these results, if contrasted with external evidence. As to the meaning of the notion 'validity', a commonplace is that a measurement is valid if it measures what is presumed to be measured. But this is, as Guilford already pointed out (1954, p. 471) "...but one step better than the definition that states that a test is valid if it measures the truth". Over the years the difficulty to come to grips with the concept of validity has resulted in a proliferation of dozens of compounds featuring 'validity'. To mention a few constituent words: conceptual (-), concurrent (-), construct (-), content (-), convergent (-), discriminant (-), external (-), face (-), predictive (validity). Within the scope of this paper it is important to state that any conclusion about (lack of) validity of measurement outcomes is bound to be based on findings extraneous to the measurement under consideration. Validity, whatever its context, can never be determined by repetition of a particular measurement, that is: the same 'rule' applied to the same 'object', which in fact would be the obvious means to establish reproducibility.

Criteria in Ergonomics/Human Factors literature
   The search in the literature has resulted in 83 papers featuring one or more of the following terms at least once (including corresponding adjectives and adverbs): reliability (in 56 papers), reproducibility (7), repeatability (3) and validity (60).
- In some 10 papers reliability is used in a non-committal way.
- In 19 papers the reference to reliability does not concern any dispersion in measurement results but the performance of human beings or systems in terms of the absence of mistakes (errors), failures etc..
- In a few papers, 'reliable' appears to be used as 'statistically significant', see for example Lintern et al. (1989), and Wogalter and Young (1991). In consulted textbooks on statistics 'reliable' or 'reliability' does not show up in this sense. The appropriate terminology centres around 'confidence' such as in 'confidence coefficient' and 'confidence interval'.
- In a number of papers that have been published in proceedings, 'reliable' seems to mean 'valid', e.g. in expressions like "a reliable prediction" and "an unreliable forecasting". Another example of confounding reliability and validity is provided by Hopkins et al. (1989), who observe that the "reliabilities" of estimated indicators to grade runners may be reduced if these indicators would be based on a test of a narrow population such as only sprinters. Presumably, the validity of indicators is meant here in the sense of generalizability. Everyday connotations of reliability such as 'trustworthiness' may have interfered in these cases.

Random variation in outcomes (measurement error)
   There are 25 papers in which the authors (intend to) address measurement error, which is referred to by 'reliablility', 'reproducibility' and 'repeatability', including some compounds. These words are used indifferently in six papers, four of which are from

the three reviewed journals. When confining the presentation of the results to these journals, there remain eight articles in which measurement error is directly addressed on the basis of empirical research by the authors. In four of these articles correlation coefficients are presented as criterion for the measurement error in a test-retest involving different subjects. Insofar as data are detailed, the objection to this approach as being inferior (see Altman and Bland (op.cit.)) seems to hold in view of large interindividual variabilities (in one case variation coefficients amount to almost 50% (DeVries et al., 1989)).

In six articles, including the four featuring correlation coefficients, conclusions about the reproducibility (occasionally indicated as reliability) of data in test-retests over different subjects rely (also) on a comparison of corresponding means. As far as specified, the statistical significance of the differences between means is established by a t-test or ANOVA. In five of the six articles no significant differences are found, which is indeed not surprising in view of considerable overlaps between standard deviations, combined with the number of subjects that is limited in most cases. However, what's more important is that, basically, these analyses of means are inappropriate for dealing with reproducibility of data, virtually for the same reason why correlation coefficients fail. In the one article (Misner et al., 1989) where statistically significant differences were found between paired means of a test (in firefighting tasks, number of subjects > 50) that was carried out twice, it is concluded that, apparently, the subjects had gained experience during the first run. But this would mean that the criterion reproducibility is not justified since differences turn out to be systematic which bears upon validity.

Amongst the articles from the three reviewed journals there are two in which the reproducibility of data actually concerns repeated measuring of the same 'object'. Grönqvist et al. (1989) adopt as criterion the standard deviation in outcomes of measurements concerning the slip resistance between shoe sole and underfoot surface in simulated conditions. Zipp (1989) takes as criterion the variation coefficient of outcomes if EMG measurements are repeated for a constant recording position.

In the papers incorporated in proceedings one more research is reported in which real repetition of a measurement serves to estimate the error (criterion: the standard deviation of outcomes). Amongst the total of selected papers no example has been found of the approach proposed by Altman and Bland as pleaded for by Burton (op.cit.).

To conclude this inventory, it is noted that the analysis of ratings in terms of interrater reliability has only been encountered in some papers in the screened proceedings. Generally, the information provided in these papers is limited and gives no reason to adjust earlier criticisms.

SUMMARY AND CONCLUSION

Within constituent disciplines of Ergonomics/Human Factors criteria for measurement outcomes are dealt with differently. In *technical sciences*, repeatability/ reproducibility are adopted as criteria for random variation in outcomes of a repeated measurement, which concerns the measurement error. A criterion for systematic differences between outcomes of measurements is virtually lacking. In *social sciences*, systematic deviation of measurement outcomes is expressed as (lack of) validity, which term features in numerous compounds. Measurement error is addressed by the term reliability in this area.

In Ergonomics/Human Factors literature, the use of the terms repeatability/ reproducibility and reliability as indication of measurement error turns out to be mixed up indeed. In addition, reliability regularly serves as criterion for the absence of mistakes or failures of human beings or systems, which has nothing to do with measurement error. Moreover, confusion occurs with the criterion validity.

As to the actual establishment of measurement error in research involving different objects, several articles have been found that present as measure the correlation between outcomes of a test-retest and/or the statistical insignificance of the difference between corresponding means. Often, both approaches can be shown to be meaningless in this respect. Appropriate alternatives as put forward in the literature have not been found applied. In view of these findings it is proposed:
* to maintain a clear-cut distinction between random variation and systematic deviance in outcomes of measurement;
* to drop reliability as criterion for measurement error, in favour of reproducibility;
* to graft the analysis of measurement error on statistical techniques that are described in the literature as conclusive rather than on current approaches that in general can be demonstrated to be insignificant.

REFERENCES

Altman, D.G. and Bland J.M., 1983, Measurement in medicine: the analysis of method comparison studies, The Statistician, 32, 307-317.

Berka, K., 1983, Measurement, its concept, its theories and problems, Reidel, London.

Burton, K., 1991, Measuring flexibility, Applied Ergonomics, 22 (5), 303-307.

Carmines, E.G. and Zeller, R.A., 1979, Reliability and validity, SAGE, London.

DeVries, H.A., Brodowicz, G.R., Robertson, L.D., Svoboda, M.D., Schendel, J.S., Tichy, A.M. and Tichy, M.W., 1989, Estimating physical working capacity and training changes in the elderly at the fatigue threshold (PWC$_{ft}$), Ergonomics, 32 (8), 967-977.

Ebel, R.L., 1961, Must all tests be valid?, American Psychologist, 16, 640-647.

Grönqvist, R., Roine, J., Järvinen, E. and Korhonen, E., 1989, An apparatus and a method for determining the slip resistance of shoes and floors by simulation of human foot motions, Ergonomics, 32 (8), 979-995.

Guildford, J.P., 1954, Fundamental statistics in psychology and education, Kogakuska Company, Tokyo.

Hancock, G.R. and Klockars, A.J., 1991, The effect of scale manipulations on validity: targetting frequency rating scales for anticipated performance levels, Applied Ergonomics, 22 (3), 147-154.

Hopkins, W.G., Edmond, I.M., Hamilton, B.H., Macfarlane, D.J. and Ross, B.H.,1989, Relation between power and endurance for treadmill running of short duration, Ergonomics. 32 (12), 1565-1571.

ISO, 1986, ISO Standard 5725: Precision of test methods, International Standard Organisation. Geneva.

ISO-IEC, 1982, Guide 36: Preparation of standard methods of measuring performance (SMMP) of consumer goods, International Standard Organisation, Geneva.

Lintern, G., Sheppard, D.J., Parker, D.L., Yates, K.E. and Nolan, M.D., 1989, Simulator design and instructional features for air-to-ground attack: a transfer study, Human Factors, 31(1), 87-99.

Lumsden, J., 1976, Test theory, Annual Review of Psychology, 27, Palo Alto.

Misner, J.E., Boileau, R.A. and Plowman, S.A., 1989, Development of placement tests for firefighting, Applied Ergonomics, 20 (3), 218-224.

Nunnally, J.C., 1978, Psychometric theory, McGraw-Hill, New York.

Saal, F.E., Downy, R.G., and Lahey, M.A., 1983, Rating the ratings: assessing the psychometric quality of rating data, Psychological Bulletin, 88 (2), 413-428.

Stevens, S.S., 1951, Mathematics, measurement and psychophysics, Handbook of experimental psychology, John Wiley & Sons, New York.

Tinsley, H.E.A. and Weiss, D.J., 1975, Interrater reliability and agreement of subjective judgements, Journal of Counseling Psychology, 22 (4), 358-376.

Wogalter, M.S. and Young, S.L., 1991, Behavioural compliance to voice and print warnings, Ergonomics, 34 (1), 79-89.

Zipp, P., 1989, The concentric anular electrode in surface electromyography, Ergonomics, 32 (7), 911-918.

Automatic Evaluation in Human-Computer Interaction

Sandrine Balbo, Joëlle Coutaz

Laboratoire de Génie Informatique
IMAG - BP 53 X
38041 Grenoble Cedex - France
E-mail: balbo@imag.fr, coutaz@imag.fr

This paper presents first steps towards the automatic evaluation of user interfaces. We propose a framework useful for classifying current and future evaluation methods. We then present our own approach to automatic evaluation with the description of a tool based on three ingredients: acquisition of end user's actions, task description, and heuristic rules that detect patterns of behavior.

INTRODUCTION

The development of interactive systems relies on an iterative process based on well identified phases such as requirement analysis, design, implementation, and evaluation. Tools are now available for supporting a number of these steps. In particular, task models such as the GOMS-like family (Card et al. 1983) and MAD (Scapin&Pierret-Goldreich 1989) provide methods and formalisms useful for structuring and conveying task analysis. Implementation is supported by even a wider variety of tools ranging from toolboxes to application skeletons and UIMSs. On the other hand, the evaluation process of user interfaces is still based on "craft techniques" (Long&Dowell 1989) and suffers from the lack of software tools. In this article, we propose a framework useful for classifying current and future evaluation techniques. We then present our own approach to the automatic evaluation of user interfaces.

A TAXONOMY FOR EVALUATION METHODS

Current taxonomies for evaluation techniques rely on the distinction between empirical and formal methods (Nielsen&Molich 1990, Senach 1990). They do not mention the automation of the evaluation process. We consider the evaluation of user interfaces from this additional perspective. The framework we propose defines a three dimensional space: automation, knowledge used, and user presence. In the following paragraphs, we discuss the dimension space and use salient existing methods and techniques to illustrate our contribution.

The three dimensional space

The automation dimension covers four distinct approaches:*non automatic methods* are performed by human factors specialists; *automatic capture methods* rely on software facilities to record relevant information about the user and the system such as visual data, speech acts, keyboard and mouse actions; *automatic analysis methods* are able to identify usability problems automatically; and *automatic critic methods* not only point out difficulties but propose improvements.

The knowledge dimension captures four distinct sources: *empirical knowledge* is used by the human factor specialist to detect usability problems and to suggest improvements. Empirical knowledge is built from the own experience of the specialist coupled with background knowledge. Knowledge about the *interface*, the *user* and the *task* is not empirical. As illustrated in the next section, a theory or a model may be the sources of such knowledge.

The user dimension covers two possibilities: the evaluation method either *works without* or *works with* the end user. Methods that do not require the presence of end users can be used as predictive tools.

Illustration

Figure 1 shows four models and techniques within our framework. Rectangles denote the values within the automation plane whereas the knowledge and user dimensions are represented as connecting lines.

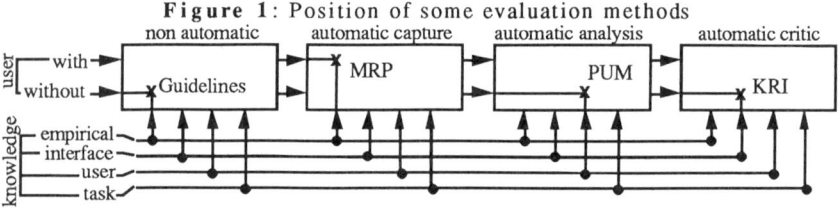

Figure 1: Position of some evaluation methods

The (Smith&Mosier 1986) guidelines result from empirical knowledge (inferred by observation and experimentation). They can be used by the human factor specialist just by "looking at the interface". The end user is not needed. Although these guidelines are easy to understand, they are difficult to use: the 480 page report includes 944 guidelines!

As a first step towards automation, MRP detects repeating patterns of behaviour by analysing a log of the end user's actions (Siochi&Hix 1991). These actions are recorded during task execution with the real system. Detected patterns can then be used by the human factor specialist as driving cues for the evaluation.

As a further step towards automatic evaluation, PUM predicts learnability or usability problems from the formal description of the knowledge the end user needs to run the user interface (Young&Wittington 1990). This description is compiled in terms of rules. These rules represent the user's ability to accomplish this particular task with that particular user interface. Based on this knowledge, the PUM cognitive architecture tries to elaborate a plan. If no plan can be generated, then the designer is notified of a potential usability or learnability problem.

KRI goes even one step further by providing the designer with an automatic critic (Löwgren&Nordqvist 1990). The tool uses a collection of ergonomic rules to evaluate a formal description of the interface. The evaluation produces a set of comments and suggestions. The improvements, however, are concerned with lexical issues only such as the minimum distance between two icons or menu items ordering.

In the next section, we present our own evaluation tool. Like KRI, it will provide the designer with a critic. Unlike KRI, it works with the user and does not rely on user interface knowledge but on a task description.

A FIRST STEP TOWARDS THE AUTOMATIC EVALUATION OF USER INTERFACES

Our system is based on three ingredients: (1) the acquisition of the end user's actions in order to observe how the user evolves within the user interface, (2) the description of a data-flow oriented task model that describes the designer's expectations, and (3) behavior rules to analyse the discrepancies between the designer's expectations and the effective use of the system. These rules can then be linked to a software usability critic. These issues are discussed in the following paragraphs.

Definition of actions

From the user's perspective, an action is a physical act than can be applied, either directly or indirectly, to an input device. For example, moving the mouse is a direct action. Looking at an icon is a direct action to the screen but an indirect, non intentional action towards an observing eye-tracker or computer vision system. From the software perspective, an action is modelled as an event whether it be direct or not.

We distinguish two types of events: (a) events that only modify the perceivable state of the system (i.e., the user interface portion of the system). These events model *syntactic actions*; (b) events that modify the internal state of the system (i.e., the functional core portion of the system). These events model *semantic actions*.

For example, window scrolling and window resizing modify the perceivable state of the system. They are performed by syntactic actions. The corresponding collection of events do not require any change within the functional core. On the other hand, the sequence of actions that results in the invocation of the delete command in a text editor implies the modification of the internal state (as well as that of the perceivable state). They are semantic actions.

Actions that are observed by the system are recorded in a file along with a date of occurrence and some type-dependent information. As described later, the semantic/syntactic typing distinction provides useful indication for organizing the behavior rules.

The task model

A task model may serve different purposes (Normand 1992). In our particular case, a task model is used as a reference to measure discrepancies between the task intended by the designer and the effective task performed by the end user.

Our task model is a tree structure where nodes represent actions. The root of the tree denotes the action to launch the application under test. A path expresses the sequence of actions involved in the execution of a particular task. The tree thus represents every possible way, at the action level, one can accomplish tasks with the system. Unlike most of task models, it does not represent a hierarchical decomposition of tasks into simpler units. It does not express the structural organisation of the task space provided by the system. Instead, it stresses the "action sequence" perspective in order to define a correspondence with the actions recorded during effective task performance.

Figure 2 shows the task model for exiting from the Macintosh Word application using the quit command.

Figure 2: An example of a task model

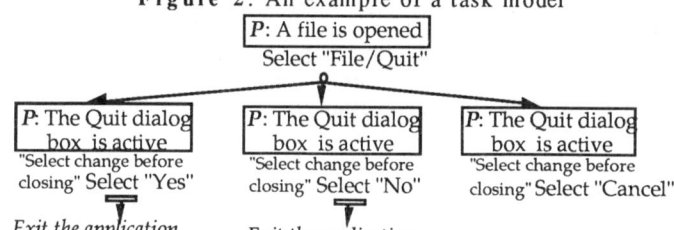

As shown in Figure 2, nodes may be decorated with preconditions *P*. A precondition must be true in order to have access to the action it decorates. For example, the selection of the menu item "Quit" in the "File" menu is possible only if a file is opened. The link between two nodes A and B may be one of the following types:

- $(A \: \square\!\!\!-\!\!\!- B)$. When A is done there is no way to return to either A or above A. B becomes the new root of the tree.

- $(A \: \circ\!\!\!-\!\!\!- B)$. This relation represents a sequence constraint: once A is performed, then the user cannot perform any other action than B.

- $(A \: -\!\!\!- B)$. This relation expresses a logical sequence between two actions: if action A is executed, then the next action to carry on the current task is B; the user however may perform other actions pertaining to another task before executing B.

The analysis stage, or how to discover interface problems by comparing the task model with recorded actions

The manual analysis of a number of log files has allowed us to identify patterns of behaviour that would reveal usability problems. We have expressed these conditions in the form of heuristic rules. Some of the rules are related to semantic actions only while others are mainly concerned with syntactic actions.

Rule 1: direction shift. A direction shift is detected from the task model as the user stops progressing along a tree path. For example, a Macintosh user may want to change the layout of a directory by selecting a) the directory icon and then b) the menu item "View/by Name". By doing so, the view of the current opened directory will be changed, not that of the intended one. To reach the goal, the user will have to open the directory of interest, make the opened directory the current window, then select the "View/by Name" menu item.Figure 3 shows the task model for this example (black arrows) as well as the user actions (grey arrows).

Figure 3: Example of direction shift

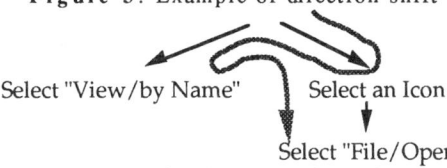

Select "View/by Name"   Select an Icon

Select "File/Open"

In the example of Figure 3, the user thought that all the options in the menubar would concern the current selected object. Although this rule is generally true in direct manipulation user interfaces, the View option applies to the current the selected window, not to the current selected object within that window! Rule 1 allows the automatic analyser to detect this sort of misunderstanding and notify the designer of potential problems.

Rule 2 immediate cancelling of an action. Cancelling an action immediately after its occurrence may denote a navigation problem within the interface.

Rule 3 re-occurring syntactic actions. The repetition of syntactic actions denote extra work due to the presentation media. For example, resizing, moving or iconifying a particular window when newly mapped onto the screen, may reveal wrong defaults options as to its initial size, location, or relevance.

Rule 4: irrelevant actions. Irrelevant actions have no meaning within the user interface under test. For example, a Macintosh user may reuse previous skill-based knowledge and use double-clicks for opening objects within a new interface where double-clicks are inoperative.

Rule 5: temporarily irrelevant actions. Temporarily irrelevant actions correspond to the execution of actions that have no meaning in the current context, but may be relevant in another situation. For example, the selection of an item that is not currently accessible is a temporarily irrelevant action. It may reflect a consistency problem.

CONCLUSION
The concepts and approach presented in this article are currently being implemented. Automatic recording has been integrated into two types of application: a system for designing user interfaces embedded within cars, and one information system to retrieve documents in a library. The user interface design system supports multiple threads of dialog and stresses direct manipulation. At the opposite, the librarian system adopts a system-driven style of interaction using

keyboard only. Heuristic rules as those presented above have been devised based on these experiments.

Although the experimental apparatus used so far covers distinct styles of interaction and different types of task, we plan to extend our observations by performing action recording for a drawing package as well as for a multimodal information retrieval system (Nigay&Coutaz 1993). The critic is under development. It is clear that our dataflow task model is appropriate to detect simple usability problems. We plan to integrate a complementary structured task model that would support the identification of anomalies related to way the system organizes the work space.

ACKNOWLEDGEMENT

We would like to thank the Ergonomics Unit, University College London, for a stimulating environment, and Alan Christiansen for his review of this paper.

REFERENCES

Card S.K., MoranT.P. & Newell A., 1983, "The psychology of Human Computer Interaction", Lawrence Erlbaum Associates

Long J. & Dowell J., 1989, "Conceptions of the Discipline of HCI: Craft, Applied Science, and Engineering", in Proceedings of the Fifth Conference of the BCS HCI SIG, A. Sutcliffe and L. Macaulay (Eds), Cambridge University Press

Löwgren J. & Nordqvist T., 1990, "A Knowledge-Based Tool for User Interface Evaluation and its Integration in a UIMS", Human-Computer Interaction-INTERACT'90, pp. 395-400

Nielsen J. & Molich R., 1990, "Heuristic evaluation of user interfaces", Proceedings of the CHI'90, Seatle. ACM New York, pp. 349-256

Nigay L. & Coutaz J., 1993, "A Design Space for Multimodal Systems: Concurrent Processing and data Fusion", to appear in INTERCHI'93.

Normand V., 1992, "Task Modelling in HCI: purposes and means", Rapport de recherche no PTI/92-02, Thomsom CSF, Division SDC, 7 rue des Mathurins, BP 10, 92223 Bagneux Cedex

Scapin D.L. & Pierret-Goldreich C. ,1989 , "Toward a method for task description: MAD", in Proceedings of the Work with Display Units Conference, Montreal, Canada, Elsevier Science Publishers

Senach B.,1990, "Evaluation de l'ergonmie des interfaces homme-machine: du prototype aux systèmes experts", Congrès ERGO-IA'90, Biarritz, France

Pierret-Goldreich C., Delouis I. & Scapin D.L., août 1989, "Un Outil d'Acquisition et de Représentation des Taches Orienté-Objet", rapport de recherche INRIA no 1063, Programme 8 : Communication Homme-Machine

Siochi A.C. & Hix D., 1991 May, "A study of computer-supported user interface evaluation using maximal repeating pattern analysis", Proceedings of the CHI'91 Conference, New Orleans. ACM New York, pp. 301-305

Smith S.L. & Mosier J.N.,1986 , "Guidelines for designing user interface software", Report MTR-10090 ESD-TR-86-278, The MITRE Corporation, Bedford, MA

Young R.M., & Whittington J.,1990, "A knowledge Analysis of interactivity", Proceedings of INTERACT'90, edited by D. Diaper, G. Cockton & B. Shackel, Elsevier Scientific Publishers B.V. pp. 207-212, Cambridge (United Kingdom)

SELECTING EARLY USER INTERFACE EVALUATION METHODS FOR USE IN PRACTICAL DESIGN SITUATIONS: A CHECK-LIST

Arnold P.O.S. Vermeeren and Mathilde M. Bekker

Delft University of Technology
Faculty of Industrial Design Engineering
Jaffalaan 9, 2628 BX Delft
The Netherlands
E-mail: a.p.o.s.vermeeren@io.tudelft.nl

The selection of an appropriate evaluation method for a specific design situation on the basis of published literature is a difficult task. By presenting a check-list of issues that are of theoretical and practical relevance, the present paper intends to aid designers with the selection of evaluation methods. The paper focuses on the use of evaluation methods at an early stage in the design process. To illustrate that there is no general answer to the question of 'what evaluation method can best be used', the check-list is applied to two example design situations; the examples show that the choice for a specific method depends highly upon the characteristics of the design situation in which it is to be implemented.

INTRODUCTION

Within the field of ergonomics it is generally accepted that in designing products one should focus on the product's prospective users and their tasks right from the beginning of the design process. In this paper, we focus on one specific group of methods for doing that, namely early evaluation methods; early evaluation methods are methods for investigating design proposals in the initial stages of their development.

In practice, designers of user interfaces seem to have problems in eliciting information from and about users (Bekker and Vermeeren, 1992). Also, the selection of user interface evaluation methods on the basis of published research that compares such methods seems to be a difficult task (Jeffries and Desurvire, 1992).

In this paper we present a general check-list that can be used for selecting an early evaluation method for use in practical situations. To minimize the chance that the issues mentioned in the check-list are misinterpreted, we describe how we applied it to two (non-existing, but realistic) example design situations. At the same time, doing this illustrates that it is no use assessing the usefulness and feasibility of evaluation methods in general; the practicality and usefulness of methods must be investigated for each concrete design situation individually. In this paper, we focus on two classes of evaluation methods: walkthrough methods and empirical methods, but first we will clarify the meaning of the term 'early evaluation' as we understand it.

Early evaluation: what is it aimed at?

One speaks of evaluation "... when some aspect of a system...is compared against some criteria...; (evaluation) is necessary to supplement the designer's model of the user, which without evaluation, is based upon knowledge derived from experience of previous similar situations, intuition and informal supposition" (Howard and Murray, 1987). This definition stresses the two-fold purpose which evaluation can serve. On the one hand it may look back, assessing the quality (usability, utility, etc.) of a design or design proposal, on the other hand it looks forward thus providing the designer with information that can help in improving the design. Especially in early design phases the forward looking perspective will usually be considered more important by designers. As no detailed design is available yet, one does not primarily seek to assess the overall performance of a design; usually one focuses on a limited set of specific design issues (such as the basic principles underlying a design proposal design or some new interaction techniques of which the usability is to be investigated).

Two classes of evaluation methods

We make a basic distinction between two classes of evaluation methods: *empirical* methods and *walkthrough* methods.

Empirical methods are methods that are based on the observation of a subject using a prototype or simulation or on subjects' opinions based on that usage. Within the class of empirical methods we see two extremes: small-scale, qualitative user studies often conducted in a rather informal set-up (from now on referred to as *'qualitative user studies'*, see for example Vermeeren and Kolli, 1993) and large-scale, quantitative user tests conducted in a very controlled experimental set-up (from now on referred to as *'quantitative user tests'*, see for example Bewley et al., 1990).

In *walkthrough* methods a system is evaluated by applying knowledge from indirect sources like theories, guide-lines or expert knowledge. Within the class of walkthrough methods we make a distinction between *unstructured methods* (based on guide-lines, heuristics or expert knowledge, see for example Nielsen and Molich, 1990) and *structured methods* (based on scientific theories); the Cognitive Walkthrough method (Polson et al., 1992) is an example of such a structured method, and will be used as a representative example.

ISSUES IN SELECTING AN EVALUATION METHOD: A CHECK-LIST

Based on the literature on user interface design practice we constructed a list of issues which are relevant for selecting early evaluation methods. Some of the literature that we investigated explicitly mentioned a number of issues that may constrain the selection of design or evaluation methods in practice (e.g., Bekker and Vermeeren, 1992; Bellotti, 1990). In other publications such issues were not mentioned explicitly but had to be deduced from what the authors said (e.g., Mills, 1986). We grouped the issues thus found into three categories: issues concerning the aim of a method, issues concerning the input required for the use of a method and issues concerning the outcomes that can be expected from the use of a method. The list of issues is presented in table 1. More details about the construction of the list can be found in Vermeeren (1993).

Two example design situations

Based on interviews conducted among user interface designers in the Netherlands (Bekker and Vermeeren, 1993) we described two example design situations (see table 2). The situations do not really exist but are realistic examples made up on the basis of descriptions of real situations. The following characteristics are used to describe the example design situations: the type of company characterised by the products or software they develop (see the header 'type of company' in table 2), educational background of the designer, team size and composition ('design team'), specific characteristics of the design process ('design issues'), project length ('duration').

Table 1. Issues relevant in the selection of an evaluation method in practice. The right column shows questions that can be asked in assessing methods with respect to a specific issue (plain text); the statements in italics are intended to clarify the questions or to suggest relevant attention points in answering them.

A. Would the method be relevant for the intended evaluation?	
1. Focus and scope	- On what aspects does the method focus? *(usability, utility, acceptability)* - How many aspects of a system or of usability does it address? *(whole system, basic UI concepts, details)*
2. System/ application dependency	- Can the method be applied to any system, application, UI-style, or only to some specific types of systems or applications?
B. What is required for using the method?	
1. Availability and accessibility of the main sources of information	- Are the main sources of information the method makes use of, available to the method user? Is it easy to access them? *Main sources of information can be: theories (e.g., cognitive), guidelines, heuristics, experts, specialists, prospective users*
2. Design models	- What type of design model is required for the use of the method? Is the type of model that is needed of the same type as the models that are used in the rest of the design process? *One can distinguish between empirical models (prototypes, mock-ups, simulations), formal models (state-transition diagrams, grammar descriptions, etc.), conceptual models (descriptions, diagrams describing general structures).*
3. Expertise and experience	- What expertise and experience is required from the evaluator for a successful use of the method? *(theoretical knowledge, experience in applying a method,...)* - What are the consequences of lack of expertise or experience? *(invalid results, not being able to use a method).*
4. Time, costs and equipment	- How much time does it take to prepare and use the method, and to analyse the data? Is special equipment needed and if so, what kind of equipment? *(e.g., video equipment, data loggers)* - Is the method perceived as tedious to conduct?
C. What can be said about the outcomes of the method?	
1. Validity and relevance	- What problems does the method have concerning the validity of the results? *(e.g., unproven, questionable, uncontrollable).* - What possible problems exist concerning the (perceived) relevance or utility of the results?
2. Direct applicability	- Are the results such that a designer only needs a minimum of interpretation and extra analysis before (s)he can use them for improvements in the design? *A method can yield various types of outcomes; it can tell something about: features of the system to be changed, (possible) user problems, working styles of users, efficiency and performance in use. It can also yield design ideas.*
3. Communication and persuasiveness	- Are the results easy to explain to others that have a say in the design decisions? *(e.g., managers, clients) .* - Can these people be easily convinced using the type of results that the evaluation method yields?

Table 2. Two example design situations (based on interviews with designers described by Bekker and Vermeeren, 1993).

Situation 1: Product Development	
Type of company	company producing and developing consumer products such as audio equipment, faxes, answering machines, etc.;
Design team	multi-disciplinary, within-company design team; user interface design is done mainly by industrial designers together with an ergonomist;
Design issues	visually oriented, industrial design approach; costs of displays and controls are serious constraints in the design process (e.g., a display of only two lines of 20 characters is available); sometimes evaluations are conducted specifically for the purpose of convincing management of the quality of design proposals;
Duration	design takes about one year; after that, the design is engineered further to optimize its manufacturing;

Situation 2: Software Design	
Type of company	software company developing computer application software that people use for their daily work (e.g., bookkeeping programs, special databases), the company is specialised in a small number of applications;
Design team	within-company design team consisting of people with a background in software development, the project is carried out for a client company that is paying for it and that will use the program in-house once it is finished;
Design issues	structured system development approach to design; the company has its own standard system development methodology that specifies all design phases in detail;
Duration	design and development takes about one year and a half.

INVESTIGATING THE METHODS' PRO'S AND CONS

First, a number of general issues concerning and related to the usefulness and feasibility of the methods are described, based on the list presented in table 1 (a more detailed description of this can be found in Vermeeren, 1993).

Second, the usefulness and feasibility of the methods are assessed on the basis of both these characteristics and the descriptions of the example situations presented in table 2.

General issues of the methods

Walkthrough methods focus on usability, and not on utility or acceptability (A1, the letter-number combinations in brackets refer to the issues in table 1.). Cognitive walkthroughs focus on learnability aspects of 'easily learned' interfaces (A2). Other structured methods may focus on a different kind of interface use (e.g., on performance speed in the case of routine use).

Structured methods require detailed design models (not necessarily implemented, B2). Descriptions of the purpose of structured methods and the way to apply them, may be hard to find. Also, these methods are extremely difficult to understand and apply for people with moderate (or less) expertise in cognitive sciences (B1,3). The aims and use of such methods are mainly described in specialised literature on human-computer interaction and psychology. In the case of structured methods lack of expertise usually leads to not being able to use a method (B3). In the case of unstructured methods the validity of the results is questionable and uncontrollable (C1). Lack of expertise may lead to even less valid evaluation results (B3). Therefore, walkthrough methods require access to one, but preferably more experts (B1).

Conducting an evaluation by using structured methods requires much administrative-like work; no special equipment is required and conducting the evaluations can be relatively quick (especially in the case of unstructured methods, B4).

All walkthrough methods tend to yield large numbers of potential problems, but usually identify only a part of the problems that are found by empirical methods; in addition, there is not much evidence for the validity of the results (C1). In walkthrough methods major problems are hard to distinguish from minor problems. Determining what the major problems are, is helped by the evaluator's expertise, but is always difficult (C2). The distinction between major and minor problems is especially important in cases where one can not fix all problems and wants to concentrate only on the major ones. Walkthrough methods give no insight into the actual use (or differences in use) of a design and do not help very much in generating ideas (C2).

Although it is not very likely that designers will use the structured methods themselves (because of their lack of expertise in cognitive science), they need to know and understand the results in order to be able to fix problems; therefore, evaluators have to communicate the detailed results to designers, which may also be difficult because of the designers' lack of expertise(C3).

The fact that structured methods are based on psychological theories, can sometimes convince people of the value of the results. Also, in the case of heuristic evaluation by experts, people are easily tempted to just believe what experts say (C3).

Empirical methods can focus on usability aspects as well as on utility or acceptability aspects (A1). Usability of prolonged use, however, is difficult to investigate empirically early in the design process (A1,2).

Empirical methods require the use of empirical models in some form (prototypes, mock-ups, simulations, screens drawn on paper). For investigating usability more sophisticated models are needed than for investigating utility or acceptability (B2). This does not imply, however, that investigating usability always requires complete, robust and working prototypes, in many cases simple simulations will do. Access to prospective users is always required for the use of empirical methods (B1).

For quantitative user tests, experience and expertise in cognitive science and in conducting research is necessary; for qualitative user studies moderate expertise or some experience may suffice (B3). Domain expertise may help in creating realistic settings for the studies.

Quantitative user tests are very time-consuming to conduct and often require special equipment. Also, processing the results of such a test can be very time-consuming. Qualitative user studies require much less time for analysing the results (B4). In addition, a smaller number of subjects can be used in many cases(B1,4).

The usability problems found in empirical evaluations are usually major usability problems (C2). Some problems concerning the validity of evaluation results can arise, namely in cases where evaluation session times are short, compared to real-life use of a system, and in cases where subjects have different motivations to use the system than in real-life (C1). Observing users and talking to them usually provides designers with many new design ideas (C2).

Showing subjects' behaviour (either in real-life or on a video-tape) is a very powerful and easy way to communicate what the problems in an interface are. Also, people are easily convinced this way, that the problems found in the evaluation are real (as they have seen subjects actually having those problems, C3).

<u>Usefulness and feasibility of the methods in the example situations</u>
Usefulness and feasibility of the methods will now be investigated for the Product Development situation and for the Software Design situation.

The Product Development situation. With respect to the <u>walkthrough methods</u> we conclude that a structured method like the Cognitive walkthrough may focus on the right issues for this situation (as do the unstructured methods). Structured methods, however, require too much expertise and will be considered too tedious to conduct in this situation. If a department with specialists in cognitive science exists within the company the required expertise may be available; in that case, however, the problem of designers having to understand the detailed results still remains. Unstructured walkthrough methods are more feasible here. The main problem with these methods is that the validity of their results is questionable (even if considerable expertise in cognitive science is available) and that they do not distinguish between major and minor problems. This makes them to a weak base for convincing people by using the results from the evaluation (this is also true for the structured walkthrough methods). The main role we see for walkthrough methods in this situation is as an initial check to filter out some very obvious design flaws before time and effort is invested in an eventually empirical evaluation.

With respect to underlined empirical evaluations we conclude that quantitative user tests may only be feasible in cases where radically new designs are proposed or in cases where management can only be convinced by 'hard figures'. Otherwise, conducting such tests as well as analysing the results will be considered too time-consuming. Also, such evaluations often require robust simulations or prototypes of integrated hard and software (which is a time-consuming and costly matter). Qualitative user studies are more useful and may be feasible in the Product Development situation. The validity problem with respect to the motivation of subjects as compared to users, may be outweighed by the persuasiveness of observing subjects actually having problems with an interface and by the amount of ideas designers get from conducting the evaluations. In sum, we conclude that qualitative user studies have a fair chance of being adopted in situations similar to the one described here; in some cases they may be preceded by the use of unstructured walkthrough methods (in exceptional cases where a detailed design and sufficient time and expertise are available, one might also consider the use of a cognitive walkthrough method for that).

The Software Design situation. Main problem with regard to walkthrough methods in this situation, is the required expertise in cognitive science; this is a significant problem in these types of situations, as in most cases no department with cognitive science expertise exists within the company. This may especially be a problem with respect to the structured methods. A specific problem with the Cognitive Walkthrough method (one of the structured methods) is that it focuses on aspects of initial learning of a system and not on aspects of prolonged use (while the company in this situation usually designs software that people have to use day-in day-out in their work).

With regard to empirical evaluations one considerable problem is that, although in this situation it is known who the users of the new software will be, designers will often consider conducting empirical evaluations not to be their work. They may also not have the expertise for such evaluations, or lack of experience with them. Designers in this situation may even consider an evaluation of little importance; since the company is specialised in a small number of applications, they are tempted to believe that they know enough about what is important and potentially problematic in a design. In addition, as the product is tailor-made for a specific client, the product usually is already sold and does not have to compete with other products. Therefore, conducting quantitative early user tests will always be considered too time-consuming and costly (unless they are required by the client). Even qualitative user studies may be considered too time-consuming. With respect to the validity of the results the main problem in this situation is that usability of prolonged use is hard to investigate in empirical evaluations, but is important considering the type of software that is being designed. Consequently, the results from an empirical evaluation will also be less persuasive.

In sum, the unstructured walkthrough methods have the best chance of being adopted in this situation. Qualitative user studies may be feasible, but only in cases where usability is considered sufficiently important by the design team.

Conclusions

We conclude that feasibility of early evaluation methods and the usefulness of their outcomes relies heavily upon the characteristics of the situation in which they are to be implemented. In the Product Development situation qualitative user studies have the best chance of being adopted, while in the Software Design situation unstructured walkthrough methods have the best chance.

The structured walkthrough methods as well as the quantitative user tests do not seem to be applicable for early evaluation in either situation. Both require too much expertise and are too tedious and time-consuming in conducting them and in analysing the outcomes. In addition, for quantitative user tests, the severe prototyping requirements are an important hindrance to their application.

DISCUSSION

We presented a list of issues relevant in the selection of early user interface evaluation methods. The list can be used as a check-list to investigate, for each design situation individually, which method can best be used. Although the list is based on an extensive literature search, it has not yet been validated or used in practice. The only application of the list so far, is to the example situations described in this paper. This application, however, clearly demonstrates the potential usefulness of the list and suggests how it can be used. To enhance the usefulness of the list, more specific information would be required on the characteristics of evaluation methods.

REFERENCES

Bekker M.M., Vermeeren A.P.O.S., 1992, Human computer interface design in practice: variability, constraints and information sources. In Proceedings of the Ergonomics Society's 1992 Annual Conference, ed. E.J. Lovesey (London: Taylor & Francis), pp. 352-357.

Bekker M.M., Vermeeren A.P.O.S., 1993, Developing user interface design tools: an analysis of interface design practice. In Proceedings of the Ergonomics Society's 1993 Annual Conference, ed. E.J. Lovesey (London: Taylor and Francis).

Bellotti V.M.E., 1990, Applicability of HCI Techniques to Systems Interface Design, Thesis Submitted for Examination for the Degree of Doctor of Philosophy, Queen Mary and Westfield College, London.

Bewley W.L., Roberts T.L., Schroit D., Verplank W.L., 1990, Human Factors Testing in the Design of Xerox's 8010 "Star" Office Workstation. In Human -Computer Interaction: selected readings: a reader, eds. J. Preece, L. Keller, H. Stolk, (Hemel Hempstead: Prentice Hall), pp. 368-382.

Howard S., Murray D., 1987, A Taxonomy of Evaluation Techniques for HCI. In Proceedings of Human-Computer Interaction-INTERACT'87, eds. H.-J. Bullinger, B. Shackel (Amsterdam: North Holland), pp. 453-459.

Jeffries R., Desurvire H., 1992, Usability Testing vs. Heuristic Evaluation: Was there a contest? SIGCHI bulletin, 24, 4, pp. 39-41.

Mills, C.B. (organizer, moderator), 1986, Usability testing in the Real World (panel discussion). In Proceedings of CHI 1986 (New York: ACM), pp. 212-215.

Nielsen J., Molich R., 1990, Heuristic Evaluations of User Interfaces. In Proceedings of CHI 1990 (New York: ACM), pp. 249-256.

Polson P., Lewis C., Rieman J., Wharton C., 1992, Cognitive walkthroughs: a method for theory-based evaluation of user interfaces. Int. J. Man-Machine Studies, 36, pp. 741-773.

Vermeeren, A.P.O.S., 1993, Towards practical evaluation methods for early user interface evaluation, Internal Report, Faculty of Industrial Design Engineering, Delft University of Technology, Delft.

Vermeeren A.P.O.S., Kolli R., 1993, Early user involvement in the design of user interfaces: a case study. In Proceedings of the Ergonomics Society's 1993 Annual Conference, ed. E.J. Lovesey (London: Taylor and Francis).

Design methods – examples

DEVELOPMENT OF A DESK TOP PUBLISHING SYSTEM: A MODEL OF GOVERNMENT-BUSINESS CO-OPERATION THAT WORKED

TAY WILSON

Laurentian University
Ramsey Lake Road
Sudbury, Ontario, Canada P3E 2C6

Many government initiatives have been aimed at stimulating businesses. Many have failed, at high cost. This study reports in detail a low cost government-business cooperation that worked both for the client and the business provider in the field of desk-top publishing of government manuals. The key to success was to effective development of an interative series of interactions in which the business provider and the client government department each concentrated on the effective performance of two tasks. The client government department, on one hand, provided, social data which successively and accurately assessed and described to the provider business, the detailed nature of their governmental task and the emerging social implications of successive technical interventions. On the other hand, the client department concentrated upon understanding in detail the capabilities/limitations of successive technical interventions provided by the business partner. The business provider, on one hand, concentrated upon understanding successive iterations of descriptions of the client task and of presented social implications of technical interventions; on the other hand, upon ensuring the client understood clearly the limits/capabilities of successive technical interventions. Having completed this intensive project, the client department found itself to have a government leading effective publishing operation and to be a major consultant for other departments wishing to achieve similar capabilities. The business partner found itself to be the only Canadian company with a personal computer software package to be listed in the top ten of North American sales.

STIMULATING HIGH TECHNOLOGY INDUSTRIES: TOWARDS A MORE EFFECTIVE MODEL

In recent years, there has been a flurry of stimulation activities by governments in countries across the world to accelerate the development of so called high technology industries. Many of these initiatives, such as the TELIDON program in Canada, based upon a new method of data compression, in which $250 million was fruitlessly spent jointly by government and industry, have been high cost failures. In an attempt to develop a model of government stimulation which might be more effective, the author, as Director of Innovation Policy for Communications Canada designed the following study to focus upon correcting two of the reasons for failure of earlier programs: over ambitiousness and failure of major players to comprehend salient properties of the technology in question.

The study model tested was as follows. First, a government division was required which had an ongoing production operation experiencing specific problems which could become a genuine, motivated customer for a particular type of new technology. Second, a provider company was required with genuine interest in developing technology of the sort required and in learning in detail from the experience. Third, the relationship between the client and provider was required to be established on a contractual basis in such a manner that included overheads due to the experimental nature of the activity. Fourth, the study was to be run on an iterative staged basis, at each stage of which, both the client government division and the business provider performed intensively two roles. The client government division, on one hand, was to provide, social data which successively and accurately assessed and described to the provider business, the detailed nature of their governmental task and the emerging social implications of successive technical interventions. On the other hand, it was to concentrate upon understanding in detail the capabilities/limitations of successive technical interventions provided by the business partner. The business provider, on one hand, was to concentrate upon understanding successive iterations of descriptions of the client task and of presented social implications of technical interventions; on the other hand, it was to concentrate upon ensuring the client understood clearly the limits/capabilities of successive technical interventions. Fifth, the experimenter was to enable and assist this four way interaction by setting up the expectations in the first place, by recording events as they transpired, by reporting them and their implications to the principal parties, by providing minimal intervention from time to time and by providing some extra monies.

TARGET TECHNOLOGY (DESKTOP PUBLISHING 1986 STYLE), CLIENT AND PROVIDER

The target technology of this study was desk top publishing. Traditional in house desk top publishing consists of groups of specialists, each contributing a small part to the production of the publication. Among the requirements for effective in house desk top publishing is the ability to create and place figures within text bodies, wrap text around them, place text within them and display the result on computer screen exactly as it will appear in the printed text (WYSIWYG - what you see is what you get). These capabilities, particularly on IBM based computers, upon which large investments had already been expended by corporate users, were not, at first, available; but, in 1986, some software was just beginning to be produced piecemeal amid considerable capability, inter-program compatibility and integration problems. Would be customers were, in general, not aware of the extent of these problems. A further stimulus to desk top publishing was the new availability, at moderate cost, of laser printers, which by writing a pattern on a drum and thus theoretically allowing, with proper software support (a big proviso), any font, figure and document layout to be produced with good resolution.

The spectrum management division of the Department of Communications, Canada was chosen as the first integrated desk top publishing client of Corel, a newly incorporated business venture, targeted at the far from desirable 1986 level of integration of personal, particularly IBM based computers, laser printers and software required for effective in house desk top publishing. The division consisted of about twenty people who write and publish each year, twenty manuals ranging from 15-200 pages containing text, tables, graphs and maps; circulars; 20,000 copies of 12 exams based upon random selection of items from a computer test bank and regulations, on an occasional basis, in the field of management and use of the radio spectrum in Canada.

PROCEDURE

A contract, or in fact, a series of contracts, were established between the client and business provider to deliver various hardware, software and training, on an iterative basis, with the aim of developing an effective in house desk top publishing capability relating to the specific requirements of the client. In this endeavour, it was understood that provider and client would each perform the two activities outlined in the fourth point in the model described above, for which cooperation, the experimenter would provide stimulus monies in the order of tens of thousands of dollars. The experimenter endeavoured to effectively maintain the two activities of each party on an iterative basis. Two devices were use to accomplish this goal. First, research branch staff in the Department were used to establish permanent accessible chronicles of the step by step introduction of desk top publishing capability, which were used as stimuli to the developing understanding of both client and provider. The chronicles were based upon a set of four types of iterative interviews with division staff: initial interviews, in which people were asked abut the nature of their work and their expectations about the changes that would occur after the desk top system was installed; post-installation interviews in which people were asked about their perceptions of the new system and the training they were undergoing; operational interviews, in which people were asked about the effects of the desk top publishing system on themselves and on the documents they publish; and maintenance interviews, about every three weeks during the course of the study with the chief of section, in which she was asked about recent events and her current perceptions of the electronic publishing system. Second, staged meetings were arranged between heads of client and provider organizations at which formal attempts were made to ensure that the client understood the capabilities and limitations of the technology, as they were in a given stage, in the same manner that provider saw them; that the provider understood, in detail, the needs/desires of the job as they were understood by the client at that stage; and that the emerging social implications for both parties, were understood, at that stage.

RESULTS AND DISCUSSION
Tribulations of Implementation

Implementation of desk top publishing began in August, 1986. In Table 1 appears a chronology covering the twelve month study period divided into four major stages indicated by dividing horizontal lines. There follows a discussion of some of the salient events. First, as is often the case in organizations that enter early into new technology, the Division was, at the time of implementation, using an outdated early 1980's micro computer

system (TurboDOS) based upon Z80 processors and an outdated word
processing system (WordStar). Moreover, although this installation
exacerbated the compatibility problems faced by the desk
publishing team, it took most of a year for the division to wean
itself from the outdated technology and, in particular, adopt
WordPerfect as the word processing program of choice. Second, it
took six weeks from program inception until the first equipment
was delivered (IBM AT compatible, 20 MByte hard disk, laser
printer based upon the Cannon engine, Wyse high-resolution
monitor, Microsoft mouse, 1200 baud modem, WordPerfect, Iprint
software for making forms and Paintbrush software for making
graphics. Initial problems included Word Perfect not being
compatible with the Wyse high resolution monitor, printer
mechanical defects and communication and operational difficulties
due to the fact that maintenance of equipment was subcontracted by
the provider to a second party and training to a third party. Due
to initial problems, it was early December before any member of
division staff began to learn IPrint and that was on her own with
the manual. A mis-understanding, in part due to the complicated,
but not unusual, subcontracting of tasks led to the refusal, based
upon inability, of the training party to demonstrate the
preparation of a complete document including graphics for the
cover with inside text and tables in an appendix. It became clear
that although principals in the study knew the experimental or
beta nature of the equipment and software, many of the staff had
expected the system to work on an off the shelf basis. Four months
after commencement of the study, the production chief used the
equipment chiefly as a word processor and was unable to produce a
simple double column French/English single page document.
        Troubles were at first attributed by staff to a single
fault, malfunctioning hardware. Staff were slow, not through any
fault of their own, to comprehend that the real underlying problem
was that each of the software packages were not compatible with
one another, not uniformly compatible with the hardware and had
unique user interfaces which introduced user complexity. One
software compatibility problem was the failure to realize that bit
mapped graphics and ASCII or code based text could produce the
same appearing pattern, yet be entirely non compatible. A
hardware compatibility problem concerned programs that were
compatible in one regard but not another. For instance, Ventura
Publisher was able to load both a WordPerfect file and an
Paintbrush file and display them at the same time but then would
only allow the user modify text in the Wordperfect part and not in
the Paintbrush part. Hardware compatible problems included
graphics produced by Paintbrush being only printed on a laser in
stair case form with dot matrix resolution and the inability to
display different fonts, bold and underlining on the monitor.
Interface problems included non consistent use of mice, pull down
menus and function keys depending upon the software used. By the
sixth month of the study, users had yet to see concrete evidence
that a desk top publishing system was capable of fulfilling their
specific document requirement needs. Additional personal
computers and an optic scanner were obtained. By the middle of
the seventh month the first real documents were produced involving
ordinary word processing, albeit with different fonts, and
manually prepared and hand pasted tables and graphs. Ventura, a
specialized program allowing the user to specify characteristics
of a document in considerable detail, was obtained at this time
but because it was very complicated to use, individuals were slow
to learn it. Moreover it further introduced additional
incompatibilities into the system. TurboDos, the outdated system,

was finally jettisoned in the seventh month of the study. By mid-May, the ninth month, at least one example of each of the major client required documents was produced with the desk top publishing system. The quality of final documents was of progressively higher and higher quality in type quality and layout. However, much of the material, particularly that requiring figures was produced by hand and pasted into the document. In fact, PaintBrush, the graphics program, failed ever to be integrated with word processing and was largely rejected.

Throughout this period, following the experimental model, client staff were iteratively updated with characteristics of the hardware/software, as they emerged in the above described detail, and provider staff were interactively updated on the implications/ expectations of the client staff. At the end of the seventh month, a meeting consisting of the President of the provider company, the Operations Division Manager and Production Section Chief from the client and the author was arranged in which the above events and related implications were discussed in detail.

Changes in Both Client and Provider

Consider the client. First, despite all the problems, satisfaction with the system as improving output was expressed by the production chief almost from the outset of the study as soon as a page was produced that looked better than any page produced by the previous system. This motivated great continuing effort over an extended time to improve the system and explain its attributes and implications. Second, over the course of the study indications of increased status of the division and its staff appeared in terms of increased staff allocation in difficult times, provision of an individual office for the chief and additional equipment purchase monies. Third, the division became sought out as consultant for other government branches wishing to engage in desk top publishing. Consider the provider. The detailed iterative nature of feedback regarding technical attribute comprehension, desired features and implications of various stages of the system development clearly motivated the President of the provider company and appeared to change significantly perceptions of technical staff. At any rate, within two years of the completion of this study, Corel systems, from a start up company, had produced "Draw" their highly successful graphics package, which was the only Canadian personal computer software to make and remain over an extended period on the top ten North American sales list. It is to be noted first, that "Corel Draw" supplanted the graphic package that failed ever to work in the experiment and, second, it is lauded for its friendliness.

In conclusion, based upon the criteria of both client and provider success, the focused low cost stimulation model tested was deemed to be a success and worth further study. However, in attributing this success, it is important to emphasize two points. First, the division was the provider's first integrated in house desk top publishing customer. As such the provider was intent on learning as much as possible from the experiment. Second, the client had great pride in producing the best documents possible in their remit and so was highly motivated to be successful with the new technology. Two final points remain. First, the training provisions for this operations, as is common to many similar computer start up operations, was far too small. The expedient of hiring a live in tutor is to be recommended where possible. Second, the failure to reject early an in-place but outmoded system (Turbodos) held back considerably the effective implementation of the desk-top publishing system. This phenomenon has been noted before by the author (Wilson, 1981) and should be avoided where possible.

Table 1. Four stage chronology of desk top publishing implementation events as separated by horizontal lines (after Whalen and Wilson, 1988)

Date	1986/87
Aug	Installation proposed
Oct	Publication clerks informed
Mid Nov	Computer, laser printer, wordperfect, IPrint delivered
	System installed in cubicle, clerks begin self-learning WordPerfect, half day training session
Ear Dec	Mouse, Modem and Paintbrush delivered, electronic publishing system moved to new room
	Authorization officer begins to learn IPrint, Control officer begins to learn Paintbrush
	Training session cancelled by Division, Production Chief fails to create single page two column document
	Officers demonstrate Paintbrush and IPrint program
Ear Jan	Control officers begin to use Iprint for work, visuals prepared for senior management
Lat Jan	New printer installed, second computer obtained, in-house tutor hired
	First circular completed
Feb	Ventura Publisher installed, index to Divisions publications completed, new clerks learn WordPerfect
	Chief and Clerk receive half-day instruction on Ventura.
Mar	More WordPerfect training, six more computers, optical scanner and new Paintbrush obtained
Apr	More WordPerfect training, exam test bank transferred to new system
	Three days training on Ventura
May	Two exams produced
Jun	Two and a half more days training on Ventura
Jul	Observation period ended

REFERENCES

Whalen, Thomas and Wilson, Tay (1988) Social Implications of Introducing a Desk Top Publishing System. Department of Communications Report. Ottawa, Canada.

Wilson, Tay (1981) Proposal for a Shared Computer Facility for the School of Automotive Studies. Technology Assessment and Applied Psychology Unit, Cranfield Institute of Technology, Cranfield, Beds, England.

USER INVOLVED REDESIGN OF A CLOCK-RADIO

C.C.C. Voûte, H. Kanis and A.H. Marinissen

Delft University of Technology
School of Industrial Design Engineering,
Department of Product and Systems Ergonomics,
Jaffalaan 9, 2628 BX Delft, The Netherlands

This paper describes a study on the usage of a new clock-radio, that
was introduced as being easy to use. Unfortunately, indications from
the field seem to imply the opposite. In a users' trial several difficulties
to operate the clock-radio could be identified as emerging under various
use-conditions and by different ways of operation. The results of the
field study have been converted into redesign requirements. These
requirements have been synthesized in a mock-up, as an illustration of
how findings in a users' trial can be directly translated into design
proposals.

INTRODUCTION

Nowadays, producers of consumer goods, such as audio-visual equipment,
often presume to pay full attention to usability during the development of new
products. However, user experience regularly seems to indicate the opposite.
Often the fast technological development of products appears to bring about an
increasing complexity and a decreasing usability. Convincing evidence for
mismatches between users and products can be obtained by observing the actual
use under everyday conditions. This approach is adopted in a case study discussed
in this paper.

The product studied is a newly developed clock-radio, introduced on the
market with ease of use as a special feature. The incentive to focus on this
particular product were indications from the field, that this clock-radio was not so
easy to use, after all.

OBJECTIVE

The use-problems, already alluded to, concerned setting and switching off the
alarm. Another problem seemed to be the illumination of the clock-dial; by some
users it was felt as irritating, whilst other users complained about a poor
readability. In order to gain a clear insight into the use, and especially into the
origin of difficulties in usage, it was decided to carry out a users' trial that should
provide answers to the following research questions:

- How and in what use-conditions are clock-radio's or alarm-clocks being used in the bedroom?
- What operational difficulties occur?
- What factors can be identified as the cause of these difficulties?

In the study, the answers to these questions were focussed on the alarm functions, rather than on the radio-part.

On the basis of the indications from the field the following aspects have been taken into account as possibly related to difficulties emerging in use:
- the lay-out of the environment of the bed, in particular the position of the clock-radio or alarm-clock, and the intensity of the ambient light;
- physical characteristics of the user, such as dimensions of the upper extremities;
- perceptual characteristics of the user;
- use-habits with other alarm-clocks or clock-radio's.

METHOD
Subjects

The recruitment of subjects proved to be troublesome, primarily since the test was considered as an invasion into one's privacy, and also because a considerable amount of time would be required on the part of the subject. Eight subjects participated, who were aged between 26 and 78 year (two of them were men). None of these subjects had had any experience with the new clock-radio under investigation.

Procedure

First, an inventory was made of the lay-out of the bedroom at the subjects' homes, consisting of establishing the position of the alarm-clock or clock-radio owned relative to the bed, and the intensity of the ambient light at midnight and after sunrise.

Subsequently, the subjects were invited to the usability lab in order to demonstrate the way they are used to operate their own alarm-clock or clock-radio. For that purpose, a simulated bedroom was built in which the alarm-clock or clock-radio was positioned interindividually according to the position relative to the bed at home. The operations were videorecorded under two levels of ambient light as perceived by the subject:
- almost dark, which was fixed at an intensity of 3 lux: the lowest level to allow videorecording;
- pitch-dark (0 lux), by blindfolding the subject.

After the subjects had demonstrated their habitual way of operating their alarm-clock or clock-radio, the new clock-radio was introduced. The subject was asked to unpack the clock-radio and to explain the different functions, while demonstrating them. This introduction was videorecorded. The subjects were then asked to use the new clock-radio at home for one week. After that week each subject was again invited to the usability lab, now to demonstrate the operation of the new clock-radio in the simulated bedroom, individually adjusted, as before.

Finally, the subjects were asked to report their experience in using the new clock-radio at home in comparison to the experience in using their own alarm-clock or clock-radio.

The research was concluded with the establishment of some antropometrics of the upper extremities, and of eyesight without spectacles or lenses on the basis of a visus-map.

RESULTS

Use conditions at home

Alarmclocks or clock-radio's have been found in all sorts of positions. As illustrated in Figure 1, the subjects generally succeed in adjusting the horizontal position according to their preferences. However, in most cases the desired height, which all subjects indicate to be at eye level, is not available.

Figure 1. Position of alarm-clocks or clock-radio's in bedrooms

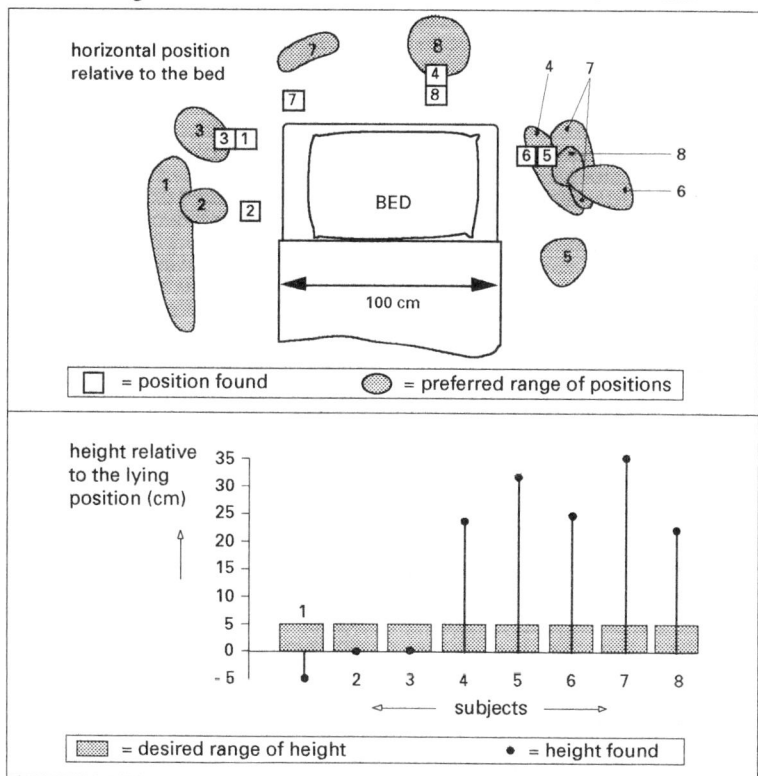

In Figure 2 the intensities are given of the ambient light in the bedroom of the subjects during sleeping hours. This intensity may vary considerably between midnight and after sunrise; in these cases, none of the subjects were bothered by a considerable intensity of light. What appears to irritate people is the light emitted by the LED's. Three subjects had a clock-radio with a LED-illuminated clock-display. Each of them complained about it, albeit that the intensity was adjusted to its lowest level. As a remedy, one subject always placed the clock-display facing backwards, while another subject put books in front of it. The origin of the difference in sensitivity for ambient light and for light emitted by LED's has not been further explored, as yet.

Figure 2. Ambient light in bedrooms during sleeping hours

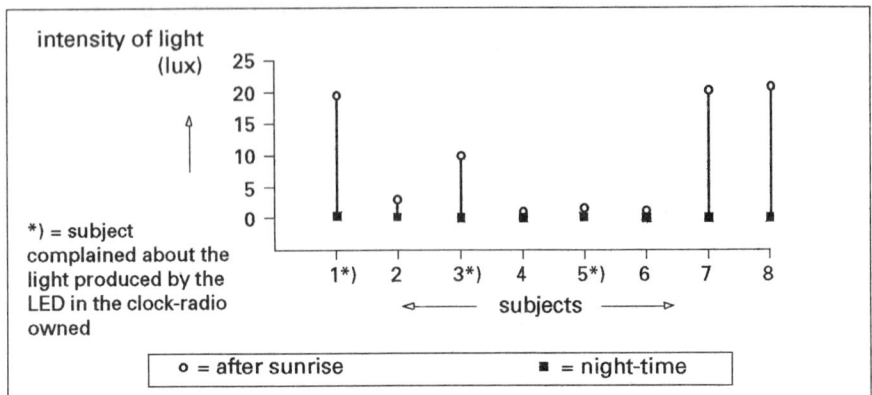

Operation

At the introduction of the new clock-radio, the subjects familiarized themselves with the operation of the product while sitting at a table, under supervision of the researcher. None of them demonstrated any persisting problems in understanding the operation. However, when using the product at home, under real-life conditions (lying in bed), various difficulties emerged. First, setting the alarm was felt as inconvenient, due to the combination of the small size of the relevant control and its position at the rear of the product. This caused several subjects, who were used to set the alarm in a lying position, to sit upright and manipulate the product with both hands. Another use-problem concerned switching off the alarm. In Figure 3 some typical examples are given of different ways of operating the relevant control. The first way shown is what seems to be assumed practice (see for example Woodson (1981), p. 610). However, several other ways of manipulation occur, which may be related to the position of the clock-radio (see Figure 1) in connection with the sleeping posture of the subject such as lying on the back or on the stomach (one subject is used to sleeping with her head under the pillow). By the other two ways of manipulation shown in Figure 3, the radio is switched on simultaneously with switching off the alarm.

Figure 3. Some ways of operation

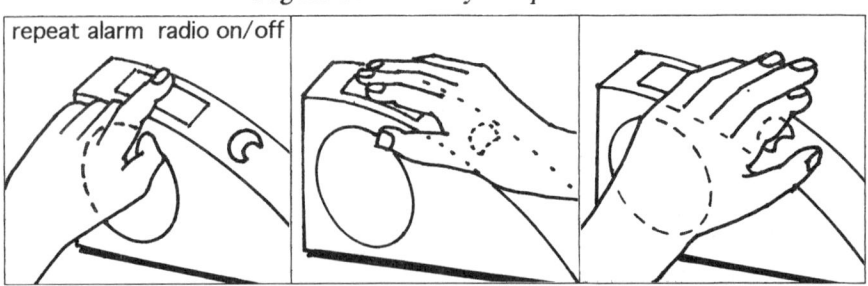

The frequency of use of various controls appears to differ considerably. For example, the alarm is set several times a week, whilst generally the time is adjusted only twice a year. Unfortunately, this is not embodied in the current design of the clock-radio: both controls are positioned next to each other on the same surface, and have to be operated in the same way.

In general, the subjects stuck as much as possible to their habitual way of use. Enforced deviations are easily felt as irritating.

None of the observed ways of use could be shown to be linked up with perceptual or physical characteristics of the subjects. For example as to eyesight, differences between subjects did not matter in operating the alarm functions. While being blindfolded, the subjects managed equally well as in dim light. In addition, both the actual position of an alarm-clock or a clock-radio and the position preferred (see Figure 1) could not shown to be related to body dimensions such as armlength or reach envelopes. What appears to be dominant here are use conditions, particularly the position and posture of the sleeper.

Functionality

The new clock-radio features an analogue clock-display. Most subjects indicated to prefer a digital clock-display corresponding to the one they own. When asked, they explained that it is more precise and easier to read; these subjects expressed some uneasiness with the analogue clock-display they had to use during the test. The other subjects prefered an analogue clock-display, because the whole time cycle, including the alarmtime, is seen at a glance. Actually, on a digital clock-display the subjects appeared to read the time quickly in figures, but apparently it took some mental effort to convert the figures into 'the time', especially for hours after 12 o'clock and minutes over 30. The time on an analogue clock-display was read faster; the minutes were mostly rounded off to a multiple of 5.

The majority of the subjects indicated that at night they were especially interested in the time left for sleeping, rather than the exact point in time. As to the indication of time, several subjects got confused by the various pointers on the clock-display of the new clock-radio.

Interindividually, preferences appeared to vary about the alarm-repeat time interval, the alarmvolume and the brightness of the illumination of the clock-display.

REDESIGN REQUIREMENTS

Any indication of inconveniences of sub-optimal use-conditions or of use-problems that have emerged from the users' trial is assumed as relevant for drafting redesign requirements (cf. Vorst et al., 1992). The following list of requirements is limited to evidence from the field study. A general consideration is that it should be possible to operate a clock-radio without the necessity of having to consult written directions of use. It is taken for granted that the clock-display will remain analogue.

Operation
1   Current functions, such as setting the alarm and setting the time, or alarm functions versus radio functions, should be readily operable with one hand, while lying in bed.
2   While operating the alarm functions, the product should not fall over or slide away.
3   Controls to activate different functions should be clearly separated.
4   Frequently used controls should have prominent positions.
5   Setting the alarm and adjusting the time should be possible both clockwise and anti-clockwise.

Functionality

6   The clock-display should be easily visible and readable for someone lying in bed.
7   The clock-display should indicate the time left for sleeping.
8   Confusion about different display information, caused by similar appearance
    such as pointers that look alike, should be anticipated.
9   The clock-display should be illuminated; this illumination should be
    continuously adjustable between 0 and 70 lux.
10 The volume of the alarm should be adjustable between 30 dB and (max.) 70 dB.
11 The time interval between a repeated alarm signal should be adjustable between 4
    and 9 minutes, in steps of no more than 1 minute.

REDESIGN

The redesign requirements have been materialized - together with some other
requirements that fall outside the scope of this paper - in a mock-up which is shown
below. This mock-up serves primarily as an illustration of how findings from a users'
trial can be synthesized in a design proposal, rather than as an elaborated redesign
that is ready to be produced as a working prototype which could be tested in practice.
The numbers in the picture refer to the redesign requirements mentioned.

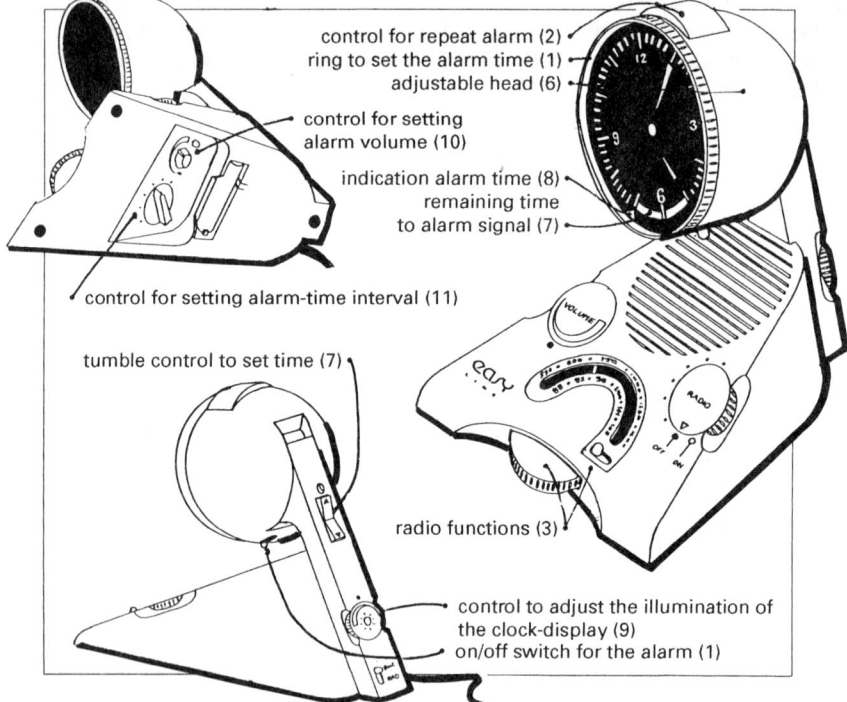

control for repeat alarm (2)
ring to set the alarm time (1)
adjustable head (6)
control for setting
alarm volume (10)
indication alarm time (8)
remaining time
to alarm signal (7)
control for setting alarm-time interval (11)
tumble control to set time (7)
radio functions (3)
control to adjust the illumination of
the clock-display (9)
on/off switch for the alarm (1)

REFERENCES

Vorst, L.M.T., Kanis, H. and Marinissen, A.H. (1992), User involved design of a
    remote control, Gerontechnology, 343-348, IOS Press, Oxford.
Woodson, W. E. (1981), Human Factors Design Handbook, McGraw-Hill Book
    Company.

A CLASSIFICATION OF DOMESTIC TASKS TO SUPPORT DEVICE PROCUREMENT AND DESIGN

V.S.JENKINS,[1] K.Y.LIM & J.B.LONG.

Ergonomics Unit,
University College London,
26 Bedford Way, London WC1H 0AP.

Domestic work is performed daily by most people. The design of new domestic devices for the home needs to accommodate human requirements and abilities, and ergonomic considerations more generally. Unfortunately there is an absence of organised, detailed information relating to domestic tasks. To help make good this deficiency, this paper suggests a means of classifying information about domestic tasks using the framework of Human Computer Interaction Engineering (Dowell & Long, 1989). To this end, a domain hierarchy was constructed, a task analysis conducted and the results tabulated to support design decisions. The tables were intended to support the procurement and design of domestic devices for the disabled. A pilot study was conducted to test the usefulness of the tables. The results of the pilot test highlighted the need to develop the tables further to provide more specific decision support for non-expert procurers. Further work needs to be conducted to extend the task classification to include devices with embedded information technology, and to study the social consequences of increased home automation for the disabled.

BACKGROUND

Domestic work in Britain in the 1990s has been defined as 'unpaid work undertaken in the household by household members for themselves or other members of the household' (Williams 1988). This definition is able to accommodate the various forms of household, as well as domestic work undertaken by people of different ages, sexes, or roles (rather than being the sole province of women who are housewives). A broad definition is required, because the nature of households is changing. Not only is there a larger number of single parent families, but there is an increasing proportion of single adult households and working spouses. Also, owner occupied dwellings increased to 61% of all households in 1981 (OPCS 1986, 1988). This change suggests an increase in the number of people responsible for the repair and maintenance of their own

---

[1] At the School of Physiotherapy, Middlesex Hospital, London W1 5PG.

homes. Taken together, these changes are likely to increase the domestic workload, because a wider range of domestic tasks will be involved, and the responsibility for the tasks is increasingly borne by individuals rather than shared. Of these individuals, some might be expected to be experts, others novices. Although experts may experience few problems performing tasks, less effort expended on them would release time for other work or leisure activities. In the case of novices, and people with special needs, difficulties with particular domestic tasks may arise. For example, persons with osteoarthritis of the hip joints are likely to have limited mobility and so may be unable to clean their homes acceptably.

Current support for domestic tasks is provided in various ways. First, there are labour saving devices, e.g. a food mixer, which reduce the domestic workload and change the nature of the tasks. Second, domestic tasks can be supported by people either in a paid or voluntary capacity, e.g. as a house cleaner. As a larger proportion of the population becomes elderly, it is likely that the demand for manual support of domestic work will exceed the supply of people available or willing to provide it. Therefore, manual support may become too expensive, particularly for public sector financing. Third, domestic support for services may be provided off-site, e.g. a hairdresser. This alternative may be less expensive than on-site provision, but would require greater effort from individuals who need the service, e.g. in the use of transport. Consequently, a wider range of domestic devices needs to be developed to reduce the domestic workload of a larger proportion of the population. In particular, new domestic devices to reduce the mental and physical workload need to be developed, e.g. domestic devices with embedded information technology. To this end, this paper reports a study that aimed :

• to develop an initial framework of tasks in domestic settings.
• to suggest a tabular form for classifying information about tasks to be performed using domestic devices.
• to illustrate how the classification may be used to organise the information to support the design of domestic devices.
• to assess the usefulness of the tables by conducting a pilot study.

These aims are addressed in the Sections that follow.

FRAMEWORK

To define explicitly the human and device behaviours of a domestic worksystem, a theoretical framework was recruited to support the classification of domestic tasks. In particular, Dowell and Long's (1989) Human-Computer Interaction Engineering framework was used, because it provided a structure that facilitates the organisation of domestic task information in a form that would support design.[2] In the Human-Computer Interaction Engineering framework, design problems are conceptualised in terms of an Interactive Work System which operates to transform the attributes of objects. The Interactive Work System consists of the human and device elements which interact to perform work expressed as task goals.

In a domestic context, the *human* element of the Interactive Work System is a user who may be a child, parent, friend, partner or carer of the person associated with the transformation of

---

[2] It is open to debate as to whether all or even some domestic work should be conceived within the ethos of work design for effectiveness assumed by the framework. This issue is not addressed here.

objects within the domain. For example, to achieve the task goal of 'clean clothes', the user may have to interact with a washing machine. The *device* in the Interactive Work System may not only be a machine, but a 'natural' entity, i.e. one which is not manufactured. For example, a dog may be a means of supplementing a security work system, in which case, it would be considered an element of the Interactive Work System. Alternatively, a dog may also be a domain object operated on by the Interactive Work System, e.g. to achieve the task goal of 'well-groomed dog'. The behaviours of these Interactive Work System components (i.e. the human and device), thus together transform the attributes of objects within the domain of application of the worksystem. For example, the behaviours of a vacuum cleaner can be both physical (e.g. collecting dirt) and abstract (e.g. indicating when emptying is necessary). Similarly, to clean a carpet, the human component of the Interactive Work System performs both physical (e.g. pushing a vacuum cleaner) and mental behaviours (e.g. deciding on a strategy to ensure complete cleaning of the carpet). Interacting together, the behaviours of the worksystem elements achieve task goals.

The *domain of application of the worksystem* comprises the physical and abstract objects whose attribute changes constitute work. In the home, the domain of work may be generally conceptualised as 'home maintenance' which involves specific domestic objects. A model of the domain of home maintenance may be decomposed and represented as a hierarchy.[3] Specifically, the whole is decomposed into hierarchical branches comprising its component parts. Thus, home maintenance is divided into two main components; household and dwelling maintenance (see Figure 1). Household maintenance is concerned with the people of the home, either as an individual or as a group. Dwelling maintenance concerns the physical structures in the home, e.g. house, furniture, etc. Dwelling maintenance can be further sub-divided into exterior and interior/room maintenance (e.g. room structures such as walls, furniture, etc.). Other components of home maintenance may include garden/vehicle/animal maintenance; although these parts may not be considered contingent rather than necessary for the existence of a home.

Figure 1. A hierarchy of the domain of home maintenance

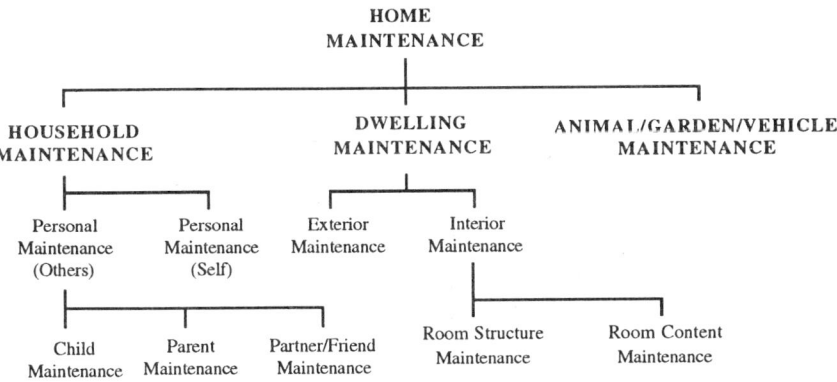

It should also be noted that domestic work, and particularly its effectiveness, may be

---

[3] For present purposes, and initially, a simple hierarchy was constructed to organise domain objects. This hierarchy is not a 'hierarchy of complexity', and so is not a domain model in the sense of Dowell & Long (1989).

viewed differently by people with different expectations. The *effectiveness* of domestic work can be thought of in terms of *quality* (e.g. how well the carpet is cleaned) and costs. The costs associated with domestic work behaviours may be viewed as comprising both human (e.g. physical and mental workload) and material costs (e.g. wear and tear of device). The concept of quality as applied to domestic work may be highly subjective. However it can be conceptualised as the difference in the perceived initial and final state of the home following the performance of a maintenance task.

Using the framework as described above, information concerning domestic tasks was classified in terms of the domain and the behaviours of the user and device of an Interactive Work System. An account of these activities follows.

A CLASSIFICATION OF DOMESTIC TASKS

To illustrate the classification scheme, information for two domestic tasks was collected and analysed. Task decomposition was supported by task analysis, which involved the observation and audio-taping of two subjects. Subjects were required to clean a carpet using different household devices: brush and pan; carpet sweeper; and vacuum cleaner. The information was then classified and is shown in Figure 2, in the form of a task classification table.

Figure 2. Example of a task classification table

PROBLEM 1 Carpet Cleaning	USER BEHAVIOURS		MACHINE BEHAVIOURS	
Devices/Means	Physical	Mental	Physical	Abstract
1. Brush and Pan	Fetch brush Carry brush Brushing movements Move obstacles	Judge need Locate device Locate task Judge frequency	Moves dirt Contains dirt	
2. Carpet Sweeper	Fetch sweeper Push sweeper Pull sweeper Move obstacles Empty sweeper	Judge need Locate device Locate task Judge frequency Judge sweeper full	Moves dirt Collects dirt Contains dirt	
3. Vacuum Cleaner	Fetch vacuum Push vacuum Pull vacuum Move obstacles Switch on/off socket Put on/off attachments Empty vacuum	Judge need Locate device Locate task Judge frequency Select attachment Sense feedback Recognise feedback	Collects dirt Contains dirt Indicate vacuum full	Judge vacuum full

PILOT ASSESSMENT OF THE DOMAIN HIERARCHY AND TASK CLASSIFICATION TABLES

A pilot study was conducted to elicit feedback for further development of the classification and to test the usefulness of the task classification tables. To this end, eight subjects were allocated to four test conditions and provided with instructions and a design problem scenario. Specifically, experts and non-expert subjects were asked to solve a design problem using either the tables or a simple list of devices. The problem comprised two scenarios concerning the selection of devices for carpet cleaning and heating control that would be suitable for an arthritic and elderly person living alone. Constraints on the choice of a device were: minimal expenditure and the person's physical limitations. To reveal the nature of their problem analysis, the subjects were required to 'talk aloud' while solving the problem, and to suggest any design modifications that might be useful. It was expected that the study would reveal differences between the test conditions with respect to the proposed solutions and scope of problem analysis. For instance, better scope of address and design solutions might be expected from experts than novices and from the use of the task classification tables than the list of devices.

To complete the study, the subjects were asked to assess the task classification and to suggest improvements that would better support problem analysis and solution. Their responses were audio-taped and then transcribed.

RESULTS AND DISCUSSION
The analysis of the transcripts and the results of the study are as follows :

1. The use of the domain hierarchy. Subjects were often uncertain about which aspects of the domain of application were implicated by the problem. For instance, the control of heating was seen both as household maintenance (5 out of 8 subjects classified it under personal maintenance (self)), and as dwelling maintenance (3 out of 8 subjects classified it under interior maintenance (room)). However, carpet cleaning was recognised by all subjects as being concerned with dwelling maintenance (the subjects classified it under interior room content maintenance). Nevertheless, 3 of the subjects were also prepared to include carpet cleaning as a part of household maintenance. This uncertainty may stem from the potential association of home cleanliness with a sense of well being.

2. The scope of the problem analysis. Appropriate address of the range of costs and constraints was used as an indication of the scope of problem analysis. There were five major constraints and costs which could be considered, namely: physical limitations of user (i.e. bending over and gripping); physical effort of user; mental effort of user; safety characteristics for the device; and financial cost of the device. Although the non-expert subjects using the task classification tables considered more constraints than non-experts using only a device list, the results of the study did not reveal significant differences in the subjects' address of design costs and constraints. Non-expert subjects, however tended to choose devices that are more expensive, i.e. over-provision of support.

3. The quality of proposed solutions. The solutions proposed by expert subjects were used as a baseline for assessing the solutions proposed by non experts. In this way, the utility of the task information tables for supporting device design was assessed. The results seem to indicate that the tables did not support the generation of solutions that are better in quality, since similar solutions were proposed by non-expert subjects regardless of whether the tables or device list was used.

Generally, the results of the pilot study were inconclusive. An analysis of the nature of *problem scope* did not reveal significant differences between the test groups. Although the subjects found the domain hierarchy and task classification tables to be useful, their impact seems to have been minimal. The primary reason appears to have been the ease with which the experimental tasks were performed. In other words, the subjects performed the tasks well regardless of whether the hierarchy and tables, or a device list were provided. Consequently, further studies need to employ domestic devices which are less familiar to the subjects. The results may then reveal the usefulness of the task classification tables more clearly.

The quality of *problem solution* indicates that the task classification tables may support the identification of more cost effective solutions by non-experts, e.g. the selection of devices that are more appropriate with respect to user requirements. Further studies should employ domestic devices which are less familiar to the subjects, so that more informative assessments may be made. A larger scale study should also be conducted to offset any experimental effects due to individual subjects, e.g. personal variables such as experience and ability in problem analysis and solution. In addition, further studies need to emphasise more design oriented tasks, since the present study only required the subjects to match devices to user requirements. Although the subjects were asked to suggest any modification to the design of the device, they were not explicitly required to do so in this study. Follow-up studies could also examine the usefulness of the tables in respect of domestic devices with embedded information technology.

CONCLUSIONS

Current sociological and demographic information indicates a growing demand for domestic devices to reduce domestic workload (both mental and physical). Although there is general agreement that appropriate ergonomic support should be provided for the development of more effective and usable domestic devices, detailed information about domestic tasks and requirements does not exist. It is argued in this paper that a domain hierarchy and a classification of domestic tasks could support the procurement of appropriate off-the-shelf devices and the development of new and/or bespoke devices. However, the task classification tables are presently incomplete and need further development and testing. In particular, the scope needs to be extended to include a wider range of domestic tasks, and the description of user and machine behaviours needs to be made more consistent. In addition, improvements to the tables suggested by subjects need to be taken into account, namely the provision of more detailed information on familiar and alternative devices such as relative cost and efficiency, and lower level design features (e.g. information about the weight of a device would support more precise assessment and design) of alternative devices. The usefulness of the tables as a design aid also needs to be tested further. It should be emphasised that related studies need to be conducted to investigate the impact of increasing automation in the home. Specifically, the studies need to examine both positive and negative effects of automation, e.g. greater independence and loss of face-to-face contact respectively in the case of the disabled.

ACKNOWLEDGEMENTS
This research was conducted in part fulfilment of the requirements for the MSc in Ergonomics, at University College London. The project was supported financially by the Bloomsbury and Islington Health Authority.

REFERENCES
Dowell, J. & Long, J., 1989, Towards a Conception for an Engineering Discipline of Human Factors. Ergonomics 32, 11.
O.P.C.S., 1986, Monitor. General Housing Survey 86/1. Office of Population Censuses and Surveys. H.M.S.O.
O.P.C.S., 1988, Monitor. General Housing Survey 88/2. Office of Population Censuses and Surveys. H.M.S.O.
Williams, C.C., 1988, Examining the Nature of Domestic Labour. Gower.

ENERGY DEVOTED TO THE ERGONOMICS OF PRODUCT HANDLING

W.P. MOSSEL

Faculty of Industrial Design Engineering
Department of Product & System Ergonomics
University of Technology Delft
Jaffalaan 9
2628 BX Delft
The Netherlands

Twelve designers of small products with handling
aspects have been interviewed about their design process to discover the part ergonomics played in it. The
used elicitation techniques are explained. In the
beginning of the design process most of the ergonomic
information comes from existing products, or is based
on designers presumptions. In the last phases of the
design process ergonomic aspects have a low priority.
A users' trial with random chosen test person to check
the ergonomics is mostly not executed.

INTRODUCTION

A product is designed to perform a certain function. The user
of the product takes action to let the product fulfil the
desired performance. The operations for the user to achieve the
desired performance vary from very simple to very complicated.
Supposing it is the intention of the designer to make the
handling of the product not unnecessary complicated the question
arises in which way the designer reaches this goal. One of the
research projects of the Department of Product & System
Ergonomics of the Faculty of Industrial Design Engineering of
the University of Technology at Delft concerns a study of the
behaviour of the designer when he(she) designs everyday consumer
products, in particular those products that require handling.
The variety of products in this field is enormous; and as some
of them are meant for the same kind of operations,
categorization of (operations with) products would be desirable,
which allows for generalization of ergonomic characteristics
within each category. Moreover, people perceive objects in terms
of their similarities and differences, and consequently, in
terms of the categories to which they belong (Lederman and
Klatzky, 1990). In this study is chosen for a category of small
products, namely domestic-, office- and do-it-yourself products

129

leaving unmovable fixed products excluded. The reason for this choice rather than looking at the handling aspect itself, is that the handling aspect of this type of products cannot be overlooked by the designer.

To get an insight in the behaviour of the designer during the design process three studies have been made:
- Analysis of twenty students' masters graduation projects at the Faculty of Industrial Design Engineering of the Delft University of Technology, (Mossel, 1988).
- Analysis of the design periodicals Design (UK), Form (FRG), Form & Zweck (GDR) and ID (USA) from the years 1983 till 1987, (Mossel, 1990).
- Interviews with designers.

This paper reports on the last mentioned study. Since the handling of the product is not the only design problem of the designer the following questions arise:
- Does the handling of the product play a role during the whole design process or only in certain phases?
- How important is "handling" compared with other product requirements for the designer?
- Are there any unprejudiced checks on "handling" during the design process?

The design process is approached -in this study- in the way it is taught at the Faculty of Industrial Design Engineering in Delft (Roozenburg and Eekels, 1991). In this approach the following phases can be distinguished: orientation, information, concept or draft, materialization, evaluation and adjustment. In a nutshell: In the orientation phase the designer builds an image of the product to be designed. In the information phase information is collected and analyzed. In the concept phase of the design ideas are generated, worked out and selected. In the materialization phase the realisation of a real product is undertaken, resulting in drawings and possibly in a working model with documentation. In the evaluation phase the first mass products are made and the product is evaluated. After the first two phases a list of requirements is composed that can be used for concept choice and evaluation.

The background of the third study is the thought that relevant information about the behaviour of the designer concerning handling of the product to be designed can be found in the way designers collect ergonomic information and the way they deal with it in the design process.

This paper describes the setup of the interview method and the results of twelve interviews with designers.

METHOD

Various techniques can be used to find out in what way designers bring in ergonomic knowledge in their design during the design process (Breuker and Wielinga, 1984). In this study a structured interview with the designers just after they completed their design project has been chosen. With the interviews the risk is run that the aim of the study is too obvious with the possible result that the importance of the

handling aspects and the attention paid to it are exaggerated. In a pilot study -with two designers- this problem seemed to arise. An effective way to avoid this is to assume a pose of sophisticated naïveté for the interviewer (Goldman, 1962). The respondent is assigned the role of educating the interviewer. The respondent is coerced into explaining aspects of the design process to the -to a certain extent- as ignorant acting interviewer. The interviewer must be able to assess the relevance of an answer, and adapt his strategy of questioning. In this way the light on the area for special attention is filtered because the questioning is about a number of aspects of the complete design process and the questions about ergonomics are not remarkably highlighted.
This leads to the followed procedure.

Procedure

Through contacts of the faculty of Industrial Design Engineering with various design offices in the Netherlands an inventory could be made up of designers active within the scope of the study. Designers were asked by mail to participate in the project with the suggestion that the researchers were interested in all aspects of the design process. (With the intention to find out any possible 'red line' that could be useful to create a design method more oriented on practice). Twelve were selected on the (aforesaid) kind of product they designed and the requirement that they worked as an industrial designer for at least five years. Then the designers (one woman, 11 men) received a questionnaire. The questions, most of which were standardized, were divided in six headings, following the phases of the design process. Under each heading each question referred to the emphasis the designer gives to aspects like marketing, ergonomics, technical constraints, styling etcetera. Other more general questions referred to the product (was it a redesign or an innovation) or to the length of time worked on the product by the designer etcetera.

After the return of the questionnaires the designers were interviewed for clarification of some of their answers and for further details. Each interview took about two hours. During the interview the product itself or a picture of it was present.

RESULTS

Twelve designers are interviewed. Eight of them have a university education (5 as an industrial design engineer, 2 as a mechanical engineer, one as an electrotechnical engineer), the four others have a college education (two industrial design engineers and two mechanical engineers). The designer worked in eight cases in a team, in two cases alone, in one case alone with the assistance of two industrial design trainees. In one case is the design is being done except by the designer by experts. The designers described their products in two cases as improvements of existing products, in eight cases as innovations and in two cases as innovations as well as improvements of existing products.

Their products were: a can with a pilferproof plastic lid
and sealed aluminium tagger; cutlery; a child resistant locked
chemical waste container for domestic use; a portable air
conditioner; a hair drier; a telescopic swivel arm for a
P.C.monitor; an ergometer (an "easy-to-handle" instrument to
adjust things in your office in a correct ergonomic way); a
plastic tablet for shampoo and soap, easy to click on the
attachment bar of the shower. In four cases the products were
lamps but different in regard of handling: a hanging lamp; the
Primopowertrack lamp (a transformer connected to a power rail.
One or two lamps can be clicked on the transformer); a standing
fluorescent lamp of which the amount of light and the direction
is controlled mechanically; an adjustable lamp that can be fixed
to spring mattresses.

From the questionnaires came out that in the orientation
phase of the design the image of the product to be designed is
mainly based on existing products, presumptions and personal
experience. (As aforesaid: in four cases the product had a
direct connection with an existing product). In two cases the
image was based on the result of a users' trial.

In the information phase of the design an analysis of use is
done by six of the designers and in one case a users' trial is
executed by a professional consumers research institute. The
product ergonomic information is taken from handbooks and/or
copied from existing products or prototypes in two-third of the
cases. The information concerns dimensions like grip width,
height for a control switch, safety regulations etcetera. In
three cases the handling aspects were concluded from simple
tests done by the designer himself on a foam model of the
product. Seven designers made a list of requirements. In six
cases handling requirements were formulated.

In the concept phase of the design ideas are generated,
worked out and selected. (Foam)models are build from the product
(or from parts of the product). As the deciding factors in the
generation of the ideas styling aspects (named 11 times) and the
constructional aspects (named 10 times) seem more important than
the ergonomics and managerial aspects (both named 5 times). At
the elaboration of the selected ideas ergonomic aspects are
taken into account 7 times, compared to styling and
constructional aspects each 11 times and managerial aspects 8
times. In almost all cases no users' trials are done, only in
two cases. The explanations of the designers for not doing
users' trials were:
"A detailed users' trial was done in the previous phase".
"Handling of cutlery is commonplace, everybody handles cutlery
most of his life, so a users' trial would be of no avail".
"The product will not often be moved, the production aspects are
more important".
"A users' trial was not included in the order. All ergonomic
aspects are adopted from the function model".
"For reasons of secrecy no users' trial has been done in this
phase".

"All ergonomic demands have been fulfilled, besides ergonomics are not a decisive factor in this product".
"We have observed the use ourself with drawings and models and also with the tealady".
"A trial is done by trainees on a foam model".
"There are no ergonomic aspects".
"About ergonomics everything has already been fixed, a users' trial would be of no avail".

In the materialization phase of the design the realisation of a real product is undertaken, resulting in drawings and possibly in a working model with documentation. In this phase constructional, managerial and styling aspects are more important than ergonomic aspects. The reason is that the ergonomic aspects are (believed to be) already defined and fixed in the beginning of this phase. At the end of this phase a kind of users' trial is done, generally by the designer himself or a not-random chosen test person. One designer thinks that a users' trial in this phase is the concern of the manufacturer.

During the evaluation and adjustment phase of the design it appeared that five designers are of the opinion that their product can be improved ergonomically. The possible improvement is an assumption of the designer and is not based on research, except in one case in which a test on child resistance failed.

Summary

The study leads to the following conclusions:
Most of the ergonomic information is taken from existing products. A part of the ergonomic information is based on the designer's presumptions.

In the end of the materialization phase the design had not been tested in a users' trial but by the designers themselves.

Generally speaking it looks as if the ergonomic aspects have a low priority. This conclusion is intensified by remarks of the designers and leads to the following picture:
- Pressure of time and lack of money limits activities. So users' trials are left out.
- Sometimes designers are called in some time after the start of a project. They then proceed in the belief that the ergonomic aspects are already studied and verified by their predecessors.
- Ergonomics are sometimes considered as a complicating factor, preferably to be avoided.
- Often the product (or the construction) is regarded as being that simple that it is pointless to do a users' trial.
- End users are expected to accept the handling of the products in the total deal of purchasing and use, although it might be uncomfortable or even dangerous.

DISCUSSION

Based on this study the following three hypotheses can be set up:
1 The manageability of a product is made subordinate to its styling and construction.

2 As far as the manageability of a product is concerned the
   designers' own presumptions and experience prevail. A
   targeted users' trial in advance is considered of low
   importance.
3 The ergonomic aspects of a product concerning handling play a
   minor role in its commercialization.

A users' trial with at random selected persons is for obvious
reasons preferable than a trial with the designer himself or
people in his near surroundings are involved. The fact that a
good users' trial not always takes place is probably due to lack
of time and money. When a users' trial is not an agreed part of
the design, the designer will not arrange for it.

The combination of the written questionnaire and the
interview (a survey and a structured interview) seams a useful
way to elicit the designers' ergonomic knowledge used in the
design process. It is useful to ask test questions in the
interview about answers in the written questionnaire. In four
questionnaires came out that the prototypes from the
materialisation phase were used in a users' trial, but in the
interviews came out that in only two cases this was done.

The choice for the category of small products (unmovable
fixed products excluded) is based on the fact that the handling
aspect of this type of products cannot be overlooked by the
designer. A further categorization of products to get a
comparison of the handling aspects of the products- has not been
done. The study is restricted to the way designers deal with
ergonomic aspects concerning handling in the design process.

REFERENCES
Goldman, A.E., 1962, The group depth interview. Journal of
    Marketing, 7, 61-68.
Lederman, S.J. and Klatzky, R.L., 1990, Haptic classification of
    common objects: knowledge-driven exploration. Cognitive
    Psychology, 4, 421-459.
Mossel, W.P., 1988, The design of products to be handled. In
    E.D. Megaw (ed.), Proceedings of the Ergonomics Society's
    1988 Annual Conference. London: Taylor & Francis,
    pp. 215-220.
Mossel, W.P. 1990, Is manageability a marketing issue for the
    designer?, Ergonomics, 4, 447-451.
Roozenbrug, N.R. and Eekels, J. ,1991, Produktontwerpen,
    structuur en methoden, 1th edn (Lemma, Utrecht).

ACKNOWLEDGEMENTS
    Thanks are due to Mary Munnik, Marike Groenendijk and Janine
Heinen for doing the interviews and their contribution in the
development of the research method.

FEASIBILITY AND USEFULNESS OF INVOLVING USERS EARLY IN THE USER INTERFACE DESIGN PROCESS: A CASE STUDY

Arnold P.O.S. Vermeeren and Raghu Kolli

Delft University of Technology,
Faculty of Industrial Design Engineering
Jaffalaan 9, 2628 BX Delft
The Netherlands
E-mail: a.p.o.s.vermeeren@io.tudelft.nl

A user interface design case study was conducted to gain first hand experience in involvement of users early at an early stage in the design process. Three methods were used: interviews, comparative user studies of specific features and studies investigating users' work strategies (user strategy studies). The feasibility and usefulness of each of these methods was assessed. The interviews and comparative studies were found to be useful and feasible; the strategy study, however, was expected not to be feasible in practical situations. In addition, it yielded less useful outcomes. This seemed to be caused mainly by the low fidelity of the prototype. It is argued that psychological theories can contribute to a better understanding of what constitutes a 'good' prototype for useful user studies.

INTRODUCTION

In the recent literature on human-computer interface design much attention is paid to methods that involve users at an early stage in the design process (see for example Bødker & Grønbæck, 1991; Grudin, 1991). In practice, however, many designers still experience considerable problems in eliciting information from users (Grudin, 1991; Bekker & Vermeeren, 1993).

Our goal was to get hands-on experience and insight into the problems and possibilities of user involvement early in the design process. For our case study, we chose a realistic design problem, a new user interface for making tables in a word processor. We decided to involve users in the design process by using three different methods. We hoped that using the methods would aid the design decisions and thus contribute to the design process.

First, the three methods that we used will be described, followed by a description of our experiences regarding their usefulness and feasibility in the design process. Second, we will use psychological theories to explain the problems with one of the methods.

Design objective

Our design objective was to develop a simple and intuitive user interface, called TableMaker, for making tables in Apple Macintosh word processing applications. The target users were primarily secretaries, researchers, students and educators, who use word processors regularly for writing letters and articles, preparing reports, lecture notes, etc. Figure 1 depicts a typical table that can be made with TableMaker. An important constraint we had in developing the interface was that the design specifications had to be developed within a period of six months (the design process is described in more detail by Kolli (1992)). The development team consisted of one designer and one human factors specialist.

THREE METHODS OF INVOLVING USERS

At the very start of the design process prospective users were interviewed. As soon as some kind of preliminary design was available, we used two other methods to get feedback from users: (a) in a series of 'comparative study' sessions users were asked to perform a number of simple isolated tasks; (b) in a series of 'strategy study' sessions users' work strategies were investigated by making users perform some more or less realistic tasks. The designer was present with the human factors specialist during all the sessions.

Figure 1. Sample table that illustrates some of the features incorporated in TableMaker.

Involving users in early design phases		
study # 1	?	**Interviews** How do users make tables now? What features do they need?
study # 2		**Comparative study** Which design solution is preferred? How do their performances compare?
study # 3		**Strategy study** Which operations do users choose? What problems do they have with the interface?

Figure 2. Illustration of a TableMaker feature used in the interviews.

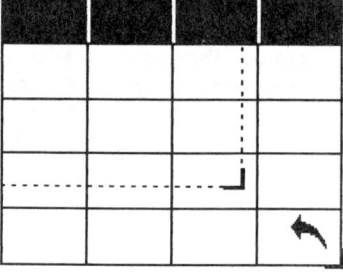

Re-sizing the table

Interviews

The objective of the interviews was to find out how prospective users make tables in their current work situation, how they structure their table making tasks and what problems they have with making tables. We wanted to find out what features are required in TableMaker programs.

We interviewed 11 subjects from various backgrounds, whom we expected to make tables in their daily work (3 graphic designers, 4 researchers, 1 secretary, 2 secretaries/bookkeepers, 1 student). After some initial questions on the subjects' background, they were asked about the frequency of making tables and the purpose of those tables. They were asked to show a typical table made recently and explain in detail the step-by-step process of construction. In addition, the subjects were asked about the frustrating and good aspects of their current way of making tables.

In the second part of the interview, subjects were confronted with a small number of tables which in complexity and style matched the tables that they usually make. They were asked to describe in detail how they, ideally, would like to make the tables (imagining they had a computer that could do anything they wanted).

In the third part of the interview, subjects were presented with paper sheets containing illustrations of the proposed features of TableMaker (figure 2). The features were explained orally by the experimenters. Subjects were asked to comment on the usefulness of each of the proposed features.

All subjects (but one) were interviewed at their regular working place. Each session lasted about one hour. The interviews were videotaped and notes were taken. The analysis was made by summarizing comments, problems and ideas of all the subjects in a large matrix, which gives an overview of the results.

Comparative study

The objective of the comparative study was to compare alternative design solutions for specific features of the TableMaker interface. These were: selection mechanisms for rows, columns and cells and selection mechanisms for lines or borders of rows, columns and cells. We wanted to gain insight into the relative quality of solutions with respect to efficiency and ease of use.

A prototype was developed in HyperCard 2.0 with a series of abstracted tables of different sizes and shapes (figure 3). A total of 16 subjects (Industrial Design Engineering students) was divided into two groups of eight subjects. Subjects were asked to make tables on the screen identical to sample tables given to them on paper. Depending upon the group to which the subjects belonged, they were asked to use simulations of, either two design solutions for the task of selecting columns, rows and cells and applying background patterns, or two design solutions for the task of selecting lines (or line segments) and applying a line thickness to the selected lines. For both tasks, the order in which the design solutions were presented, was balanced between the subjects. In total, every subject carried out at least 15 different tasks with each design solution.

A data logger was built in the prototype to record the sequence of actions and to time them in order to be able to analyse the sessions in detail later on. Building the prototype took about one week. Each session lasted about one hour.

Figure 3. Sample screen of the prototype used in the comparative study

Figure 4. Sample screen of the prototype used in the strategy study

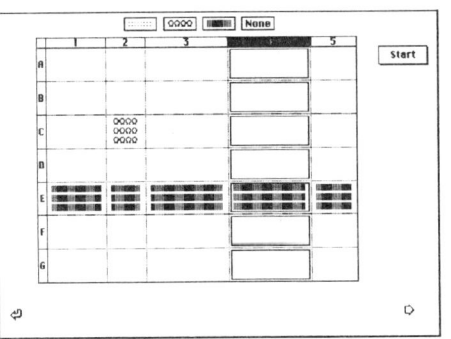

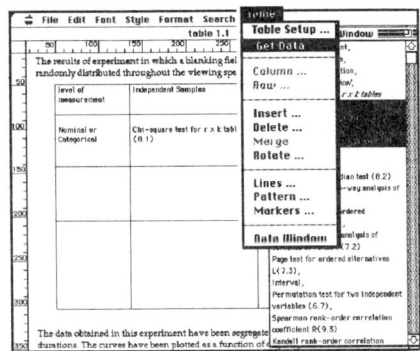

Strategy study

The objective of the strategy study was to find problems concerning the way in which the user interface accommodates the various strategies users adopt in creating new tables or editing existing tables (e.g., starting with a default number of rows and columns and then expanding the table, or counting the number of rows and columns and then directly making a table of the right size).

The prototype was built using HyperCard 2.0. An extra menu called 'Table' was added to a program that looked like a word processor (figure 4). Only those menu commands necessary for making or editing a table and entering or importing text were made active. The other menu commands and dialogues for changing text style, adding graphics etc. were available for browsing but were deliberately made inactive . Some features were temporarily modified for the sake of prototyping due to limitations of HyperCard 2.0 and programming skills of the designer. It took about four weeks to build the prototype.

Five subjects (secretaries who normally work with MS-Word on an Apple Macintosh) worked with the prototype to create and edit sample tables given to them on paper. Each subject session lasted about one hour and a half. During the sessions the computer screen was videotaped. The videotapes were not analysed during the design process, however, as this was thought to be too time-consuming and was not expected to reveal much extra information as compared to the information the designer would obtain by just being present at the sessions.

RESULTS, EXPERIENCES AND DISCUSSION

In the interviews subjects were clearly able to articulate their requirements for TableMaker. This provided us with a number of realistic ideas for new system features such as a 'marking tool' for graphically highlighting important data items in a table. Also, the results showed that various groups of prospective users adopt radically different strategies in making tables. These strategies determined the basic design concepts for the interface. For the interviews we used no special expertise or equipment (besides a video camera). We found that conducting the interviews was (both time and costwise) a very effective way of determining essential features of the application.

The comparative study also showed clear results. Both for the task of selecting columns and rows and for the task of selecting lines, the solutions that incorporated direct selection were preferred by most subjects. However, in some specific instances the other solutions (selection of columns and rows through keyboard commands and selection of lines as borders of cells, columns or rows) were considered more useful. Hence, the final design was based mainly on the 'direct selection' solutions, but incorporated aspects of the other solutions as well.

Prototyping was easy and prototype response times were short enough so as not to hinder task performance. Absence of features other than the studied features made it easy to build the prototype. The original plan of comparing the solutions on the basis of a careful analysis of the logged data was considered to be too time-consuming; just observing the subjects and talking to them proved to be sufficient for the designer to get useful information about the quality and effectiveness of solutions. Also, for both design solutions (for line selection and for column and row selection) the designer felt that the number of subjects in the study could have been smaller. The additional value of observing another subject after having seen already five was perceived as being very low.

The strategy study showed us that in general, users did not have any problems in comprehending the various features and the functioning of the interface. The main objective of this study, however, was to find possible problems in the interface with respect to the various strategies users might want to adopt in making tables. We found that in many instances subjects chose (and stuck to) suboptimal strategies and we had clear indications that simulation artifacts might have been an important cause for this. For example, dragging a line to change the width of a column was found to be cumbersome in the prototype as it required very precise positioning of the cursor in selecting the line. One subject mentioned that she used that feature as much as possible because she found it a challenge to try and 'grab' the line.

In order to find out what aspects of the prototype might have distorted the results, we searched for theories that could tell us which are the critical aspects that make an application a good tool. We assumed that if those aspects were prototyped realistically, the prototype would have been good enough for evaluation of the application with regard to the most important usability aspects.

We used the human activity theory described by Bødker (1991) in combination with theories and principles summarized by Hammond (1987). The key aspects of the human activity theory can be summarized as follows (based on Bødker (1991)): To conduct a certain activity, a person has a repertoire of *operations* that are applied in conscious *actions*. *Operations* are thought to be triggered by material conditions (and are not chosen consciously), while *actions* take place with specific intentions and are conducted consciously. Through learning users can transform actions into operations (this is called *operationalisation*). In some conditions a user may have to reflect on them consciously again (like in cases where something in the interaction goes wrong unexpectedly). In such cases, former operations are *conceptualised* once more into conscious actions. An important role of the user interface now is to support the users' *development and use of operations* toward all the objects (in our case the tables) and artifacts (TableMaker) of an activity (table making). Hammond (1987) summarizes principles and theories of skill acquisition and determines conditions influencing the extent to which a sequence of dialogue transactions will become automated (i.e., operationalised). He also describes user performance characteristics of transactions performed as *actions* and of transactions performed as *operations*.

From these theories, a list of design issues can be derived that are critical for the functioning of a user interface as a proper tool. On the basis of that list, a list of key components of a system can be deduced. We assume that those key components are the components that have to be prototyped realistically (this way the prototype can also function as a proper tool). The list of design issues derived from the theory described above include: (a) the repertoire of transactions that can be used in the task, (b) response times for required feedback (feedback that is necessary for transactions, as opposed to redundant feedback), (c) transactions that require very precise perceptual analysis or motor control, (d) the number of transactions (and redundant transactions) in a sequence of transactions and (e) default settings in the user interface.

Now, if we investigate the prototypes used in the comparative and strategy study, we find that the prototype used in the comparative study simulated the key components of the system quite realistically. The prototype we used in the strategy study, however, did not match this requirement in a number of ways: (a) some actions that were important in the task, did not have the required functionality; the repertoire of transactions for the task was not complete (e.g., deleting and inserting columns was only possible at the right and bottom side of a table, whereas in the actual design it should be possible to insert or delete columns anywhere in the table); (b) concerning the response time for required feedback there were a number of situations in which the prototype was too slow: adding a column while typing text took too long; in case of dragging a line it took too long before the cursor changed its form to indicate that the system was ready for dragging and it also took too long for the line to start following the cursor; (c) selecting a line for dragging required very precise positioning of the cursor; (d) changing the number of rows and columns by using the insert or delete commands required more transactions than in the actual design; (e) in some situations defaults were used, in other situations this was not the case; also, defaults turned out not to be the optimal ones for efficient use. We assume that the differences between the prototype and the intended design as described above were so important and affected subjects' strategy choices to such an extent that the study lost much of its usefulness.

Most of the differences between prototype and design described above find their origin in the limitations of the prototyping tool available (HyperCard).

CONCLUSIONS

We conclude that conducting interviews with potential users can be an extremely useful, both time and costwise, effective way of investigating users' current way of working. It can provide designers with fresh and relevant ideas for features to be incorporated into a new application and with requirements for the application as a whole as well as for the various features to be incorporated. We believe that the presence of the designer during the interviews is crucial for its usefulness.

A method such as the one we used in the comparative study can be useful and feasible for investigating the use of specific interaction features. We found that in the early phases of design, a small number of subjects (e.g., five subjects) can be enough to get a clear picture of what the possible problems of an interaction feature are and how people will use it. This is in line with research findings described by Virzi (1992). Observation of subjects using a particular feature (either in real-life or on a videotape) may provide the designer with enough information to continue with the design process confidently. A formal analysis of the data does not seem to be necessary and may not always be feasible in practical situations.

We expect the method which we used in the strategy study, not to be feasible in practical situations. In addition, the usefulness of the outcomes of the method is limited. Better prototypes are required to improve the usefulness of the method. Creating such prototypes, however, requires new, more powerful and easy to use prototyping tools. In the absence of such tools, other methods might be considered; for example, methods that make use of a number of prototypes, each one specifically made for studying one of the work strategies that can be adopted. Such prototypes will be much easier to construct. Subjects can then get acquainted with the various prototypes and the usability of the prototypes with regard to the strategies can be discussed in interviews.

At a more abstract level, we conclude that psychological theories like the human activity theory (Bødker, 1991) might contribute to a better understanding of what constitutes a 'good' prototype. The system components that relevant psychological theories consider as the key components of an interface may be the same components that have to be prototyped realistically on order to conduct useful (early) user studies. Future research will be aimed at investigating this assumption.

REFERENCES

Bekker M.M., and Vermeeren A.P.O.S., 1993, Developing user interface design tools: an analysis of interface design practice. Proceedings of the Ergonomics Society's 1993 Annual Conference, ed. E.J. Lovesey (London: Taylor & Francis).

Bødker S., 1991, Through the interface-a Human Activity Approach to User Interface (Hillsdale, NJ: Lawrence Erlbaum)

Bødker S., Grønbæck K., 1991, Cooperative prototyping: users and designers in mutual activity. Int. J. Man-Machine Studies, 34, 453-478.

Grudin J., 1991, Systematic sources of suboptimal interface design in large product development organizations. Human-Computer Interaction, vol. 6, 147-196.

Hammond N., 1987, Principles from the psychology of skill acquisition. In Applying Cognitive Psychology to User-interface Design eds. M.M. Gardiner & B Christie (Chichester: John Wiley), chapter 6, pp. 163-188.

Kolli R, 1992, TableMaker, an ergonomic user interface for making tables. In Proceedings of the EUC'92 Conference (Brussels: SA Apple Computer NV), pp. 91-98.

Virzi R.A., 1992, Refining the test phase of usability evaluation: How many subjects is enough? Human Factors, 34 (4), pp. 457-468.

Computer assisted tasks

A Computer Based Tool for Accessing Anthropometric Databases

N.I.Beagley, R.A.Haslam & K.C.Parsons

Department of Human Sciences,
University of Technology,
Loughborough,
Leics. LE11 3TU

Based on the requirement of the Vehicle Design and
Systems section of the Army Personnel Research
Establishment for improved access to their
anthropometric database, a computer based tool has
been developed to assist data retrieval. The
utilisation of the improved graphical capability of
desktop computers using a Macintosh based "Hyper-
media" environment, presented the potential for novel
approaches to the presentation of information, as
compared with traditional database query systems. The
opportunities and pitfalls of this extended freedom in
interface design as encountered in the design of this
anthropometry tool are discussed.

INTRODUCTION

In response to the requirement of the Vehicle Design and
Systems section of the Army Personnel Research Establishment for
enhanced access to the anthropometric data required for their work
in the area of Vehicle Ergonomics, a project was initiated to
develop a bespoke computer based tool to assist the section in
their work. The scope of the project dictated a modular structure
for the information system, based on prioritised information
categories (Beagley,Haslam,Parsons 1992). This paper discusses the
design, development and implementation of the Anthropometry module
of the D.A.V.E. system (Database Application in Vehicle
Ergonomics).

BACKGROUND

Work on the development of this tool has been undertaken by a single developer working amongst the eventual user group. The approach to the system's development is based upon a user driven evolution of the system through iterative prototyping.

This work has been undertaken using SuperCard on the Macintosh computer. This is a high level development environment which provides a method for non-programmers to create bespoke hypermedia applications. Continuing advancements in computer technology has provided affordable computers, capable of providing the processing levels required to adequately support high level programming code and colour output on a large screen. This opens the door to a large number of potential developers who now have a greater range of choices in the design of their interface.

As a result of technological growth there is an inevitable lag in the standards and guidelines needed to support the decisions these designers must make in the development of their systems. The incremental prototyping approach to in-house system development reinforces system design and provides a method of identifying problems involved in interface design which are, as yet, undocumented.

INFORMATION SELECTION

An important component of system design is to ensure that the information to be contained in the system is actually required by the user. Redundant information inevitably impedes user's access to the information they require.

This system was to be designed primarily for the use of the Vehicle Design Section. As a result it was possible to restrict the spread of data to be included in the initial version, to that which was specifically relevant in the work of the section. It was decided that the primary source of data would be the Anthropometric Survey of the British Army (Gooderson 1982). This however contains measures such as skinfold thickness which were unlikely to be required during the normal course of work of the group. Similarly, surveys of certain populations have more bearing on the work of the group than others. Through the application of a questionnaire to the members of the user group, a list of required measures and populations was compiled. The motivation for determining relevant data was to enhance the future user interface through the removal of redundant information. Whilst the Anthropometry module has been designed around data appropriate to vehicle design, the database itself was left intact to allow expansion of the interface to include omitted measures, should a user requirement for these measures be seen.

SYSTEM INTERFACE

The combination of Hyper-media development software and computer support for large screen (19"+) colour graphics expands the potential of system interface design. Command line interfaces appeared as a consequence of the limitations in the ability of

early machines to display information. In the absence of a true
natural language interface the command line interface requirements
for a user knowledge of commands and syntax acted as an obstacle
to many potential computer users. The move to support full screen
interfaces whilst not displacing the command line interface did
open the door to the development of graphical interfaces. These
interfaces, in the form of the WIMP interface of Windows and the
Macintosh Operating System are set to dominate the computer market
by opening it up to the large population of potential users.
Hardware development has not stopped there. In order to safeguard
the usable lifetime of a new system the developer should review
hardware trends and consider employing recently released output
devices for which guidance in usage is unavailable. The diversity
of computer output devices presently available raises numerous
questions for the interface design of future computer tools.

The 19" colour monitor output format was chosen for the
development of the D.A.V.E. system in preference to a more usual
13" screen due to the potential for presenting, either more
information on screen, or the same amount of information more
clearly.

The appearance and function of the interface was influenced by
the following principles in addition to more specific interface
design guidelines.

1. To minimise the depth of the system (i.e. Reduce the number
of levels through which the user is required to navigate in order
to find the required information)

2. To provide an intuitive interface through the adoption of a
3D metaphor for the appearance and function of the interface
components.

3. To provide the user with "on-line" help through
demonstration and/or description depending on the user's
preference.

4. To tailor the information specifically to the user group,
thereby avoiding redundant information where possible.

5. To maintain consistency within the interface in order to
provide a blueprint for future modules.

In the design of the Anthropometry module's interface it was
decided to use the extra screen space to present more information
than would have been legible on a 13" screen. This decision was
based on the aim of minimising the system's depth as a trade off
against potential screen clutter. By reducing the depth of the
system the goal is to simplify navigation within the system and
provide rapid access to the relevant information.

Fig. 1 The Anthropometry Module Interface

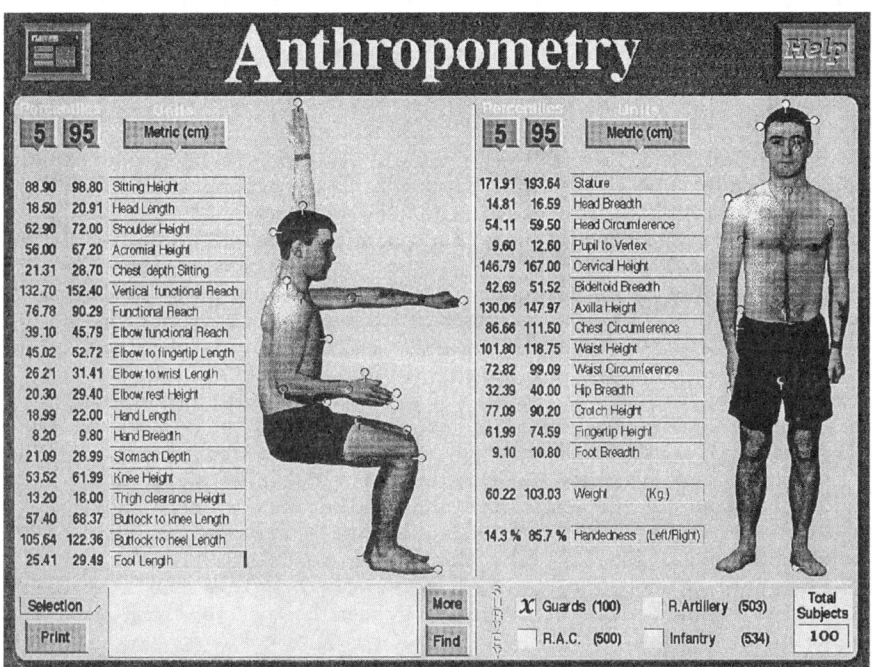

It was found that the 5th & 95th percentile measures for populations were most regularly required in the work of the section. Therefore an attempt was made to present these values for all the measures the group specified on the first screen. Consequently the user is only required to interact with the system when they require data different from that normally used. In order to assist the visual search through the 35 measures presented on screen, a pictorial index was devised with which the user can highlight the list dimension by pointing to the measure of interest on a representative photograph. The concept of on-line help was expanded through the provision of a demonstration option for the user in addition to the usual descriptive paragraph. The help menu is revealed by pressing the help button which will be located in the top right of the screen of all D.A.V.E. modules. The descriptive paragraph is displayed when the user passes the cursor over the function's description in the menu. If the user clicks on the descriptive line of the menu, the actions required to complete the queried function are demonstrated on screen in conjunction with a verbal description of what is being carried out. This mimics the way an experienced user might demonstrate a function to a naive user. The parallel verbal description helps to set an acceptable pace to the demonstration.

Guidelines and standards were unavailable to support all the decisions made in these design features of the interface. As a result, it has been necessary to base any measure of acceptability on the ability of the group to use the system. This usability assessment is integral to the development process due to the incremental prototyping approach taken.

INCREMENTAL PROTOTYPING

In order to allow maximal user involvement in the specification and design of the system the approach to development has been through incremental prototyping. Using this technique, the designer's early concepts of the system are quickly cobbled into a semi-functional program which will eventually evolve into the "final" product. The designer is able to use an early prototype to test whether the design is capable of meeting the functional requirements of the system, as well as conforming to interface guidelines and standards. At this stage, evaluation of the prototype by experts in Human Factors can provide important criticism concerning usability considerations.

Informal involvement of the members of the user group throughout the system's development provides them with an understanding of the abilities and limitations of the computer system. From this they were able to form a stronger concept of what they would ideally like the system to do for them. Incremental prototyping using Hyper-media provides the flexibility to allow the specification of the system to evolve along with the developing system.

Using the incremental approach, the limited resources available for developing the system can be channelled into development of the system's functionality or interface, depending on which factors are highlighted in the evaluation of the previous prototype.

EVALUATION

In addition to the informal evaluation of the evolving system carried out through user group involvement in the module's design, the module has been been formally evaluated at two points in it's development life, to date. An early prototype was evaluated by a group of Human Factors experts who commented on the functionality and interface of the prototype. Although possessing the same skills as the developer it was hoped that they could highlight problems missed by the developer as a result of his close involvement with the development process. Problems highlighted at this stage included interface consistency, measure list grouping, omitted measures, etc. The criticisms received were either countered by a rationale for the approach taken or implemented as far as possible within the development time scale. The iterated prototype was then enhanced to full functionality for formal presentation to the user group.

The user group, whilst encouraged to use the system, were not required to undertake a tutorial in the operation of the system before it's evaluation. Instead, as a method of introduction and

evaluation they were given a task sheet to be completed in their
own time and at their own pace. The tasks required the users to
find numerical values for anthropometric questions similar to
those encountered in the section's work. The task sheets tested
whether each individual was able to make efficient use of the
system's functionality with time taken and accuracy as indicators
of the usability of the system. The aim of the interface is that
it is ultimately to be intuitive. To test this, the evaluation did
not follow instruction in the use of the system but rather each
individual's approach to retrieving information from a basically
unfamiliar interface. In addition to the task sheet results,
background logging of the user's actions provided a picture of the
different approaches taken by the members of the group to
interrogate the system. By highlighting specific difficulties
encountered by the users it has been possible isolate usability
considerations which may be peculiar to this restricted user
group. By tackling these group specific problems it is hoped to
ensure the suitability of the system to provide the information
required by the section.

CONCLUSION

By taking a user centred approach to development through the
technique of iterative prototyping, it has been possible to
develop a novel interface for the access of anthropometric data.
The success of this technique is dependent on the ability to
quickly and easily remodel the program to meet the iterative
requirements of the user group.

REFERENCES

Beagley, N.I. and Haslam, R.A. and Parsons, K.C., 1992, Designing
an Information System for Experts in Vehicle Ergonomics, In
Contemporary Ergonomics 1992 (London: Taylor & Francis) pp

Def.Stan. 00-25 (Part 7), 1986, Human Factors For Designers of
Equipment: Visual Displays

MIL-STD-1472D, 1989, User-Computer Interface, Chapter 5.15

Maddix, F., 1990, Human-Computer Interaction Theory and Practice
(Ellis Horwood)

NASA-STD-3000/Vol./REV.A, 1989, User-Computer Interaction Design
Considerations, Chapter 9.6

Gooderson, C.Y., 1982, Final Report: The APRE Anthropometric
Survey of the British Army 1972-1977, Unpublished MoD Report

This work was carried out with the support of the Army Personnel
Research Establishment, Ministry of Defence.

CRAMP - THE DEVELOPMENT OF A GRAPHICAL USER INTERFACE

Martin Bontoft, Ergonomics Consultant
Jane Dillon, Ergonomics Manager, Royal Mail Research and Development

This paper describes the development of a Graphical User Interface (GUI) for CRAMP, the software that amends the automated sorting and routing of mail in the United Kingdom. The design development of the system placed emphasis on the involvement of users. Substantial ergonomics research was conducted before the technical specification was drafted. A GUI was specified with the aim of allowing users to develop sorting information easily, intuitively and with minimal training or documentation support.

Paper prototypes of screens were followed by two iterations of testing with software prototypes. By these means, Royal Mail was able to develop a simple and usable interface to a complex and important system.

INTRODUCTION

This paper describes a case study of the development of a graphical user interface. The application is software that produces and amends the plans which are needed to operate automatic sorting machines within Royal Mail. Without sort plans sorting machines would not work. In simple terms, sort plans contain information about which letter goes in which box on a sorting machine. It is essentially a database, application software and a sophisticated human computer interface. The new system is called "Computerised Routing for Automated Mail Processing", (CRAMP).

The development process can be characterised as "user-centred". At each stage of the development the involvement of users was actively sought. Two phases of prototype testing were undertaken; the results of these were fed back into the design baselines.

BACKGROUND

Royal Mail is in the process of acquiring new high speed sorting machines designed to improve its mail processing capability. The current method of amending sorting plans requires each processing centre to embark on a complex process of form filling. These forms are later despatched for central data input and processing.

This system is slow and error prone and would not be able to support the demands of the new generation of sorting equipment.

It was recognised that data manipulation, not data entry, should be the major user activity. This should be carried out locally—where people have the necessary local information, where data integrity is of paramount importance, and where the control over mail processing needs to be exercised.

A central facility does, however, act as a repository of information: a contact point where procedures can be explained and questions answered. Decentralising control over sorting plans would mean that users would not be able to rely on such expertise. It was therefore apparent that the new system must be inherently easy to use and the development of a usable Human Computer Interface (HCI) became priority in the early stages of the project.

INTERFACE DEVELOPMENT

CRAMP can be considered to be constructed of two parts—the data processing engine and the user interface. Traditionally, in the development of such systems, all importance is placed on the data processing engine and user interfaces have been subservient to its requirements.

For an application like CRAMP, with usage distributed among a large number of sites and naive computer users forming the bulk of the user population, the user interface becomes a significant factor. Indeed, it became clear that if either the data processing engine or the user interface is less than adequate, there would be a high probability that CRAMP will fail.

The model chosen for the CRAMP HCI development is somewhat informal, though nonetheless uses well-proven and established techniques. It is characterised as being "user-centred" and may be represented by the diagram below (see Figure 1).

The model has the advantages of being iterative and comprehensive, while remaining flexible enough to fit into a tight development schedule. Many of the techniques run in parallel with the development of entity relationships and data flow diagrams. As such, they helped forge collaborative links between the database developers and the HCI team.

Fig 1: User centred design cycle (adapted from Alger et al, 1990)

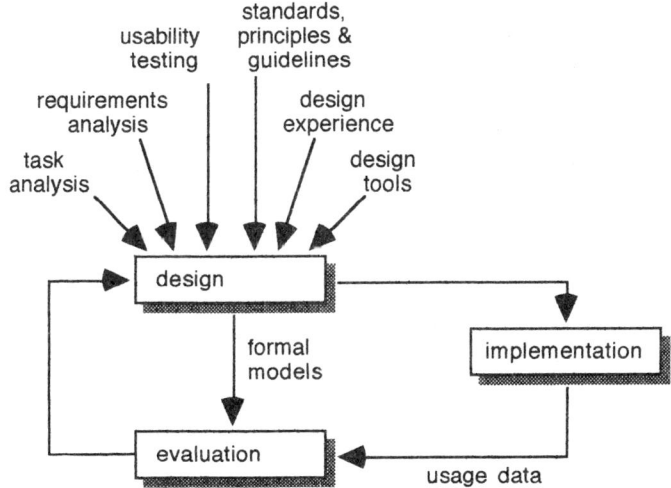

Task analysis

Users of the current system (POSP) were visited and invited to describe the high level tasks that form their jobs.

This was aimed at the elicitation of the users' conceptual model of the system that is served by the tasks they carry out. In other words, how does the user think the system works, what is the users' rationale for their activities? Whilst this was inevitably coloured by the system on which they had gained their experience, it was possible to pull out key factors. These formed the basis of what the eventual HCI should support.

Requirements analysis

Clearly, the database designers have to analyse the users' functional requirements, they must know what their product should do. The HCI team need to know this also—the HCI has to be designed to support that functionality—but they also need to know the users' interface requirements. The interface style must be matched to the users' needs and abilities if the system is to be easy and satisfying to use.

In this particular instance the interface style was easy to decide. The requirements analysis and the task analysis had identified a number of factors that indicated the way forward.

There were clear indications that a simple, visual metaphor could be established between the postcode (ie, letters) and the "pigeon hole" or "box" on the sorting machines. The user might place a postcode in a box, virtually, on a graphics screen. CRAMP's software would then interpret that to mean that all letters with that postcode should be directed to that box.

Usability testing

This premise was tested with users to establish its worth. Paper prototypes of screens were produced from Apple Macintosh screen dumps mounted onto card. These were introduced to users in the form of a game. The rules were explained, and all the constituent parts were provided for the users to be able to construct part of a sorting plan.

The results of this 'quick and dirty' usability testing were encouraging. The design appeared to be intuitive and readily acceptable. The problem then became one of specifying this design basis to the system integrators.

Standards, principles & guidelines

At this time the choice of software and hardware was beginning to be decided. Processing, portability, and compatibility all dictated a UNIX® operating system on a RISC workstation. The decision was made to use the X-Window™ graphics window system and the OSF/Motif™ interface style.

Motif was able to support the design basis that had been developed. Moreover, the style is based on human factors and behavioural research, (OSF, 1991).

Design experience & Design tools

The interface style had been developed and tested, a suitable graphics window system had been specified. The design was turned over to the system integrator; they had the design experience and the necessary tools to construct workable prototypes and, finally, the system itself.

DESIGN, IMPLEMENTATION & EVALUATION

These three areas formed a cycle of events as depicted in figure 1. Royal Mail had requested of the system integrator that they provide two realistic prototypes for user testing. Each of these was delivered, tested with the future users, appraised and given back for revisions.

Prototypes

The aim of the first prototype was to validate the design in general terms. Hence, as much as possible of the final system was prototyped; not all of it was what might be termed 'functional', however. Much of the software was a mock-up of the final screens, only basic functions were available: just enough for navigation purposes.

Users were invited in pairs to the Royal Mail's test facility within the Research & Development Centre. The users spent either a morning or an afternoon in front of the prototype. They were accompanied by a member of Royal Mail's HCI team and a member of the system integrator's team.

Each pair was fully briefed on the purpose of the session. The software was demonstrated to show the capabilities of window systems. The session moved onto task scenarios drawn from the envisaged use of the system. The points that arose were noted for later analysis.

The comments from both sessions were documented and discussed. A value judgement had to be made about some user comments, it was not feasible simply to accept everything they said. The majority of comments were either evidently well-founded or simple to implement. Some, however, were neither and they had to be justified against their assessed impact to the development programme.

The first prototype had proved broadly acceptable, this gave confidence that the design approach was reasonable. It also meant that the second prototype could be targeted at validating the core functionality of the final system.

The procedure for testing this second prototype was exactly the same as for the first. The same users were invited to attend and give their comments.

DISCUSSION

The end result of this procedure was an agreed and documented HCI description. This fully described the objects and processes that together formed the CRAMP application.

The development schedule was tight in order to meet pressing operational requirements. As a consequence, some of the agreed HCI functionality had to be trimmed. The involvement of users during the design development provided a sound basis for deciding which features could be delayed and which were essential for the early implementation. The final mix of features could not have been adequately predicted without user involvement. CRAMP should has been constructed, tested and accepted. The system is evidently acceptable to users.

The user input to this project has been carefully managed to maximise its benefit. It was scheduled to coincide with opportunities for change within the software development programme, and the input was assessed (as was all design input) to ensure it contributed to the quality of the design.

The process of gaining the users' design input must be carefully managed. Users need to be educated about the aims of the system, and led through different test scenarios using the appropriate tools. Indiscriminate and unstructured involvement is likely to lead to requirements that cannot be incorporated; this is not just unproductive, it may actually be counter-productive.

ACKNOWLEDGEMENTS

The assistance of the Royal Mail Data Systems Group and TRW Financial Systems Inc. is gratefully acknowledged.

REFERENCES
Alger et al, 1990, " A Guide to Usability", The Open University, Milton Keynes, UK
OSF, 1991, "OSF/Motif Style Guide", revision 1.1, Prentice Hall, New Jersey, USA

USING TASK NETWORKS TO MODEL ERROR CORRECTION DIALOGUES FOR AUTOMATIC SPEECH RECOGNITION

K.S. HONE and C. BABER

Industrial Ergonomics Group.
School of Manufacturing and
Mechanical Engineering,
University of Birmingham,
Birmingham.
B15 2TT

Automatic speech recognition does not yet allow error free interaction between human and computer. The prevalence of recognition errors is cited as a major impediment to the success of the technology. The design of error tolerant systems has become increasingly important. Despite the range of approaches to error correction, it is often difficult to objectively compare different approaches. This paper reports comparison of four error correction approaches, using task network models. The data obtained compare favourably with those from experiments using human subjects.

INTRODUCTION

Given the present state of automatic speech recognition (ASR) technology, it is clear that the design of error tolerant systems is an important area of study. Furthermore, current technology does not allow for fully automatic detection and correction of errors, so there exists a need for error correction dialogues which remain under human control. While errors can arise in both the device and the user (see Baber et al., 1990), in this paper we assume that user errors have been minimised in order to focus our attention on recognition errors. Of the three categories into which recognition error are generally classified, viz. insertions, classifications, substitutions, the most common form is substitutions (Brown and Vosburgh, 1989). Not only are errors arising from substitutions the most common, but also these present the biggest problem in terms of prevention and correction. This type of error is, therefore, the main focus of efforts to develop error detection and correction strategies.

There are many different strategies that can be adopted to allow user controlled error correction and there is much debate among researchers as to which is the optimum strategy. Recent papers by Murray et al. (1992) and Ainsworth and Pratt (1992) illustrate strategies relying on the ASR devices next best match or strategies relying on user repetition.

Murray et al. (1992) compared three different error correction strategies. Two of the strategies involved user repetition of the original word when a misrecognition occurred, while the third used the nth. choice offered by the device. In both repetition strategies, templates for rejected recognitions were deleted from the 'active set'; in one strategy (template removal with immediate reinstatement - IR) the full template set was reinstated immediately after the next recognition; in the other method (serial removal of templates - SR) templates for rejected recognitions were serially deleted and only reinstated when the user said "o.k.". The nth. choice strategy successively used the device's next best match, when the user said "no" to a misrecognition, until the correct match was found and the user said "yes". Murray et al. (1992) found that the nth. choice technique produced faster transaction times that either IR or SR, while SR performed better than IR in terms of percentage of errors corrected in two attempts. However, due to limitations in the device used and in order to maintain subject confidence in the performance of the device, Murray et al. (1992) did not progress beyond two attempts for each misrecognition. Thus, it would be interesting to see whether the results held when $n > 2$. Furthermore, although they do not report a recognition accuracy, Murray et al. (1992) imply that the task employed produced a very low recognition accuracy. It would be interesting to see whether variations in recognition accuracy affect the results.

Ainsworth and Pratt (1992) also compared an error correction strategy using the device's next best match ("elimination without repetition") with a method involving repetition with serial template removal ("repetition with elimination"). In contrast to the Murray et al. (1992) study, Ainsworth and Pratt (1992) found that the strategy involving repetition produced superior performance to the nth. choice strategy. It is proposed that the differences between the two studies can be explained to a large extent by such factors as vocabulary used, subjects used, experimental conditions, ASR device used and the particular types of errors the devices made. These confounding factors make it difficult to decide which of the error correction strategies can be considered optimal.

The aim of the work reported in this paper is to overcome these problems by using a computer modelling technique to which can allow variables to be held constant across trials of different strategies. A task network modelling technique was used to model the strategies used in Murray et al. (1992) and Ainsworth and Pratt (1992), with the addition of a fourth, 'simple repetition', strategy in which misrecognised words were repeated until recognition was successful.

METHOD

Task networks can be used to model the performance of sequences of events or tasks within a specific segment of activity (Meister, 1990). Each task in the network can be described in terms of a completion time and a probability of progression to a further task. In this study we used the student version, i.e. 25 nodes, of MicroSAINT running on an IBM PS / 2. The four error correction strategies were represented as task networks, and entered into MicroSAINT. Figure one shows the network for IR.

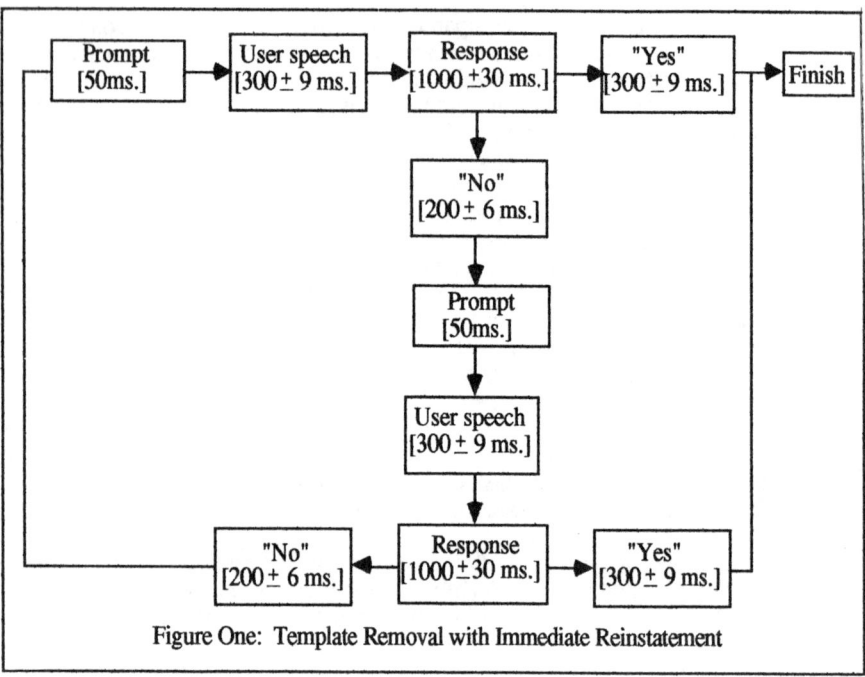

Figure One: Template Removal with Immediate Reinstatement

We assumed that the words to be recognised would be three phonemes long,
with equal confusability, and that there would be ten words in the vocabulary set.
The timings used were derived from a pilot study and based on one phoneme
lasting 100ms. with a 3 ms. standard deviation per phoneme. The probability of
successfully passing between tasks was assumed to be related to recognition
accuracy; so we used different levels of probability to correspond to recognition
accuracy levels of 50%, 85%, 90%, 95% and 99%. Each task model was run 500
times for each recognition accuracy level.

A number of assumptions were used in the derivation of the task network
model. This was primarily the result of using a highly limited number of nodes.
We assumed perfect human performance because it would be difficult to accurately
predict and model errorful performance. The "yes" / "no" responses to prompts
were assumed to be correctly recognised at all times, in order to reduce the
problems of getting caught in loops. The probability of a word being recognised
on repetition was assumed to be the same as its probability of initial utterance, i.e.
we assumed that the 'user' would make no attempt to alter speaking style when
using the device. Further, we assumed that all words had an equal probability of
being recognised, after the initial misrecognition; while this may seem rather
extreme, if we assume that we are using the 'e' set, then the assumption is
conceivable (if not wholly accurate). We assumed that all errors which occurred
would result in substitution and require correction. If a substitution is made then
all the words in the set have an equal chance of being recognised. Thus, the
recognition accuracy should alter with changes in set size. We describe this
assumption using equation one:

$$Nra = 100 - (100 - Pra) . (C - 1) / (V - 1) \qquad \text{Equation one}$$

[where Nra: new recognition accuracy; Pra: previous recognition accuracy; C: new vocabulary size; V: initial vocabulary size].

RESULTS

The results obtained show a variation in mean transaction time as a function of recognition accuracy, as one might expect. Of more interest is the similarity between these data and those obtained by Murray et al. (1992).

Strategy	50%	85%	90%	95%	99%
Repetition	3257.7	1948.1	1817.7	1718.8	1669.1
IR	3109.1	1923.5	1814.7	1718.8	1669.1
SR	2988.5	1923.5	1814.7	1718.8	1669.1
Nth. Choice	2598.7	1858.8	1784.9	1704.9	1657.1

Table One: Mean transaction times for strategies

In all conditions, nth. choice produced the fastest mean transaction time, probably because the is no requirement for users to repeat words in this condition. No differences between the strategies are apparent above a recognition accuracy of 90%. At the lower recognition accuracies, differences start to appear, with simple repetition decreasing in effectiveness and at 50% SR is superior to IR.

We also analysed the results in terms of percentage of errors corrected at each attempt at correction. In the interests of brevity we only present the data for the two template removal strategies, 50% recognition accuracy (see figures two and three).

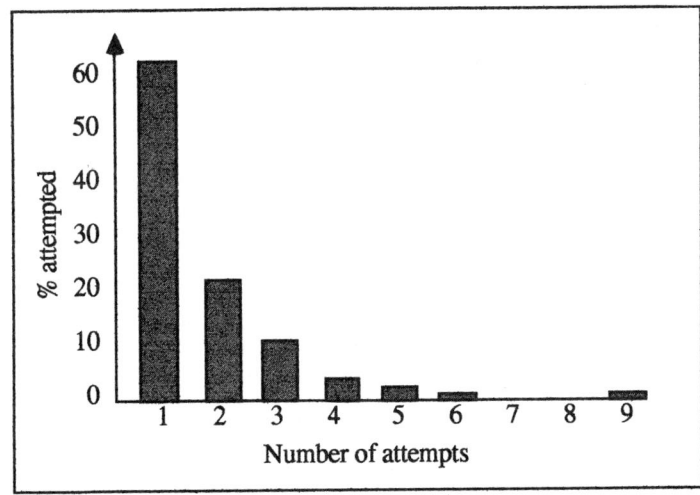

Figure 2: Attempts at error correction for strategy IR

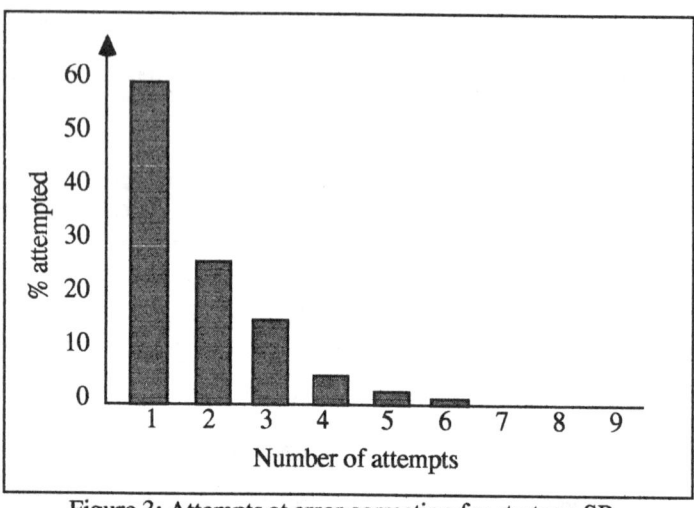

Figure 3: Attempts at error correction for strategy SR

The value of 50% recognition accuracy was chosen in order to compare our data with those of Murray et al. (1992), who stated that their ASR device produced very low recognition accuracy.

In terms of the number of attempts at error correction needed for each strategy, the nth. choice and SR strategies were superior to the other strategies, with more errors corrected using fewer attempts.

CONCLUSIONS
The results of this study have several points in common with those of Murray et al. (1992). The success of the nth. choice strategy, in terms of fast transaction time, was shown in both studies, and appears to offer superior performance by removing the need for users repeating words; an activity which is both time consuming and irritating. The pattern of attempts at error correction are also markedly similar between these data and those of Murray et al. (1992). Comparison between the studies is limited by the use of a ceiling in the Murray et al. (1992) study. However, there were sufficient similarities for us to assume that the modelling technique used in this study can produce valid data.

DISCUSSION
While the primary aim of this study was to validate the use of task network modelling as a means of investigating error correction strategies in ASR, the results produce some useful points which can, we feel, be generalised to a range of ASR applications. The fact that there was little difference in performance at high recognition levels implies that all strategies would be equally effective. Previous work has shown that, when confronted with a misrecognition, many users tend to repeat the word (Baber et al., 1990). Thus, at high levels of recognition accuracy, simple repetition could be sufficient to ensure smooth progression of the interaction. As recognition accuracy decreases, then more involved error correction strategies may be needed. This observation, although intuitively appealing, seems to be somewhat at odds with the speech community; it is often

the case that sophisticated error correction "dialogues" are appended to high performance ASR devices, perhaps to maintain a perception of 'high - tech', perhaps because the designers have an interest in modelling human error recovery. What this study suggests is that the sophisticated error correction in high recognition accuracy systems is redundant. Previous work suggests it will be highly confusing to users.

REFERENCES

Ainsworth, W.A. and Pratt, S.R. (1992) Comparing error correction strategies in speech recognition systems   In Ed. C. Baber and J.M. Noyes **Proceedings Interactive Speech Technology** [Loughborough: Ergonomics Society]

Baber, C., Stammers, R.B. and Usher, D.M. (1990) Error correction requirements in automatic speech recognition   In Ed. E.J. Lovesey **Contemporary Ergonomics 1990** [London: Taylor and Francis]

Brown, N. R. and Vosburgh, A. m. (1989) Evaluating the accuracy of a large vocabulary speech recognition system **Human Factors Society 33rd. Annual Meeting** [Santa Monica, CA: Human Factors Society]

Meister, D. (1990) Simulation and modelling   In Ed. J.R. Wilson and E. N .Corlett **Evaluation of Human Work** [London: Taylor and Francis]

Murray, A., Frankish, C.F. and Jones, D.M. (1992) Data entry by voice: facilitating the correction of misrecognitions   In Ed. C. Baber and J.M. Noyes **Proceedings Interactive Speech Technology** [Loughborough: Ergonomics Society]

Auditory warnings

ALARMS IN A CORONARY CARE UNIT

Neville Stanton

Human Reliability Associates Ltd
1 School House
Higher Lane
Dalton
Wigan
Lancs
WN8 7RP

This paper presents the findings of a study that investigates alarm systems in a coronary care unit. Data collected over a three days of observations suggest that the design of such systems have severe shortcomings in ergonomic terms. Explanations of the shortcomings and proposed solutions are given.

Alarm handling in the CCU

Alarm handling forms a small part of the nursing staff duties which are mainly connected with patient care. These duties include assistance with washing and dressing, administration of drugs, arranging meals to be delivered, preparing patients for theatre, keeping patients and family informed of their progress, and generally seeing to the patients' needs. Although the staff/patient ratio is quite high (normally three nursing staff in the Coronary Care Unit (CCU) for each shift) the patients' status means that they are kept busy, particularly if one of the patient's condition deteriorates. The monitoring systems mean that the nursing staff are relieved of monitoring each patient continuously. The alarm systems acts as an interruption and call for attention if any of the parameters falls outside tolerance.

Alarm systems in the CCU

This paper considers the design of alarms in a CCU. Alarms take two forms in the CCU: auditory alarms and auditory-visual alarms. The syringe pumps and blood pressure meter are alarmed by auditory media, whereas the patient's heart rate is alarmed by auditory and visual media via the ECG monitor. In addition there is a panic button, to call for assistance, which initiates an audible alarm and a red light outside the CCU.

Different types of equipment are used within the CCU, some of it performing the same type of function, i.e. the different types of syringe pump. The ECG monitoring system, manufactured by Hewlett Packard, displays each patients ECG with a 30 second trace via a VDU. This information is accompanied by the patient ward and bed number, heart rate (beats per minute), and alarm status. There were two VDUs in the CCU and four beds, two either side of the VDUs. The VDUs also displayed the ECGs of some patients in the cardiac wards.

Brief description of the study

The study was undertaken over three working days, from 9.00 a.m. to 4.18 p.m. This was to determine if the data collected had been representative of a 'typical' day in the CCU. Whilst it is accepted that all three of these days may be atypical, it could be viewed as a useful exercise to see if they were similar. The approach taken and detailed findings are presented in Stanton (1992).

Main findings

The three days of observation could be considered homogeneous as there were no statistical differences in the number of alarms that were presented. This provides a basis for confidence in interpreting the other data. It was reported that the ECG alarms could be linked to patient activity which accounted for 37.6% of all the alarms. This figure would have undoubtedly been higher if the facility of suspending alarms and changing thresholds had not been available. Therefore this degree of control needs to be regarded as an important feature of the alarm system and should be included in future systems design.

Very few of the signals were identified as 'alarms', i.e. satisfying all of the criteria proposed in that they:

* attract attention;
* were not predicted;
* called for intervention, were inaction would have been detrimental patient.

Between two and five percent of the signal satisfied all three criteria. All of the signals attracted attention. Most of the signal were not predicted 'a priori', but they were not surprise events at the same time in that their presentation could be explained by the circumstance of their onset, for example the patient moving. To have predicted the alarm, one would have to predict the degree of patient movement that would be sufficient to trigger the alarm. Approximately twenty seven percent of the signals resulted in some level of intervention, but most of these would not have led to any deterioration of the patient if the intervention had not taken place. Two thirds of the interventions were checking activities. In this respect the alarms may be seen as useful in prompting the nursing staff in checking the patients and the monitoring system. However, this needs to be carefully balanced as the alarm system should not falsely call for attention too frequently if it is to be trusted.

Alarm presentation is characterised by busy periods followed by intervals of relative calm. During the busy periods the nursing staff are put into a 'fire-fighting' role, dealing with the high priority demands. Whereas in the calmer periods they are able to carry out their normal nursing duties.

Design considerations

The ECG monitor has some shortcomings in ergonomic terms and could benefit from redesign. There were five main problems that were recognised in the observational studies.

(i.) Multiple Alarms

There could potentially be problems if there was more than one alarm at a time on the same VDU. This is because the audible call and the 'red/amber/green' light system may only draw attention initially to one patient. However, it is recognised that the likelihood of this ever occurring is very small.

(ii.) Suspending Alarms

It was also noticed that if the staff suspended alarms on a particular patient, the "ALARMS SUSPENDED" tag covered the patients heart rate. This is of particular concern because of the recognition that the nursing staff are often able to use the heart rate as an indication of potential problems, which may be the very reason that the alarms are suspended.

(iii.) Identifying Patient

The nursing staff had obviously had difficulties in the past in identifying the bed of the patient that related to the ECG presented on the VDU. Evidence of this lay in the adhesive label stuck to the bottom of the VDU that indicted the bed position of patients. A possible means of resolving this is to configure the screen layout in a more meaningful way and to present some means of identifying the bed of the patient the alarm refers to.

(iv.) Monitoring Patients in Other Wards

Another problem relates to the difficulties of being unable to see the patients in other wards that are being monitored in the CCU. In one example the nurse noticed that a patient had 'flat lined'. This could have been a very urgent situation. The first course of action was to telephone the ward and ask the nurse to check the patient. Given the serious nature and possible consequences of this incident, this kind of procedure could waste valuable seconds in recovery time. The provision of CCTV (closed circuit television) could significantly reduce the time taken to ascertain the status of the patient as well as reducing the incidence of false alarms.

(v.) Prioritisation

Finally the system attempted to provide a prioritised auditory warning, yet there was no evidence from the observations that this was warranted. The purpose of the auditory system is to attract the attention of the nurses. There did not appear to be any useful additional information provided by the urgency rating. Very few of the alarms that were categorised as high urgency actually result in any action. Therefore the merit of this auditory distinction must be questioned. Once attention has been drawn to the event, the maintenance of a visual distinction between the different types of event may help in determining the appropriate type of response, if any is to be taken.

Syringe pumps

The syringe pumps also have shortcomings in ergonomic terms as was highlighted in the introduction. The Welmed goes a long way to resolving these problems. However two problems still remain: identifying the source of the alarm and integrating the alarms. During the observations it was noticed that the nurses did not always immediately recognise which bed the syringe alarm was coming from. This is a problem of an omnidirectional signal in the absence of a clear visual indication. Although there was a red LED on every syringe pump this was far too small to be seen from a position at any distance away from the bed. It is clear that the auditory signal needs to be accompanied by some clear visual reference to the bed that it is emitting from. As was also indicated in table 1 in the introduction there are a variety of syringe pumps which appear to be largely set up for the patient based on availability than any other criteria. This results in a variety of different audible signals, all which essentially carry the same message, i.e. "the syringe is running out". Human factors principles call for consistency in the transmission of information, clearly then this elementary guideline is broken.

Blood pressure
   The blood pressure monitor has the same shortcomings in ergonomic terms as the syringe pumps in identifying the source of the auditory signal. However, there may only ever be one patient in the ward whose blood pressure is being monitored, so this is less of a problem than that identified on the syringe pumps.

Panic
   Finally the panic alarm has some shortcomings in ergonomic terms and could benefit from redesign. Although there is both auditory and visual alarm outside the ward there is only visual indication within the ward of which patient the alarm refers to. However the procedure of the nurse staying with the patient is designed to overcome the shortfall in the alarm system.
From the shortcomings in the alarm system is is possible to offer a suggestion of how future systems might offer substantial benefits.

Audible distinction
   In principle at least the maintenance of some audible distinction between calls from the panic alarm, ECG monitor, syringe pumps, and blood pressure monitor is not undesirable. It allows for the nursing staff to make some preliminary assessment of what is calling for their attention. However, it is desirable that some effort be put into an integrated approach for presenting this information. A methodology for designing auditory symbolic information has been proposed elsewhere (Stanton, 1993). Such that the auditory channel could contain 'symbolic' reference to the source of reference.

Integration
   The visual alarm system could be integrated also. For instance, each bed could have a panel above each patient similar to that suggested in figure 1. This would not only indicate the type of alarm, but the source also. More detailed information could be provided at the bedside or central monitoring unit if necessary.

Assistance	ECG	Syringe pump	Blood pressure

Figure 1. Proposed visual alarm unit.

Conclusions
   Given that only a small percentage of signals require intervention, it must be difficult for the staff to treat each alarm as if it were an emergency. In some cases the protocol was to telephone the ward and request a nurse to check the patient. In

a real emergency this procedure might waste important seconds that could be devoted to recovery techniques. Further it was noted that the spatial representation of patients on other wards was inadequate for them to be identified from the VDU, and that a sticker system had been developed. Thus the nurses had to translate the screen representation to the ward position before telephoning the ward. Clearly there is the need to reconsider screen layout to mimic ward layout to save time in this translation.

USING PSYCHOPHYSICS TO PREDICT PERCEIVED URGENCY

J EDWORTHY & E J HELLIER

Department of Psychology
University of Plymouth
Drake Circus
Plymouth
DEVON PL4 8AA

The effects of auditory warning speed, frequency, number of units and the degree of inharmonicity on perceived urgency were scaled using an application of Steven's Power Law. Exponents of 1.35, 0.38, 0.50 and 0.12 were obtained respectively, showing that changes in warning speed has a greater effect on perceived urgency than the other parameters tested. The results have a number of applications in the future design and development of auditory warnings, as well as suggesting areas for more theoretical research.

INTRODUCTION

The design and implementation of auditory warnings systems has received a significant amount of attention over the last few years. There are many human factors and ergonomic aspects of such systems requiring attention, but this paper focuses on the issue of perceived urgency of sound, a topic which our research group has researched in detail over the last few years (Hellier and Edworthy (1989); Edworthy and Loxley (1990);   Edworthy et al., (1991)).

In practical terms, one of the prime motivations for carrying out this research is that warnings, and the situations for which they are designed, are poorly matched; this makes them difficult to identify and therefore potentially unergonomic.

There are a number of ways in which this mismatch can be addressed and reduced. One way is to capitalise upon already-learned associations between particular types of sound and particular types of situation. For example, research has shown that listeners tend to associate sirens with danger situations, and to some extent this phenomenon extends across cultures (Lazarus and Hoge (1986); Hoge et al., (1988)). This knowledge has clear application in the area of alarm design and implementation.

In many situations, the receiver of a warning or alarm does not know its meaning. This applies not only in situations and places frequented by the general population, but in areas where warnings are heard regularly by professionally-trained operators. A study which demonstrates this clearly is a recent one by Momtahan et al (1993), for an operating room and an intensive care unit in a Canadian teaching hospital in which 23 and 26 alarms respectively were available at the time of the study. The results of this study showed not only that many warnings were potentially maskable by other warnings and other sounds in the area, but that of the 26 alarms in the operating room, on average only 10 to 15 were recognised; in the intensive care unit, only 9 to 14 of the total of 23 alarms were recognised. Thus operators are regularly put in the position whereby they do not know what an alarm or warning is signalling. One potential improvement to this situation might be to indicate to the listener the degree of urgency of the situation so that at least the urgency or importance of the situation is conveyed through the alarm itself. Under these circumstances, the operator receives some indication of the urgency with which he or she should react direct from the warning itself. Studies show that alarm systems are generally devoid of such 'urgency mapping' (Momtahan & Tansley (1989); Finley & Cohen (1991)). In practice, the urgency of an alarm situation is not generally signalled through its acoustic characteristics, so the operator is not normally able to gauge the urgency of a medical situation on the basis of the acoustic qualities of the warning itself. In situations where the meaning is not known, appropriate urgency mapping may therefore be an advantage. Of course, the operational urgency of some situations will be context-specific, but with increasing use of artificial intelligence in medical monitoring, preliminary screening of potentially dangerous situations can be gauged by computer systems, and alarms produced accordingly.

For urgency mapping to be feasible, however, we need to know the relationship between sound, and judgement of urgency. For this reason the systematic exploration of the relationship between acoustic parameters and perceived urgency has been the main focus of our work, and a set of studies is reported in this paper.

Many acoustic parameters have been shown to affect subjective judgement of urgency (Edworthy et al., (1991)). Increases in the speed, pitch, length and so on of an acoustic stimulus increases the urgency of that stimulus, and other parameters which are not as readily quantifiable as these three can also have quite clear effects on perceived urgency, such as musical structure and rhythm. However, it would also be useful to know the relative strengths of these effects for both practical and more theoretical reasons. In practice, a manufacturer wishing to alter the urgency of a particular alarm, or set of alarms, needs to know not only how an acoustic parameter affects the urgency of a sound, but by how much. If he or she wishes to make a large change in the urgency of a warning, then it is best done so 'economically'. That is, by selecting a parameter for alteration which will have a striking effect on the urgency of the sound - one for which a relatively small change in physical level will have a large effect on the subjective judgement of urgency. More theoretically, the relative strengths of individual parameters might provide insight into the psychological correlates of urgency.

One well established method through which the relationship between objective, physical quantities and subjective judgement can be described is that of Steven's Power Law (1957), where the relationship is expressed in the terms

$$S = kO^m$$

where S is the subjective judgement (in this case, urgency), k is a constant, O is the objective value (the speed, frequency, loudness or other characteristic of an auditory stimulus) and m is the exponent derived from the gradient obtained when the objective and the subjective data are plotted against one another. The larger the value of m, the more powerful is the physical stimulus in its ability to alter urgency. If m is high, then relatively small changes in the physical parameter will produce large changes in urgency; if m is small, then larger changes will be required in the physical parameter values in order to produce a similar change in urgency. The feasibility of using Steven's Power Law as a descriptor in this way has been explored in some of our earlier work (for example, Hellier and Edworthy (1989)). In the paper presented here, we derive exponents for four of the most significant acoustic parameters known to be associated with perceived urgency, and compare their relative strengths.

One approach to the design of ergonomic auditory warnings has been advanced by Patterson (1982). Warnings made according to these guidelines consist of a short pulse of sound, lasting typically about 200ms, with a full harmonic structure and a shaped onset envelope to avoid startle. This pulse of sound is repeated several times to form a warning burst, which may last two seconds or so. The frequencies of each pulse and the time intervals between them may be different from one another. This method of construction lends itself well to the manipulation of perceived urgency, and so the stimuli tested in the studies presented here are constructed in the same way. The four parameters tested in the experiment are warning speed, which is indicated by the inter-pulse interval; frequency, which is indicated by the fundamental frequency of the pulses; number of units, which is the number of times a small unit of sound repeats; and degree of inharmonicity, which is indicated by the number of harmonics in the pulse which are not integer multiples of the fundamental frequency. All of these parameters are known to affect perceived urgency, but the relative strengths of these effects were not previously known. Subjects were played acoustic stimuli based on these four parameters and asked to rate their urgency by matching their subjective judgement to line length. A similar procedure has been adopted by Kuwano and Namba (1990), who asked subjects to match helicopter noise to line length.

METHOD
Subjects
Twelve subjects participated in the study. Each subject was screened for potential hearing loss using a Peters AP250 Audiometer, and none was found to have any such loss. Subjects varied in age from 18 to 50 years.

Apparatus
Stimuli were presented to subjects through a Cambridge 1401 interface and 1701 filters using a Tandon PC. The low-pass filters were set at a cut-off of 4kHz. Subjects heard the stimuli over air, in a sound-attenuated booth.

Stimuli
Four sets of seven stimuli were constructed. The SPEED stimuli varied only in their interpulse interval, which ranged from 0 to 500ms. The FREQUENCY stimuli varied only in their fundamental frequency, ranging from 210 to 680Hz. The NUMBER OF UNITS stimuli ranged from a 4-pulse sound, which was a 2-pulse sound repeated once, to 6 such units, resulting in 12 pulses. The INHARMONICITY stimuli varied in the number of inharmonic components in the harmonic structure of the

pulses, ranging from 0 to 14 inharmonic components.

All stimuli consisted of a set of 6 200ms pulses varying only along the parameter under investigation. In the case of the NUMBER OF UNITS stimuli, both the length and the number of units varied, as this was the parameter under investigation. Further details of the stimuli can be found elsewhere (Hellier, (1991)).

Procedure

Subjects were given the following instructions:

"This experiment is concerned with your subjective experience of urgency. You will be presented with a series of sounds in random order. You are requested to match the length of a line to the urgency of each sound so that its subjective length is equal to the subjective urgency of the sound. Let short lines represent low urgency and longer lines represent high urgency. Do not try to be consistent, it is only your immediate impressions that are of interest. Any questions?"

Each stimulus was then presented twice to the subject, in a block randomised order. The subjects were given practice trials prior to the start.

RESULTS

The stimuli for each of the four acoustic parameters were ranked in the direction predicted by earlier work, such that stimuli were judged to be more urgent as they became faster, higher in pitch, more repetitious and more inharmonic. Engen's data transformation (1971) was performed on the data in order to eliminate inter- and intra-subject variability. The method of least squares was use to fit the data for each parameter to a straight line. An exponent was derived for each of the four parameters, which can be seen in Table 1.

Table 1. Urgency exponents for speed, frequency, number of units and inharmonicity, and the change required in physical parameter in order to produce specified changes in urgency.

Parameter	Exponent	50%	Double	Treble urgency
Speed	1.35	1.3	1.6	2.2
Frequency	0.39	2.8	6.0	17.4
Number of units	0.50	2.2	4.0	8.9
Inharmonicity	0.12	28.5	307.0	8773.0

The percentage variance accounted for in each of the plots was 98%, 93%, 98% and 84% for speed, frequency, number of units and inharmonicity respectively.

DISCUSSION

Table 1 shows that each acoustic parameter produced a different exponent, suggesting that they each have differing strengths of effect on perceived urgency. The higher the exponent, the stronger the effect. Thus, of the four stimuli tested, speed has the highest exponent and therefore is the parameter which has the most influence in altering the urgency of a stimulus. The power law equation can also be used to predict the amount of physical change in a stimulus which would be required to produce a particular estimate of urgency. Table 1 shows the predicted amount of change in the levels of the physical stimuli which should be required in order to produce specified changes in urgency. For example, in order to double the urgency it is necessary either to increase the speed of a stimulus by a factor of 1.6, the frequency by a factor of 6, the number of units by a factor of 4 or the inharmonicity by an unworkable factor of 307. Thus the urgency of a stimulus is more economically altered through changing its speed than by altering the other parameters tested here. The parameter with the smallest exponent, inharmonicity, would produce only very small changes in urgency even if the whole range, from completely harmonic, to completely inharmonic, was covered. In this sense, it is the least useful of the four in altering the urgency of an acoustic stimulus although some reservations must be expressed about the quantifiability of this parameter, and the appropriateness of the application of the power law in its description.

The results obtained have a number of applications. Aside from showing that the power law can be used to relate objective quantities to subjective judgement, the results will be useful in future auditory warnings work because they show more clearly the strength of some of the acoustic parameters already known to affect perceived urgency. They show that some parameters are more powerful than others, and thus the manufacturer wishing to alter the urgency of a warning, or set of warnings, will be able to predict the degree to which specific alterations might change the urgency of those warnings.

Perhaps most useful is the possibility that the results can be used to predict theoretically equal levels or urgency between acoustic parameters. For example, the equations can be used to predict the value for each of the four parameters which would lead to an urgency judgement corresponding to a specific line length, and therefore equal urgency. This information is essential if we are to identify the major influences on perceived urgency, or indeed any other subjective attribute of sound, because in order to establish the salience of individual acoustic parameters it is important that they represent comparable ranges so as to make comparisons meaningful. This is important for research purposes (Freed & Martens (1986); Hellier (1991)).

The ability to predict units of equal urgency is also important practically because it begins to show us how we might construct warnings that are different from one another, and yet approximately equally urgent. This is useful if a set of warnings is required for situations which are already operationally prioritised, and where there may be several warnings required for a single priority. Environments where this is the case include aviation and hospitals.

There are many issues yet to be covered by this research. One line of research which needs to be explored is that of the relationship between perceived urgency and performance. It is important to establish that warnings which are coded appropriately in urgency produce responses which vary in some way in relation to the urgency of those warnings. This area of research is being explored by our group in the hospital environment, an area in which there is much research and standards interest at present. Another, more theoretical, line of research is that of the effects on urgency when many acoustic parameters are combined, and the potential additivity of such effects.

REFERENCES

Edworthy, J., Loxley, S., & Dennis, I., 1991, Improving auditory
warning design: Relationship between warning sound parameters
and perceived urgency. Human Factors, 33(2), 205-231.
Edworthy, J., & Loxley, S., 1990, Auditory warning design: the
ergonomics of perceived urgency. Contemporary Ergonomics 1990,
Ed. E. J. Lovesey (London, Taylor & Francis).
Engen, T., 1971, Scaling methods. In Experimental Psychology, ed. J.
Kling & L. Riggs (London: Methuen & Co.)
Finley, G. A., & Cohen, A. J., 1991, Perceived urgency and the
anaesthestist: Responses to common operating room monitor
alarms. Canadian Journal of Anaesthesia, 38(8), 958-964.
Freed, D., & Martens, D., 1986, Deriving psychophysical relations
for timbre. In Proceedings of the International Computer Music
Association, pp 393-405
Hellier, E. J., 1991, An investigation into the perceived urgency of
auditory warnings. Unpublished doctoral dissertation, University of
Plymouth.
Hellier, E. J., & Edworthy, J., 1989, Quantifying the perceived urgency
of auditory warnings. Canadian Acoustics, 17(4), 3-11.
Hoge, H., Schick, A., Kuwano, S., Namba, S., Bock, M., & Lazarus, H,
1988, Are there invariants of sound interpretation? The case for
danger signals. In Noise as a Public Health Problem, 2 (Hearing,
Communication, Sleep and Nonauditory Physiological Effects), ed.
Berglund et. al. (Stockholm: Swedish Council for Building Research),
pp 253-258
Kuwano, S., & Namba, S, 1990, Continuous judgement of loudness and
annoyance. In Proceedings of the 6th Annual Meeting of the
InternationalSociety for Psychophysics, pp129-139.
Lazarus, H., & Hoge, H., 1986, Industrial safety: Acoustic signals for
danger situations in factories. Applied Ergonomics, 17(1), 41-46.
Momtahan, K., Hetu, R., & Tansley, B, 1993, Audibility and identification
of auditory alarms in operating rooms and an intensive care unit.
Ergonomics, in press.
Momtahan, K., & Tansley, B., 1989, An ergonomic analysis of the
auditory alarm signals in the operating room and recovery room. Paper
presented to the Annual Conference of the Canadian Acoustical
Association, Halifax, Nova Scotia, Canada, October 1989.
Patterson, R. D., 1982, Guidelines for auditory warning systems in
civil aircraft. Civil Aviation Authority report No. 82017.
Stevens, S. S., 1957, On the psychophysical law. Psychological Review,
64, 153-181.

Health and safety

HUMAN DEPENDENT FAILURES:
A SCHEMA AND TAXONOMY OF BEHAVIOUR

P.D. Hollywell

Electrowatt Engineering Services (UK) Ltd
Electrowatt House, North Street
Horsham, West Sussex RH12 1RF

This paper describes a recent study that examined the nature of Human Dependent Failures (HDFs). It determined that HDFs are those system failures which result from the coexistence of two factors: one that provides a susceptibility for humans to fail in a set of actions due to a particular Root Cause and a Coupling Mechanism that creates the conditions for multiple human actions to be affected by the same root cause. A schema and taxonomy which shows how root causes and coupling mechanisms can lead to HDFs was developed and have been used to enhance Human Reliability Assessment (HRA) techniques for coping with HDFs in Probabilistic Risk Assessments (PRAs).

INTRODUCTION

Over the last decade, human error has come to be recognised as a potential major contributory cause of serious accidents in a wide range of hazardous industries. In addition, there has been a growing appreciation that the systematic consideration of human error in the design, operation and maintenance of highly complex systems can lead to improved safety and more efficient operation.

The comprehensive and systematic identification, prediction and reduction of potential human errors within a system are the basic elements of Human Reliability Analysis (HRA). HRA is most often used to provide quantitative estimates of Human Error Probabilities (HEPs) for a Probabilistic Risk Assessment (PRA) in order to achieve declared safety targets. HRA can also be used qualitatively, to identify and minimise potential human errors. Thus HRA can make a significant contribution to improving total system safety. Whichever HRA approach is adopted, however, there needs to be a full and proper appreciation of the potential for Human Dependent Failures (HDFs). Unfortunately, HDFs are not well understood and including them realistically in HRA can be problematic.

Not only are HDFs important in the application of HRA, but a good understanding of them is vital for the design, operation, maintenance and management of systems in which redundancy is employed as a means of achieving high reliability requirements; e.g. in nuclear power stations. This is because HDFs have the potential to compromise the assumed independence of redundant sub-systems, which can reduce the reliability of the system.

HUMAN DEPENDENCY IN HRA/PRA
Definition
     Human dependency occurs in the context of two or more human
actions, where the probability of failure (or success) of the
outcome of one action is different depending on whether a failed
(or successful) outcome has occurred for a previous action. It is
a special case of conditional probability, i.e. when two events
are not statistically independent, applied to human actions.
Human dependency can have both beneficial and detrimental effects.
It is normal in the context of HRA/PRA applications to give most
consideration to the detrimental effects.
     The above definition suggests that almost all linked human
actions are likely to exhibit dependency between them. However,
the importance of human (and other forms) of dependency in PRA
arises from the fact that the combined probability of the failure
of two distinct events is the product of the probability of
failure for each event only when the events are completely
independent. When some form of dependency exists then the
combined probability of failure is different from that obtained
from the simple product. Thus, the serious possibility arises
that, for safety assessments in particular, a failure to take
account of dependencies within a situation could result in an
underestimation of the actual risk. This is clearly
unsatisfactory.

Forms
     It is generally accepted that there are three forms of human
dependency:-
i)   Within - Person Dependency: Human dependency arising due to
the same person performing a set of related or different actions.
     For example: During refuelling on a nuclear power plant the
same person carried out scheduled maintenance on a series of Motor
Operated Values (MOVs). Due to a repeated wiring error, it was
discovered that two of the MOVs stroked in the reverse direction.
ii)   Between - Person Dependency: Human dependency arising due to
different persons working as a team performing a set of related or
different actions. In general, very small teams, particularly two
person ones, are considered.
     For example: During an outage on a nuclear power plant a small
team of maintenance staff carried out some scheduled maintenance
in a cramped location out on the plant. Each team member was
allocated work in a similar plant area; each area being very close
to a MOV. An inspection following the work revealed that due to a
common working practice, the motor housings of two containment
spray MOVs were found to be broken because they had been stepped
on.
iii)   Global Dependency: Human dependency arising due to a common
factor, or set of factors, peripheral to the task itself which
influences: (a) the same person performing actions, (b) different
persons working as a team performing actions, (c) different
persons working independently performing actions.
     For example: Some months after a period of shutdown on a power
station, three safety valves developed flange leaks. These leaks
were caused by using an incorrect gasket and nuts of the wrong

material. These parts were given to the maintenance staff by the
station stores for the task of maintaining these particular
valves. Note that it is irrelevant who actually carried out the
maintenance of the valves; the common factor was the provision of
the wrong parts for the job by stores staff.

NATURE OF HUMAN DEPENDENCY
Human Dependent Failures

Human Dependent Failures (HDFs) are those system failures which
result from the coexistence of two factors; one that provides a
susceptibility for humans to fail in a set of related or different
actions due to a particular Root Cause and a Coupling Mechanism
that creates the conditions for multiple human actions to be
affected by the same root cause. The existence of Defences
against HDFs will greatly reduce their occurrence. These
different contributors to human dependency will now be considered.

Root Causes and Coupling Mechanisms

Root causes are those changes of state that greatly increases
the likelihood for humans to fail in a set of related or different
actions. Coupling mechanisms are those links which make the
impact of the root cause common to multiple human actions.

As an illustration of both factors, let us consider a human
activity which is carried out according to a common specification,
instruction or procedure (e.g. design specifications, drawings,
test specifications and maintenance procedures). If any
documentation errors exist within the common specification, etc,
then it is extremely likely that there will be failures in the
multiple human actions. In this example, due to the strong
coupling that exists between the multiple human actions it would
be appropriate to assess a level of dependence which is close to
complete dependence. Therefore the stronger the link between the
root cause and the set of actions due to the coupling mechanism,
the higher will be the Assessed Level of Dependence. The
strongest coupling should be assessed as complete dependence; no
coupling should be assessed as zero dependence.

Defences

Defences against HDFs, which prevent root causes affecting
multiple human actions, can be considered as being restricted to
only two basic types: quality and diversity (Edwards, 1989).
Thus, in order to defend against HDFs, either the potential root
causes of human dependency have to be reduced (i.e. improve the
quality of the features common to multiple human actions) or the
coupling mechanisms have to be weakened (i.e. increase the
diversity between multiple human actions). Historically the
quality approach to defending against HDFs has mainly been used,
except for a few situations where exceptional reliability
requirements have been specified when some degree of diversity has
been applied.

In order to apply the quality approach to defence, one needs to
take into account the basic quality control that can be achieved
and the quality assurance methods applied to detect and eliminate
errors and deficiencies in common features which could form

possible root causes of dependent failures. In order to apply the
diversity approach to defence, one needs to take into account the
adequate diversity between multiple human activities that can be
reasonably implemented in an appropriate form. The degree of
independence achievable by the defences can therefore be expressed
in terms of the quality of those features common to multiple human
actions and in terms of the diversity between multiple human
actions. The degree of independence achievable by the defences
will therefore have an impact on the assessed level of dependence.

SCHEMA OF HUMAN DEPENDENCY
    A diagrammatic representation, or schema, of human dependency
in line with the above description was developed. The schema
clearly shows how root causes and coupling mechanisms can lead to
HDFs. Defensive techniques, aimed at reducing HDFs, can be
directed towards the potential root causes (i.e. quality) and/or
towards the coupling mechanisms (i.e. diversity). It is the
strength of the coupling mechanisms and the effectiveness of the
defences that will determine the assessed level of dependence.
Such a schema is supported by Ballard (1988).

Figure 1.   Schema of Human Dependency

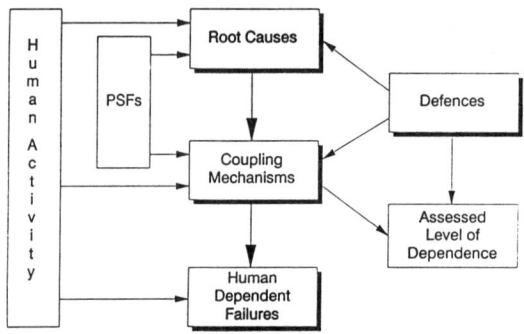

It should be noted that the precise nature of the human
activity being considered will determine the potential root causes
and the coupling mechanisms as well as identifying the credible
HDFs. Additionally, it will be the Performance Shaping Factors
(PSFs) which influence human performance within these two factors.
    It is interesting to note that often human activities or jobs
are "designed" in order to minimise excessive task demands, either
physical or mental, on a worker. This is especially critical
where there are significant consequences should the worker fail to
perform the task correctly and therefore a high degree of human
reliability is required. In order to overcome possible problems
with high workload, whether it be physical or mental, a task can
be broken into sub-tasks which can be carried out by more than one
person. In order to overcome possible problems in achieving high
reliability a task is often performed by one person and is then
checked by a second person. Although the above job redesign
strategies are able to alleviate the problems of high workload and
high reliability, they do introduce their own dependency problems.

TAXONOMY OF HUMAN DEPENDENCY

An extensive review of the human dependency and human error literature (e.g. Rasmussen and Pedersen, 1982) has enabled the production of a classification, or taxonomy, which is based on the above schema. The taxonomy is not exhaustive but is indicative of the nature of the factors which comprise the schema.

Root Causes

Since HDFs arise from errors and deficiencies in some form of human performance, the full list of their root causes should be very similar to that for human error.
* External Events
  - Distractions (from systems, persons, etc)
  - Interfering tasks (from systems, persons, etc)
* Excessive Task Demands
  - Inadequate manual skills (force, speed, precision, etc)
  - Inadequate memory
  - Insufficient or incorrect knowledge
* Intrinsic Human Variability
  - Variability in manual skills (force, speed, precision, etc)
  - Variability of attention
  - Lapses of memory
  - Erroneous recall of knowledge
* Person Incapacitated
  - Sickness
  - Injury
* Organisation Deficiencies
  - Corporate
  - Management
  - Group

Coupling Mechanisms

Coupling mechanisms link the impact of the root cause to multiple human actions. Therefore it can be said that any human activity that employs one or more common resource is likely to involve some dependence. Note that a 'resource' should be considered in its widest possible meaning.
* Information
  - Spoken (instructions, supervision)
  - Written (drawings, documents, written procedures)
  - Displayed (control panels, VDUs, signs, warnings, tags)
* Equipment
  - Tools
  - Facilities
  - Support services
  - Work environment
* Intrinsic Human Capabilities
  - Physical (size, weight, reach, strength, precision, workload)
  - Mental (detection/discrimination, attention, memory, workload)
* Skill
  - Training and qualifications
  - Experience and mental models (understanding, assumptions, beliefs)

* Organisation
  - Corporate policies, standards and procedures
  - Management attitudes and behaviour
  - Group values and norms

Human Dependent Failures

Since HDFs arise from errors and deficiencies in some form of human performance, the full list of HDF modes should be very similar to that for human error.
* Human Activity Not Performed To Completion
  - Multiple actions not carried out
* Human Activity Performed Erroneously
  - Multiple actions carried out inaccurately
  - Multiple actions carried out in wrong sequence
  - Multiple actions carried out too early/late
  - Multiple actions carried out too little/too much
* Human Activity Performed With Extraneous Actions
  - Unnecessary multiple actions carried out

Defences

As already stated, defences can be directed towards: (i) the root causes of dependent failures and/or (ii) directed towards the coupling mechanisms, in order to defend against HDFs. With the first approach, the quality of the common features affecting human actions will determine the incidence of root causes and will therefore affect both dependent and independent human failures equally. With the second approach, the diversity of human actions will affect the strength of coupling between them and the conditional probabilities relating dependent and independent human failures.

Defence strategies for the 'quality approach' against HDFs tend to focus on the systematic application of good ergonomics. Therefore, the systematic application of ergonomics would appear not only to be effective in reducing human errors in general, but would also serve to reduce the likelihood of HDFs caused through human related root causes. This conclusion is consistent with that arrived at in the study by Lucas (1987).

Defence strategies for the 'diversity approach' against HDFs tend to focus on the following points in order to achieve success of the activity:
i)   The use of redundant information sources for the multiple actions
ii)   The avoidance of relying on the same essential piece of 'equipment'
iii) Not relying on unreasonable human capabilities; better to support person(s) or redesign activity
iv)   Ensuring that persons undergo extensive, wide-ranging training and are given an opportunity to develop a variety of contrasting operational experiences
v)   Not relying on too great a level of imposed consistent behaviour, either throughout an organisation and/or over the period of time it takes for the activity to be completed.

It is very important to note that there are some obvious practical problems in applying all the defence strategies listed above as there are potential conflicts between maintaining quality whilst attempting to obtain diversity. Some of the diversity approaches tend to go against current standardisation practices, designed to enhance quality. The applicability of such defence strategies will need to be determined on a case by case basis.

FUTURE DEVELOPMENTS

It should be appreciated that the above taxonomy is based on a wealth of knowledge taken from the human reliability, cognitive psychology and risk assessment literature. Although such knowledge provides extremely useful insights into human dependency, and thus enables the continued enhancement of HRA techniques, much more real world human-orientated dependency data is required in order to validate the above taxonomy.

The schema and taxonomy detailed in this paper can be usefully employed as a tool in order to assist in raising the general awareness of the nature of human dependency with design, operations, maintenance staff as well as with project management. It could also be used as a method of generating defence strategies for specific situations identified as being vulnerable to human dependency.

ACKNOWLEDGEMENTS

Thanks is due to the Health and Safety Executive, Nuclear Safety Research Management Unit, who funded the research on which this paper is based. The views expressed in this paper are solely those of the author.

REFERENCES

Ballard, G.M., 1988, Dependent Failure Analysis in Probabilistic Safety Assessment. In SRS Quarterly Digest, April, pp. 3-13.

Edwards, G.T., 1989, Dependent Failure Assessment: Principles, Engineering Structures and Data Requirements, SRS/GR/83 (SRD Association).

Lucas, D.A., 1987, Human Errors Causing Dependent Failures in Nuclear Power Plants: A Database, Taxonomy and Analysis. In Reliability '87, Volume 2.

Rasmussen, J., and Pedersen, O.M., 1982, Formalised Search Strategies for Human Risk Contributions: A Framework for Further Development. Riso Report No. M-2351.

RISK HOMEOSTASIS IN A NON-TRANSPORTATIONAL DOMAIN

*THOMAS W. HOYES and **CHRIS BABER

*Human Factors Research Unit,
Organisation Studies & Applied Psychology Division,
Aston Business School, Aston University,
Birmingham B4 7ET

**Industrial Ergonomics Group
School of Manufacturing & Mechanical Engineering
University of Birmingham,
Edgbaston, Birmingham B15 2TT

The application of RHT to non-transportational physical risk is considered with reference to two cases: the nuclear industry and air-traffic control. It is suggested that neither case provides strong support for a homeostatic mechanism in risk-takers, although each case may still be reconciled with RHT, and evidence exists for compensation that does not negate intrinsic risk changes. It is suggested too that risk homeostasis is not a context-free theory of risk taking, and that the ways in which it might be extended to non-transportational risk therefore need clarifying.

INTRODUCTION

Risk homeostasis and other so-called compensation models have been postulated in terms of road-user behaviour. These theories share an emphasis on the level of risk populations are prepared to accept, rather then the absolute level of intrinsic risk, as the unique determinant of accident loss. Here, intrinsic risk refers to the risk inherent in the environment. The theory suggests that time-unit 'actual' risk can only be reduced by motivational interventions, and that non-motivational interventions aimed at improving the intrinsic level of risk are negated by the behaviour adjustments of those affected by them. As stated already, the theory has, in the main, been applied to road-user behaviour. Only recently (Wilde, 1986) have writers begun to consider the possibility that the theory might have wider implications encompassing other cases of physical risk.

The question of whether the homeostatic principle can be extended outside the road traffic environment it was postulated to explain to more general cases of risk does rather presuppose that it can explain even road-users' behaviour after some change in intrinsic risk. This somewhat basic assumption has certainly not met with universal agreement (see, for example, McKenna 1985). In this paper, however, we will deal with issues relating to physical risk outside the road traffic environment. By briefly examining two other physical risk-taking environments - nuclear process control and air traffic control - we hope to show that the theoretical terms in which RHT has been stated go some way to preventing its wider application. We intend to show also that the limited empirical evidence we have available would appear to oppose the notion of a complete homeostatic mechanism operating in nuclear process control and in air traffic control.

UTILITY

Wilde's (1988) concept of utility is, in the main, a time-referenced one. RHT suggests that utility is necessary for behaviour adjustments to be made in response to changes in intrinsic risk. In other words, RHT is not about risk *for its own sake*. Out of this emphasis on time comes the consideration of the behavioural pathway to homeostasis of speed.

The replacement of a time-referenced with, say, an attention-referenced, concept of utility, is difficult to consider seriously. The possibility that attention-related pathways might provide utility in cases where intrinsic risk changes has been examined (Hoyes, 1990; Hoyes, et al., 1992; Hoyes and Stanton, 1992); no evidence for any attention effect capable of restoring previously existing risk levels has so far been found. A study by Hoyes and Stanton (1992) in a simulated control room showed that performance on a spatial secondary task was unaffected by changes to intrinsic risk, where manipulations were made to presentation rates of alarms, and to the ratio of alarms to non-alarm. Hoyes (1990) found that although attention differences may exist over different levels of intrinsic risk, the effect did not *negate* intrinsic risk benefits. A study by Hoyes et al. (1992) showed that where attention is inferred from position-on-track in a validated driving simulator, no change is seen to follow from manipulations to intrinsic risk. In short, where time-referenced utility cannot be applied (as in most process control tasks), its replacement is far from clear.

ACCIDENT LOSS

In Britain, records for accident rates on our roads exist from 1927. In that year, no fewer than 2,774 pedestrians were killed on British roads. Wilde (1988) suggests that for homeostasis to work, perceptions of accident loss must enter a closed loop. If accident loss for road traffic maps on to a 2-year adjustment period, one imagines that for 'safer' environments, the adjustment period is a multiple of this. Thus, in a non-transportational domain, RHT may be true, but almost trivial.

RISK HOMEOSTASIS IN THE NUCLEAR INDUSTRY

HM Nuclear Installations Inspectorate (NII) has long recognised that human error is important in plant safety (NII, 1982; Whitfield, 1987). The major topics considered by the NII

under the heading of ergonomics include: allocation of operator
tasks; control room design; cognitive/conceptual errors; and
maintenance errors (Whitfield, 1987). Other writers have
discussed human factors relating to safety in the nuclear
industry without reference to any compensatory actions on the
part of operatives (Marshall, Duncan and Baker, 1981; Marshall
and Baker, 1985). Marshall and Baker, for example, see the
contribution of human factors to nuclear safety in terms of
automation, information presentation and training. Other
researchers have focused attention on the processing of alarm
information, modelling alarm-initiated activity in the nuclear
power plant control room (Stanton, Booth and Stammers, 1992).

What is clear from research activity here is that the emphasis
in nuclear regulatory control is very much on the intrinsic side.
This is not to say that the human factors perspective is ignored
in nuclear risk, for it is not; rather, through selection and
training, system design, operator-plant interface, etc., effort
is made to reduce operational, as well as intrinsic, risk.

The President's Commission report on the Three Mile Island
incident (Kemeny, 1979) illustrates the nuclear industry's
emphasis on human factors in accident loss, but in a way detached
from any compensation theory. Stanton, et al. (1992) identify
four human factors problems that contributed to the accident: so
many alarms were activated in the environment that the operators
could not hope to process them all in the time scale of the
incident; the position of controls and indicators was badly
thought-out; some instruments went off-scale during the incident,
making diagnosis difficult, or impossible; and there was a 2.5
hour time lag between event and alarm. What is missing from this
description is any attribution of responsibility: the *environment*
was to blame for the accident, not the operators, who were
powerless to do anything.

An aspect of the Three Mile Island incident was simulated in a
mock control room by Hoyes and Stanton (1992). The study
involved asking participants to process scrolling alarms by
taking decision action in a matching task. Three presentation
rates were used in the study: an alarm every one second, every
four seconds, and every eight seconds. What Hoyes and Stanton
found, perhaps not surprisingly, was that the one-second
presentation rate was associated with very high levels of errors
that could be expected to lead to an equally high probability of
accident. Participants did not compensate for a level of risk
they found unacceptable, perhaps because they *could* not
compensate, since to restore some previously existing level of
accident loss was quite simply beyond their control. In these
circumstances, then, there was no risk homeostasis effect.

Whether this finding can be generalised is, of course, another
matter (see Hoyes, 1992a, 1992b, and Hoyes & Glendon 1992). What
is clear is that little evidence exists at present for RHT in the
control room. Perhaps there are two reasons for this. One may
well be the limited feedback already discussed that follows from

low levels of accident loss. Another may be that when behaviour adjustments within the physical risk-taking environment are not possible - as in the Hoyes and Stanton study - paricipants are reluctant to tend towards homeostasis through avoidance of the environment.

APPLYING RHT TO AIR-TRAFFIC CONTROL

Air traffic control has frequently come under the spotlight of ergonomic attention, often with a view either to improving the intrinsic safety characteristics of the ATC environment or to examining the human factors involved in the task (see Ball & Jackson, 1988; Langan-Fox & Empson, 1985; Stager and Hameluck, 1990; and Wiener, 1980, for a interesting cross-section of this work). Two cases of intrinsic risk changes in ATC are examined here within the framework of RHT: workload variation and computer-based interventions.

On the workload side Stager & Hameluck (1990) analysed 301 operating irregularities by ATCOs (air traffic control operators) over a one year period. They found that such errors were not related to rated workload. However, all this tells us is that high workload is not necessary for ATCO error to occur. Since the relative occurrence of low, moderate and high workload is not reported by the authors, the finding does not tell us whether high workload - translating to high levels of intrinsic risk - increases the *actual* risk of that environment. However further support for the negation of the intrinsic risk in minor workload variation has been well-documented. Sperandio (1971) described a system of feedback in which workload variation serves to regulate the operator's choice of action: in conditions of high workload, a more efficient strategy will be adopted. The consequences of this strategy switching are such, according to Sprerandio, as to negate the intrinsic risk effects of workload variation.

If we look at intrinsic risk as reflected in computer-based operator support, it can be seen that safety interventions can be expected to lead to improvements in actual risk (Hoyes, 1990; Thackray and Touchstone, 1989). However, this is not the whole story. Hoyes (1990) looked at the effects of attention of ATCOs when a computerised decision-support system was introduced. Using a secondary task, within a few seconds, operators had, it would seem, reduced their attention on the primary task of ATC. Perhaps then, when improvements of sufficient magnitude are made to the intrinsic safety of an ATC environment, operators do change their behaviour, but not to such an extent, at least in the short-term, as to negate the intrinsic risk benefits.

THEORETICAL IMPLICATIONS OF APPLYING RHT TO NON-TRANSPORTATIONAL RISK

In suggesting that risk homeostasis can be achieved through the three pathways (see Wilde, 1988) wider implications arise. In physical risk-taking environments other than road traffic, with what is the concept of migration from one mode of transport to another to be replaced? Are we, in non-transportational risk,

to be left with the alternatives of behavioural adjustments
within the physical risk-taking environment, or avoidance of that
environment?

The three-pathway theory brings a further problem. RHT makes
its predictions at the level of a given jurisdiction. So far as
individual transportation is concerned in hypothetical
jurisdiction, the implications of the theory are clear enough.
Suppose an individual gets up for work one morning and looks out
to see ice or fog. Three options (the three pathways) are
immediately available. First, the individual might consider that
by driving more carefully, actual risk may be brought back to
within target parameters. Second, the individual may decide to
travel into work by train, supposing the train's actual risk to
be in line with target risk. Third, if no form of transport is
available through which actual risk can be matched to the
individual's target risk, the individual may elect to stay at
home, thus maintaining the best correspondence of actual with
target risk. How this three pathway model can be applied to non-
transportational risk, where 'avoidance' is, one imagines, met
with recruitment, has never been specified.

A further conceptual difficulty in generalising RHT is the
important distinction Wilde makes between *per capita*, *per km*
and *per time unit* risk. How this distinction translates to non-
transportational physical risk is also unclear.

CONCLUSIONS
The terms in which compensation theories have been proposed
are so related to the road-traffic behaviour they seek to explain
that it is sometimes difficult to extract from them
quintessentially context-free concepts and predictions.

In the application of RHT to transportational research, Hoyes
and Glendon (1992) point out that they can envisage no finding,
either existing now, or potentially existing in the future, that
could falsify RHT. They suggest that RHT's status as a theory
would be much improved if its proponents would spell out what
would constitute its falsification. In non-transportational
risk, the same would seem to apply.

REFERENCES
Hoyes, T.W., 1990, Risk homeostasis in a simulated air-traffic
    control environment, Unpublished Masters Thesis, The
    University of Hull.
Hoyes, T.W., 1992a, The investigation of alarm media in a
    laboratory environment, Paper presented at one-day conference
    in the Human Factors of Alarms (Ergonomics Society), Aston
    University, Birmingham, UK.
Hoyes, T.W., 1992b, Risk homeostasis in simulated environments,
    Unpublished PhD thesis, Aston University, Birmingham, UK.
Hoyes, T.W. and Glendon A.I., 1992, Risk homeostasis theory:
    issues for future research. In press, Safety Science.

Hoyes, T.W. and Stanton N.A., 1992, Risk homeostasis theory in a simulated alarm-handling task, Paper presented at one-day conference in the Human Factors of Alarms (Ergonomics Society), Aston University, Birmingham, UK.

Hoyes T.W., Dorn L. and Taylor R.G., 1992, Risk homeostasis: the role of utility. In Contemporary Ergonomics, ed. E.J. Lovesey (London: Taylor and Francis), pp. 139-144.

Kemeny J., 1979, The Need for Change: The legacy of THI. Report of the President's Commission the the Accident at Three Mile Island (Pergamon, New York).

Langan-Fox C.P. and Empson J.A.C., 1985, 'Actions not as planned' in military air-traffic control, Ergonomics, 28: 1509-1521.

McKenna, F.P., 1985, Do safety measures really work? An examination of risk homeostasis theory, Ergonomics, Vol 28, No. 2: 489-498.

Marshall, E.C., Duncan, K.D. and Baker, S.M., 1981, The role of withheld information in the training of process plant fault diagnosis. Egonomics, 24,: 711-724.

NII, 1982, Ergonomics/human factors in the design and safe operation of pressurised water reactors. Report NII - ONSWG (92) P.1.

Sperandio, J.C., 1971, Variation of operator's strategies and regulating effects on workload, Ergonomics 14(5): 571-577.

Stager, P. and Hameluck, D., 1990, Ergonomics in air traffic control, Ergonomics, 33(4): 493-499.

Stanton, N.A., 1992, The Human Factors of alarm handling, Aston University, Birmingham, UK.

Stanton, N.A., Booth, R.T. and Stammers, R.B., 1992, Alarms in human supervisory control: a human factors perspective, International Journal of Computer Integrated Manufacturing, Vol 5(2), 81-93.

Thackray, R.I. and Touchstone, M.R., 1989, Detection efficiency on monitoring task with and without computer aiding, Aviation, Space, and Environmental Medicine, Vol 60(8), 744-748.

Whitfield D, 1987, Human reliability from a nuclear regulatory viewpoint, Contemporary Ergonomics, ed. E.D. Megaw: 58-63.

Wiener, E.L., 1980, Midair collisions: the accidents, the systems, Human Factors, 22(5), 521-533.

Wilde, G.J,S., 1988, Risk homeostasis theory and traffic accidents: Propositions, deductions and discussion of dissension in recent reactions, Ergonomics (UK), 31,4, 441-468.

Wilde, G.J.S., 1986, Beyond the concept of risk homeostasis: suggestions for research and application towards the prevention of accidents and lifestyle-related disease. Special Issue: Risk, Accident Analysis and Prevention, Volume 18(5), 377-401.

PASSENGER CONTAINMENT ON AMUSEMENT DEVICES

J A JACKSON

Health & Safety Executive
Broad Lane
Sheffield
S3 7HQ

This paper describes the results from the pilot study of a project aimed at providing guidance for designers of amusement rides to enable them to build systems which contain the passengers in a more secure manner. Initial work has concentrated on Chair-O-Plane type rides for young children. The results showed that the method of restraint selection developed in the pilot study was a useful guide to the suitability of particular restraint designs and that there was scope for improving the present restraint designs, particularly the ease of restraint adjustment to ensure a good fit.

INTRODUCTION

Accidents have occurred on rides where the poor design of passenger containment systems may be considered to be a major contributor. To investigate the problem the Health & Safety Executive (HSE) is currently undertaking a study with the main aim of producing design guidance for passenger containment on amusement devices based on sound ergonomics principles. A literature survey showed that the majority of work carried out on restraint systems has been for the automobile and aviation industries. None of the papers found related directly to the amusement industry. The automobile and aviation work proved to be of little use in this context mainly because the forces involved, the duration of containment and the economics are very different.

The study first needed to gather some basic information on ride types and existing containment systems and their use from field studies. These studies showed that many restraints had been designed to fit the carriages rather than accommodating a range of passengers. The possible consequences of not designing for passengers would be poor fit and inadequate security leading to an increased risk of accidents occurring. The child's Chair-O-Plane type rides are typical examples and were considered suitable for detailed examination in a pilot study. The Chair-O-Plane is a roundabout with chairs hung from the rotating canopy. The pilot study consisted of four phases the first of which was the field studies. In phase two, information gathered in phase one was used to model the Chair-O-Plane ride on an ergonomics CAD system. Phase three involved the selection of an alternative restraint system, suitable fasteners and the development of techniques for taking anthropometric measures. The fourth phase used a full scale mock-up of the Chair-O-Plane seat to test the selected restraint designs on three user groups.

METHODS
Phase one : field studies

Data on two child's Chair-O-Plane type rides were gathered during visits to travelling fairs. One was a hand cranked ride and the other was powered. Other than this they were of similar construction with similar seat and restraint designs. The seat and restraint on each ride was measured, the dimensions of which are shown in Table 1. The seats on the motorised ride were slightly wider, but less deep, than those of the other ride. The restraint system on each ride was the same, a length of chain attached to the front of the upper edge of the seat with a spring tensioned clip on the loose end. In both cases the length of chain was greater than the width of the seat.

Table 1 : Seat dimensions

DIMENSION (mm)	HAND CRANKED	MOTORISED
seat width	300	330
seat depth	300	255
backrest height	205	215
backrest depth	30	30
chain length	450	570

   Video recordings were made whilst members of the public were using the rides. Several observations were made when the video recordings were analysed. It appears to be common for parents, as well as the ride operator, to fasten small children into the seat and for older children, estimated to be eight years and over, to fasten themselves in. Most children were observed sitting in what was considered to be the correct position although one child was observed sitting at the front of the seat, tilting it forward and increasing the risk of slipping under the chain, as shown in Figure 1.

Figure 1. video images of (a) child in correct seating position and (b) child in forward position

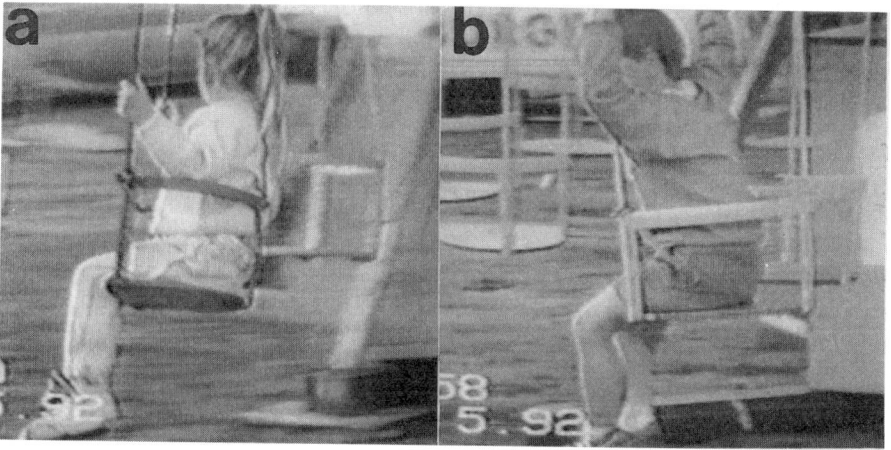

<u>Phase two : computer modelling</u>
   The measurements from the motorised child Chair-O-Plane ride were used to model the seat using Mannequin Human CAD software. A three dimensional scale model of the Chair-O-Plane seat was built up in the computer and male child models of between three and nine years of age were sat in the seat.

Phase three : restraints, fasteners and anthropometry

The fundamental problem with the present design was that the restraint was fitted to the seat and not the passenger. It was decided that to fit the restraint directly to the passenger a waist belt was needed. The belt could then be attached to the rear of the seat in a suitable manner. This would provide a system which was fitted directly to the passengers and pulled them back into the seat. A suitable fastener would need to be fitted to the waist belt.

Eleven common types of fastener, including the existing design, were selected for testing. Each fastener was attached to a length of strap so that it could be made into a loop when fastened. The fastening and unfastening times were recorded. Subjects assessed the ease of fastening, ease of strap length adjustment, security of the fastener and the ease of unfastening using a five point rating scale running from very easy (secure) to very difficult (insecure). The trials were carried out in the laboratory with the fasteners being presented in a random order. No instruction on how to use the fasteners was given prior to, or during, the tests. The aim was to select four fasteners for use in user trials.

It is reasonable to assume that the size of the passenger will affect the fit of the restraint. Each subject had eleven anthropometric measures taken which relate to this particular seat and restraint. The eleven measures were as follows:

1. weight, 2. stature, 3. sitting height, 4. abdominal depth, 5. horizontal forward reach, 6. thigh clearance, 7. buttock-popliteal length, 8. knee height, 9. bideltoid width, 10. hip breadth, 11. abdominal girth. All the measures were made according to the descriptions given by Pheasant:1990 with one exception. It is recommended that thigh clearance is measured on the uncompressed thigh. On the Chair-O-Plane ride the legs are unsupported and hence the thigh is compressed, therefore, thigh clearance was measured on the compressed thigh. Weight was measured using a Salter model 109 scale and stature was measured using a Holtain stadiometer. Sitting height was measured using a Holtain sitting height table which was modified by attaching a flat piece of board to the upright in order to measure abdominal depth and horizontal reach using a Harpenden anthropometer. All other measures were made with the Harpenden anthropometer with the exception of abdominal girth which was measured with a flexible tape measure. The measurement techniques employed were tested for repeatability.

Phase four : user trials

The dimensions of the hand cranked design were used to build a full scale replica of the Chair-O-Plane seat which was then suspended from an A frame. The chain restraint, as used on the original seat, was attached to brackets on the front of the chair. Provision was also made for using the chain with a webbing crotch strap. The remaining three restraint belts were attached to the rear upright of the seat by a removable fastener. Samples were taken from three user groups which were ride passengers, parents/guardians and trained operators. The trained operators were adults who had been instructed in the proper use of the restraints. The age range of the ride passengers was 4 to 11 years old. These were split into four groups, each spanning two years. Each child was measured and then tested in each of the restraint systems. Children under 8 years old were only asked to unfasten the restraints. Fastening and unfastening times were recorded. For the parents and trained operators, the fastening and unfastening times were recorded and the ease of fastening, the ease of adjustment, the security of the restraint and the ease of unfastening were recorded using a five point, balanced rating scale.

RESULTS
Computer modelling

The relationship between seat depth and buttock-popliteal length was shown to be a factor which could explain why some children sit forward in the seat. Using 50th percentile models, a 5 year old American male would not be able to sit back in the seat without the front edge of the seat impinging upon his popliteal.

Fastener selection

The subjective data was used to develop a method of selection. Histograms from the rating scales were plotted, an example of which is shown in Figure 2, and the modes for each of the restraints were determined. The first stage in the selection process discounted any fastener which any subject failed to close correctly. The second stage rejected any fastener which had a mode score of less than 3 for any of the properties measured. This process rejected 7 of the fasteners. The 4 selected for use in the user trials were an automobile seat belt type, a staple and eyelet buckle (attached to a leather belt), a snap clip and the existing chain restraint.

Anthropometry techniques

The largest errors occur when measuring fleshy parts of the body, notably abdominal depth and thigh clearance. These errors were less than 10% and were therefore considered to be acceptable for these measurements. The techniques developed using the board attached to the upright of the sitting height table appear to allow reproducible measures to be made within acceptable error limits.

User trials

Body measurements are given in table 2. The passenger restraint trials produced two main findings. Of the children given the option to use the crotch strap only 47% did so. Although the unfastening times were recorded, it was considered to be more significant that children failed to unfasten some of the restraints. The percentages were 9% for the chain, 25% for the chain with the crotch strap, 9% for the seat belt and 13% for the snap clip. None of the children failed to unfasten the leather belt.

Neither fastening nor unfastening times were significantly different, at the 5% level, between restraints for the parents though the unfastening times were significantly different for the the trained operators. Ease of adjustment was significant for both groups.

Applying the same selection criteria to these data as for the fastener selection survey the passenger's and parent's results would eliminate all but the leather belt. In the trained operators survey the seat belt and snap clip would be eliminated and the leather belt would be preferable to the chain.

One problem which came to light during the trials was that the webbing in the snap clip had a tendency to slip through the clip when fastened round a child's waist.

Figure 2. Example of fastener trials histogram.

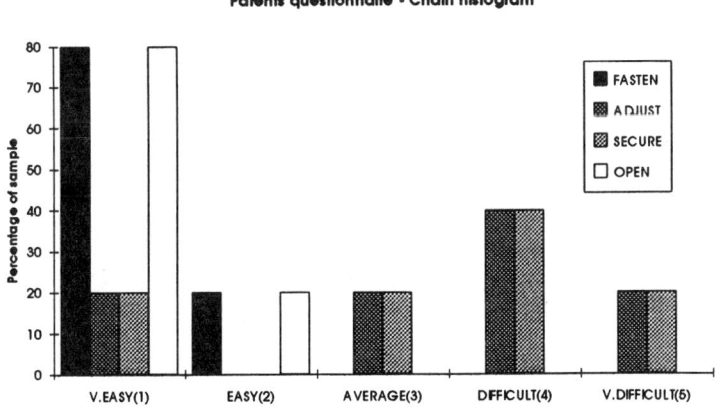

Table 2 :Anthropometric measures (weight is in kg all other measurements are in mm)

	4 & 5	year	olds	6 & 7	year	olds
	5th	**50th**	**95th**	**5th**	**50th**	**95th**
weight	17	21	25	17	26	35
stature	1042	1128	1214	1116	1210	1304
sitting height	594	630	666	605	658	711
abdominal depth	155	175	196	145	180	214
horizontal forward reach	477	538	599	504	554	604
thigh clearance	75	85	95	82	97	112
buttock - popliteal	254	299	344	285	329	373
knee height	291	329	367	329	371	413
bideltoid	247	279	311	258	301	344
hip breadth	190	214	238	191	228	264
abdominal girth	502	569	636	527	616	704
	8 & 9	year	olds	10 & 11	year	olds
	5th	**50th**	**95th**	**5th**	**50th**	**95th**
weight	20	29	38	24	38	51
stature	1236	1340	1444	1299	1422	1545
sitting height	654	707	760	680	739	798
abdominal depth	127	162	197	144	177	210
horizontal forward reach	563	619	674	544	647	751
thigh clearance	77	95	114	88	106	123
buttock - popliteal	325	368	411	349	402	455
knee height	367	417	468	394	437	480
bideltoid	285	316	347	297	335	374
hip breadth	203	232	262	218	263	307
abdominal girth	522	604	685	528	657	785

DISCUSSION

The computer modelling showed that buttock-popliteal length was an important measure in terms of the fit of the seat. Further work needs to be done to investigate other body measures which affect fit.

A problem with the experimental design of the fastener selection trials was that the fasteners were judged out of context. Subjects just made a loop, adjusted the size of the loop, tugged on the fastener to assess security and then unfastened the loop. It would have been better to have had the restraints attached to a chair, or simply fixed around an upright, and for them to be fastened round some object of similar girth to a child. The problem of the snap clip slipping when at an angle to the strap would have been identified at this stage if this had been done. This set up would have restricted access to the fasteners which could also have an effect on how they were rated. However, the selection and rejection criterion were considered to be valid.

The subjects for the user trials were selected from the relevant populations except for the trained operators. Actual ride operators were not available so colleagues from the HSE laboratories were trained in the use of the restraints and then used in the survey. The cut off point for children fastening themselves in would not be precise in age terms but for the purpose of these tests it was considered reasonable to assume that children aged eight and above would normally be capable of fastening themselves into restraints. With hind sight, this was a fairly accurate judgment when fastening and unfastening failure rates were assessed. The measurement of fastening and unfastening times for children was not as useful as the measurement of failure rate for unfastening the restraints. Below 8 years old failure to unfasten restraints was a problem. A small child who feels trapped by a restraint may attempt to defeat the restraint in some unconventional way so endangering themselves. From the observations made during the trials, the leather belt, ie the staple and eyelet buckle system, was the only restraint which children adjusted when fastening themselves in. Also, the use of the crotch strap with the chain caused problems.

With the exception of the leather belt, parents made little or no attempt to adjust restraints until prompted to do so by being asked how easy it was. Although the first instruction asked them to fasten the restraint, it was interesting to note that few subjects took this to mean fasten the restraint so that it fits. The seat belt was generally judged to be secure or very secure but this may be due to the design being associated with automobile seat belts so biasing the subjects perception of security.

CONCLUSIONS

In concluding this pilot study, the most fundamental error in the present design of Chair-o-plane rides is that the restraint fits the ride and not the passenger. From the above results it has been shown that ease of adjustment is one of the most important considerations and that the most preferred type of restraint, for use on the Chair- o- plane type ride, would be one which utilised a staple and eyelet type buckle on a waist belt. To put this into practice suitably durable belt materials and secure methods of attachment need to be investigated, but the belt and buckle was the only restraint which was correctly adjusted and fastened by everyone who used it.

REFERENCES

PHEASANT, S., 1990, Anthropometrics an introduction. BSI.

HOW AND WHY CHILD SAFETY RESTRAINTS IN CARS ARE MISUSED

J.A. RAINFORD[1], M. PAGE[2] & J.M. PORTER[1]

[1] Department of Human Sciences, Loughborough University of Technology
[2] ICE Ergonomics, Swingbridge Road, Loughborough

The relationship between the misuse of child safety restraints and user knowledge was examined. Thirty-six cases were studied and an informal interview was carried out with each supervising adult on aspects of use, correctness of installation, and the implications of misuse. Correct installation examined the amount of restraint movement once secured to the car, the position of seat-belt buckle in relation to the restraint and the existence of any twisted webbing. The usage aspect addressed the height of harness shoulder straps, the amount of slack present in the harness, the position of the harness buckle, use of crotch strap, position of seat belt across child, twisted webbing and the use of any webbing guides. The greatest areas of concern identified were excessive harness slack and seat movement, incorrect height of shoulder straps and incorrect placement of the harness buckle.

INTRODUCTION

There is a known problem regarding the misuse of child safety restraints in cars as cited by Shelness & Jewett (1983), Cynecki & Goryl (1984), Kahane (1986) and Prior (1991) amongst others, all of which were observational studies. Misuse is defined as being any deviation from correct use, as outlined by the manufacturer. This reduces their crash protection and may result in unnecessary injury or death.

Whilst it is known that car restraint incompatibility, restraint design and poor instructional information all contribute to the problem, no one was found to have addressed the issue of how much users actually know about aspects of correct installation and usage nor the likely consequences of misuse. It was decided to try to establish causes of misuse from the angle of how much users know about correct use, the hazards of misuse and to compare this with observed practice.

METHOD

Thirty six cases were surveyed during June 1992 at locations where the child, a range of restraints, car and the adult subject would all be found together (for example, playgroups and shopping centres). The sample covered the models of 6 different manufacturers and included 7 infant carriers, 11 convertible framed restraints (9 forward and 2 rearward facing), 7 booster seats, 4 booster cushions and 1 harness.

Subjects were asked to fit the child into the restraint and a checklist was used to assess aspects of installation and usage. This was followed in each instance by a structured personal interview in an attempt to ascertain how much was known about these areas, and the relevant consequence of misuse in each instance. Most questions were multi-choice. The assessment of the correctness of the installation examined the amount of child seat movement once secured to the car, the position of the vehicle seat belt buckle in relation to the restraint and whether the webbing was twisted. Usage addressed the height of harness shoulder straps, amount of slack present in the harness, position of the harness buckle, use of the crotch strap, position of the seat belt across the child, twisted webbing and the use of any webbing guides.

Correct usage was derived from discussion with experts and consultation of instruction leaflets and the literature (CAPT, 1989; Arnberg, 1978; Burdi, Huelke, Snyder & Lowery, 1969). The recommended practice is as follows:
i) Harness slack
If increased force is not to be applied to the thorax and to the neck as a result of the child's chest hitting a slack harness, only one finger (flat of a hand) should be able to be inserted between the shoulder straps and the child. This allows a slow controlled deceleration from the harness webbing stretching.
ii) Height of the shoulder straps
These should be level with or just above the child's shoulders. Any higher and they may slip off, possibly leading to ejection during rapid vehicle deceleration. Any lower would increase the forces applied to the child because there would be excess slack present. No force should be applied to the abdomen as the liver, kidneys and bladder are relatively unprotected.
iii) Position of the harness buckle
The crotch strap should be short enough to keep the harness buckle down low, i.e. over the crotch. If the lap straps of the harness are allowed any higher, either the buckle or the straps may cause internal injuries. Crotch straps should always be used as they prevent the child from submarining, i.e. slipping under the lap straps.
iv) Seat belt position
The diagonal belt should lie over the shoulder, i.e. a load-bearing structure. Lying against the neck could cause injuries. As children do not have well developed anterior superior iliac spines (the bony "hooks" on the pelvis), the lap belt can ride up onto the abdomen, so causing injury in a collision. It thus needs to be kept in position around the bony part of the hips.
v) Twists in the webbing
These should be eliminated otherwise if a child is forced against them they will cause localised pressure points which increase the possibility of injury.

RESULTS AND DISCUSSION
Results were analysed through the use of observation on percentages and significance testing where appropriate using Fishers exact test of probability and the Chi Square test. However, as all the results proved non-significant (p>0.05) it was deemed appropriate to look at the trends in the data as the lack of statistical significance could partly be due to the small sample sizes.

Background information on sample
19 male and 17 female children and babies were observed with a mean age of 2 years 2 months and a standard deviation of 1 year 5 months. Of the cars investigated 14 were two door and 22 were 4 door models and all of them were

fitted with rear seat belts. 36% of the subjects had owned the restraint for 1 year or less and 61% had for 2 years or less.

Five questions common to all subjects for infant carriers and framed restraints were asked, these related to their views on the direction in which the infant carriers should face, twisted webbing, height of shoulder straps, harness slack and harness buckle position. Overall experience of child restraints did not appear to have had much effect on how much subjects know about their correct usage as those with 1 year or less experience fared only marginally worse than the other subjects in answering the questions (mean correct of 2.9 : 3.0 out of a maximum of 5.0). Regarding the experience of a particular type of child restraint, those with greater than one years experience fared slightly better (mean 2.7 : 3.2).

72% of the child restraints investigated were bought new, 20% were bought second hand and 8% were borrowed. This highlights the need to educate the public who acquire second hand restraints on the importance of obtaining instructions and the restraint's history as those involved in a crash should always be replaced.

All child restraints were fitted by the subject or their partners, possibly because of the high number which use the seat belt for installation (ie. no special fitments are required). As none were fitted though a garage (or retail outlet) the subjects interviewed had all missed out on a source of professional advice. This was most appropriate for the 4 subjects using tethers.

19 subjects transferred their child restraint between cars, therefore increasing the possibility of car-restraint incompatibility whilst 9 of the 19 who did this once a week or more would have greater familiarity, there would also be an increased risk of error.

Instructional information

40% of the adult subjects using second hand or borrowed restraints did not receive a full set of instructions, so missing vital information on usage and warnings. However, whilst 75% of all subjects still claimed to possess instructions, 59% of them reported that they never referred to them. Within this particular sample, referral to instructions did not appear to be affected by their clarity, as only 2 out of 11 subjects who did refer to the instructions found them difficult/very difficult although a larger sample may present a different picture.

The high proportion of adult subjects not referring to instructions may be attributable to the fact that 56% of those who never referred to them all possessed booster seats/cushions which are generally very easy to fit.

Aspects of installation

94% of adult subjects knew that infant carriers should face rearwards, though 62% did not know that this was to help prevent head/neck injuries.

The 4 restraints using tethers had all 4 straps connected, though 2 subjects did not know that these should be attached to the mounting point of adult belts, other parts of the car being potentially too weak to withstand crash forces.

17% of those subjects using their restraint on the rear seat felt it acceptable for the buckle to lie on the restraint's frameworkand 56% did not appreciate that this could lead to the buckle becoming undone in an accident (buckle crunching).

Only 61% of subjects said that they would always untwist webbing. In practice 53% had twists present in the belt, tether or harness, even though the avoidance of twists was stressed in the manufacturers' instructions. As this is not an obvious point, users may choose to ignore the problem, especially in the absence of instructions.

Whilst 82% of subjects felt that up to 5 cm of seat movement was acceptable, only 44% had installed their restraint within these limits.None of the instructions reviewed stated the likely consequences of excessive seat movement (ie. a child striking the car interior during a crash), but 86% of subjects were aware of this hazard. Part of the problem of excessive seat movement may actually have arisen from car-restraint incompatibility, a point which needs to be stressed by retailers when advising the public on their selection of a suitable restraint.

Aspects of fitting the child into the restraint
83% of subjects knew that the shoulder strap height should be level with or just above the childs shoulder, but in practice 24% had them well above or below this. Of those who were correct in practice, 9% did not know the correct height, and so had positioned them by chance. 74% of subjects did not appreciate the dramatic increase in restraint force associated with the rapid deceleration of a crash, and even the possibility of ejection.

32% of subjects quoted one finger width and 40% quoted two finger widths of harness slack as being acceptable. However, only 20% had one and 24% had two finger widths present. 56% had three finger widths or more present, thus 80% of the harnesses were too loose if one finger width is considered correct. 48% of subjects did not appreciate the danger of neck or thoracic injury associated with a loose harness. Knowledge clearly does not necessarily encourage correct practice as 43% of those who knew the hazards were incorrect in practice. It is felt that this aspect of misuse is due to a combination of ambiguous instructions and the difficulty often encountered in adjusting the harness.

56% of subjects did not know that the harness buckle should lie over the crotch and in practice 36% of buckles were positioned over the abdomen. However, crotch straps are often a fixed length which predetermines the buckle position which may not be suitable for certain sizes of child. 72% of subjects did not realise that a buckle positioned over the abdomen can cause internal injuries in the event of a crash. All 12 restraints with crotch straps used them, though 2 subjects did not realise that its purpose was to stop 'submarining' i.e. the child slipping under the lap belt.

Regarding booster seats/cushions, all used inertia reel 3 point lap and diagonal belts. 18% of subjects thought the lap portion should lie around the abdomen and 9% wherever it was comfortable. 44% did not realise the hazard of submarining associated with positioning belt above the pelvis. However, all correctly lay around the pelvis in practice as a result of using the belt guides on the restraints. This design clearly helps to reduce the risk of misuse.

91% of subjects correctly placed the diagonal portion over the shoulder, though 18% were found across the neck and 9% under the childs arm. 22% did not appreciate the risk of neck injury from a wrongly positioned belt.

Some restraints have a fixed diagonal belt guide, as opposed to an adjustable one, which may not allow for correct positioning. None of the instructions reviewed stressed the importance of the correct position, possibly leading to some accepting an inappropriate height. In light of this, there appears to be a need for making guides adjustable and for providing clear guidelines for their use.

GENERAL DISCUSSION
To varying degrees, errors were found in all aspects of installation and usage of child restraints (bar the observed position of lap belts around children). Similarly, there is an apparent lack of knowledge in all areas, which is by no means restricted to one sector of the sample. There were often high proportions who did not realize the implications of misuse and some of those that did were

still incorrect in practice. Most people did not appreciate the magnitude of the forces associated with a crash and how the extent of injury is related to certain aspects of misuse.

Having looked at aspects involving a number of subjects (some groups contained as few as 4 subjects), it was felt that the greatest cause for concern appears to be harness slack, as high proportions were wrong subjectively, (68% gave incorrect answers regarding amount, 48% did not appreciate the implications of misuse and observation revealed that 68% of restraints had excessive slack present). Other major problem areas were the height of the shoulder straps (16% incorrect answers, 74% unaware of misuse hazard and 64% incorrect from observation) and seat movement (18% incorrect answers, 14% unaware of misuse hazards and 56% incorrect from observation).

The findings from this study are in agreement with previous work. For example, Prior (1991) noted that 38% of harnesses were maladjusted overall (n = 2779) and Knott (1991) found 68% of harnesses were too slack (n = 121). Hodgson (1988) found 50% of seats (n = 36) had slack in seat belts or tethers, but for harness slack (n = 12) only 22% were unacceptable and only 16% (n = 19) had badly positioned shoulder straps. Misrouting of seat belts was not examined in the present study, though Shelness & Jewett (1983) found a 75% error rate in 2323 devices. Cynechi & Goryi (1984) found errors were lower through an enclosed frame with a belt slot (23% of 1006 compared with 41%).

Why then are such misuse problems occurring? Possibly because of the difficulty in understanding instructions and obtaining specific guidance and information relating to the implications of misuse. Difficulty in using the seat, such as adjusting the harness length, may also be a cause. The harness may be adjusted to fit a child wearing bulky clothes and not readjusted when this is not the case. Whilst not investigated here, it is known that car-restraint incompatibility can prevent a secure installation. For example: the seat belt may not be long enough for some rearwards facing restraints; the rear seat cushion profile may prevent the restraint from lying flat; and some car manufacturers have recently moved the rear seat belt mounting points to improve the positioning of the lap belt for adult passengers whilst making it worse for many current designs of restraint.

CONCLUSION

As only a relatively small sample was considered, any results should be interpreted with caution. However, it does appear that there is a need for the following:

i) Clear restraint instructions should be provided which contain all the relevant information and stress the hazards of misuse. These instructions should be able to be affixed to the restraint so that they can be read during installation or adjustment and are not so prone to being separated from the restraint when it is used by other people throughout its serviceable life.

ii) The design of restraints should be improved to encourage the correct and easy adjustment of the harness. Some of the more expensive current models offer a single strap to tension the harness on each trip which seems a good design solution although its cost may preclude it from the cheaper models.

iii) Public education literature on correct usage and its importance should be made more available as there are many restraints in use now that have been separated from their manufacturer's instructions or came with poor quality instructions.

iv) Ideally, a list of those restraints and cars that are incompatible should be clearly displayed by every retailer and retailing staff should be able to advise on correct installation and adjustment.

REFERENCES

Arnberg, P.W., 1978, "The Design and Effect of Child Restraint Systems in Vehicles".Ergonomics, Vol 21, No. 9, pp 681-690.

Burdi, A.R., Huelke, D.F., Snyder, R.G. And Lowrey, G.H., 1969, "Infants and Children in the Adult World of Automobile Safety Design: Pediatric and Anatomical Considerations for Design of Child Restraints".J. Biomechanics, Vol 2, pp 267-280.

Child Accident Prevention Trust, 1989, "children in Cars: Principles of Safety and the Law" rev.Occasional paper No. 12, CAPT, August 1989.

Cynecki, M.J. And Goryl, M.E., 1984, "Incidence And Factors Associated With Child safety seat misuse".US Department of Transportation National Highway Traffic Safety Administration. DOT HS - 806, 676.

Hodgson, J.R., 1988, "The Misuse Of Child Restraints".Unpublished Final Year Project Report.Department Of Human Sciences, Loughborough University Of Technology.

Kahane, C.J., 1986, "Evaluation Of Child Passenger Safety. The Effectiveness And Benefits Of Safety Seats".Us Department Of Transportation. National Highway Traffic Safety Administration, Technical Report. DOT HS 806, 889.

Knott, N.G., 1991, "The Fitting And Use Of Child Restraints In Cars, September 1991".Hertfordshire County Council, Transportation Department, "Goldings", North Rd., Hertford, SG14 2PY.

Prior, D., 1991, "Tightening Their Belts - An Essay On The Misuse Of Child Restraints In Cars".Educational Occasional Paper, 5 June 1991.British Standards Institution, Education Section.

Shelness, A. And Jewett J., 1983, "Observed Misuse Of Child Restraints".SAE/NHTSA Child Restraint and Injury Conference Proceedings, San Diego 1983, Warrendale SAE Paper 8 31665.

SAFE SURFACE TEMPERATURES OF DOMESTIC PRODUCTS

K C PARSONS

Department of Human Sciences
Loughborough University
Loughborough, Leics
LE11 3TU.

Surface temperatures of 'non functional parts' of domestic products should not achieve levels that cause burns to the skin. To design safe products, designers require information about skin reaction to contact with surface temperatures in terms of relevant factors such as user group, user behaviour, material type and condition and likelihood of contact and its duration. A review of literature is presented and a recently proposed European standard (PrEN 563) described. The standard is concerned with developing limiting values for the surface of machines and if adopted would provide greater protection for industrial workers than for users of domestic products.

INTRODUCTION

People come into contact with surfaces around them throughout their lives either directly with bare skin or through clothing. The temperature of the surface, along with other factors such as material type and condition, will influence the skin reaction to any contact in terms of thermal sensation and comfort or in extreme heat or cold, in terms of pain or skin damage. Some products are heated as part of their function. For example it is necessary for the hob on a cooker to become very hot to cook food. It is not the function of other parts of products to be hot, for example the front of an oven door or control knobs. These non-functional parts however, can become hot when the product is in operation. To design products with surface temperatures that are 'safe' it is useful to know which surface temperature would produce a burn, for example, when touched by bare skin. It is also useful to know the likelihood and nature of contact and by whom. These are considered below.

BURN THRESHOLDS
   A burn threshold is the temperature of skin above which a superficial partial thickness burn would occur and below which it would not. The most notable attempts to determine burn threshold data for human skin were carried out in the 1940s. Leach et al (1943) and Sevitt (1949) exposed anaesthetised guinea pigs and some rats with shaved backs to a heated brass cylinder at a number of temperatures. Henriques and Moritz (1947) conducted a series of experiments with anaethetised pigs and humans using flowing water across the skin. They plotted curves of average skin temperature, to produce a burn, against exposure time and provided the main source of data still used today. Siekmann (1989,1990) used a thermesthesiometer ('artificial finger') to measure contact temperatures of heated discs on a range of materials. Using the data of Henriques and Moritz he then proposed burn thresholds in terms of surface temperatures of typical materials. This work was used in the development of a proposed European Standard PrEN 563 (1992).

The principle of contact temperature is that when two surfaces of different temperature come into contact heat will flow from the hotter to colder surface. Although temperatures will change, there will be a temperature at the interface that will be achieved instantaneously and will not change. This is contact temperature Tc and is given by:

$$Tc = \frac{Tsk.bsk + Th.bh}{bsk + bh}$$

where
   Tsk = temperature of skin
   Th = temperature of the hot body
   bsk = thermal penetration coefficient of skin
   bh = thermal penetration coefficient of hot body

The thermal penetration coefficient is the square root of the product of the density, specific heat and thermal conductivity of a material. The contact temperature demonstrates the effects of different material types. For example if skin touches wood at 100 °C a low contact temperature would be calculated and it would show less severe reaction than if it touched metal at 100 °C where a high contact temperature would be calculated. Parsons(1992) considers the use of a modification to the contact temperature equation for practical use. He demonstrates how it could be used to calculate an equivalent contact temperature index (Tceq) and lead to a method for determining a reliable scale for establishing safe surface temperatures. This technique has potential but requires development and is as yet untried. No assessment method has used it to date.

ESTABLISHMENT OF SURFACE TEMPERATURE LIMITS

PrEN 563 (1992) allows the establishment of temperature limits for machinery in support of the European Machinery Directive (89/392/EEC and amended 91/368/EEC). Although restricted in application to machinery this standard is based upon data concerned with skin reaction and is not specific to a particular context such as machines. It would be just as valid to use it to assess domestic products whether machines or not. Other standards however provide limit values for domestic products. For example Harmonization Document CEN HD 1003 considers the front surfaces of domestic cooking appliances burning gas, that can be touched accidentally.

As an Illustrative example, consider a cooker assessed as if it were a machine used in industry and also as if it were a domestic product.

i) Assessed as a machine: PrEN 563 would apply and the following method used.

1) Identify persons who may touch the surface. Include those who will use the appliance (eg. adults) and those who will not use it but may still come into contact with it (eg. adults and children in the home or cleaners and on-line maintenance workers at work). Perform a task analysis to see who would come into contact with the surface and the likelihood of contact.

2) Identify materials from which the surface is made. (eg. smooth enamel in this example)

3) From the task analysis establish likely and maximum contact periods (eg. 4 seconds in this example).

4) Establish burn thresholds from figures 1 and 2. Figure 1 presents burn thresholds for bare (uncoated) metal. For a contact time of 4 seconds, these range from 58°C below which a burn would not be expected to occur, to 64°C, above which a burn would be expected to occur. Figure 2 provides the increase in burn threshold if they were coated in 160μm enamel. For a contact period of 4 seconds, this increase is 2°C. The burn threshold spread in this example is therefore 60 °C to 66°C.

5. The surface temperature limit value is between 60 and 66°C. A judgement must be made taking into account the context and area of application. For domestic use and risk to children a limit close to 60°C would seem reasonable. For industrial use a limit closer to 66°C may be reasonable.

ii) Assessed as a domestic gas cooker: HD 1003 would apply and the following method used:

The cooker would be operated under standard test conditions in a room at 20°C. To conform to the 'standard' the INCREASE in temperature measured in contact with frontal surfaces of the appliance which can be touched accidentally, must not exceed the following limits:

- Metal and painted metal : 60 K
- Enameled metal : 65 K
- Glass and ceramic : 80 K
- Plastic : 100 K

That is for a starting temperature of 20 °C, after 60 minutes of operation, the surface temperatures must not exceed 85 °C for enameled metal. According to PrEN 563 this would produce a burn.

SAFE SURFACE TEMPERATURES
    A safe product is one that will not cause injury. A safe surface temperature could be defined as one that would not produce damage to the skin (eg. a burn). This does not mean that it may not contribute to an accident. For example reaction to contact with a surface causing pain or simply surprise may lead to an accident. What is safe must therefore be considered in the context of a particular product, its use and its environment. All products will carry with them an associated risk. Risk assessment is a developing subject and has not been used in the context of safe surface temperatures. Thomas(1992) cites the UK Health and Safety Executive in providing the following equation:

Risk = Hazard Severity x Likelihood of occurrence

This can be used as a basis for design decisions. For example if the likelihood of occurrence is low then depending upon consequences, higher surface temperatures may still provide an acceptable risk. The question for the design of domestic products therefore is what is acceptable? It would seem unlikely however that acceptable risk should be greater for domestic products than that for industrial machines.

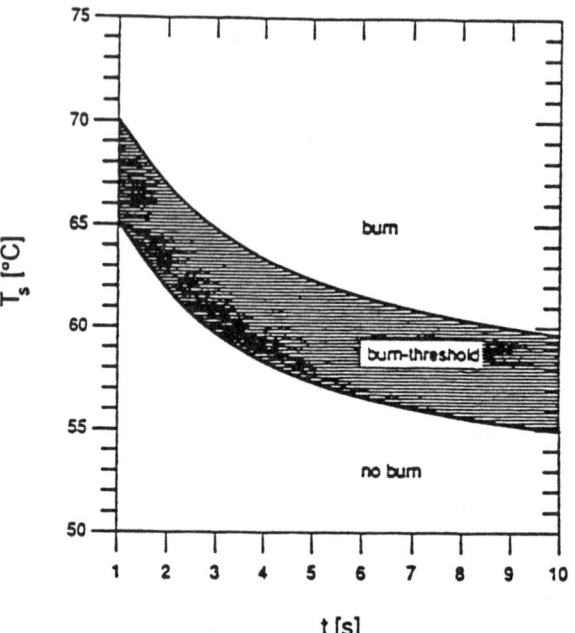

Figure 1: Burn threshold range for skin contact with
bare uncoated metal.

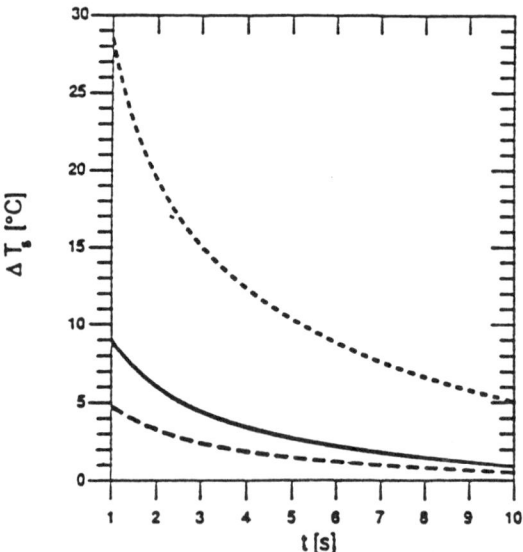

Figure 2: Increase in burn threshold for metals coated
with 400μm Rilsan(.....),90μm powder(_____)
and 160μm porcelain enamel(-----).

The appropriate design of products, involving safe surface temperatures, requires fundamental ergonomics research and knowledge about the reaction of human skin to contact with surfaces. It also requires methods in Applied Ergonomics, both to represent estimates of hazard severity, such as likelihood of burning (eg. Equivalent Contact Temperature - Tceq) and to quantify likelihood and overall consequences of contact leading to risk assessment, acceptable risk and hence design criteria in terms of safe surface temperatures.

REFERENCES

CEN PrHD 1003 (1988) Heating in contact with the front of the domestic cooking appliances burning gas. European Committee for Standardization, CEN, Brussels.

CEN PrEN 563 (1992) Draft for the safety of machinery: Temperatures of touchable surfaces:Ergonomics data to establish temperature limit values for hot surfaces. European Committee for Standardization, CEN, Brussels.

Henriques F C and Moritz A R (1947) Studies of thermal injury I. The conduction of heat to and through the skin and the temperature attained therein. A theoretical and experimental investigation. Am.J.Pathol.23,531-539.

HMSO (1992) The single market : Machinery. Department of Trade and Industry, March 1992, London.

Leach E H, Peters R A and Rossiter R J (1943). Experimental thermal burns, especially the moderate temperature burn. Quarterly journal of experimental physiology, 1943,32,67-89.

Parsons K C (1992) Contact between human skin and hot surfaces: Equivalent Contact Temperature (Tceq). In proceedings of the fifth Int. Conf. on Environmental Ergonomics, W A Lotens and G Havenith (EDs), pp144-145. TNO, Netherlands. ISBN 90-6743-227-X.

Sevitt S (1949) Local blood-flow changes in experimental burns. J. Path. Bact, 61. 427-442.

Siekmann H. (1989) Determination of maximum temperatures that can be tolerated on contact with hot surfaces. Applied Ergonomics, 20,4,313-317.

Siekmann H. (1990) Recommended maximum temperatures of touchable surfaces. Applied Ergonomics,21,1,69-73.

Thomas N (1992) Legal use of personal protective equipment and assessment of risk. In proceedings of the fifth Int. Conf. on Environmental Ergonomics, W A Lotens and G Havenith (EDs), pp48-49. TNO, Netherlands. ISBN 90-6743-227-X.

AN ERGONOMICS APPRAISAL OF THE PIPER ALPHA DISASTER

W. H. GIBSON   and   E.D. MEGAW

Four Elements Limited
Greencoat House
Francis Street
London SW1P 1DH

School of Manufacturing and
Mechanical Engineering
The University of Birmingham
Edgbaston
Birmingham B15 2TT

The following study is based on a review of the transcripts of the Public Inquiry into the disaster on the Piper Alpha offshore oil platform. The analysis is from an ergonomics perspective and provides ergonomics information about both the platform and the disaster. Four main areas of concern were identified from the analysis. These were the management of safety, interface design, training, and human behaviour in response to fire and evacuation. Inadequacies were found in the ergonomic consideration given to these areas on Piper Alpha prior to the disaster. These areas provide examples of the possible impact that poor ergonomic design can have on the safety of offshore installations. Finally this paper discusses the validity of using the transcripts from the inquiry as a method for ergonomic evaluation.

INTRODUCTION
Overview of the Piper Alpha disaster event sequence
Because of the death of key personnel and destruction of a major section of the platform during the disaster, there is uncertainty as to the causes and precise progression of the accident sequence. The following summary is therefore based around the most likely scenario presented in the Cullen report (1991).

On the 8 June 1988 an operator attempted to start a pump which was an integral part of the main production process. A failure in communication meant that the operator was unaware that a valve directly linked to the pump had been removed for maintenance and therefore the pump should not have been started. Prior to this a blind flange was used to cover the open pipework left by the valves removal but this had probably been inadequately secured by an operator. When the pump was started highly flammable hydrocarbons leaked past the inadequately secured blind flange and into a production module .

The hydrocarbons exploded, various engineered safety systems failed and further explosions and fires ensued. The primary cause of death to personnel was a large pool of flaming oil and gas fires which gave off large quantities of smoke, engulfing the platform within fifteen seconds of the initiating event. This made evacuation almost impossible. The majority of personnel died through smoke inhalation. This is the largest offshore disaster to date and lead to the death of 165 platform personnel.

Literature concerning the Piper Alpha disaster

The Cullen inquiry raised a number of issues which clearly deserved consideration form an ergonomics perspective. In particular chapters of the final report covered the permit to work system and shift handovers, training for emergencies and the management of safety. Also the design of evacuation equipment and evacuation procedures were key issues. Despite this input there was a surprising lack of expert evidence by human factors specialists during the inquiry. Literature directly relating to ergonomics and the disaster is briefly summarized below.

Tombs (1991) discusses the accident in terms of 'distorted communications', in particular, management's failure to act upon relevant warnings available prior to the disaster. Fitzgerald et al (1990) present a human factors approach to the effective design of evacuation systems supported by information from the inquiry transcripts. Munley & Williams (1990) discuss the failure of a senior manager on Piper Alpha to effectively direct the evacuation of personnel as an example of cognitive failure.

It is clear that ergonomics considerations were central to the disaster but, unlike such incidents as Three Mile Island, the extent of published human factors literature is relatively small.

METHOD

The study was centred around a review of the transcripts of the Piper Alpha inquiry. They are a written account of all that was spoken at the inquiry, numbering roughly 1500 pages. This study was concerned with the evidence given by the 61 survivors and those who were directly involved with managing, and working on, the platform prior to the disaster.

RESULTS

From the analysis, the areas found to contain sufficient information for further consideration were, the management of safety, training and interface design a further area, human behaviour, fires and evacuation, is treated in detail in Gibson (1992).

Management of safety

One of the central features of the management of safety is the importance of feedback to management as to the status of the system under their control. Also feedback must be processed in the right manner such that the information can be used effectively in improving systems safety. On Piper Alpha examples of problems with effective feedback and the failure to act on safety information prior to the disaster were found.

A technician was working under the permit to work system on a valve prior to the disaster. As he began to remove the valve, he was hit by a strong blast of flammable gas which blew an operators hat off but luckily did not ignite. This leak occurred because the equipment had not been adequately isolated. On reporting this incident to the maintenance superintendent, he was told that he had been hit by a blast of water and that no action would be undertaken. The information was thus suppressed even though the operators, who were completely dry after the incident, knew the diagnosis to be incorrect (39, 83)[1]. Feedback on an event with clear safety significance was thus suppressed. Possible problems with both the permit to work and isolation procedures could have been brought to the attention of the relevant safety authorities if feedback channels had been working effectively. Significantly both the permit to work system and the isolation procedures played a key role in the disaster.

It is clear that information on the potential safety problems on Piper Alpha were available but no action was taken to eradicate or reduce these problems. Cullen (1991, pp 231 - 234) discusses a previous accident on the platform in which an operator died. This had clear implications for the inadequacy of the permit to work and shift handover systems, again key issues during the disaster. The incident was not adequately considered by management and, therefore, recommendations arising from the incident were not adequately formulated to fully tackle these problems. It should also be noted that adequate feedback was not established to monitor the recommendations which had been proposed.

Similarly, a safety report on the platform which highlighted certain large scale design weaknesses was not considered adequately. One scenario presented in the report mirrored 'very closely what happened on Piper...' during the accident (110, 83) but this knowledge was not adequately incorporated into the safety management system and subsequently acted upon. The Piper Alpha disaster had, in some senses, been predicted and knowledge of this was available but ineffectively used by management.

Interface design

When a fire or gas alarm was identified in the control room the precise location of the source of the alarm could only be found from walking behind a panel to read further indicators (48,33). Thus, an ergonomic design principle was violated and functionally related equipment was not placed close together. This layout has important implications for safety because the precise location of any gas or fire problem needs to be quickly discovered. This is highlighted in the accident sequence where the control room operator did not have sufficient time to go around to the back of the panel to check the more precise location of the initial leak.

The frequency with which spurious fire and gas alarms occurred on the platform meant that they had become viewed as untrustworthy. This meant that, during the accident sequence, an operator was despatched to plant to see if he could identify the leak (48, 33). Using human 'leak detectors' is both

---

[1] Transcripts are referenced as ('Day of trial', 'page number')

unsatisfactory and will severely increase the time taken to identify a leak. The impact that a lack of consideration of ergonomic design may have is clearly highlighted in this case and has direct relevance to the disaster.

During the event sequence, operators were attempting to start a pump. In order to achieve this, controls in three different positions around the pump had to be used. Clearly functionally related equipment which was required to be operated in sequence was not located close to each other. This violates a basic ergonomics design principle. This lead to four important consequences:

1. Considerable time was required for a single operator to restart the pump.
2. If, as during the incident, one operator was at each control area, noise and lack of a line of sight between them produced severe coordination problems.
3. Operators could not check if erroneous actions were being carried out by another operator.
4. Possibly due to (2) and (3) above, an identical pump was unable to be started by operators. If this had been achieved the initial leak would not have occurred.

Due to a lack of evidence as to the precise actions at the pumps, the reasons why the identical pump was not started are unknown. From an ergonomics perspective this poor design has a large potential significance in terms of the disaster. During the trial, lack of human factors input into the inquiry meant that witnesses were not questioned as to the problems involved in restarting the pumps.

Training

Examples from the inquiry illustrated the lack of adequate training given for specific jobs. The control room operator stated that he had received no specific training for his job (49,4). A senior platform manager had not had specific training until three years after taking up his position (110,21). He also stated that he viewed the most essential part of his training as 'experience on the platform' (110,4). It should also be noted that both the control room operator and the manager had no official qualifications (49,4 & 110,4) and their knowledge was based around experience offshore. Personnel in these two positions played a vital role during the disaster.

One of the key features of the initial leak was that a blind flange covering open pipework had not been secured properly. The supervisor for this job had newly been promoted and he failed to check that the flange had been fitted securely. Also it is possible that he did not use the permit to work system effectively. Failure to transmit plant information through the permit to work, as to the status of this valve, was also a key factor in producing the initial leak. Clear inadequacies in the training of this supervisor were presented in the transcripts (50,87). As a new supervisor, with inadequate knowledge of safety procedures, the only option was for him to learn from his mistakes. In the context of the disaster these mistakes were to be an important causal factor.

DISCUSSION

Possibly one of the most important aspects of this study was the insight gained into the usefulness of public inquiry transcripts to allow a deeper understanding of the causes and background to major disasters. For this reason the final section will discuss the usefulness and validity of a consideration of the transcripts as a form of interview data.

Interviewers

These are the Representatives of Parties who questioned the witnesses. Inherent within the legal framework is the different emphases that representatives will place on factors considered during questioning. Representatives will act on behalf of parties to support their parties interpretation of events. For example, representatives of Occidental, the owners of Pipe Alpha, supported their clients interests by stressing the lack of corporate responsibility for the disaster. Each witness can be cross-examined by any Representative. This system allows Representatives to select the aspects of witnesses evidence to be viewed from various perspectives. It provides a range of views biased to different parties interests but hopefully provides a more objective overall view because the evidence will have been considered in a variety of ways.

Prior to the Inquiry the Representatives had access to a large body of data on the disaster such as witnesses' statements and the preliminary Petrie Report (1989). Thus many issues were already considered in detail before the Inquiry began. For example the accident sequence described above had already been assessed as the most likely course of events in the Petrie Report. Deeper level causes such as the inadequacy of the permit to work system, training for emergencies and shift changeover had already been identified.

Due to this fact and the nature of the Inquiry format, the questioning was highly directed, that is, the interviewer came with a clear idea of the types of questions to be asked and commonly lead witnesses to give a specific response. This is a useful technique to ensure that interviews are conducted efficiently, but also leads to bias in the data collected because there will be a tendency to reinforce previous findings rather than discover new ones.

In summary the present study was limited by the directed nature of the Inquiry in discussing issues that had already been shown to be important rather than covering a broader range of causal factors. Thus as broad a range of ergonomic issues were not presented as were first anticipated.

Interviewees

The interviews were conducted in public and within the Inquiry format. These factors produced bias in the evidence given by interviewees. For example, because those involved in the accident may feel that they could be made accountable for the accident they may be unwilling to present information which might implicate that them. One instance of this is that a supervisor, whose actions proved central to the most probable accident sequence, was often evasive in his answers to questions. Secondly many of the witnesses were still involved with the offshore industry. This again may bias the data because they were less likely to publicly criticise their employers. Despite aggressive questioning a senior platform manager resolutely refused to implicate the company's management as a causal factor in the disaster despite the evasive answers this commonly required. It should be noted that those who had left the offshore industry may also have been biased in their attitudes against their former employees.

The evidence should also be viewed in terms of its relationship to the disaster. Survivors were still suffering from the adverse psychological effects of the disaster. They were often confused about events both prior to and during the accident sequence and some had to retire form questioning due to the stress created by recounting these events.

Another important issue is that there was no expert witness who gave evidence on the role of ergonomics in the disaster. It is surprising that despite the central importance of human error in various forms and the key focus on such ergonomic issues as training, safety management and procedures that no witness was called to give specific evidence on the role of human factors in the disaster or for recommendations on this issue in the future. This can be contrasted with the high profile of ergonomics within the Nuclear Industry, found, for example, in the analysis of the Three Mile Island incident.

The Inquiry as an interview form is clearly not an ideal source form which to collect data for objective ergonomic study. Bias is clearly an issue in terms of the interviewer and interviewee, despite the broad objectivity and legally enforced honesty that was required. These methodological problems were outweighed by the detailed information available from the Inquiry and the lack of availability of other sources.

REFERENCES

Cullen, W.D. (1991) The Public Inquiry into the Piper Alpha Disaster (2nd edition) London: HMSO.

Fitzgerald, B. P., Green, M.D., Pennington, J.and Smith, A.J. (1990) 'A Human Factors Approach to the Effective Design of Evacuation Systems' in Piper Alpha Lessons for Life-cycle Safety Management Institute of Chemical Engineers Symposium Series Number 122, pp. 167-180.

Gibson, W.H.(1992) An ergonomics appraisal of the Piper Alpha disaster MSc Thesis: School of Manufacturing and Mechanical Engineering, the University of Birmingham.

Munley, G.A. and Williams, J.C. (1992) 'Cognitive failures: Experiences and remedies' in Human Factors in Offshore Safety - Seminar Documentation Business Seminars International.

Petrie, J.R. (1989) Piper Alpha Technical Investigation Interim Report London: Department of Energy.

Tombs, S. (1990) 'Piper Alpha - A Case Study in Distorted communication' in Piper Alpha Lessons for Life-cycle Safety Management Institute of Chemical Engineers Symposium Series Number 122, pp. 99-111.

SHIPBOARD ACCIDENTS IN THE ROYAL NAVY

P.B. JENKINS

Human Factors Department, Institute of Naval
Medicine, Alverstoke, Gosport, Hants.
PO12 2DL

Three hundred and seventy seven accidents that took place on Royal Naval surface
ships were analysed and classified according to their main causal factor. The most
common classes of accident involved falls (31 %), burns (9 %) and moving
machinery (8 %). A Priority Model was developed to ascertain the most important
classes of accident according to their frequency and consequences. The only classes
of accident that had high consequences were those that had a very low frequency
e.g. impacts with moving objects. The classes of accident that were considered to
be the most important, were falls, both on the same level and between different
levels, especially those involving ladders. It was concluded that there is a need for:
further research into accidents involving ladders and hatches; a laid down criterion
under which <u>all</u> accidents should be reported and modifications to the injury
reporting form, to improve the collection and subsequent use of accident data.

INTRODUCTION

Injuries suffered at work are a common and serious problem, representing a major cause of
mortality and morbidity in Western countries. In Great Britain during 1979 for example, one
in 200 workers in manufacturing and construction industries were killed or suffered severe
injuries. As with many other similar places of work, ships are often associated with high
accident rates. Due to the nature of the ships' design and the conditions in which personnel
on board must operate e.g. confined space and motion, risk is high and accidents and injuries
are likely.

Each year the Marine Division of the Department of Transport publishes accident figures in
'Casualties to Vessels and Accidents to Men'. Data from all vessels over 100 tons gross
tonnage that are registered in the UK are included. The most recent return from 1988 (DOT
1990), indicates that there were 603 notifiable accidents. Slips and falls were most common,
with those occurring on the same level accounting for 23 % of the total, and those occurring
between different levels for 14 %. Accidents involving manual handling (16 %), and those
involving machinery and tools (16 %), were also major categories.

The only military navy that has published detailed work on accidents and their causation is
that of the United States. Helmkamp & Bone (1986), found that shipboard accidents involving
falls and the use of machinery are amongst the most frequent, accounting for 18 % and 24 %
respectively of all shipboard accidents. Although it is difficult to determine the financial cost,

it has been estimated that accidental injuries accounted for almost 25 % of all days lost because of hospitalisation of American Naval personnel (Melton and Hellman (1977)).

The collection of accident data
The first step in taking effective action in the interests of personnel safety at sea, is to collect data on how and why accidents occur. After the collection of accident data, steps can then be taken to prevent accidents recurring, and/or attempts made to reduce the seriousness of any injuries sustained. To do this effectively requires reliable and accurate accident data, which can then be used to identify potential areas for accident prevention strategies.

Accident procedure in the Royal Navy
Although little direct accident data has been collected in the Royal Navy, data on all injuries received by personnel have been routinely collected for many years using MOD Form 298 (Report on injuries or immediate death resulting from other than natural causes - service personnel). In practice all shipboard accidents are briefly recorded in an accident log, and in addition a MOD Form 298 is raised if the accident results in anything more than minor injuries. MOD Form 298's provide a way of estimating the number of accidents occurring on board ships. A survey of the 377 records generated during the period 1986 - 1991, revealed that numbers reported have increased steadily since 1986.

METHOD

Development of the Priority Model
A priority model was developed in order to rate the relative importance of each class of accident. A priority model was developed in terms of two factors: Frequency and Consequences i.e. unavailability for work of injured person. The injured person's statement, provided the 'class' of the accident according to the main causal factor e.g. fall, burns etc. Inspection of all available forms produced the absolute frequency of each class. Examination of the sick - lists provided by each ship, revealed the consequences (in terms of time off duty) of the accident.

Procedure
Part 2 of the MOD Form 298 asked the injured person to give a statement '... explaining circumstances in which injury was received'. Each statement provided at least some genuine information on which a decision as to the main causal factor could be based. In some cases there were a number of factors involved, in such instances the main causal factor was identified. The frequency of each class of accident was then taken to be the number of MOD Form 298's identifying each main causal factor. Frequency of accident was also enumerated according to the part of the ship in which it occurred.
To quantify the consequences of the injuries sustained by the personnel, MOD Form 298's were matched with entries on the weekly sick lists supplied by relevant ships. The injuries sustained were classed into one of three groups. Injuries that required time off duty for examination and slight treatment only, were considered to be of low consequences and for the purpose of the model were arbitrarily assigned a score of 0. Injuries that required further time away from their normal duty, but did not result in the person being medically downgraded were classed as being of medium consequences and assigned a score of 2. Finally cases that involved the person being medically downgraded were classed as having high consequences, and were given a score of 4.
A matrix was constructed with each of the two variables having 3 elements - low, medium and high. An absolute frequency (as indicated by the number of MOD Form 298's) of above 30 was taken to be high, 15 to 30 medium, and less than 15 to be low. Similarly a mean consequence of less than 1 was considered low, between 1 and 2 was considered medium and

a mean consequence above 2 was considered high. The mean consequence value was calculated for each class of accident, according to causal factor and according to the location of the accident. Only accidents for which the eventual consequence of the injury sustained was known were included. A summary of the criteria by which each class / location of accident was categorised is shown in Table 1.

Collection of accurate accident data

To find ways of improving the collection of accident data in the RN, MOD Form 298 was carefully scrutinised for deficiencies and was also compared with standard checklists of the type suggested by Adams et al. (1981). Items included on such checklists include: exact time and date of accident; exact location of accident; age; sex; rank or rate; time into duty watch; causes of accident (e.g. from the Materials Handling Research Unit classification of causes of accident); nature and bodily site of injury.

		ABSOLUTE FREQUENCY (F)		
		Low	Medium	High
MEAN C O N S E Q U E N C E S (C)	Low	F = < 15 and C = 0 - 0.99	F = 15 - 30 C = 0 - 0.99	F = > 30 C = 0 - 0.99
	Medium	F = < 15 C = 1 - 1.99	F = 15 - 30 C = 1 - 1.99	F = > 30 C = 1 - 1.99
	High	F = < 15 C = > 2.0	F = 15 - 30 C = > 2.0	F = > 30 C = > 2.0

Table 1: Summary of the criteria by which the seriousness of the accidents were classified

RESULTS

Frequency by main causal factor

The total breakdown of the 377 incidents by main causal factor is shown in figure 1. Nearly a third of all the accidents were reported as involving falls, with 68 being from one level to another, the majority (62 %), of which involved ladders. Forty - eight involved falls on the same level, these included tripping over objects on the deck, and slipping and falling on greasy decks. Burns, moving machinery, lifting, chemicals and hatches were also each responsible for between 5 and 10 % of the accidents.

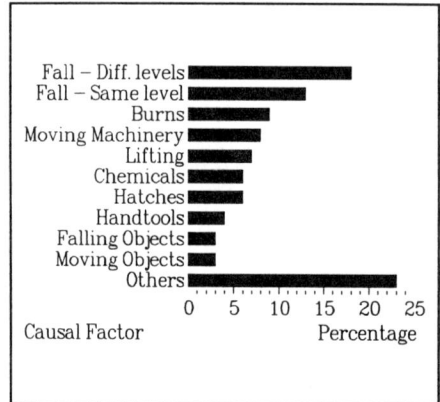

Figure 1: Frequency of accidents by main causal factor

Frequency by location

Location was often not easily identifiable as the MOD form 298 had no specific place to enter this information. Of the 377 cases only 192 could be definitely assigned to a particular location, of these ladders accounted for 23 %, with passageways and hatches, mess and dining areas, and the hangar and flight deck area accounting for about 18 % each.

Consequences

Table 2 provides a summary of the consequences of the 377 accidents. One hundred and thirty-eight of these resulted in time being taken off for examination and minor treatment only. The other 99 accidents all resulted in time off work or restrictions being placed on the availability of the injured person for certain duties.

Consequences (C)	No.
a) Time off for treatment only	138
b) Time off for treatment and put on light duties	4
c) Treatment and sick leave	57
d) Treatment, sick leave, and light duties	3
e) Treatment, medically downgraded, and sick leave	29
f) Unknown	140

Table 2: Summary of consequences of the 377 reported accidents

Matrix of frequency (by main causal factor and location) v consequences

		ABSOLUTE FREQUENCY (F)		
		Low	Medium	High
MEAN CONSEQUENCES (C)	Low	HANDTOOLS FALLING OBJ.	CHEMICALS	BURNS MOVING MACH.
	Medium		HATCHES LIFTING	FALLS (DIFF) FALLS (SAME)
	High	MOVING OBJ.		

Table 3: Matrix of accident frequency versus mean consequences - classified according to causal factor of accident

DISCUSSION

Frequency by main causal factor

Slips, trips and falls are the most common type of accident reported in many situations both at home and at work. With the added dangers encountered on board (e.g. ship motion and deck wetness), it is perhaps not surprising that this type of accident accounted for nearly one third (31 %), of all those reported at sea. What was perhaps more unexpected was the fact that of these, 36 % (11 % of the total), were due to falls from ladders. Ladders are used frequently by everyone on board so all personnel do have a high exposure to this particular hazard. Further, such familiarity may result in under-perception of risk. Users may become more careless therefore, resulting in a high frequency of ladder accidents.

Frequency by location

Unfortunately the MOD Form 298 does not specify the exact location of the incident to be entered. Thus answers were rarely specific and often just gave the name of the ship, making this a difficult task. However, the one factor that all the most commonly identifiable locations exhibited is that their use is not limited to any particular section of the ship's crew. Therefore accident frequency in these locations would be expected to be higher than in locations that are inherently more dangerous, but are used primarily by limited sections of the ship's crew e.g. engine room.

Consequences

Due to missing sick-lists the consequences to the injured person were identified in only 237 cases. Ninety-five of these accidents resulted in the injured persons taking time off work, either being confined in the ships sick bay, or being evacuated from the ship and placed in a shore hospital. Of these 95, thirty-five had further consequences in that the injured person was placed in a reduced medical category for up to 4 months. This means that they were unfit for duty at sea for the period stated, and as such was of the most serious consequence to the ship.

Matrix of frequency versus consequences

Accidents that resulted in injuries of a high consequence, occurred only infrequently. As indicated, the classes of accident that have a high frequency were those involving burns, moving machinery, and falls both on the same level and involving different levels. Of these 4 groups it is the falls that have the highest consequences. Falls on the same level mainly involved tripping over objects, and slipping on substances that made the deck hazardous e.g. rain, grease or oil.

Comparing locations on board ship with respect to accident frequency and their consequences, ladders were again evident. Accidents here were of a high frequency and medium consequences. Other accident locations that yielded similar results were those occurring in the mess and dining areas.

Limitations and assumptions of the priority model

Two assumptions must not be forgotten when interpreting the results of the priority model. Firstly, in calculating the frequency of each class of accident from the completed MOD Form 298's, no account has been taken of any differences in frequency of exposure to the hazard. To produce an effective comparison of the hazards on board, would have meant calculating rates of accidents per man day exposure. This is impossible due to the non-existence of the exposure data. In the absence of this, frequency was calculated from the number of completed MOD Form 298's as outlined earlier. Simplifying the consequences to time off work is also not the best measure of the consequences of the accident. This does not take into account rank or trade of the injured person. To compare the consequences of an injury to a chef and a similar injury to a radar operator is difficult if not impossible.

MOD Form 298 as an accident data collection tool

MOD Form 298 was not primarily designed to act as an **accident** reporting form, however it can provide useful accident data. It includes many of the essential points that are essential to a good accident data collection system. Experience showed that many of these details were completed rather too generally to be of use in accident data collection e.g. giving just the name of the ship as the location, or forenoon or afternoon as the time of the accident.

MOD Form 298 does have a use as an accident data collection tool, but much of the information it provides is not as specific or as definitive as it could be. Minor changes (e.g. request the exact location of the incident), would probably improve the quality of the data obtained. In addition analysis would be easier if the person involved was asked to indicate the cause of the accident from an extensive list.

There are two further items which if added would increase the value of the form. Firstly if the accident occurred on duty, then the time into the duty watch needs to be recorded. Secondly it would be useful to have a section that allows the final implications of any injury sustained to be recorded i.e. in terms of time off duty. Although this may prove time consuming in the short term, it would lead to much easier accident analysis in the future, with all the essential information being present on one form.

An alternative to modifying the MOD Form 298 in this way, is to introduce a completely new form which would be completed only after an accident. This form would be completed in addition to the MOD Form 298, but would be considerably shorter. Its big advantage would be that it would only provide data relevant to accident research. It could also be used to provide information on so-called 'near misses' and 'trivial events'.

GENERAL DISCUSSION AND CONCLUSIONS

A review of MOD Form 298's revealed that the number of reported shipboard accidents has increased since 1986, reaching a peak in 1989 and since levelling out at about 120 per year. Due to an increase in health and safety awareness and the need to prevent future negligence cases, it is likely that this is effectively an increase in the reporting level rather than an actual increase in the number of accidents.

The most important classes of accident identified by the priority model involved the following factors : falls, hatches and lifting. Of these, falls, both on the same level and between different levels were considered to have the most serious consequences. The most important locations of accidents as identified by the priority model occurred on ladders, in passageways and in the mess and dining areas.

The validity of the priority model used would have been improved if the accident reporting system had been more detailed. At present it is the judgement of the senior medical person on board each ship as to whether the injury sustained requires reporting i.e. a MOD Form 298 to be completed. Thus the seriousness of the accidents reviewed (in terms of injury received), differed quite considerably. Completion of a MOD Form 298 needs a clear criterion to be laid down under which this should be made mandatory.

The information requested on the MOD Form 298, if made more specific would benefit accident research. Recommended additions to the form are a section that allows the precise location to be reported and the selection of the cause from an extensive list. Alternatively a separate accident reporting form could be developed.

REFERENCES

Adams, N.L., Barlow, A. and Hiddlestone, J. (1981). Obtaining Ergonomics Information about Industrial Injuries: A Five - Year Analysis. Applied Ergonomics, 12, 2, 71-81.

Helmkamp, J.C. and Bone, C.M. (1986) Hospitalizations for accidents and injuries in the US Navy: environmental and occupational factors. Journal of Occupational Medicine 28, 4, 269 - 275.

Department of Transport, (DOT) Marine Division (1990). Casualties to Vessels and Accidents to Men: Vessels registered in the UK - return for 1988. HMSO.

Melton, L.J. and Hellman, L.P. (1977). Causes of Hospitalisation of Active-Duty Personnel. U.S. Navy Medicine 68, 18-20, 1977.

Ergonomics in the energy industry

AN INTERNATIONAL SURVEY OF HUMAN FACTORS INVOLVEMENT IN THE NUCLEAR INDUSTRY

LINDA HERMAN

Electrowatt Engineering Services (UK) Ltd.
Electrowatt House,
North Street
Horsham
West Sussex RH12 1RF.

Following a number of nuclear incidents, e.g. Three Mile Island, nuclear regulatory policy has been concerned increasingly with enhancing the safety and reliability of nuclear power plants. The policy entails the prediction of "human error" (and hence prevention) to ensure safety in the design, operation and maintenance of nuclear facilities. In particular, the implementation of highly automated plants has led to an awareness of failures that may be attributed to human error. Thus, human factors input is increasing in importance in the nuclear industry. This development resulted in greater interest in human factors research and development to ensure that advanced automation is not compromised by poor human reliability due to inadequate design. National and international attention on such concerns, is reflected by an increase in support for collaborative research and development programmes. This paper reviews the human factors role, research activities and impact for the nuclear industry in Sweden, Canada, France, UK and Germany.

INTRODUCTION

In recent years, human factors have gained increasing national and international prominence in the nuclear industry. This development is due largely to the occurrence of a number of nuclear accidents; such as Three Mile Island in 1975 and, more recently, Chernobyl. Invariably, a major contributor to the accidents is human error arising from poor safety management and operator-interface design, i.e. a system that does not support adequately the requirements of operators.

Consequently, nuclear licensing authorities have focused increasingly on human factors to better ensure the safe design and operation of nuclear power plants. For example, in response to accidents such as Three Mile Island, the United States Nuclear Regulatory Commission (USNRC) has sponsored extensive human factors research concerning the safety of nuclear power plants. Similar research which impacts nuclear licensing policy is ongoing in Europe. The initiatives include benchmark exercises and collaborative research funded by the International Atomic Energy Agency (IAEA) and the Organisation for Economic Co-operation and Development (OECD).

The research funded by USNRC is published widely and will not be described in this paper. Instead, the paper focuses on human factors research and developments in the nuclear power industry of other countries, namely Sweden, Canada, France, Germany and the UK.

In the account below, the nuclear regulatory approach of the particular country is outlined, followed by a review of the role and impact of human factors with respect to nuclear licensing. In addition, areas of international collaboration are highlighted where appropriate.

SWEDEN

The human factors department of the Swedish Nuclear Power Inspectorate, Statens Kraftinspektion (SKI), is known as "MTO", i.e. Man, Technology and Organisation. The goal of the department is to ensure safe and efficient interaction between the operator, the Technology and the Organisation; focusing in particular on the safe design, operation and maintenance of nuclear power plants. This goal is to be achieved by a policy comprising dual regulatory roles, namely the Regulatory and Supervisory Role, and the Safety Promotional Role.

The Regulatory and Supervisory Role is concerned with issuing formal regulations and guidelines and with the analysis of failure incidents. In addition, it is involved with the inspection, enforcement and licensing of nuclear installations and procedures. The Safety Promotional Role addresses the research and development required to support improvements in safety of nuclear utilities. Thus, they are concerned with the definition of scientific and technical issues and with the development of tools. To this end, the research areas investigated include the following : competence and training (both individual and team); control room simulation systems and computer-assisted control systems; the impact of work organisation, training and testing on human reliability in the context of plant maintenance and operation; methods for reporting and analyzing failure incidents, and for safety assessments.

Thus, the dual role approach facilitates the direct incorporation of human factors research into regulatory procedures which are implemented in collaboration with plant management. Such an approach engenders an effective safety culture that would improve individual and team training (hence the capabilities of operators in handling emergencies), leading to an overall reduction of human error in both plant maintenance and operation.

CANADA

The Atomic Energy Control Board (AECB) is the Canadian governmental body responsible for regulating the development and use of nuclear energy. The objective of AECB's human factors section is to establish human factors criteria that can be implemented in a regulatory licensing policy, and to define methods for assessment and evaluation of licensee programmes. Thus, a utilities' capability is assessed continuously against standards established by the AECB. A licence would be granted to a utility only if standards have been satisfied.

In summary, the focus of the human factors section of AECB is on the development of principles and criteria that apply to the human sub-systems, e.g. conditions of employment and work. To this end, research programmes have been commissioned to investigate the following concerns : human performance analysis and human error; organisational and management aspects such as quality reviews, supervision, operations communications, personnel fitness monitoring; plant upgrades such as better automation and more appropriate allocation of functions; working conditions (e.g. shift schedules, manning, training, etc.); operating procedures in both normal and emergency situations.

FRANCE

France's approach towards human factors activities in nuclear licensing has similarities with the approach adopted in the UK. In particular, the nuclear inspectorate specifies what safety criteria must be achieved, but emphasizes that the onus is on the nuclear utility to demonstrate how the criteria would be achieved.

The Commission a l'Energie Atomique (CEA), i.e. the Atomic Energy Commission in France, is responsible for specifying the human factors goals and research requirements, while Electricité de France (EDF), France's only utility, is responsible for carrying out the human factors evaluation and design activities in collaboration with the CEA and FRAMATOME (primary consultant).

France's safety concerns are focused primarily on the human operator since human error is regarded as an important contributor to system failure, i.e. the human operator is considered to be potentially unreliable (Colas, 1992). Thus, the research activities are directed mainly at uncovering how and why human errors occur, and how they can be prevented. In this way, appropriate design improvements may be made to minimise and eliminate (if possible) the probability of human error. To this end, particular emphasis is placed on improving the design of the human-machine or human-computer interfaces. To support these improvements, a major proportion of human factors analysis of normal and failure conditions is carried out with actual operators on large scale simulations (Carnino, 1988; Mosneron-Dupin and Lars, 1991; Mosneron-Dupin, 1988). The scope of such research activities comprises human reliability analysis (HRA), probabilistic safety analysis (PSA), accident investigation and analysis, risk management, analyses of operating procedures and interactions between operators as individuals, a team and an organisation, in the context of a complex system.

The results of EDF's investigations are usually analysed and discussed with CEA and FRAMATOME. Methods and recommendations are then agreed and incorporated subsequently into CEA's licensing policy. In addition, EDF's human factors research may impact nuclear licensing policy internationally through the IAEA, e.g. recommendations for incorporating human reliability analysis into probabilistic safety analysis.

These activities have benefited EDF in the following areas: more appropriate allocation of tasks between human and machine, improved designs of human-machine and human-computer interfaces, better operating procedures, training and organisational design. The likelihood of human error is thus minimised.

UNITED KINGDOM

Her Majesty's Nuclear Installations Inspectorate (NII) is responsible for nuclear licensing in the United Kingdom. The NII, in consultation with the major nuclear utilities such as Nuclear Electric Plc and British Nuclear Fuels Plc (BNFL), is responsible for specifying safety assessment criteria. To support such specifications, these nuclear utilities have comprehensive human factors research programmes that are ongoing. Since the human factors research activities are published widely (e.g. Ackroyd, 1991; Kirwan, 1989; Kirwan & James, 1989), they will not be reviewed here. It suffices to say that, the main research areas comprise human reliability analysis, safety management, personnel selection and training, system design, task analysis and design and operator-plant interface (Whitfield, 1990).

In summary, BNFL and Nuclear Electric are involved in a broad range of human factors activities that impact nuclear licensing directly. In particular, the onus is placed on the utility to demonstrate that the required safety criteria have been satisfied, e.g. for the safe operation of THORP and Sizewell B (see Whitfield 1987a, 1987b, 1988, 1990). Thus, the research undertaken by the utilities forms a part of the acquisition of a license.

GERMANY

Germany is, surprisingly, lagging behind its Western neighbours in integrating human factors in its nuclear regulatory policy. In particular, the contribution of human error to system failures have not been emphatically acknowledged or recognised. However, with the unification of East and West Germany, the position is shifting in favour of human factors.

The Kerntechnische Ausschuss (KTA) is the German body responsible for defining nuclear safety standards. The KTA comprises a standing committee of members from various organisations, e.g. technical supervisory agencies, nuclear utilities and universities. Current KTA Safety Regulations address instrumentation, control room design and installation. At the present time, specific human factors input to these regulations are under consideration, i.e. have not been finalised (see below).

The Federal Ministry for Radiological Protection (BfS) is responsible for initiating and procuring research and development programmes which are carried out primarily by nuclear utilities such as Gessellschaft für Reaktorsicherheit (GRS) mbH; technical supervisory agencies of each state, namely the Technischer Uberwachungs Verein (TUV); and commercial organisations (e.g. Elektrowatt). These organisations investigate and analyse potential and actual accidents, e.g. risk analysis (German Risk Study, Volume 2). The results of these investigations are used by the Federal Ministry for the Environment, Nature Protection and Reactor Safety, as the basis for licensing and operational assessment.

The imminent KTA human factors regulations together with the research outputs, will lead to : a better understanding of operator behaviours and potential types of human errors; improved operator-machine interfaces; the enhancement of existing techniques for human reliability analysis and probabilistic safety analysis. In other words, the research outputs enhance the safety regulations implemented by the KTA.

INTERNATIONAL COLLABORATION

The human factors and safety concerns of the countries surveyed are similar because international collaboration has been ongoing for many years under the auspices of the International Atomic Energy Agency (IAEA) and the Organisation for Economic Co-operation and Development (OECD).

The IAEA is an independent, inter-governmental organisation within the United Nations. IAEA has been involved extensively in human factors research (referred to generally as human-machine interface research) in the following areas : human reliability analysis and probabilistic safety analysis; expert systems; operational safety management (e.g. human performance indicators and incident analysis); safety culture and balanced automation and human action. These research activities are carried out in collaboration with the nuclear utilities of various countries. One outcome of the activities and international meetings is the international application of recommendations generated by a particular country, e.g. EDF's recommendations for incorporating human reliability analysis into probabilistic safety analysis.

The OECD is internationally funded to undertake human factors research activities in the following areas : the incorporation of human reliability analysis into probabilistic safety analysis; incident analysis; cognitive error analysis; personnel training; the application of information technology and expert systems in control room (OECD Report, 1990). The emphasis of OECD is on longer range research, i.e. it is generally concerned with theoretical development and prototype testing, rather than operational research which has immediate implications for the current nuclear industry.

CONCLUSION

The paper reports a survey of current human factors research and its contributions to nuclear licensing. The survey revealed common areas of human factors concern across countries. It is likely that these similarities may be a result of international collaboration led partly by the IAEA. However, slight differences in human factors emphasis remain due to the peculiarities of the nuclear licensing authorities of each country.

In particular, to reduce human error, the Canadian and Swedish licensing authorities seem to advocate the centralised research, development and specification of human factors criteria, which are then incorporated into regulatory procedures and safety standards. In contrast, France and the United Kingdom seem to devolve research to the individual nuclear utility. The research findings are then discussed and agreed with the licensing authorities and appropriate human factors criteria are thus specified explicitly. Consequently, the emphasis appears to be on self-regulation in the sense that the utility is responsible for demonstrating how safety requirements will be satisfied.

The French view of human reliability is less optimistic, and may be more realistic than the view of the Germans and Americans. In particular, the French authorities consider the probability of human error to be high, and thus focus the research on improving human reliability via better designs of the operator-machine interface (Chevallon, Colas and Ellia-Hervy, 1990).

REFERENCES

Ackroyd, P., 1991, The Role of Human Factors in Fault Studies - A Model for Design. In : Fault and Hazard Analysis for Nuclear Plant, IMechE seminar organised by the NEC of the Power Industries Division, 7th March 1991.

Carnino, A., 1988, An EDF Perspective on Human Factors. In : IEE Fourth Conference on Human Factors and Power Plants, June 5-9, 1988, Monterey, California.

Chevallon, Colas A. and Ellia-Hervy, A., 1990, Human Factors and Safety, Ten Years of Experience (Les Facteurs Humains et la Surete, 10 Annees d'Experience), SFEN, 6-7 December 1990.

Colas, A., 1992, Socio-Professional Culture and Operating Quality and Safety in EDF Nuclear Power Plants, EDF, France, 1992.

German Risk Study: Nuclear Power Plants -- An Investigation into the Risk due to Accidents in Nuclear Power Plants, Volume 2 Reliability Analysis (In German).

Kirwan, B.,1989, A Human Factors and Human Reliability Programme for the Design of a Large UK Nuclear Chemical Plant. In : Proceedings of the Human Factors Society 33rd Annual Meeting, Denver, Colorado, October 16-20, Volume 2.

Kirwan, B. and James, N., 1989, The Development of a Human Reliability Assessment System for the Management of Human Error in Complex System. In : Reliability '89, Volume 2, 5A/2/1 5a/2/11.

Mosneron-Dupin, F., 1988, Human Factors Data and the Use of Simulators, EDF France, Human Factors Seminar, 14 March 1988, Bournemouth, UK, ESRRDA Seminar.

Mosneron-Dupin F. and Lars, R., 1991, Probabilistic Human Reliability Analysis: The Lessons derived for Plant Operation at EDF, In : International Symposium on the Use of Probabilistic Safety Assessment for Operational Safety, PSA, 1991, IAEA-SM- 321/57, Vienna, Austria.

OECD Report, 1990, Institute für Energiteknikk, Annual report 1990, OECD Halden Reactor Project.

Whitfield, D., 1987a, A Regulatory Perspective on Human Factors in Nuclear Power. In : Human Reliability in Nuclear Power, 22-23 October 1987, Regent Crest Hotel, London.

Whitfield, D., 1987b, Human Reliability from a Nuclear Regulatory Viewpoint. In: Proceedings of the Ergonomics Society's 1987 Annual Conference, Swansea.

Whitfield, D., 1988, A Regulatory Perspective on Human Factors in Nuclear Power, London, IBC Technical Services Ltd. In : Human reliability in nuclear power: A state of the art evaluation of the nuclear power industry's position on human reliability.

Whitfield, D., 1990, An Overview of Human Factors Principles for the Development and Support of Nuclear Power Station Personnel and their Tasks. In : Proceedings of the IMechE International Conference, Quality Management in the Nuclear Industry, The Human Factor, 17-18 October.

THE DEVELOPMENT OF AN INTERACTIVE GUIDELINE DOCUMENT FOR ADVANCED CONTROL ROOM DESIGN REVIEW

Daniel L. Welch   Clifford C. Baker
Thomas M. Granda   Patricia J. Vingelis
Carlow International Incorporated
3141 Fairview Park Drive, Suite 750
Falls Church, VA 22042
John O'Hara   William S. Brown
Brookhaven National Laboratory
Department of Nuclear Energy
Human Factors & Performance Analysis Group
Upton, NY 11973

Next generation Nuclear Power Plant (NPP) Control Rooms are currently being developed in the United States. This paper summarizes an effort to develop a computer-based, interactive document of design review guidelines, intended to assist reviewers in the evaluation of the new control rooms.

INTRODUCTION

Advanced control room (ACR) concepts are being developed in the commercial nuclear industry as part of future reactor designs. The ACRs will utilize advanced human-system interface (HSI) technologies that may have significant implications for plant safety in that they will affect the operator's overall role (function) in the system, the method of information presentation, the ways in which the operator interacts with the system, and the requirements on the operator to understand and supervise an increasingly complex system. The U.S. Nuclear Regulatory Commission (NRC) reviews control rooms (including their HSI) to ensure that they are designed to good human factors engineering (HFE) principles and that operator performance and reliability are appropriately supported in order to protect health and safety.

The principal guidance available to the NRC, however, was developed more than ten years ago, well prior to these technological changes. Accordingly, the human factors guidance needs to be updated to serve as the basis for NRC review of these advanced designs. The purpose of this paper is to discuss the development of an NRC Advanced Control Room Design Review Guideline, hereafter referred to as the "Guideline." The term "Guideline" (with a capital G) refers to the entire document, while the term "guideline" refers to the individual guidelines within the document.

Brookhaven National Laboratory and Carlow International Incorporated, with the sponsorship of the NRC, are currently involved in the development of the Guideline. The overall objectives of this effort are:
• to develop an approach for the evaluation of advanced HSI;
• to develop a Guideline to support the review of advanced HSIs, based upon accepted HFE principles, standards, and guidance, from both within and outside the nuclear community;
• to develop an interactive, computer-based version of the

Guideline to facilitate guideline access and to provide user aids
to support the conduct of reviews;
• to perform tests and evaluations of the Guideline in order to
support its technical validity, scope, content, and functionality;
• to identify those areas important to performing reviews of
advanced NPP HSIs, for which available guidance is inadequate.

This paper will summarize the development of the interactive
Guideline.

REQUIREMENTS ANALYSIS

Based on the number of guidelines available and the potential
need for frequent updates due to rapidly evolving technology, it
was determined that an interactive, computer-based document could
provide improved access to the guidelines, provide user aids for
the compilation of guidelines needed for a specific review, and
permit easier modification/update.

An analysis of the inspection task and the variety of ways the
document could be used by NRC personnel was performed to identify
initial interactive document requirements. Four general categories of requirements were developed, as discussed below.

Review and Inspection Task Requirements revolved around supporting all aspects of NRC NPP control room review elements and
phases, modes of document use, and review-by-system efforts.

Usability Requirements reflected the belief that the design
should reflect the guidelines contained in the database, i.e. the
Guideline document should practice what it preaches. This let to
the following principles:
- Easy Learning & Retention
- Minimal User Errors
- Minimal Memory Load
- On-Line Help
- Efficient Performance
- High User Satisfaction
- Minimal User Input
- Meaningful Error Feedback

Electronic Document Functional Requirements included:
- Graphics Support
- Rapid Search Initiation
- Exact and Approximate
  Word Search
- Location Landmarks
- Evaluation Function
- Automatic "GOTO"
- Restricted Field Search
- Keyword List
- Browse
- Placemarker
- Note Taking

Hardware Requirements included:
- Storage Capacity
- Size
- Battery Life
- Weight
- Design
- Display Screen
  Input Devices

HARDWARE AND SOFTWARE

For the near-term purposes of mocking-up and evaluating a
prototype of the interactive Guideline, the current implementation
is in HyperCard™ on an Apple Macintosh™ portable computer.

HyperCard satisfies the functional requirements of the document
very well. It has the capability to display text in multiple,

scrolling fields which can be simultaneously displayed. This
provides users with a high degree of control over access to,
manipulation of, and display of information. Hypercard also has
the facility to implement such features as "book-marks" and an
embedded notebook. It supports "note links", whereby the user can
access notes pertaining to a specific topic or field. Graphics
support is included whereby non-textual, bit-mapped images can be
displayed in association with the text. In addition, HyperCard
offers a powerful help facility, allowing users to seek context
sensitive help such as clarification of a navigation problem, or
more detailed information pertaining to a specific area or field
in the document. A decision about the hardware platform for
Guideline implementation will be made over the next several
months.

FUNCTIONS AND USER INTERFACES
General
   Figure 1 presents the general design of the Guideline review
screen. The screen is divided into two sections; the upper section
displays the guidelines and related information; the lower section
provides reviewer support functions. The upper section contains
two "Areas". The "Guideline Area" (to the left) displays the
hierarchical context of the guideline, the guideline title, and
the text of the guideline. The "Additional Information Area" (to
the right) contains a window which provides further information,
clarification, or examples related to the guideline. Below this
window are three soft buttons. The "Source" and "Methods" buttons,
when activated, display the name of the source document for the
guideline and the suggested method of evaluation, respectively.
The "Show Figure" button will display any figure, table or graphic
associated with the guideline.

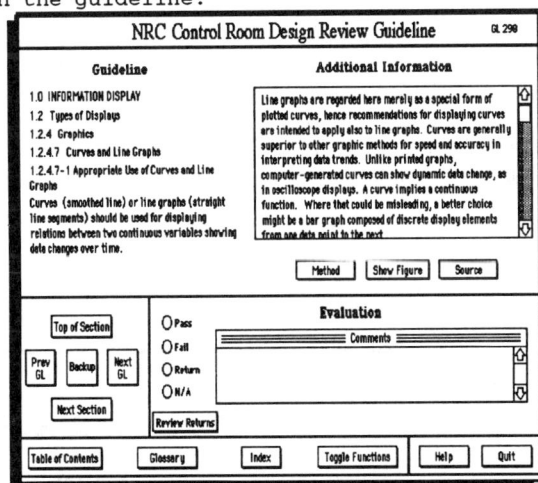

Figure 1. General Guideline Presentation and Review Screen

   The reviewer support section (the lower part of the screen) is
divided into three zones: guideline evaluation, navigation, and
document support. The guideline evaluation zone contains the
evaluation buttons and the reviewer's comments window. The navigation zone contains buttons for moving around in the Guideline
document. The document support zone (located across the bottom of

the screen) contains buttons that invoke functions frequently
used during a review; i.e., Table of Contents, Glossary, and
Index. The Toggle Functions button causes the document support
zone functions to alternate between those functions just mentioned
(the default display) and other less frequently used functions --
Clear All Evaluations, Search, and Report. At the lower right
corner of the screen are the Help and Quit buttons. When the Help
button is activated, the user can point to any field with the
mouse and click to display permissible operations from that field.
Activating the Quit button exits the Guideline.

Navigation Functions

The Next GL or Prev GL buttons cause the next/previous guideline
in the current section to be displayed. The Top of Section button
causes the first guideline of the current section to be displayed.
The Next Section button causes the first guideline of the section
following the current section to be displayed. The Backup button
allows the user to retrace his or her path through the Guideline
(a different function from Prev GL).

Evaluation Functions

The evaluation buttons permit the reviewer to record assessments of whether the system being reviewed conforms to the intent
of the guideline (Pass) or not (Fail), or whether the guidance is
not applicable to the system (N/A). The Return button allows the
reviewer to tag a guideline, indicating that it was applicable but
that insufficient information was available at the time to make an
evaluation. When the Return button is activated, a window appears
allowing the reviewer to indicate the reason for tagging the
guideline by choosing from a pre-defined list. Guidelines tagged
as "Return" can be reviewed at a later time by using the Review
Returns button. The reviewer selects the category of "return"
items to be reviewed; the Next button is used to move through the
selected items. The Next button appears on the screen (next to
the Review Returns button) only when returns are being reviewed.

Document Support Functions

Activation of the Table of Contents button opens a pop-up
scrollable window on the screen listing sections, subsections,
areas and subareas of guidance, which can be accessed by a double-
click (Figure 2).

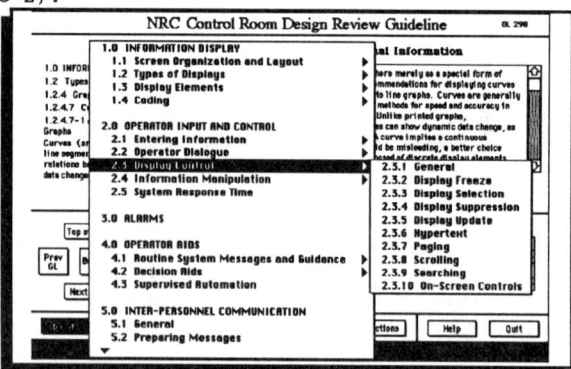

Figure 2. Table of Contents Screen

Activating the Index button causes the Context Index screen to be displayed. The index lists all words contained in the Guideline in alphabetical order, along with the count of their occurrences (Figure 3). Double-clicking on a word in the index displays all occurrences of that word in guideline titles or text; essentially a Key-Word-In-Context (KWIC) display. Each line in the window displays the selected word in the center and the surrounding text as it appears in the displayed section of the Guideline. Clicking on the desired text transfers th display directly to the guideline from which the selected text came.

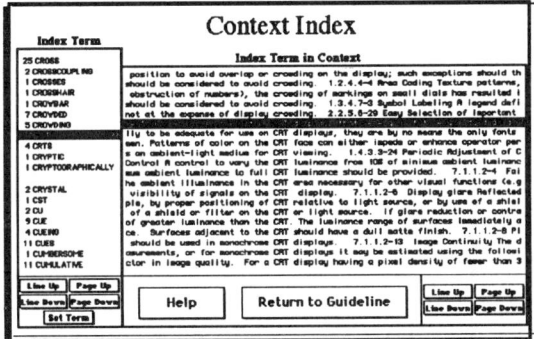

Figure 3. Context Index Screen

A generic HFE glossary, alphabetically arranged, is available through the "Glossary" button (Figure 4). Definitions can be saved by pushing the "Add to Holder" button and then the "Save" button. The definition can also be printed by activating the "Print" button.

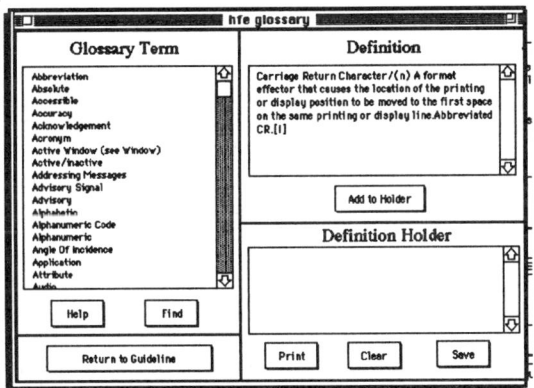

Figure 4. Glossary Screen

Secondary Document Support Functions

Searching the Guideline for a given term may be accomplished in three ways. Activating the Search button causes a dialogue box to be displayed that allows the reviewer to enter a search term. Pressing the OK button initiates a field-independent search for the term. Clicking on the current search term (shown to the right of the Search button) also initiates a field-independent search. Clicking in a specific field while holding down the Shift key limits the search to the selected field.

The "Clear All Evaluations" button causes all reviewer evaluations and comments to be removed. Because of the consequences of such an action, the reviewer is asked three times to confirm this operation.

When the Report button is activated, the report specification screen is presented (Figure 5). The screen displays the number of guidelines passed, failed, not applicable, or marked for later review. The reviewer may choose to include any or all of these evaluation categories in the summary. Activating the "Build Summary" button causes the specified summary to be generated and the summary report screen to be displayed. The report can contain all or as few as one guideline. The content of the summary can be viewed using navigation buttons. From this screen it is also possible to print the summary to a text file, send the summary to a printer, or return to the report specification screen.

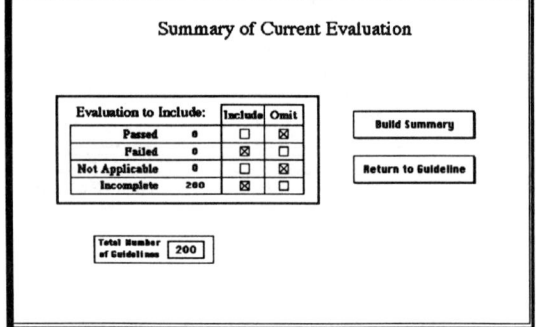

Figure 5.   Report Function Screen

System Help

On-line help is available from the main screen through the Help button. When activated, the help window is displayed and the user is provided instructions of how to utilize the help function. When in help-mode, the reviewer can point to anything on the screen and the help window describes the section of the screen and indicates how the selected function operates. Help is also available for the glossary and index functions.

FUTURE DEVELOPMENT

The approach established to review, develop, and integrate additional information into the Guideline will be an ongoing task in order to ensure that the document always has the most up-to-date and valid HFE review guidance. Since the technology is rapidly changing and the nuclear industry's experience (as well as that of other application areas) will be increasing, the Guideline will need to remain a living document in order to meet NRC needs.

ACKNOWLEDGEMENT

This research is being supported by the US Nuclear Regulatory Commission. The NRC Project Manager, Jerry Wachtel, provided many valuable recommendations on the development of this project.

Human error and process safety

HUMAN ERROR INCIDENTS IN ELECTRICITY SUPPLY

A.I. GLENDON

Human Factors Research Unit
Organisation Studies and Applied Psychology Division
Aston Business School
Aston University
Birmingham B4 7ET

While errors are essential for learning, when they occur in unforgiving situations they can produce undesired outcomes. The paper describes an analysis of 52 human error incidents from the electricity supply industry. Based upon the skills-rule-knowledge performance classification devised by Rasmussen and human error categories developed by Reason, five categories of human error incidents are revealed. A number of these involved errors at more than one level. The study briefly identifies types of interventions designed to reduce risk to system integrity including: managerial controls, human resources and communications as well as ergonomic controls. The S/R/K model is evaluated.

INTRODUCTION

The skills-rules-knowledge (S/R/K) classification of human performance developed by Rasmussen (1980, 1982) is a widely-used typology of human performance. Combined with human error categories (eg Reason, 1990), it is useful in diagnosing and analysing human error incidents.

While errors are essential for learning to occur, when they happen in unforgiving situations they can produce undesired outcomes. This paper presents an outline diagnosis of 52 human error incidents from a number of organisations in the electricity supply industry (ESI) over a number of years.

There is no central data collection system for the collation of human error incidents in the ESI and therefore no national details or aggregated data are available for analysis or investigation. Few studies exist in this area and no recent systematic study has been undertaken of incidents in this field. An earlier study is that of Hale (1970).

Incident outcomes typically vary from no effect upon plant or personnel - ie merely the recording that an erroneous action occurred, to temporary loss of supply to customers and plant or equipment damage - which may be slight or severe. Very rarely indeed do such incidents result directly in personal injury.

The sample of incidents analysed for this paper is not claimed to be representative of the population of ESI incidents. Without access to complete industry records, incidents could not be randomly selected and those analysed are an opportunity sample

only. Most of the material was obtained from reports which were
written as a result of inquiries into the incidents. In some
cases, this information was supplemented by personal interviews
with staff who were involved in the incidents and by visiting
incident locations for more detailed explanation.

The objectives of this paper are to:

1. Present a systematic analysis of a sample of human error
incidents within the ESI.
2. Outline a range of interventions to reduce the risk of ESI
incidents.
3. Evaluate the utility of an S/R/K based classification of human
error in analysing ESI incidents.

ANALYSIS OF 52 HUMAN ERROR INCIDENTS WITHIN THE ESI

The categories or levels used for this analysis are:

1. Skill-based errors - for example, actions not as intended.
2. Errors at the skills-rules (S/R) interface - for example, while
carrying out a sequence which has been carried out previously.
3. Rule-based errors - for example, when following a new but
predetermined sequence, such as a series of switching
instructions.
4. Errors at the rules-knowledge (R/K) boundary - for example,
following a sequence of actions in a wrong location.
5. Knowledge-based errors - for example, not having required
information.

In a number of cases, errors at more than one level were
associated with incidents. However, the basis for the
categorisation is the level of initiation of the error sequence.
In the case of some incidents, this categorisation is rather
arbitrary because human errors associated with the incidents were
identifiable at more than one performance level. Thus, the
following categorisation is indicative rather than definitive.

Skill-based errors

There were 8 incidents in this category. Three of these were
literally slips of a physical nature in which the individuals had
lost their balance or moved awkwardly and knocked against
equipment which had triggered the incident. In all three cases, a
circuit had tripped as a result of the slip. A further 3
incidents occurred during sequences involving actions repeated on
a number of successive occasions - for example, operating
selectors on a series of panels on which the wrong action was
carried out - ie were slips in the sense of actions not intended.
Of the remaining 2 incidents, one was a lapse error which involved
a control engineer overlooking a symbol on a wall diagram and the
other could have been due to either a slip or a lapse which
resulted in an incorrect protection setting.

Errors at the skills-rules boundary

Seven of the incidents were identified as having resulted from
errors at the S/R boundary. The individuals involved in the
incidents classified under this heading were all engineers
carrying out sequences of events which were the same or very
similar to sequences which they had previously carried out. In

two of the cases, engineers were working with equipment which
performed the same function as that of very similar standard
equipment elsewhere but which had a different (in one case
contrary) operational feature. These errors are likely to have
arisen through the equipment being contrary to stereotype, thereby
confounding user expectations. The remaining errors under this
heading were slips and lapses which typically resulted in capture
errors - for example, a step in a sequence being omitted. In one
case there was a suspicion of a rule violation. It is also
pertinent to record that a few background (or 'performance
shaping') factors were mentioned in the incident reports,
including:
   - person's wife involved in a road accident earlier that day;
   - large number of excess hours worked prior to incident;
   - subsequently found to be ill at time of incident.

Rule-based errors
   Ten of the incidents were classified under this heading. As
much of the work of engineers and other staff involves switching
sequences, it is unsurprising to find a number of errors of the
'pure' rules type. However, 2 of these - both cases of incorrect
selection of an action - might have been classified at the S/R
boundary. The most common error type under this heading was steps
omitted from a sequence (5 cases) and one incorrect sequence
followed. There was one violation and one case of equipment being
modified - which might have been classified at the R/K boundary.

Errors at the rules-knowledge boundary
   Seventeen of the incidents were categorised as resulting from
errors at this intermediate level. It is unsurprising to find the
largest number of incidents at this boundary because much of an
engineer's work involves the application of well learned sequences
of actions (many enshrined in rule books) in situations which are
to an extent novel and which usually require some problem-solving
(ie knowledge-based) activity. Of particular note are situations
where responsibility for solving a problem (for example, how to
carry out a particular switching operation) is shared between two
(or more) people (for example, a field engineer and a control
engineer). The risk which may be accepted under these
circumstances could be greater than in the case of an individual
acting alone. Incidents categorised at the R/K boundary were
invariably more complex than those under other headings and
involved errors at more than one level. Each incident classified
under this heading is unique in respect of its precise combination
of error factors. Those which emerged were: shared responsibility
between individuals (10 cases), information missing/unclear (5),
rule violation (4), occurred early or late in shift (4), went to
wrong location (4), also involved skill-based error (3), equipment
modified/non-standard/unfamiliar (3), distractions (2), work
pressure (2), misperceived risk (2), step omitted from sequence
(1), personal factors (1).
   The total of 41 factors indicates that each of the incidents
classified at the R/K boundary had around 2.5 factors associated
with it, the most common being shared responsibility for an

operation and information unclear or missing. In some cases, an earlier error gave rise to a subsequent error (for example, switching at a wrong location).

Knowledge-based errors

The remaining 10 incidents were classified as being of the 'pure' knowledge (K) type. These were all unique and although there were typically many factors associated with each, the main factors which a number had in common were: shared responsibilities between two or more parties (8 cases), communication difficulties (6), lack of/ambiguous information (6), planning errors (3).

Overall analysis

The small sample size precludes valid cross-tabulated analyses, for example age or work experience of those involved with error category. In any case such analysis would be compounded by those cases, mainly at the K and R/K levels, in which more than one party was directly involved. There was no association between age or experience and the incidents studied - these tended to reflect the age distribution of the staff population.

In 18 of the 52 incidents, there was evidence of some form of work pressure, while in the remainder there was no such evidence. This suggests the possibility of portraying incidents on a bi-modal distribution in which around one-third are associated with individuals being particularly high in arousal (perhaps experienced as stress) and the remaining two-thirds occurring when arousal levels are low. A number of the incidents occurred near the beginning or end of a work shift - when arousal levels may be particularly high or low, depending upon circumstances. However, this type of allocation to time is problematic in the case of some incidents, particularly those at the knowledge level and at the R/K boundary, as some of these took place over a period of time rather than at an instant.

INTERVENTIONS

A number of approaches to the control of human error incidents are possible. Various forms of human reliability analysis, for example based upon error likelihood estimates, are described by ACSNI (1991). Another approach, also based upon systematic risk appraisal is task analysis for error identification, developed by Stanton and Baber (1992).

Once detailed diagnostic analyses have been completed, a variety of control measures may be recommended. While there is some irony in seeking technical solutions to what are classified as *human* errors (van de Kerckhove, 1992) humans cannot operate as reliably as machines, however much a sense of 'discipline' is instilled into them. Thus, if it is possible to operate something incorrectly, sooner or later it will happen somewhere. Various estimates of the frequency ranges of different error types have been published (eg Rasmussen, 1980, Embrey 1988).

Taken together, errors in the three categories covered by the skills- and rules-based levels, some of the same technical, design and ergonomic controls are likely to be common to their reduction and include:

1. Impose physical restraint - forcing function to make impossible things which are illegal (eg violations) - an example would be an interlock or cut off switch.
2. Remove human from system - machine does task.
3. Inform user of system state - ie make system state explicit.
4. Make system state more transparent - ie make present condition obvious to user.
5. Show affordances - make explicit what future states are/not available.

Control measures appropriate to dealing with the types of errors described under the K and R/K headings may be listed under the following headings:
1. Managerial - including: supervision, work scheduling, safety auditing, safety inspections, contractor selection, post incident procedures.
2. Human resources - including: training, selection and recruitment, authorisation, promotion, staffing levels, rewards.
3. Communications - including: reporting/recording systems, rules and procedures, safety documentation, information provision, interfaces within the organisation and with other organisations, employee involvement and consultation.
4. Ergonomics - including: design, engineering controls, procurement procedures, work practices and patterns.

While control measures targeted at incident antecedents are vital in seeking to reduce numbers of incidents, these are unlikely to be completely eliminated and therefore post-incident strategies are also likely to be required. In addition to considering some of the items listed under the managerial, human resources and communication control measure headings, often some form of counselling is required for those whose professional self-esteem may have been reduced by being involved in an incident. In a small number of cases, individuals have been involved in successive incidents within a short time period and it is possible that anxiety levels associated with the first incident have been a factor in a second incident.

EVALUATION OF THE S/R/K MODEL

A great strength of the S/R/K model, its simplicity and elegance, is also a weakness because it is hard to represent the complexity of factors in many incidents. For example, for a complete understanding of an incident it is necessary to know how the parties involved came to be in the positions they were in. Very few of the incidents were exclusively associated with errors of a 'pure' type; nearly all involved errors at more than one level and often involved more than one person. This is not a criticism of the theory but an observation which does not invalidate it nor preclude its use. It is also acknowledged that S-R-K should not be seen as a strict tripartite classification but as representing ranges on a continuum. In this study 46% of the incidents were attributed to errors at the interfaces of two of these levels. The Rasmussen classification usefully sets outer limits for human behaviour (and human error) although some categories within those limits may be problematic. It is difficult to classify some incidents, even in respect of the 'main' errors and distinguishing performance shaping factors from other issues is also difficult. The model may be unsuitable for classifying complex incidents - in which errors at the knowledge level are invariably involved. However, supplemented with more detailed data, it could be useful for identifying appropriate interventions. In the cases described here, given the complex and complete range of error types, a correspondingly comprehensive set of interventions will be required.

A final issue is that of sampling adequacy. How representative of human errors are those which lead to incidents in the ESI? One methodological issue here is that of data capture. There are signs of a revival of interest in 'near miss' reporting as a tool in the armoury of safety and loss control professionals (eg van der Schaaf et al., 1991). Other issues in this area are both technical (eg definition, practicability, awareness) and organisational (eg forgiveness, organisational culture, organisational learning). In the long term, this is likely to be an important approach both for monitoring the effectiveness of interventions and also seeking prevention strategies in place of waiting for serious outcomes.

REFERENCES

Embrey, D. E., 1988, Assessment and prediction of human reliability. In Health, Safety and Ergonomics. ed A. S. Nicholson and J. E. Ridd (London: Butterworths), pp. 33-47.

Hale, A. R. 1970, Accidents occurring during high voltage electrical circuit switching. Report to Electricity Council, (London: National Institute of Industrial Psychology).

Health and Safety Commission, 1991, Advisory Committee on the Safety of Nuclear Installations, Study Group on Human Factors, Second Report: Human Reliability Assessment - a Critical Overview (London: HMSO).

Rasmussen, J., 1980, What can be learned from human error reports? In Changes in Working Life. ed, K. D. Duncan, M. M. Gruneberg and D. Wallis (Chichester: Wiley).

Rasmussen, J. 1982, Human errors: a taxonomy for describing human malfunction in industrial installations, Journal of Occupational Accidents, 4, 311-335.

Reason, J., 1990, Human Error (Cambridge: CUP).

Stanton, N. A. and Baber, C,. 1992, Task analysis for error identification: a methodology for reducing human error by 'machine' design. Health and Safety Regulation, Assessment and Control. Health and Safety Unit, Aston University, Birmingham.

van de Kerckhove, J, 1992. Is human behaviour calculable? A few notes on 'human reliability'. Paper presented at Conference on Safety and Well-Being at Work: A Human Factors Approach, November (Loughborough).

van der Schaaf, T. W., Lucas, D. A. and Hale, A. R., (eds) 1991. Near Miss Reporting as a Safety Tool (Oxford: Butterworth-Heinemann).

THE DEVELOPMENT OF A METHODOLOGY TO STUDY MAINTENANCE SAFETY IN CHEMICAL AND NUCLEAR POWER PLANTS.

JANE CARTHEY,
INDUSTRIAL ERGONOMICS GROUP,
DEPARTMENT OF MANUFACTURING
AND MECHANICAL ENGINEERING,
UNIVERSITY OF BIRMINGHAM,
BIRMINGHAM, ENGLAND . U.K.

KEITH REA,
SAFETY DEPARTMENT,
BRITISH NUCLEAR FUELS PLC,
RUTHERFORD HOUSE,
RISLEY, NR. WARRINGTON ,
CHESHIRE, ENGLAND, U.K.

The paper describes a five stage methodology which was developed to study the impact of maintenance upon safety within nuclear power and chemical process plants. The five stages of the methodology are outlined and the implications of findings from a study using the methodology are then discussed. Central findings were that maintenance cannot be treated as an entirely generic problem and that both direct and indirect safety effects impact upon maintenance safety.

Maintenance involves the organisation and performance of tasks which have the ultimate aim of preserving the original design of the system. It is a multi-faceted area covering several distinct types of activity. Two types of maintenance activity are scheduled and breakdown maintenance. Scheduled maintenance involves defined intervals for maintenance activity which are adhered to in order to satisfy regulatory requirements and ensure system safety and availability. Breakdown maintenance can be defined as involving the restoration of a failed system to a functional state.

While scheduled and breakdown maintenance activity refer to task performance at the sharp end of the system, i.e. at the maintenance task interface, the management of maintenance should also be considered when analysing maintenance-safety relationships. Maintenance management involves decisions as to the structuring and prioritisation of maintenance tasks. These are separated in temporal and physical proximity from the maintenance task interface, i.e. where maintenance operations occur.

The need for human factors intervention in maintenance activities is supported by research which points to maintenance as a root cause of system incidents (Rasmussen, 1980). Areas of maintenance activity where problems have been identified include design, procedures, communication, training, and maintenance management systems. Literature in these areas is briefly reviewed below :

It has been stated that failure to consider maintenance requirements in the design stage leads to problems for maintenance personnel, such as constrained physical access (Christensen, 1981), and the need to adopt unergonomic body postures to carry out the maintenance task (Haslegrave, Tracy and Corlett, 1987).

Recent research on maintenance procedures has shown that personnel not following procedures or deficient procedures for maintenance lead to a significant proportion of reportable incidents (Badalamente et al. 1982). Similarly, recent research by Law (1981) has shown that communication problems impact upon the maintenance safety relationship by causing maintenance delays and extended outages.

As far as training is concerned, relationships have been found between plants with good maintenance training programmes and above average safety performance (Young, 1987). Finally, research on maintenance management system impacts upon the maintenance safety relationship has shown that failures can lead to difficulties, such as communication problems with the work order process, inefficient delineation between different maintenance functions, shortcomings in organisation and a lack of effectiveness.

While previous research has identified several areas where maintenance related safety problems occur, there is a need to study the complex interplay between these factors within a real life maintenance system. The methodology described in this paper permits such analysis. The methodology was developed for a study of chemical and nuclear power plants and it comprised of the following five stages; maintenance data collection, the development of a framework of maintenance throughout the plant life cycle, the specification of performance shaping factors influencing maintenance safety performance, a qualitative error analysis of the maintenance functions specified in the framework of maintenance functions throughout the plant life cycle, and error reduction measure prioritisation. Each of these stages is discussed below ;

STAGE ONE ; MAINTENANCE SAFETY DATA COLLECTION.

The first stage of the methodology was data collection using semi-structured interviews with plant personnel. Personnel involved at various phases of the plant life cycle and different levels within the organisation were interviewed, for example, designers, safety assessors, plant managers, health physics personnel, maintenance foremen and maintenance operators.

Interview questions were constructed which probed traditional areas where human factors problems with maintenance have been identified, for example, training, procedures, communication, maintenance management systems and design. The semi structured interview technique was adopted to allow opportunity for respondents to detail problems with the particular maintenance system that they work in and also to recount any specific maintenance safety problems that they had experienced. Also, interview questions aimed to get an insight into the performance shaping factors which affect maintenance, by identifying variables such as time available for task performance, environmental conditions and so on, which may have had either positive or negative effects upon task performance.

Information verification across respondents was sought, i.e. facts from one interview were verified in subsequent interviews with different personnel. The information was further verified by reference to company documentation such as incident reports, procedures and training manuals.

This data collection stage provided a database of information on maintenance safety problems within specific plants and across the organisation as a whole.

STAGE TWO; DEVELOPMENT OF A FRAMEWORK OF MAINTENANCE FUNCTIONS.

From the information collected in stage one, a framework of maintenance functions throughout different phases of the plant life cycle was developed. The plant life cycle refers to the plant evolution, it's infra structure and operational requirements over time, beginning at the point of preliminary design ideas. In the present study, the phases of the plant life cycle isolated for analysis were design, safety case, commissioning and operations.

Plant life cycle phases were then subdivided to facilitate a more detailed representation of maintenance operations. For example, in the study the operations stage was subdivided into maintenance management, maintenance foremen, maintenance operators, Health Physics and automated maintenance management systems. Each of the maintenance functions for these areas was then listed.

An example of the maintenance functions listed for a particular stage (the design stage) is as follows; equipment design, tools for the job, shielding requirements (against radioactivity), access to equipment to be maintained, technological advances, level of knowledge / skill of the maintenance operators and anticipation of future regulatory requirements.

STAGE THREE; SPECIFICATION OF PERFORMANCE SHAPING FACTORS FOR MAINTENANCE.

The interview data was used to define performance shaping factors (P.S.F.'s) which influenced maintenance performance and safety at each phase of the plant life cycle that had been delineated in the previous stage. Interview responses where subjects were asked what factors they felt influenced maintenance performance and safety were the primary source of information in this part of the methodology.

The maintenance performance shaping factors, selected on the basis of information verified across respondents, were then listed for each phase and maintenance function. For example, the performance shaping factors listed for the maintenance function "equipment design" in the design stage were early design meetings, cost versus safety compromises, feedback of information from maintenance staff, past experience of the designer and the complexity of the designers decision making task.

STAGE FOUR; QUALITATIVE ANALYSIS OF MAINTENANCE FUNCTIONS.

Qualitative analysis of each of the maintenance functions specified in the framework of maintenance throughout the plant life cycle was carried out. This analysis was presented in tabular format with the plant life cycle phase listed in the first column and the specific maintenance function under analysis listed in the second column (see Figure 1:0).

The performance shaping factors influencing each maintenance function were then allocated by the researcher, with reference to the interview data. These were then listed in the third column of the table.

In the next stage of analysis, the direct and indirect safety impacts of failures for each maintenance function were determined by the research. Once again reference was made to the interview data and the results listed in the fourth column of the analysis table.

Finally, possible error reduction measures aimed to prevent each of the direct and indirect safety impacts outlined were compiled. The interview data is the primary source of information for this, although examples from previous human factors research may also be applicable to the maintenance context under examination.

The product of this stage of the methodology is therefore a number of tables which detail maintenance safety problems at various stages of the plant life cycle and possible modes of error reduction for each problem identified.

FIGURE 1:0 ; TYPICAL ANALYSIS TABLE.

STAGE	FUNCTION.	P.S.F.'S.	SAFETY IMPACTS.	ERROR REDUCTION
Design.	Equipment	1.Cost versus safety compromises 2.Early design meetings. 3.Feedback of information from maintenance personnel. 4.Past experience of designers. 5.Complexity of designers decision making task.	<u>Direct:</u> design does not support safe maintenance. <u>Indirect:</u> Redesign is needed later in plant life cycle ; Hazards due working on a built system. Operational delays.	1.Computer database of maintenance 2.Inclusion of maintenance staff in early design meetings 3.Use of designers with commissioning experience.

STAGE FIVE ; ERROR REDUCTION MEASURE PRIORITISATION.

In the final stage of the methodology the identified error reduction measures were prioritised. Three separate categories of error reduction measure prioritisation were used. Firstly, plant specific error reduction measures for particular problems identified during interviews were prioritised. Secondly, generic

error reduction measures to counter maintenance safety problems
common in several plants were prioritised. These may be indicative
of wider organisational maintenance safety failings. Thirdly,
generic error reduction measures for different plant life cycle
phases, outlined in the maintenance framework, were prioritised.

These three categories of error reduction measure
prioritisation ensured that all levels of the maintenance safety
problems are addressed, i.e. plant specific, organisational and
plant life cycle problems.

DISCUSSION.

The usefulness of this methodology has been shown in a study
to examine maintenance safety in two chemical and one nuclear
power plant. Although the findings of this study are confidential
its broader implications for the use of the methodology and
maintenance safety research in general can be discussed.

Support for the use of semi - structured interviews as a
means of gathering data about maintenance safety was found in the
study. For both identification of precise maintenance safety
problems and performance shaping factors which influence
maintenance safety performance the semi - structured interview
approach was found to provide useful information.

The study also provided support for the usefulness of a
differentiation between generic and plant specific maintenance
safety problems in the methodology. Maintenance problems cannot be
treated as entirely generic across an organisation. Treating them
as such would mean that attention is diverted away from important
plant specific problems. Exclusive focus towards ergonomic
intervention in one area at the expense of other equally important
areas may result. Thus, there is a need for maintenance safety
research to separate out plant specific from more generic
maintenance safety problems.

The study also highlighted the fact that both direct and
indirect safety impacts affect maintenance safety performance.
Direct safety impacts are those factors which directly effect the
ability to carry out the maintenance task, for example, poor
design, inadequate procedures, training and communication.
Indirect safety impacts are those factors which have their effects
primarily through operational delays which, in turn, have knock on
effects for maintenance safety performance. For example,
operational delays cause maintenance backlogs which have safety
consequences for the performance of the maintenance task. Indirect
safety impacts are often found to arise from attempts to counter
the effects of direct safety impacts, for example, operational
delays caused by redesign of equipment or rewriting of procedures.

A final finding from the study which has implications for
future research is that maintenance safety problems often arise
from decisions separated in both temporal and physical proximity
from the hands on performance of the maintenance task. For
example, poor management , designer , commissioning and safety
assessor decisions or practices. Therefore there is a need for
future research on maintenance safety to examine these areas and
devise ergonomic interventions to counter their safety

consequences. The methodology presented here allows such analysis. The major limitation with the methodology is that it relies upon the subjective assessment of only one assessor. Thus, the methodology is open to the criticism that it may be subject to assessor bias. Future research should aim to use a team of assessors and ensure that interviewees are consulted before the prioritised recommendations are put into operation.

CONCLUSION.

It is concluded that the present methodology is useful for examining maintenance safety since it provides insights into generic and plant specific safety problems and both direct and indirect safety impacts. It also allows the study of problems which arise away from the maintenance task interface.

REFERENCES.

Badalamente, R.V.,Ashton, W.B., Chockie, A.D., Imhoff, C.H., Truby, D., and Matsumoto, S.W., 1982, Recommended Program for the Development of Maintenance Guidelines in Nuclear Power Plants. Prepared for the U.S. Nuclear Regulatory Commission, P.N.L.-4475, September, 1982.

Christensen, J.J.,1981, The Human Element in Safe Man Machine Systems. Professional Safety, March, p.31.

Haslegrave, C.M., Tracy, M. and Corlett, E.N.,1987, Industrial Maintenance Tasks Involving Overhead Working. Contemporary Ergonomics, E.D. Megaw (ed.),p. 197-202.

Law, T.M.,1981, Workshop Proceedings : Power Plant Maintenance and Maintainability. E.P.R.I., N.P.-8-3-LD. Prepared by Pickard, Lowe and Garrick Inc., Palo Alto, California,Electric Power Research Institute, March.

Rasmussen, J.1980, What Can be Learnt from Human Error Reports ?, In ; Changes in Working Life, eds. K.D. Duncan, M. Gruneberg and D. Wallis. John Wiley and Sons, London.

Young, M.E.,1987, The Use of Simulations in Nuclear Station Field Skills Training. In; Proceedings of the C.S.N.I. Specialist Meeting on Training of Nuclear Reactor Personnel, Held at Orlando, Florida, April 21-24, U.S Nuclear Regulatory Committee, Washington D.C., U.S.A., p. 193- 203.

ERGONOMICS IN NUCLEAR POWER AND PROCESS SAFETY

Daniel L. Welch, Ph.D.
Carlow International Incorporated
4131 Fairview Park Drive
Falls Church, Virginia, 22042

The purpose of this presentation is to define the problem of accidents in the nuclear and petro-chemical energy industries, to very briefly trace the application of Ergonomics within the US nuclear power industry, to give an overview of on-going and near term developments within that field, and to draw what parallels are available to current events in the petro-chemical industry.

Ergonomics became a major player in the US nuclear power industry immediately after the Three-Mile Island accident. Since that time, Ergonomics has had major impacts within US nuclear generating plants. Like the Pre-TMI nuclear industry, the US Petro-Chemical industry has been generally unaware of Ergonomics and somewhat resistant to it's application. Yet experience indicates that human error can be identified as a cause of a large number of accidents within both industries.

Stating the problem is simple. "Accidents in the nuclear and petro-chemical industries can be catastrophic, involving major loss of life, vast numbers of injuries, the necessity to evacuate hundreds of thousands of people, billions of dollars in property loss, and enormous, adverse impact on the environment."

There are reasons for a public perception of threat, as shown in Table 1.

Table 1. Accidents involving death, injury, and evacuation lead to a public perception of threat.

• Mar 79 TMI	--0 Dead, Panic Evacuation, $1B Loss
• Dec 84 Bhopal India	--3500 Dead, 200000 Injured
• Nov 84 Mexico City	--650+ Dead, 2700 Injured, 200000 Evac
• Aug 85 Institute WV	--135 Injured
• Apr 86 Chernobyl USSR	--31 Dead, 500 Injured, 115000 Evacuated
• Jul 88 North Sea	--167 Dead, $1B Platform Loss
• Oct 89 Houston TX	--23 Dead, 130 Injured
• Dec 89 Baton Rouge LA	--1 Dead, 7 Injured
• Jul 90 Houston Tx	--6 Dead, 17 Injured

Interestingly, TMI was an engineering success in that there was no loss of life and no significant release of radiation. Thousands of people did evacuate their homes, however, and a billion dollar plus plant was lost. Bhopal is the worst case example. The Aldicarb Oxime and Methylene Chloride leak in Institute West Virginia eight months later unfortunately occured at a facility owned by the same company which owned the Bhopal facility. This did not enhance their public image.

Chernobyl was the worst nuclear power generating accident to date. It was indeed a major human, economic, and ecological disaster. However the use of safeguards which are employed as standard practice in the U.S. and a containment facility, which is also common to all U.S. commercial nuclear plants, would have avoided or greatly changed the nature of the event.

Two back-to-back chemical explosions in Houston Texas in 1989 and 1990 focused US public and government interest on safety in the petro-chemical industry. Since that time both the American Petroleum Institute and the American Institute of Chemical Engineers have published new guidelines on safety in the chemical process industry and the management of process hazards. The US Occupational Safety and Health Administration (OSHA) has published new rules for the management of highly hazardous chemicals. The combination of highly visible incidents and public and governmental response creates a parallel in the evolution of Ergonomics safety consciousness between the nuclear and the petro-chemical industries, which we will now pursue.

Three Mile Island was the most serious nuclear incident in the US. It was an Ergonomics failure in every sense of the word. For example, at the height of the incident, over one hundred annunciator alarms were flashing within the control room, simultaneously and without prioritization, and a new annunciator alarm was going off every second. The fact that critical emergency feedwater valves were closed was not recognized for critical minutes, because a maintenance tag from another control blocked the valves' display lights. What radiation was released, was released because the position display for a Pressure Operated Relief Valve was a demand versus a response display. The display showed what the valve was being commanded to do, not the actual status of the valve. Hence the display said "Closed" but the valve was actually stuck open, permitting radioactive steam to be released.

A number of actions followed TMI and many issues arose (Table 2). First, the Nuclear Regulatory Commission mandated that the industry undertake a detailed review of their control rooms to identify and correct Ergonomics related design problems.

Table 2. Post-TMI activity within the US nuclear power industry.

- Detailed Control Room Design Reviews (DCRDRs)
- Advanced Control Room Design
- Procedures Verification
- Maintenance Enhancement
- Training Enhancement

The DCRDR process (Table 3) was initiated in 1980 and was successfully completed in 1989. Hundreds of millions of dollars were spent by the industry to perform the reviews and to implement the improvements suggested by the reviews. However, the DCRDRs only solved the glaring problems created when plants and control rooms were designed and built with no consideration

given to the human factor. Enhancing and perhaps optimizing
control room design awaits a new generation of **advanced plant
and control room designs**.

Table 3. The post-TMI DCRDR process was composed of six elements.

> • Review of Operating Experience
> • Review of System Functions and Task Analysis
> • Control Room Inventory
> • Human Engineering Guideline Survey
> • Verification of Task Performance Capabilities
> • Validation of Control Room Functions

In the control room of a modern US nuclear plant, the controls
and displays are primarily manual and analog, with some digital
instrumentation and computer applications. The advanced control
room, however, is a truly revolutionary step in terms of Ergonomics. Digital instrumentation and computer generated displays
have replaced analog devices and computer control is fundamental. Artificial intelligence applications are massaging data and
taking much of the processing and interpretation burden off the
operator, freeing him for advanced supervisory activities. Hard-
wired, backups are available in case of seismic or other incidents which might disrupt computer functioning.

Ergonomics improvements in advanced plants are not confined to
control room instrumentation and controls, however. **Procedures**
for operating the plant can also be optimized. A great deal of
this effort has been accomplished as part of the DCRDR process
in terms of Emergency Operating Procedures, obviously a critical
consideration in that we want to insure that emergency procedures are effective and efficient. Yet the same consideration
should be given, and currently is being given, to the nature and
development of <u>standard</u> operating procedures, to insure that
safety and efficiency are not degraded as a result of boredom,
mis-directed attention, conflicting procedural demands, the
neglect of low-probability events, etc.

Outside of the control room, Ergonomics issues are being
actively investigated in the area of **maintenance**. As our
current nuclear generating plants reach the half-way point in
their expected lifetimes, the issues of preventive maintenance,
re-fitting, and retro-fitting are becoming more and more critical. Operations has developed a "safety culture" in response to
TMI – maintenance must now undertake an equivalent effort,
including the development and implementation of appropriate
training; the addition of external safety "eyes" for advising on
maintenance activities; and the employment of specific maintenance operating experience feedback.

Finally, Ergonomics concerns in **training** are currently being
investigated by a number of sources. A major issue in training
is the potential use of advanced simulators. Current US simulators permit utilities to teach operator candidates how to work
the reactor. This is equivalent to using a flight simulator to
teach someone how to fly. A much more useful application of the

simulator, however, is to teach an experienced pilot how to fly when things go wrong with his plane. While this is currently being accomplished in aviation, the present crop of nuclear control room simulators is generally incapable of training operators how to handle worst case accident situations and additional research and investment in this area is required.

The outcome of all these efforts is that much has been accomplished in terms of instrumentation and control effectiveness, work space layout for efficiency, the limitation of equipment demands to fall within human potentials, the availability and efficacy of protective gear, and equipment and procedure design to keep mental and physical workload within human capabilities under stressful conditions. However, Ergonomics research and development efforts must be continued post-DCRDR, in a number of new directions. The goal of these efforts is the eventual development of a standardised plant, based on optimized designs for the control room and other equipments, enhanced maintainability, and safe, effective, and efficient procedures for normal, non-normal, and emergency operations.

This prior experience of the nuclear power industry corresponds to current happenings within the petro-chemical industry today. The explosion in Houston Texas on October 23rd 1989 may have acted as the petro-chemical industries' TMI. Unlike TMI, 23 workers were killed in that accident and more than 130 were injured. Like TMI, Ergonomics considerations seem to have played a large role in the accident.

The OSHA Report on the explosion indicates the proximal cause of the explosion.

> The tests showed that the air hoses that supplied the
> air pressure (by which the actuator mechanism opened
> or closed the valve) **were improperly connected in**
> **a reversed position.** The hoses, connected in that
> way, would open a closed . . . valve even when the
> actuator switch was in the closed position. (US
> Department of Labor, 1990)

There is no more clear example of a Ergonomics deficiency. It simply should not have been physically possible to connect safety critical air hoses in a reversed position.

Like TMI, this incident was reported in the popular press. Unlike TMI, the coverage lasted for no more than three days. The government, on the other hand, has responded to this incident. As discussed above, OSHA conducted an investigation of the accident and published a report detailing the nature and causes of the accident, the industry accident history, and OSHA's program to prevent petrochemical accidents.

In summarizing the industry's accident history, the OSHA report references an earlier report by Charles Rivers Associates (US Department of Labor, 1988) which estimates that 25% of accidents within the industry as a whole are caused by equipment failure. Another 25% is attributed to "Human Error" and 33% is attributed to "Unknown Origin". The OSHA report notes that this analysis did not go beyond the obvious; that the underlying factors that contributed to the accidents were not examined.

An effective root cause analysis would undoubtedly place a good proportion of those unknown origins into the human error column. Additional analysis would probably show some Ergonomics issue at the root cause of the human errors.

The report goes on to estimate that the petrochemical industry can expect approximately 100 incidents per year, resulting in an estimated 53 fatalities, 985 injuries, and the evacuation of 18000 people. Public response to similar estimates for the nuclear power industry could be easily predicted.

OSHA's report to the president concluded with an assessment of the causes of accidents in any industry sector and 8 proposed OSHA actions to reduce the likelihood of accidents. Table 4 presents OSHA's view on the causes of petrochemical accidents (Garrison, 1989).

Table 4. Causes of loss in 150 industry accidents.

- Insufficient Recognition of Hazards
- Ageing and Poorly Maintained Equipment
- Unsafe Conditions or Procedures
- Poor Planning
- Improper Risk Management
- Unsafe Engineering Practices
- Inadequately Trained Personnel
- Disproportionate Attention to Production

Similar problems can be defined as the cause of many incidents within the nuclear power industry and Ergonomics is well experienced in analyzing and solving these types of issues.

Of the 8 proposed OSHA actions, 2 are of special importance. Action I calls for OSHA to:

expedite completion of its rulemaking requiring employers to implement comprehensive chemical process safety management plans for hazardous chemical processes. (US Department of Labor, 1990)

The new rulemaking, known as "Process Safety Management of Highly Hazardous Chemicals", became effective on 26 May 1992. The basic elements of the new regulations are shown in Table 5.

Table 5. OSHA Action I is composed of 10 elements.

1. Develop Management System to Identify and Address Hazards
2. Communicate That Information to Employees
3. Perform a Process Hazard Analysis
4. Develop Procedures to Accommodate Changes
5. Develop Safe Operating Procedures
6. Train Employees in Those Procedures
7. Develop a Preventive Maintenance Program
8. Institute a Hot-Work Permit System
9. Develop a Facility Emergency Action Plan
10. Make Contractors Aware of Hazards

These recommendations are very close to the mandated actions laid upon the nuclear power industry after TMI. Ergonomics should play an integral part in all of these efforts, especially the process hazard analysis, procedures development, employee training, and maintenance practices. (In fact, Ergonomics, or more specifically Human Factors Engineering, is specifically listed as a required element of the process hazard analysis.)

Action V of the OSHA report is as follows:

OSHA will employ all the means at its disposal to ensure that every establishment in the petrochemical industry implements technologies and safe work practices that are widely accepted and generally used by the industry and its contractors. The agency will encourage the petrochemical industry to incorporate new technologies into chemical processes to decrease the likelihood of a workplace accident. (US Department of Labor, 1990).

It can be argued that Ergonomics is both new to the petrochemical industry and an accepted, effective technology which should be implemented in chemical process safety.

What, then, should be directions in reaping the benefits of Ergonomics for the petrochemical energy industry? First, emphasis must be placed on the human operator as *the critical* safety system. This mind-set is basic to all Ergonomicss related system improvement and is a necessary first step in developing a culture of safety in industry. It requires placing the human factor on an equal plane with equipment integrity, management approach, and other elements viewed as important by upper management.

Second, all levels of organizations within the petro-chemical industry (especially upper-management) need to become familiar with the existence of the Ergonomics discipline and its applicability in this area, and to commit to implementing Ergonomics related safety programs within their own field of responsibility.

Third, the industry must clearly define and proceduralize the role of Ergonomics in the design and improvement of facilities. While the applicability of Ergonomics is obvious from the Ergonomics practicitioners' side, it is necessary to use the specific term and to direct specific actions, much as the nuclear power industry did after TMI, in order to insure the effective application of Ergonomics to the problems.

REFERENCES

Garrison, W.G., 1989, Large Property Damage Losses in the Hydrocarbon-Chemical Industries -- A Thirty-Year Review. 12th. Ed. (Chicago: March & McLennan Protection Consultants).

U.S. Department of Labor, Occupational Safety and Health Administration, 1990, The Phillips 66 Company Houston Chemical Complex Explosion and Fire: A Report to the President. (Washington, DC).

U.S. Department of Labor, Occupational Safety and Health Administration, 1988, Industry Profile and Cost Assessment for a Proposed OSHA Standard Covering Process Hazard Management in Chemical and Petrolium Industries. (Boston: Charles Rivers Associates.).

Memory and cognition

DOES REAL-TIME MEMORY PROCESSING CAPACITY DECLINE WITH AGE?

Neil Morris and Isobel Lamb

University of Wolverhampton
School of Health Sciences
62-68 Lichfield Street
Wolverhampton
WV1 1DJ

A number of recent studies have suggested that memory updating, a real-time memory processing task, can be an important element in the mental workload associated with managing complex, constantly changing tasks. These studies produce patterns of data similar to those found by Morris and Jones (1990) using a memory updating task. This study employed the latter task to examine the changes in real-time memory processing across age groups. Three groups, aged 16-21, 38-43 and 60-65 respectively, were compared. All three groups showed the expected effects of manipulating real-time processing but the groups did not differ in the magnitude of their response. All members of the two older groups were graduates and the youngest group were selected from a population of undergraduates and potential undergraduates. It is concluded that real-time processing is not impaired towards the end of the normal working life in a population of highly educated people.

INTRODUCTION

It is tempting to view human memory as an archive that the individual draws upon to perform a range of activities many of which are concerned with procedures relating to work. Indeed some researchers have acknowledged that our long-term memory specifically stores such procedural memory and this type of memory tends to remain intact even when brain damage leads to profound amnesia (Lishman, 1987). However, material withdrawn from long-term memory, and fresh input from external sources, is represented in working memory, i.e. real-time memory, and the contents of this real-time memory change second by second unless we retain material by some type of rehearsal. Because our working memory has a very limited capacity (within a given time period) and it is subject to a perpetual influx of new material while we are conscious its usefulness is limited by the rate at which we can update its contents (Morris, 1991). Memory updating involves the extraction of useful input and its synthesis with other material held in consciousness in real-time. For example, if one is reading a book then sentences are parsed and ones understanding of the text is updated but the verbatim text is not held. One extracts meaning, in real-time, and what is extracted is largely dictated by ones current understanding of the story at the time of reading. Someone who can do this more rapidly than another individual while attaining the same level of comprehension could be said to be a more efficient real-time processor. However such real-time data capture and synthesis need not be limited to the process of reading; the role of memory updating has been stressed in air traffic control (Yntema,1963) and within command and control scenarios

(Morris and Jones, 1988; Morris, Milne, Jones and Quayle, 1991). This literature is reviewed, from an ergonomic standpoint, by Morris (1991) and in theoretical terms by Morris and Jones (1990). Indeed it could be argued that memory updating may be a limiting factor in the comprehension of what is happening in the workplace (or elsewhere) at any given moment. Our own simulations of complex tasks, described in the above papers, suggest that information overload occurs largely because human real-time processing rate lags behind the rate of input into working memory.

We have developed a procedure that taps at least one facet of this updating process and allows us to statistically separate updating from passive memory storage. This task is an adaptation of the running memory task employed by Pollack et al (1959) and it involves presenting subjects with lists of consonants of varying lengths. Thus one list might have eight consonants and a different list might have ten consonants. The critical features of the procedure are that 1) the subject always recalls **only the last six items**, irrespective of the list length and 2) **the subject never knows how many letters there are in a given list**. This means that the subject must hold the first six items in memory and, if there are more than six presented on a given trial, they must drop items from the beginning of the list and add new ones to the end until presentation is completed. Earlier research by Morris and Jones (1990) has shown that this is not a passive process of just forgetting 'old' items; the six items are actively rehearsed. The experimenter can control both the rate of presentation and the number of updates that must be made.

Clearly this task is somewhat artificial but it does produce data that resembles that collected using a fairly sophisticated simulation of a freight service with a real-time updating requirement (Morris et al, 1991). This suggested to us that it would therefore be a useful tool for investigating real-time processing capabilities that are relevant to complex task performance.

In particular, the recent literature in cognitive gerontology has stressed that memory loss with aging, in a healthy aged population, may be restricted to the processing of complex information. Furthermore, it has been proposed that deficits are associated with executive level impairments to memory rather than loss of memory capacity/substrate. Executive functions are usually understood to be strategic deployment of memory resources by an individual with an intact memory. Thus one may have some memory resource but fail to take advantage of this. For example, Burke, Worthley and Martin (1988) note that older adults are less likely to engage in retrieval strategies, but rely instead upon spontaneous retrieval of information. This could be a motivational problem as Kausler (1970) argued that the aged are aware of the possibility of using retrieval strategies but do not **spontaneously** use them. This implies amelioration by incentive but other theorists suggest that there is an actual degradation of the executive apparatus.

Craik (1977) suggested that the aged show deficits only when the task confronting them involves active manipulation of information (as opposed to simple retention of material) or when divided attention is required (see Craik, Morris and Gick, 1990 for recent empirical demonstrations of this in a population aged 70+). It is therefore of some concern that aging individuals might show very serious cognitive impairments in the latter part of their working lives when they are engaged in tasks that involve creative thinking or abstraction of general principles from complex data sets. This would be particularly true for individuals who have followed a career path leading to greater responsibility for decision making after many years of work experience. Such 'executive' losses would be especially serious because a modest loss in simple memory capacity might be ameliorated by regular reference to a diary or other memory aid but loss of data manipulation and comprehension skills would be much more difficult to compensate for.

Our memory updating task is particularly taxing with respect to executive functioning because it requires real-time data transformation for successful performance. In addition, it allows us to test an additional, and simpler, model of aging memory loss. Welford (1958) argued that the aged are more susceptible to interference i.e. that activities that normally interfere with the memorial process have a relatively greater effect on older people. To test this possibility we have included an articulatory suppression condition in this study. Articulatory suppression consists of the repetition of a single word, e.g. "the", while trying to memorise material. It prevents the use of repetition as a rehearsal strategy and interferes with the use of verbal memory (see Baddeley, 1976 for a detailed account) and we know, from our previous studies, that it does not interfere with the updating (executive) component of our task. Thus by including this condition,with the memory updating task, we can examine the possibility of both executive and memory interference components of memory loss.

These two possibilities have clear implications for the predicted outcome. The task, plus articulatory suppression, yields three factors which could be impaired in an aged population. In addition, if the task is performed 'normally' a specific patterning of results not associated with aging should be produced. Age could interact with memory updating, indicating executive impairment, with articulatory suppression, supporting the interference hypothesis, or with serial position which would suggest loss of a memory substrate (serial position data provides a reasonably pure measure of memory in this task). This task normally produces main effects of all these factors and one particular characteristic finding is that there is a deficit associated with memory updating that does not vary in magnitude with the number of updates that must be made. Thus the main effects establish that the task is performed by the components of working memory suggested in Morris and Jones (1990) and any interactions with age should shed some light on the nature of the memory deficits found in aging, if any.

Finally, our studies of memory updating have been concerned with particular professional groups, namely those involved in the operation of complex real-time systems and our choice of a 'professional' population selected across most of the normal span of employment reflects this. Most studies of aging have used an elderly population up to 10 -15 years senior to our group.

METHOD

Thirty six subjects were recruited from three age groups, 1) 16-21; 2) 38-43; and 3) 60-65 years respectively. Each group consisted of an equal number of males and females. All members of the two older groups were graduates as were the parents of the younger group.

A BBC model B microcomputer was employed to visually present forty randomly generated lists of consonants of varying lengths. Each list contained either six, eight, ten or twelve items. The presentation order of trials of each list length was randomised, with the restriction that not more than two lists of the same length should be presented consecutively. The first eight trials were used for demonstration and practice, and contained two lists of each list length presented in the experiment. The remainder were split into two blocks of sixteen trials, containing an equal number of trials of each list length.

Response sheets provided for subjects contained six separate boxes for serial recall of the last six consonants presented on each trial. No indication was given as to the list length on any trial.

The procedure was identical for all three age groups. They each participated in two treatments, articulatory suppression and control (no suppression), and four conditions (four different levels of updating). During the articulatory suppression treatment, subjects were required to whisper the word "the" at a rate of not less than twice per second during both presentation and recall. The control treatment was identical except that articulatory suppression was not required.

The four levels of the updating factor were concerned with the number of items in each list. The task required the subject to recall, in forward serial order, the last six items presented on any given trial. Thus the number of updates required was a function of list length. When six items were presented 0 updates were required and when eight items were presented then +2 updates were required. Presentation of ten items required +4 updates and twelve item lists required +6 updates. All subjects completed an equal number of trials on all four levels of the updating factor, under both treatments. The order of treatments was counterbalanced to reduce the likelihood of carry over effects. Subjects controlled the instigation of each trial by pressing the space-bar to initiate a trial. A tone was heard half a second before and after a stimulus set presentation. The latter tone was used as a recall signal.

All instructions were given to subjects verbally, prior to the practice trials. During the experiment subjects were reminded of the condition they were in at the commencement of a block. They were also instructed to guess any items they could not remember and the necessity for forward serial recall was emphasised. Subjects were not informed how many trials of each list length were to be presented but it was stressed to them that some lists would contain only six items and that they should not therefore ignore any items. The experimenter was present throughout the session and subjects were run individually to ensure compliance with instructions.

RESULTS

The mean number of items correctly recalled (with serial position collapsed) is shown in Figure 1. A four-way, mixed design ANOVA was used to analyse the raw data. There was no main effect of age ($F(2,33) = 2.04$, $p>0.05$) but there were main effects of articulatory suppression ($F(1,33) = 22.37$, $p<0.0001$), updating ($F(3,99) =9.94$, $p< 0.0001$) and serial position ($F(5,165) = 23.40$, $p< 0.0001$). Articulation and updating did not interact ($F(3,99)$ 2.09, $p>0.05$) but both articulation ($F(5,165) = 3.92$, $p<0.01$) and updating ($F(15,495) = 6.30$, $p<0.0001$) interacted with serial position. Age did not interact with articulation ($F(2,33) = 2.27$, $p0.05$), updating ($F(6,99) <1$) or serial position ($F(10,165) = 1.04$, $p>0.05$). None of the higher order interactions were significant.

A simple main effects analysis was carried out to examine the interactions. Serial position affected both treatments except at serial position six. Further analysis revealed that serial position affected all four levels of updating but the decrement with updating was confined to serial positions 1 -3). Newman-Keuls comparisons on the number of updates showed significant decrements (all p <0.01) of all levels of updating (+2, +4 and +6) compared to zero updates. However, all comparisons amongst non-zero updates were non-significant.

To summarise, all the main effects and interactions found in earlier studies are present in this data. However there was no main effect of age and it did not interact with any of the critical manipulations (updating, articulation or serial position). Indeed age was the only factor that failed to account for a significant amount of variance in any comparison.

DISCUSSION

These data suggest that there is no impairment to memory updating across ones normal working life. In addition, there is no indication of any apparent loss of memory capacity or particular susceptibility to interference in the older population. The lack of these effects is unlikely to be due to a lack of statistical power given that all the F-ratio's associated with aging were very small and that the main effects associated with within subject factors were of considerable magnitude. It is likely, therefore, that there is indeed no impairment **in the population studied.**

A number of caveats need to be considered. First, we cannot be sure that these findings will hold for a more general population. Our population may, for example, be particularly healthy and well nourished relative to a broader sample of the elderly and there is no doubt that they are better educated and, perhaps, more highly motivated. Nevertheless, one can make a positive point i.e. that the finding that memory inevitably deteriorates with age probably does not hold for all sections of the population until well into retirement. The older employee may well be as mentally capable as a fresh graduate. However a second point should be borne in mind. This study tested memory across one hour only. It may be that older people have less 'staying' power that younger individuals. We cannot say more about this from our data other than to point out that much of the literature purporting to show aging deficits has been based on even briefer testing. The final point is rather obvious. Our memory task may not tap any ability that is employed in the work place or it may tap some ability that is of little

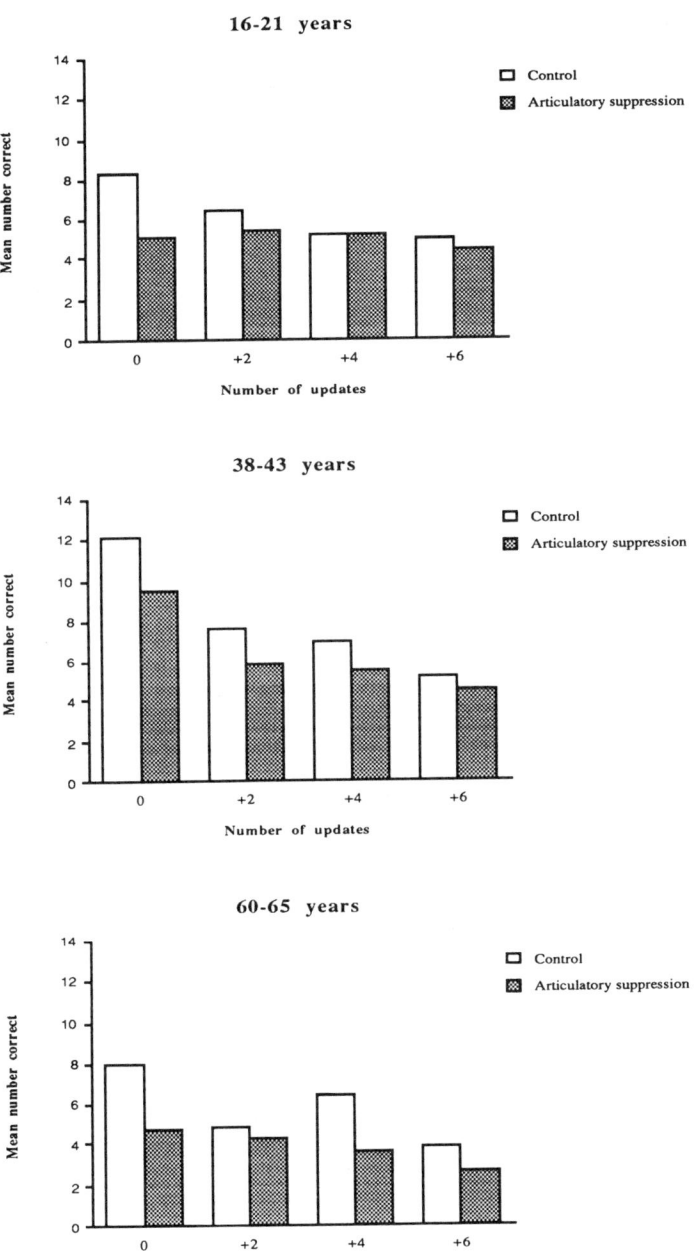

Figure 1. Mean number of consonants correctly recalled by 16-21, 38-43 and 60-65 year olds with and without articulatory suppression.

significance. We think that this is unlikely given the demanding nature of the task and its similarity to elements of many complex tasks. Given the catalogue of physiological deficits that Comfort (1964) suggests awaits the 'over-forties' we find it comforting that there is at least the possibility of retaining a high level of intellectual efficiency.

REFERENCES

Baddeley, A.D. (1976). *The Psychology of Memory.* New York: Harper and Row

Burke, D.M., Worthley, J. and Martin, J. (1988). I'll never forget what's her name: Aging and tip of the tongue experiences in everyday life. In Gruneberg, M.M., Morris, P.E. and Sykes, R.N. (Eds.). *Practical Aspects of Memory: Current Research and Issues, Vol. 2, Clinical and Educational Implications.* Chichester: Wiley.

Craik, F.I.M. (1977). Age differences in human memory. In Birren, T.E. and Schaie, K.W. (Eds.) *Handbook of the psychology of human aging.* New York: Van Nostrand Reinhold.

Craik, F.I.M., Morris, R.G. and Gick, M.L. (1990). Adult age differences in working memory. In Vallar, G. and Shallice, T. (Eds.) *Neuropsychological impairments of short-term memory.* Cambridge: Cambridge University Press.

Comfort, A. (1964). *Ageing: The biology of senescence, 2nd. Edition.* London: Routledge & Kegan Paul.

Kausler, D.H. (1970). Retention-forgetting as a nomological network for developmental research. In Goulet, L.R. and Baltes, P.B. (Eds.) *Life-span developmental psychology: research and theory.* New York: Academic Press.

Lishman, W.A. (1987). *Organic psychiatry: The psychology and consequences of cerebral disorder.* Oxford: Blackwell.

Morris, N. (1991). The cognitive ergonomics of memory updating. In Lovesey, E.J. (Ed.) *Contemporary Ergonomics 1991.* London: Taylor and Francis.

Morris, N. and Jones, D.M. (1988).The effect of relevant and irrelevant transcription on performance on a freight service simulation. In Megaw, E.D. (Ed.) *Contemporary ergonomics 1988.* London: Taylor and Francis.

Morris, N. and Jones, D.M. (1990). Memory updating in working memory: the role of the central executive. *British journal of psychology, 81,* 111-121.

Morris, N., Milne, A., Jones, D.M. and Quayle, A.J. (1991). Updating one's knowledge about the current status of vehicles in a freight service simulation. *Applied ergonomics, 22,* 401-408.

Pollack, I., Johnson, L. and Knaft, P. (1959). Running memory span. *Journal of experimental psychology, 57,* 137-146.

Welford, T.A. (1958). *Aging and human skill.* London: Oxford University Press.

Yntema, D.B. (1963). Keeping track of several things at once. *Human Factors, 5,* 7-17.

DOMESTIC ENERGY MANAGEMENT AND BILL DESIGN

D.I. WILLIAMS, J. YOUNG AND C.M. CRAWSHAW

Department of Psychology,
The University of Hull,
HULL, HU6 7RX

To conserve energy and to save money, householders
need to be able to control domestic energy consumption efficiently.   Control depends upon using
feedback, which could easily be provided by information on the bill.   But empirical studies show
understanding of electricity and gas bills to be
very poor.   It is demonstrated that a bill designed to make clear the necessary cognitive steps
involved can produce significant increases in
comprehension.

INTRODUCTION

Why study fuel bills?   Their manifest inadequacy as a
universal form of communication would be reason enough.   The
standard gas bill for example consists of a line of numbers,
the last of which is the cost of gas used - it would take an
above average member of MENSA to see any progression or link
between each set of digits.   In fact comprehension is impossible without reference to the back of the bill and the use of a
pocket calculator!   The main focus of our research is the
problem of domestic energy management and the role of the bill
in it.

Energy costs form a substantial part of any householder's
budget and for lower income groups it is frequently the most
substantial part.   It is apparent, therefore, that we should
aim to use energy efficiently and to get the most heat we can
for our money.   Efficient behaviour by individuals would have
two other advantageous consequences.   First, it would help to
conserve what is a natural resource and second, it would have
beneficial effects on the nation's economy.   Successive governments have varied in their approach to energy matters, as
motives, issues and attitudes have changed.   Policy has had
two dimensions, conservation and efficiency.   Following the
oil crisis of the 1970's the finite nature of energy resources
was acknowledged and conservation was promoted as illustrated
by the government's "Save It" policy.   But to save energy in
absolute terms might imply limitation of economic growth hence
the later move to concentrate on the economic benefits of using
energy efficiently as represented by the British Gas slogan
"Use It - Carefully".   There is a hint this year, with discussions over coal/gas stocks, that conservation thinking is re-

entering the political arena.   The policy of the present
government and preceding Tory administrations is to control
energy use by price.   This is a blunt and not very efficient
instrument.   For the poor, and particularly the elderly poor,
it produces a situation where even the most efficient energy
strategies do not allow resources to provide adequate heat and
lead to the annual crop of excess winter deaths through hypo-
thermia (Boardman, 1988).   But the basic problem with the
pricing policy is that consumers cannot react to price if they
do not know what their actions cost them.   The current situation is like shopping in a supermarket with no prices on the
goods, where shoppers put in their baskets what they need to
live and then queue at the check-out to see if they can afford
to pay.   And the next week, maybe another tin or two might be
added, but with no possibility of distinguishing between the
cost of a tin of beans and the cost of a tin of meat.   No-one
would be willing to shop like this, but it is the way that most
of us buy our energy.   We know something of how people cope.
There is evidence that on moving to a new house householders
keep consumption to a minimum, and then wait - with concern -
for the bill to arrive.   Subsequent bills are seen as bigger
or smaller (in terms of money, not energy consumed) and consumption is adjusted accordingly.
        Feedback provided by bills is impoverished in that it is
delayed and aggregated.   There is little link between leaving
a three kilowatt fire on all day and the total amount payable
on a bill which arrives three months later.   This situation
is exacerbated by the increasing use of estimated readings.
In fact, the fuel bill at present is generally used only as a
notification of charge.   Yet there is an opportunity to learn
from the bill if the way it was calculated could be understood.
At the very least, consumers ought to be able to find out
whether a bigger bill was due to an increase in tariff or to an
increase in consumption.   We know that people tend to judge
expense by the absolute cost on the bill rather than by the
unit price (Electricity Consumers' Council, 1985).   And there
is good evidence that people can learn to use information from
the bill to produce real economies (Ester, 1985).   This is
especially true if they learn to check the bill against meter
readings and then to keep track of consumption through the
meter, although it must be acknowledged that meters pose problems of their own and are very difficult to read (Gaskell and
Pike, 1983).
        So to conserve energy and to save money, householders need
to use energy efficiently.   To do this they need knowledge of
the amount and cost of energy used.   Bills should be able to
provide this information.   But they seem to be designed by
accountants who put their system needs ahead of those of the
customer.   This design failure also costs the energy utility
money.   For there is evidence that poor understanding of bills
produces expensive problems for the supplier.   The majority of
customer queries are about bills.   And all energy utilities
have to employ large numbers of staff to deal with queries and

complaints which could be reduced with adequately designed
bills.

In practice, fuel suppliers are complacent, despite evidence
from within their own organisations that bills are difficult to
understand (e.g. Attan, 1985) and despite criticism from independent bodies like the National Association of Citizens'
Advice Bureaus and the Consumers' Association.   Utilities
consistently deny that there are any difficulties (personal
communications) and spokespersons for the Consumer Councils do
not regard it as a serious issue.   As usual, queries are
usually attributed to the "idiot consumer" rather than the
inadequacies of the information systems.   Yet there clearly
are major problems.   Our research suggests that there is no
consumer who fully understands their gas bill - probably the
only public document that cannot be understood by one hundred
per cent of the target population!

The increasing number of fuel disconnections and the ever
present excess of winter deaths points to a failure of the poor
at least to maintain comfort levels within their income and,
for the rest of us, large fuel bills are unpleasant and restrict expenditure on other preferred items.   At least some of
our inefficiency can be attributed to the lack of learning
opportunities which bills could provide.   Our theoretical
model views the householder as a process controller (Dale and
Crawshaw, 1982; Williams et al., 1985) operating a house rather
than a plant or machine, but having to adjust various control
devices (thermostats, time switches, etc.) in line with tariffs
and anticipated demand, all under conditions of delayed feedback.   In this model, adequate feedback is essential if the
controller is to learn.   It is evident that the feedback given
by bills is inadequate and the information that they contain is
difficult to extract.   Let us consider the research evidence
for this statement.

EMPIRICAL STUDIES

In the first study (carried out by Nicola Browse) a strati-
fied sample of forty householders was asked questions about the
standard pre-metric gas bill {This gives a line of digits
across the bill indicating - Date of reading; present meter
reading; previous meter reading; cubic feet (100's); therms;
charges,£'s}.   Everyone could state how much had to be paid,
but only seventy per cent could locate the cost of a therm.
Only twenty-two per cent had any idea of how to check that the
bill was correct and nobody could explain how to check that the
therm reading was right.   And only half the sample could
actually make the necessary calculations when they were set out
for them in numerical form.   Using a points score for understanding each item on the bill, an index of comprehension was
calculated of fifty per cent.   In a parallel study, using
an Economy 7 electricity bill the index of comprehension similarly calculated was fifty-seven per cent.

So bills clearly provide comprehension problems for the
householder.   New bills were designed by defining the cogni-

tive steps which are required to translate the meter reading
into the final charge, and then setting out the steps on the
bill in the appropriate order.   This resulted in the bill
being set out as a progressive sum, as illustrated:-

present meter reading	3539
– previous meter reading	2299

= gas used (in cubic feet – 100's)	240

The new bills were then tested on a new sample of householders in a duplication of the original studies.   Results for the
gas bills showed a fifty per cent increase in people able to
understand how the charge was arrived at and a fifty-four per
cent increase in those who were able to follow the calculation
and check that the charge was correct.   The comprehension
measure rose to eighty-five per cent.   Similarly, in a parallel study, the Economy 7 electricity bill comprehension rose to
ninety per cent.   Attention to the cognitive ergonomics of
bill design clearly pays dividends.
    The advent of metrication gave promise of a new bill format
which it was hoped might be more user friendly.   Once the bill
had been in use for a while we carried out a second study to
see if the bill was indeed easier to understand.   The booklet
produced by British Gas – "Understanding Your Metric Gas Bill"
– received the "crystal mark" of the Plain English Campaign but
does not suggest that the new bill will be any easier to understand than the old {the new metric bill has a line of figures
across the bill, indicating – date of reading;   present meter
reading;   previous meter reading;   gas supplied – 100's cubic
feet – cubic meters;   kWh;   charges}.   To be fair, the information on the bill is complex.   Meters continue to be non-
metric and measure gas used in cubic feet which then needs to
be converted to cubic metres which in turn has to be translated
into calorific value to calculate the unit price which is in
kilowatt hours.   The British Gas booklet outlines seven steps
required to check the bill.   Their steps are confused, as the
single task of calculating the volume of gas used is separated
into two functions while the two steps of working out the
calorific value and then calculating the kilowatt hours is
collapsed into one step.   We have defined the steps thus –

1.   The volume of gas used is calculated by subtracting
      the present meter reading from the previous one.
      This gives an answer in hundreds of cubic feet.
2.   Cubic feet are converted to cubic metres by multi-
      plying by 2.8.
3.   To calculate how much heat was provided by this
      volume of gas the number of cubic metres is multi-
      plied by the calorific value (given on the front
      of the bill).

4. The number of megajoules (which is not shown on the bill) is divided by 3.6 to give kilowatt hours.
5. The number of kilowatt hours is multiplied by the price per kilowatt to give the cost of gas used.
6. The standing charge is added to the cost of gas used to give the total cost.

It would not be very surprising if householders were confused by this task when attempting to do it without the booklet and bearing in mind that most of the relevant information is in faint, small print on the back of the bill! We therefore took a small sample (thirty householders) half from detached and half from terraced houses and gave them the same questionnaire that was used in the first study.

Everyone could locate the amount to be paid. Eighty-seven per cent could locate the cost of a kilowatt hour but only seventy-four per cent had any idea what it was. Sixty per cent of the sample were unable to say what the figures meant or how the cost was calculated and nobody could do the calculation without prompt cards.

Overall the level of understanding was very poor with nobody able to do the calculation required and only thirty-eight per cent having any idea what it was about. An obvious way to improve the situation would be to set out the calculation together with all relevant information on the face of the bill, as demonstrated in our first study. But there is a more radical solution. Gas bills are more complicated than they need be because British Gas does not quote a fixed price, although of course the gas price is, in fact, fixed for each billing period. Price varies with calorific value and presumably British Gas feel that if the price was seen to rise because of "good quality gas" this would be seen as a real price rise and lead to public criticism. But by the same token, the price would be seen to fall if the calorific value was lower. Surely, this accuracy of information together with political expediency is not sufficient reason for confusing the consumer by expecting them to make these calculations in order to understand the bill. As all the calculations between the measure of cubic feet and the fuel cost are the same for each consumer in each quarter, it would be a simple matter to quote a unit price for gas on the bill. Not only might this make the bill easier to understand, it would provide a simple unit metric to allow consumers to check their consumption over successive quarters. A group of students at the University of Hull (M. Busfield, N. Cairns, G. Clayton, E. S. Patricio, S. Rodgers, and E.Rowley) devised this strategy and designed a mock-up of a unit cost bill. {This new design has bold headings of meter reading (present - previous); units used; unit price (pence); amount £'s}. But before testing it, they conducted a survey of twenty households investigating attitudes to, and understanding of, the metric bill. Results were broadly in line with our previous study. But in addition they were able to examine comprehension of each step in calculating the total gas used.   Results were as follows:

Step 1 (calculating cubic feet used)   : 85% successful
Step 2 (converting to cubic metres)   : 58% successful
Step 3 (calculating calorific value)   : 31% successful
Step 4 (calculating kilowatt hours)   : 31% successful
Step 5 (calculating cost of kilowatt hours)   : 50% successful
Step 6 (adding standing charge for total cost) : 58% successful

Then they obtained a sample of sixty students with no previous experience of paying gas bills;   half were shown the metric
bill and half the new unit cost bill for thirty seconds, then
allowed to keep it before them while filling in a questionnaire
to test their comprehension.   Three measures were used:
a) a user friendly rating;   b) a comprehension score;   c)
time taken.
The unit cost bill was judged nearly twice as user friendly as
the standard metric bill, and was significantly easier to
understand ($F = 26.25$, df 1/58, $p < .001$) and significantly
faster to use ($F = 33.3$,   df 1/58, $p < .001$).   The new design
is clearly effective.

CONCLUSION
     Many documents, fuel bills, bank statements, sales invoices,
appear to be designed to meet the accountancy system of a
company rather than the needs and limitations of the consumer.
This straightforward project with fuel bills demonstrates the
utility of a cognitive ergonomic approach to design which
benefits not only the consumer but should, by reducing the
number of queries, also benefit the company.   Better bills
should enable householders to use the information on them to
manage their domestic energy and finances more efficiently.

REFERENCES
Attan, J., 1985. Research Report 13, Electricity Consumers'
     Council.
Boardman, B., 1988.   Economic, Social and Technical Considera-
     tions for Fuel Poverty Policy, D.Phil., University of
     Sussex.
Dale, H.C.A. and Crawshaw, C.M.,   1982.   Some characteristics
     of the human controller.   CIB/CIE Workshop: Persons not
     People. (Capenhurst).
Electricity Consumers' Council, 1985. New Metering Technology.
     Information Paper 11.
Ester, P., 1985.   Consumer Behaviour and Energy Conservation.
     (Dortrecht: Nijtroff).
Gaskell, G. and Pike, R.,   1983.   Residential Energy Use: An
     Investigation of Consumers and Conservation Strategies.
     Journal of Consumer Policy, 6, 285-302.
Williams, D.I., Crawshaw, A.J.E. and Crawshaw, C.M., 1985.
     Energy Efficiency and the Domestic Consumer.   Journal of
     Interdisciplinary Economics, 1, 19-27.

VDUs in process control

SCREEN SEARCH EFFICIENCY AND SPATIAL FREQUENCY

D. SCOTT

Department of Psychology
University of Strathclyde
155 George Street, Glasgow
G1 1RD

Previous HCI studies show that icons are more
readily located on VDUs than verbal labels.
One explanation involves spatial frequencies
and peripheral vision. This paper describes a
visual search study in which spatial frequency
grids were employed to examine the role of the
variables cycle-frequency, high/low contrast,
and high/low target-distractor similarity. Eye
movement analyses of fixation durations and
scan paths were then conducted to determine if
these parameters could account for the
observed differences.

INTRODUCTION

Reducing search time for computer menu displays and
other VDU work has considerable value in terms of
potential financial savings. The energy saving
potential of human-computer interaction (HCI) is
perhaps exemplified by Tullis' (1983) report that
employees of a single insurance company will view over
4.8 million displays per year. Employees in the Bell
System using only one particular software package will
extract information from over 344 million displays per
year. The reduction of only a fraction of a second in
the time it takes users to process each display could
lead to large savings in both time and money. Tullis
further estimates that 55 labour-years can be saved by
accomplishing just a one second reduction in the time
which workers spend analysing information on each
screen of a computerised service system accessed in a
particular year.

Icons represent a fundamental aspect of today's
interfaces but we need to know exactly when and why
icons might be preferable to their verbal counterparts.
Previous studies (e.g. Scott & Findlay, 1991) have

reliably shown that the locatability of computer-type icons is faster than for more conventional verbal labels. One suggestion (e.g. Scott, 1992a) for this is that text involves a large number of lines closely spaced together; i.e. it has a high spatial frequency which would be harder to resolve in peripheral vision. It should be noted here that one of the most important characteristics of visual search is the distinction between what is in <u>foveal</u> as compared with <u>peripheral vision</u> at any particular moment. The decision as to where to look next is presumably based on information available in peripheral vision.

This study therefore investigated the role of spatial frequencies in icon versus word locatability. Due to the difficulty of controlling for the multivariate nature of icons, it was decided to use "pure" spatial frequency grids rather than "real" icons as experimental materials.

EXPERIMENT 1

The first part of the study employed spatial frequency grids to examine the role of variables such as cycle frequency, high/low contrast, and high versus low similarity of targets and distractors. It was hypothesised that a) visual search would be quicker for items of lower spatial frequency and that b) search times would be quicker where contrast was higher and where distractors (non-targets) were less similar to the target.

METHOD

Forty-eight subjects participated. Each presentation contained a central reference grid surrounded by eight grids in a circular arrangement equidistant from the centre. The peripheral grids consisted of one target and seven distractors.

Arrays were prepared with MacDraw and laser-printed ready for photographic slide preparation.

1: **Number of cycles** Number of black/white stripes within each target varied from 1, through 1.5 (one black stripe, one white, one black), 2, 2.5, 3, 3.5, 4, 4.5, to 5. Thus, there were nine different patterns involved.
2: **High versus low contrast** Bands were either (a) pure black and pure white or (b) dark grey and light grey.
3: **High versus low similarity** This referred to the similarity between target and non-targets (or distractors). The high similarity arrangement with a target of two cycles, for example, would have non-targets of 1, 1.5 (two) 2.5 (two), 3, and 3.5. The low similarity arrangement with a target of 5 cycles

would have non-targets of 1 (two), 1.5 (two) 2, 2.5,
and 3 cycles. Figure 1 shows an example of the
arrangements used.

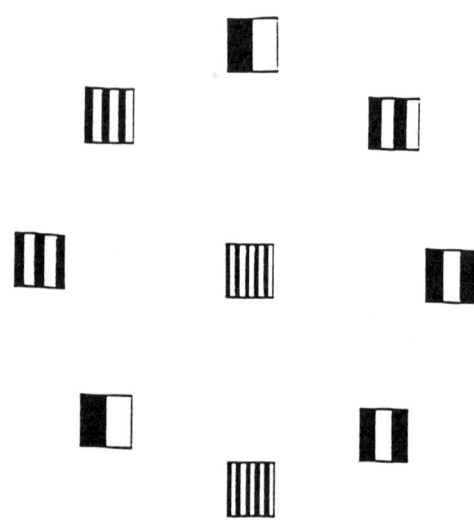

Figure 1: 5-cycle, low similarity, high contrast
arrangement

   In addition, targets were presented at each of eight
positions of the compass. There were thus a total of 9
x 2 x 2 x 8; i.e. 288 arrangements. Search arrays were
presented to subjects via slides displayed from a
projection tachistoscope. Subjects were required to
search for the grid corresponding to the central grid
and to press a key when found.

RESULTS AND DISCUSSION
   With the exception of the extreme high frequencies,
the pattern followed that hypothesised. A three-way
repeated measures ANOVA was conducted which revealed
highly significant effects within all variables and all
interactions.
   Analysis by omega-square showed the proportion of
the total variance accounted for by the similarity
factor was 63 times that of the contrast factor, and
that for the cycles factor was 54.5 times that of the
contrast factor. Thus, the contributions of the
similarities and cycles factors are not greatly
different, but are far more important than the
contribution of the contrast factor.

It seems that the greater is the number of items to be inspected, the longer is the search time. So, when the target consists of a low number of cycles, peripheral vision is able to convey this information to the visual system and the search is rapid. On the other hand, the similarity between the target and the distractors is also important. When the similarity is high, an increase in the number of candidate locations requiring viewing can be expected, with consequent longer search times. The third variable, contrast, also effects search times, although it neither increases nor decreases the number of grids to inspect. It is supposed that this variable modifies the required time to process each grid.

The prediction that search times would be lowest to targets which were grids of lower spatial frequencies was confirmed. One likely determinant of search speed is the ease with which the target can be located in peripheral vision. Rapid search occurs when a target is defined by a single and conspicuous feature. In such a case, the target 'pops out' (Treisman, 1991). It has been shown by Sagi (1988) that pop-out can occur on the basis of spatial frequency properties. Could the difference in search times among the different targets used in the experiment be related to different degrees of pop-out? If a target is readily detectable in the visual periphery, the subject will be able to direct his gaze immediately to the location and make an appropriate response. The pattern of eye movements made by the subject should reflect the extent to which the search is of this nature.

EXPERIMENT 2

To further explore this effect, a second study involving eye movements was performed. If the aforementioned rationale is correct, particular eye movement measures would be expected to be modified by the contrast, similarity and cycles variables. For instance, the more useful is peripheral vision, the more frequently a first search saccade would be directed to the target, and the smaller would be the number of grids inspected before the target is located.

The scleral search coil method has proved useful in analyses of visual search and displays (e.g. Scott & Findlay, 1990, 1991a,b,c,d,e; Scott, 1992b) and was chosen due to the need to monitor both the vertical and horizontal dimensions. On the assumption that peripheral vision is important in the search task and that a target detected in peripheral vision might attract the eyes, then a prediction is that the proportion of <u>first saccades to target</u> will be related to search speed, and conversely the <u>number of items</u>

fixated before the target will show a negative
correlation with search speed.

RESULTS AND DISCUSSION

1: Search paths A phenomenon commonly seen was that the
subject's first saccade is to a non-target just a
half-cycle removed from that of the target. It would
appear that peripheral vision incorrectly draws the eye
to the non-target. Within such cases, the subject seems
to quickly realise their error, presumably as the
pattern being observed does not match the pattern
currently held in visual memory, and then the subject
would re-view the reference item and immediately locate
the target.

2: Fixation durations There was no discernible
variation in fixation durations across the cycle
frequency variable. Despite the fact that a clear
pattern was found for search times in Experiment 1,
this does not appear to have a correlate in this eye
movement parameter.

3: First saccades to target By chance alone an index of
12.5 would be expected (one target amongst eight
items). All the nine frequencies yielded indices higher
than 12.5. The correlation between the index and the
combined search times was almost significant and
negative.

4: Number of items fixated before target This index was
found to change across the cycles variable. The pattern
of these changes was not greatly different from that
observed between cycles and search times in Experiment
1.

GENERAL CONCLUSIONS

   Experiment 1 showed a significant effect of the
three variables employed. However, their relative
effects are not equal. Contributions to the total
variance of the similarity and cycles variables was
stronger than that of the contrast variable. The
explanation put forward for these results suggested
that some eye movement measures can be related to these
variables.

   Experiment 2 tried to explore whether this was the
case. The principle results were that: (a) contrast is
the only variable related to fixation durations; (b)
similarity is related to first saccades to target and
the number of items fixated before the target; and (c)
the cycles variable is also related to these two eye
movement measures.

   It was concluded that the search times found for the
three spatial frequency variables associated with the
icon superiority have no correlate in eye movement
parameters. Icons are certainly more varied in terms of
area covered and shape than are words, but such

differences are unlikely to account for their consistent and reliable visual search superiority. It would appear though that there is an attribute of icons which results in them being located perhaps "just a saccade away", and that attribute may well be the ease in peripheral perception of their lower spatial frequency composition.

REFERENCES

Sagi, D. (1988). The combination of spatial frequency and orientation in effortlessly perceived differences. Perception and Psychophysics, **43**, 601--603.

Scott, D. (1992a). Peripheral vision, conspicuity and VDUs. In: E.J. Lovesey, (ed.), Contemporary Ergonomics 1992. London: Taylor & Francis. pp. 41-46.

Scott, D. (1992b). Do antialias fonts enhance displays? Technical Report HF153 Hursley: IBM (UK) Laboratories.

Scott, D. & Findlay, J.M. (1990). The shape of VDUs to come: A visual search study. In: E.J. Lovesey, (ed.), Contemporary Ergonomics 1990. London: Taylor & Francis. pp. 353--358.

Scott, D. & Findlay, J.M. (1991a). Visual search, eye movements and display units. Technical Report HF 147 Hursley: IBM (UK) Laboratories.

Scott, D. & Findlay, J.M. (1991b). Visual search and VDUs. In: D. Brogan, (ed.), Visual Search II. London: Taylor & Francis.

Scott, D. & Findlay, J.M. (1991c). Optimum display arrangements for presenting status information. International Journal of Man-Machine Studies, **35**, 399--407.

Scott, D. & Findlay, J.M. (1991d). The role of peripheral vision in computer VDU text editing. In: Y. Queinnec, (ed.), Designing for Everyone. London: Taylor & Francis. pp. 686--688.

Scott, D. & Findlay, J.M. (1991e). Future displays: A visual search comparison of computer icons and words. In: E.J. Lovesey. (ed.), Contemporary Ergonomics, 1991. London: Taylor & Francis. pp. 246--251.

Treisman, A. (1991). Search, similarity, and integration of features between and within dimensions. Journal of Experimental Psychology: Human Perception and Performance, **17**, 652--676.

EVALUATION OF VDU SEQUENCE-BASED MIMIC DISPLAYS FOR PROCESS CONTROL

P.J.Thelwell

Industrial Ergonomics Group, School of Manufacturing and
Mechanical Engineering, The University of Birmingham.

J.Reed

BNFL Engineering, Risley, Cheshire.

B.Kirwan

Industrial Ergonomics Group, School of Manufacturing and
Mechanical Engineering, The University of Birmingham.

This experiment aimed to increase the human factors
adequacy of sequence-based VDU mimic display design,
increasing the operability of the user-interface and
decreasing the likelihood of error. An important area
for investigation was the presentation of alarms
through the use of a "Fetch alarm" facility. To
ratify which type of display this facility should
default to, real process control operators using a
simulation based on a real plant control system were
tested under different configurations of the "Fetch
alarm" for two levels of task complexity.
Recommendations were derived from the results.

INTRODUCTION

There is a need for human factors (HF) input into the
design of the displays and controls utilised in the Central
Control Room (CCR). Great use is being made of computerised
systems and VDU-based interfaces for simple and complex process
control operations. It is of foremost importance therefore,
that in order to guard against human error in these new
technology systems, sufficient human factors attention is given
to the design of the operator-machine interface.

This study aimed to derive guidelines for the use of
sequence-based VDU mimic displays in process control. In
particular it considered how to configure the "fetch alarm"
facility. This facility does just what it says - when an alarm
occurs and the operator is alerted, the operator is able via a
single action to access an appropriate display by means of a
"fetch alarm" key. HF guidelines are available for the design
of status-based VDU mimic displays which are based upon plant
flow diagrams to represent plant status. There is, however, a
deficiency of guidelines for the design of sequence-based (or
task-based) VDU display systems - displays which provide

selected information and control facilities for particular
operations rather than all monitoring information and control
facilities for the entire part of the process.

This study was based on the New Oxide Fuel Complex (NOFC)
currently under construction at BNFL's Springfields site. The
plant process will be monitored and controlled primarily from a
CCR utilising a Sequence Control and Data Acquisition (SCADA)
System. Normal operation will largely consist of sequences
(such as heating up an autoclave, transferring a product etc.),
each sequence being controlled automatically by the control
system. The operator-interface consists of VDU mimics linked in
a functional tree hierarchy comprising status and sequence
displays as well as trend displays, menus and alarm lists.
Sequence (or task-based) displays are used to initiate and
monitor sequences, with status displays used to provide
information for general plant area monitoring. The operators
will be provided with all the relevant information (and only
the relevant information) for a particular sequence on the
sequenced-based display.

In both status and sequence-based mimic displays alarm
indications are "embedded". In addition, there are alarm
listings where all alarm occurrences are listed with time and
date tagging. This investigation looked at the configuration of
the "Fetch alarm" facility in order to determine the most
appropriate display format the "fetch alarm" should access;
either a status-based mimic display, a sequence-based mimic
display or an alarm listing display. If the "fetch-alarm" is
configured to go to sequence-based displays, there may be
times when the alarm is not related to the accessed or any
sequence. In such cases it would be more appropriate to access
the context-independent status mimic or alarm listing display.
If the alarm does relate to a particular sequence, then
diagnosis may be assisted if the "fetch alarm" is configured to
present the sequence-based mimic.

The objective of this study, via an experimental
investigation, was to provide guidance for use by designers on
the appropriate configuration of the "fetch alarm" facility.

METHOD
Subjects

Eighteen trainee BNFL process control operators were used
(twelve male, six female), aged from eighteen to mid-thirties.
All subjects had a similar amount of process control
experience, were familiar both with the use of simulations and
the hardware used in the experiment, and were also adept with
the process plant symbols.

Equipment

A static simulation based on part of the NOFC control system
was developed on a PC. The simulation was basically a
simplified version of the real system.

Figure 1. Simplified diagram of a typical Sequence-based Mimic.

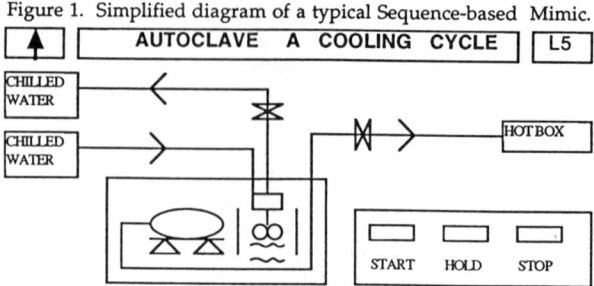

The simulation consisted of ten VDU display screens linked in a functional tree hierarchy. This comprised a plant area overview mimic, followed by a status-based mimic showing a detailed (but not over-complicated) overview of part of the process. From this, the Sequence menu can be accessed, displaying six sequences associated with the process. Each of the Sequence control mimics could be accessed from the sequence menu. An alarm listing display could be accessed from any display.

Experimental Design

Three conditions of the main independent variable of "Fetch alarm" configuration were tested — these being either: a Status-based mimic display, a Sequence-based mimic display, or an Alarm listing display. Two conditions of the second independent variable of "task complexity" were incorporated into the experiment; these were "simple" or "complex" tasks.

A within-subjects experimental configuration was adopted. To reduce any learning effects between the experimental conditions all trials were given to each subject in a randomised order. There were three tasks for each of the two levels of task complexity ie. six different tasks per "Fetch alarm" condition. Thus, there was a total of eighteen different tasks per subject. However, the complexity of each task was standardised within each level of complexity for each configuration of the "Fetch alarm".

Experimental Procedure

There were three main procedural phases: the training phase, the experimental phase and the subjective assessment phase.

The training phase involved an introduction and demonstration, followed by six pilot trials, providing each type of task which would appear in the experiment.

In the experimental phase, subjects started at a particular screen, and were given a scenario describing the position of various plant sequences. They would be required to select the "Fetch alarm" target with a mouse which would access the display, from where diagnosis of the alarm would begin.

Each subject performed eighteen tasks in a randomised order. Subjects' search strategy was recorded — which screen the subject first chose to move to from the "Fetch alarm" configured screen. Timing of the tasks commenced when subjects activated the "Fetch alarm" target and was stopped when the subject gave a diagnosis of the fault condition.

A structured questionnaire gained subjective assessment of the subject's preferences for the different "Fetch alarm" configurations.

DISCUSSION OF RESULTS
Task performance times

The analysis of variance showed a significant difference between the various configurations of the fetch alarm with respect to the time taken to complete the tasks $(F(2,34)=3.400, p=0.0451)$. There was a highly significant difference $(F(1,17)=191.466, p<0.0001)$ between the two levels of complexity. This is not surprising since complex tasks would naturally take longer than simple ones. There was also a significant interaction $(F(2,34)=3.721, p=0.0346)$ between both independent variables. This is illustrated in the graph below:

Mean time taken to complete tasks for each configuration of "fetch alarm" at the two levels of task complexity.

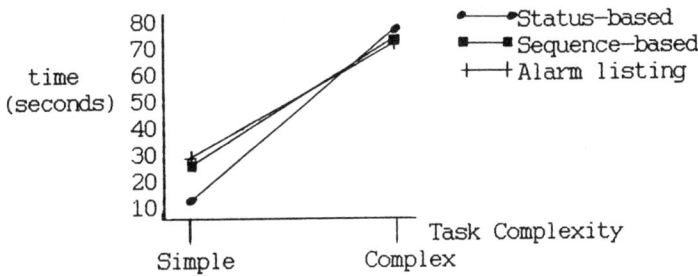

The significant results were analysed in greater detail. There was a significant difference $(p<0.0001)$ for task complexity between some or all conditions of the "Fetch alarm" configuration. At the "complex" level of task complexity no significant difference $(p=0.937)$ among the "Fetch alarm" configurations was detected.

To further explore the results, there were significant differences between the "simple" level and "complex" level tasks for all conditions of the "Fetch alarm" configuration at both the $p=0.05$ and $p=0.01$ levels of significance. However, there was no significant difference between different "Fetch alarm" configurations at the "complex" level. Thus, findings confirmed what is shown in the graph of means above. To summarise, in "simple" tasks the use of a "Fetch alarm" configured to invoke a status-based display appears to result in a quicker diagnosis than where the "Fetch alarm" is configured to call up a sequence-based or alarm listing display. There appeared to be no significant difference between these latter two configurations for the "simple" level of task complexity.

Operator preference

The questionnaires were analysed both quantitatively and qualitatively.

The quantitative analysis indicated that that there were no overall significant differences between the subjects' rankings of preference for the three configurations of the "Fetch alarm" (chi-squared-r(2)=3.44 < critical value of 5.991 for p<0.05). There was a significant trend found (L=227, z=1.83, p=0.0336) in subjects' order of preference: most preferred was the status-based (mean rank = 1.67), next came alarm listing (mean rank = 2.05) and least preferred was the sequence-based (mean rank = 2.28) configuration.

Most subjects (88%) noted that sequence-based VDU systems were suited to certain control systems.

Subjects were asked whether it would be better for the operator to be able to choose which type of display the "Fetch alarm" calls up – 61% of subjects thought it would be.

In situations where an alarm occurs in more than one detailed sequence display subjects were mixed over where the "Fetch alarm" should be configured to take an operator.

In situations where an alarm is not directly associated with any sequence, 61% of subjects thought that a "Fetch alarm" should be configured to go to a status display and 39% thought an alarm listing display would be preferable.

An insight into the subjects' search strategy can be gained from the qualitative analysis of the first type of display invoked by subjects after the "Fetch alarm" configured display.

In the "simple" task condition, the majority of subjects diagnosed the cause of the alarm straight-away when the "fetch alarm" was configured to invoke the status-based display. When it was configured to a sequence-based display, only approximately half of the subjects solved the diagnosis problem from that display alone. When it was configured to an alarm listing display, only about a quarter of the subjects could diagnose the alarm's cause from that display.

In all "simple" tasks it was possible to give a full diagnosis from the "Fetch alarm" configured display. However, when the "Fetch alarm" was configured to a sequence-based mimic or alarm listing display, subjects tended to confirm their diagnosis by going to another display. The minority that did not give an answer straight away in the status-based configuration went to the alarm listing first.

In all "complex" tasks, subjects could not give a full diagnosis from the "Fetch alarm" configured screen displayed. In the status-based display "fetch alarm" configuration, most subjects invoked a sequence-based display next. In the sequence-based configuration about half of the subjects invoked another sequence-based display, the other half being equally split between the status-based and alarm listing displays.

In complex tasks subjects' chosen strategy was to use the status-based display as an overview, but anything that was not on it would guide them to the sequence that was running, which they knew from the scenario. When they were only provided with an alarm in a sequence it was less definite whether the cause was in another sequence or not, and so some subjects used the strategy of always checking the status display, others checked on the alarm listing. With the alarm listing fetch alarm configuration there was an even split between those that went to the sequence display and those that went to the status display, this being because there was inadequate information given in the alarm listing for them to be sure where the cause of the alarm was displayed.

Although some subjects checked the alarm listing for further information, most realised it was only going to inform them what they already knew, so they would then concentrate on the sequence-displays. The only risk with such a strategy is that if they had failed to diagnose the cause of the alarm on the status-based mimic, just concentrated on the sequence displays, they could not diagnose the cause of the alarm, which could have serious implications for plant safety.

Implications from subjects overt preferences tended to be that status-based displays and alarm listing displays are preferred to the sequence-based ones. It was interesting to note that although the majority of subjects thought sequence-based VDU systems seemed a good idea for some control systems, they were not particularly keen to have to use them. Some subjects experienced difficulties in navigating around the simulation commenting that it seemed a long-winded process to have to go through the sequence menu. However, if subjects were adequately trained this menu could be removed from the screen or other access mechanisms employed.

In summary, for the "simple" tasks, the "Fetch alarm" should be configured to the status-based mimic displays. For "complex" tasks, it would appear from subjective assessment and analysis of search strategy, that subjects used the status-based display primarily as an overview to obtain information on the status of items and, as one subject put it, to visualise whereabouts in the process the alarm has occurred.

Conclusion

In conclusion, the following guidelines, for alarm system design and sequence-based VDU mimics were derived:
(1) In simple tasks the use of a "Fetch alarm" facility configured to invoke a detailed status-based display is recommended for the quickest means of accessing information for alarm handling tasks.
(2) In more complex tasks, the "Fetch alarm" facility should be configured to invoke a detailed status-based display. However, further research is required to assess the effect of other task factors not covered in the present investigation.
(3) Future research should endeavour to explore operators' search strategy in more detail for wider ranging tasks in a dynamic simulated environment or through a task analytical approach.

OPERA - THE OFFICE PERFORMANCE EVALUATION & RATING AID

N. Coleman

EPP Human Factors
168, Cheltenham Rd
Bristol
BS6 5RE

Recent changes in the health and safety legislation mean that employers now have to ensure that all display screen workstations in their organisation meet certain minimum standards of ergonomics. This will be a costly and on going exercise. OPERA has been developed in order to promote comprehensive self assessment and training to minimise associated costs.

OPERA is an interactive software database of ergonomics criteria and standards covering all aspects of the modern office environment. Designed as an easy to use application, OPERA is able to guide the user step-by-step through an ergonomics evaluation of a display screen workstation. At the end it specifies which aspects of the workstation must be changed to comply with new regulations concerning work with visual display units, and suggests other modifications which, although not mandatory, would lead to improved ergonomic design.

This paper describes the design and development of OPERA which is based on the quantitative approach to VDU workstation design adopted in Australia.

INTRODUCTION

Recent changes in legislation mean that employers now have to ensure that all display screen workstations in their organisation meet certain minimum requirements. They must also ensure that every employee who regularly uses display screen equipment is informed of the requirements of the new regulations and trained in the effective use of their adjustable office furniture.

Few organisations however, will have the knowledge and expertise required to carry out the assessments and training in-house, and the relevant ergonomics standards, codes of practice and literature that could be used remain relatively inaccessible.

The Office Performance Evaluation and Rating Aid (OPERA) is an easy to use, structured data base of ergonomics criteria and standards covering all aspects of the modern office environment. It has been developed to provide non-experts with the information they require to carry out workplace assessments, product evaluations and user training in line with the new legislation.

The OPERA package

The OPERA package consists of three main parts: OPERA Part I (Evaluation); OPERA Part II (Checklists); OPERA Part III (Training).

OPERA PART I - EVALUATION has been designed primarily for use by employers or designated Health and Safety Representatives. It takes the user, step-by-step, through a detailed evaluation of a display screen workstation. At the end it specifies which aspects of the workstation must be changed to comply with the new legislation, and suggests other modifications which, although not mandatory, would lead to improved ergonomic design.

OPERA PART II - CHECKLISTS contain detailed lists of the ergonomic criteria associated with particular items of office equipment. It has been designed to assist purchasers to select items or products of high ergonomic quality. Manufacturers of office equipment could also benefit from referring to OPERA Part II.

OPERA PART III - TRAINING has been designed as an easy and fun to use system, informing display screen users of everything they need to know about the new regulations. It also provides the training that display screen users require for a healthy workstyle, enabling them to set up their workstations so that they are able to work in comfort with minimal risks to their health.

The OPERA Strategy

To help reduce the costs associated with the volume of workstation assessments required in large organisations, the training package in OPERA Part III includes a very basic user evaluation of the display screen workstation. In direct contrast to the type of evaluation designed for Health and Safety Representitives in OPERA Part I, the questions are very general and rely heavily on the subjective opinions of display screen users. Once the user has completed the training package, a final certificate report sheet is printed summarising the responses to the questions. This information sheet can then be used by a Supervisor or Health and Safety Representitive to shortlist workstations requiring a more indepth analysis using OPERA PART I. Depending on the severity and nature of the problem identified, and the knowledge and experience of the Health and Safety Representitive, the information within OPERA may be sufficient to help generate a solution to the problem. Alternatively, further ergonomics expertise may be required.

BACKGROUND

OPERA, Parts I and II were designed with reference to research carried out by the National Health and Safety Commission in Australia (Worksafe Australia). In 1991 Worksafe produced a suite of ergonomics checklists for the evaluation of office equipment.

The criteria used in the checklists were originally developed from first principles using relevant anthropometric data and ergonomics standards and guidelines as part of a research / consultancy project for the Commonwealth Government of Australia. As the need arose they were transferred into the more usable, highly quantitative checklists that could be used by non-experts to assess the ergonomic quality of various items of office equipment.

When the Health and Safety Executive issued the first drafts of the regulations for work with display screen equipment in the UK, the requirement for a suitable, more quantified approach to display screen workstation design was identified. The strengths and benefits of the Worksafe checklists were recognised and EPP Human Factors sought permission to use the basic concept, format and data in the Australlian checklists as the basis for an interactive software package which could be used for comprehensive workstation assessment. OPERA was then developed as an in-house tool to increase the efficiency and quality of EPP's services in office ergonomics. It soon became apparent that clients were interested in OPERA as a package that they could use themselves, and OPERA was developed as a proprietary application.

OPERA DEVELOPMENT

The original in-house version of OPERA was developed using Filemaker Pro[1] software and the decision was taken to market final versions of OPERA as three Filemaker Pro templates. The advantages associated with this approach - minimal development costs; compatibility with both Apple Macintosh[2] and Windows[3] environments; consistent, easy to use interface capabilities - were considered to outweigh the disadvantages - limited capabilities of Filemaker Pro software; and the requirement of Filemaker Pro software to run OPERA. Additionally, marketing OPERA as three separate templates had the important benefit of allowing employers to control the access of the different user groups to the three separate parts.

OPERA Parts I and II: Design and Development

OPERA parts I (Evaluation) and II (Checklists), were developed together and consist essentially of similar information presented in different ways. They both include detailed lists of ergonomics criteria for the evaluation of the work chair; fixed height desks; adjustable height desks; split level desks; footrests; lighting and the visual environment; various types of document holder; display

---

[1]. Filemaker Pro is a registered trademark of the Claris Corporation.
[2]. Apple and Macintosh are registered trademarks of Apple Computers Incorporated.
[3]. Windows is a registered trademark of the Microsoft Corporation.

screen characteristics, position and adjustability; and the general office environment (eg. thermal, mechanical, safety). Whereas OPERA Part I systematically guides the user through all sets of relevant criteria depending on the type of tasks being performed, the workstation design and the display screen equipment used, OPERA Part II provides for immediate access to the set of data required.

Format

The information in OPERA parts I & II is presented to the user on different levels depending on the depth of information required. Figure 1 shows a small section of OPERA part I, illustrating how users can assess the different levels of information. Screens 1 and 2 are examples of top level screens providing basic information on the importance of good posture and working practice associated with a particular aspect of the display screen workstation (in this case, the worksurface). On screen 2 the user is asked whether that particular aspect of the workstation (eg. the fixed height desk) meets ergonomics criteria. If the user is not qualified to make that judgement, they have the option of clicking on the 'Don't Know' button which takes them to the level of detail shown on screen 3. These screens show examples of the highly quantitative yes/no questions which form the basis of OPERA parts I and II. Users requiring a more indepth understanding can then access the 'help' screens (e.g. screen 4) associated with individual questions for further information on why the measurements are important, what the measurements are based on, and how the measurements should be made.

Criteria and Rating Scales

The criteria which form the basis of OPERA parts I and II are very similar to those originally developed for the Worksafe checklists. Where appropriate, however, British anthropometric data have been used instead of Australian data, and British standards have been more closely adhered to than the Australian counterparts. The criteria used in OPERA are aimed at ensuring that the workstation is suitable for at least 90% of the British workforce.

As the user moves through the lists of criteria (in the form of Yes/No questions) within OPERA, their responses to questions are recorded. 'Yes' is considered a favourable response, 'no' is considered unfavourable. Having completed all the questions associated with a particular aspect of the display screen workstation, OPERA returns to the appropriate high level screen (screen 2 in fig. 1) where the user is presented with an approximate rating of that aspect of the workstation based on the responses to the previous questions.

Some questions are considered 'mandatory', some are considered 'highly desirable' and others are considered 'desirable' (indicated by the presence of a ●, ◑ or ○ in front of the question, Fig. 1, Screen 3). Mandatory criteria must be met in order for the workstation to pass the new legislation. Highly desirable criteria are very important in terms of worker satisfaction and comfort but cannot be classified as mandatory given the openness of the regulations and possible task dependency.

For each aspect of the workstation, the relationship between the ratings assigned by OPERA and the answers given to the questions are summarised in Table 1 below:

RATING CATEGORY	REQUIREMENT
'Meets highest standards'	All mandatory criteria and over 80% of highly desirable criteria met.
'Exceeds minimum requirements'	All mandatory criteria and over 50% of highly desirable criteria met.
'Meets minimum requirements'	All mandatory criteria met.
'Fails to meet minimum requirements'	Some mandatory criteria NOT met.

Table 1. Rating categories used in OPERA

OPERA Part III (Training): Design and Development

As well as providing training for employees, OPERA part III acts as a good introduction to the new legislation. It contains information on who might be classified as a 'user', and what types of display screen equipment are included. The four main sections of OPERA Part III cover the users rights to eye tests, appropriate work organisation and workstation design, and training.

Sections covering workstation design and training are by far the largest. The training section goes through the possible health hazards associated with display screen work and provides users with information on how to avoid physical, visual and mental fatigue. The section on workstation

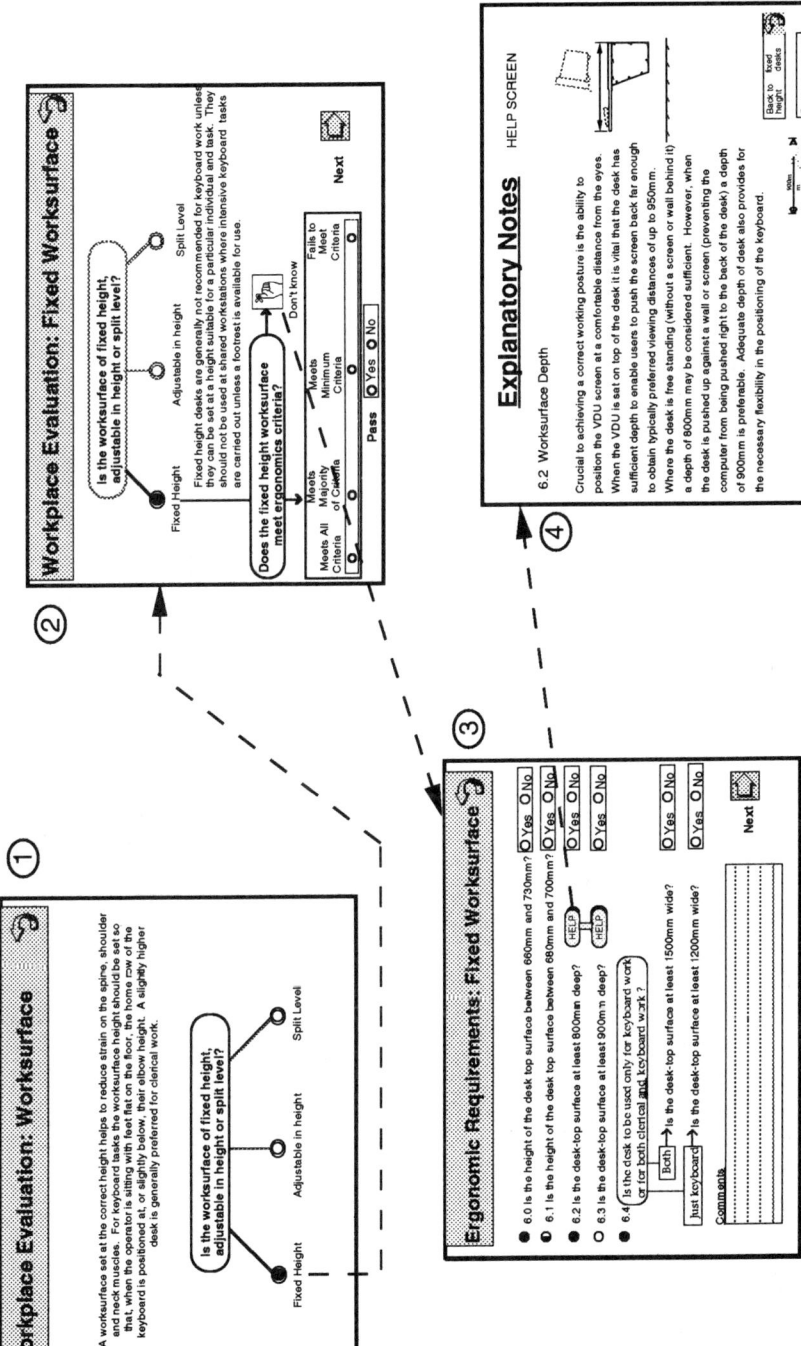

Figure 1. A selection of screens from OPERA part I

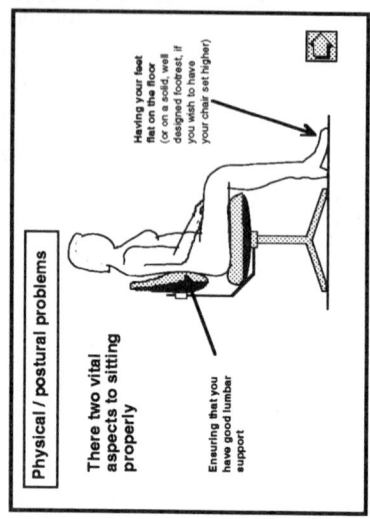

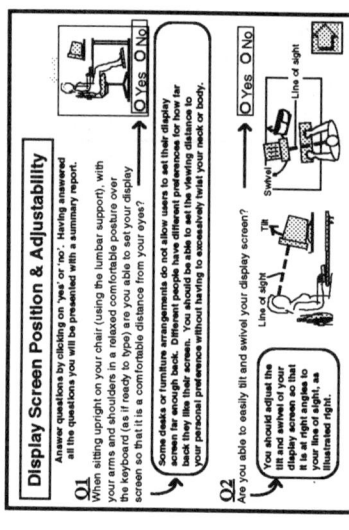

Figure 2. A selection of screens from OPERA part III

design allows users to check that their workstation meets the most fundamental requirements and provides further information on appropriate and inappropriate posture. A selection of scaled down screens from OPERA part III have been included in figure 2 to illustrate the graphical nature of the training package.

OPERA: BENEFITS AND CAUTIONS

The principle benefit of OPERA is that it enables employers to become, and remain, more self sufficient in the administration of the new legislation. The detailed lists of criteria in Parts I and II allow for focused examinations of display screen workstations by users who have little experience of ergonomics. The different levels of information allow users to expand their knowledge and understanding of ergonomics while using the package.

Despite the obvious benefits of OPERA there are a few cautionary notes. There is a danger that Health and Safety Reps., designers of office equipment, or other users of OPERA will blindly apply the criteria without understanding the underlying rationale. This may lead to inappropriate decisions concerning the office environment or the suitability of a products design.

It is possible that the OPERA criteria will become the only criteria against which display screen workstations are assessed. It is important that users are consulted and that specific task requirements are given full consideration. Where there is any doubt, professional ergonomists should be consulted.

OPERA can only be used to assess traditional office environments. Non conventional designs (e.g. kneeling chairs) are not catered for.

The criteria within OPERA should not be seen to limit designers' scope for creativity. However, a creative design should still fulfil its basic function.

CONCLUSION

In view of the new legislation concerning work with visual display units there is a strong demand for a practical, easy to use, readily available data source of ergonomic criteria for the evaluation of the office environment. OPERA Part I (Evaluation) was developed in response to this demand. Combined with parts II (Checklists) and III (Training), the OPERA package offers a comprehensive approach to the new regulations. Without eliminating the need for the ergonomist it is hoped that OPERA will increase the general level of understanding and practical knowledge of office ergonomics.

REFERENCES

LONG, A.F. and COLEMAN, N., STEVENSON, M.G., LUSTED, M., CARRASCO, C.,1991, Ergonomics checklist: The first step in assessing office equipment. Ergonomics and Human Environments, Proceedings of the 27th Annual Conference of the Ergonomics Society of Australia.

STEVENSON, M.G., 1991, Ergonomics issues in contemporary office workstation design. Ergonomics and Human Environments, Proceedings of the 27th Annual Conference of the Ergonomics Society of Australia.

HAMILTON, W.I., 1992, Health and Safety Requirements for Work with Display Screens. EPP Human Factors Document.

HEALTH AND SAFETY EXECUTIVE. Guidance on the Health and Safety (Display Screen Equipment) Regulations 1992. European Directive, No. 90/270/EEC.

THE USE OF COLOUR IN CRT DISPLAYS FOR THE PROCESS INDUSTRY

H.J.G. ZWAGA

Psychonomics Department, Utrecht University
2 Heidelberglaan, 3584 CS Utrecht,
The Netherlands

The prevailing approach to CRT-display design , especially to the use of colour in these displays appears to result mainly from personal preferences of display designers and preconceived opinions about the effectiveness of display design options such as colour. It is argued that task-oriented display specifications and the compilation of application specific design rules instead of general ergonomics guidelines and standards is a way to improve CRT-display design.

INTRODUCTION

In the process industry, especially the use of mimic displays in VDU-based process control systems is still increasing. As to the introduction and implementation of Distributed Control Systems (DCSs), it seems appropriate to make a distinction between industries such as the petro-chemical industry and industries providing a public service, such as the nuclear power industry. In the first a DCS is implemented almost as a standard piece of equipment. In the latter kind of industries, the implementation is a slow and cautious process. Since the first introduction of DCSs in the beginning of the seventies, the display design flexibility has expanded, and is still expanding. The increased flexibility has, however, not resulted at the same time in better displays, i.e. displays supporting performance in a more efficient way. Especially in the petro-chemical process industry this has resulted in display design that is technology-driven and only loosely linked to the different aspects of the task of the operator.

Of the display design options, the use of colour is the most prominent but also the most disputable. In the petro-chemical industry the colour display options are often used to their full extent, while the mono-chrome options such as size, intensity, spacing, etc., are only sparingly used. The result of this 'design process' are displays with detailed flow-scheme-derived representations of the process, crammed with information and with an excessive use of colour.

This practice has at least three related causes. The first is that design engineers make an ad hoc use of different display design principles: display embellishment, display layout options, and the coding of information. How and when information is presented according to one principle or another seems mainly guided by personal preferences and preconceived ideas about the best way to present process information. The second reason is that the independent project approach with a short time path common in the petro-chemical process industry, makes design engineers hardly inclined to take ergonomists on the project team. They are seen as another potential cause of delays. At best they are prepared to consult ergonomics standards and guidelines. The last reason for the prevailing display design practice lies in these very standards and guideline documents. In the opinion of design engineers these documents are barely useful. The information provided is too general and requires interpretation.

These three aspects determining the design of many displays in practice, i.e. preference for colour as a display design tool, avoiding the involvement of ergonomists, and the fact that ergonomics guidelines and standards do not provide the kind of answers the design engineers expect from them, will now be discussed in more detail. It will appear that the way colour is used in DCS displays is just one of the symptoms of inadequate display design and the outline of a procedure will be presented how to convert the standards and guidelines into application specific display design tools.

THE USE OF COLOUR IN PRACTICE
Cosmetics
All design engineers seem to be convinced that, because colour is available as a design option, it has to be used. For most display designers this appears to be self-evident and unquestionable. Most overview publications on the use of colour in CRT-displays mention this preferred use of colour (e.g. Travis, 1991). Also Hopkins (1992) mentions this aesthetic appeal of colour and stresses that it is only a thin line that separates an attractive display from an ugly and garish display because of the unrestrained use of colour. It is a remarkable phenomenon, that people who usually would not dare to try their hand at graphic design, show no reluctance to play the role of 'graphic' designer when specifying CRT-displays. Cosmetics and embellishments in displays are potential clutter. It is not overacting, to warn against making gaudy displays. Commercially available DCSs usually have a standard and fixed set of colours (eight or sixteen) most of which are highly saturated. Objectively speaking, this leaves

the display designer with only very few options to specify
aesthetically attractive displays. Considering the displays in
use, one has to conclude, that users apparently get accustomed to
displays, if these are aesthetically not too offending.

Layout
    Deciding how to arrange and present the information in a
pleasing way on the screen, always interacts with decisions about
how to format the information in a display, i.e. the layout. The
layout of a display is intended to guide and direct the user's
attention when he or she is scanning a display, or to assist in
locating specific information in a display. Of the many formatting
aids available, display designers have a strong preference for
colour. In most cases colour is used as an additional aid to
format a display. Location and colour is often used for the
presentation of the 'house keeping information' in a display. Date
and time, display title, plant name, etc. have, for example, been
located in the top lines of the screen and are also given a
specific colour. Spacing and colour is, for instance, used in
alarm summary displays consisting of a chronological listing of
alarms. Time, tagname, and descriptor are not only arranged in
columns, but each column also has its own colour. In mimic
displays, the measured value of variables is presented in one
colour and the associated engineering units in another.
    When formatting a display, two observations can be made with
regard to the use of colour. The first is that apparently colour
is seen as a means to enhance the appearance of a display, as well
as to increase the effect of mono-chrome formatting aids such as
size, spacing, brightness contrast, inverse video. In addition it
might also be that the use of colour is preferred because it is so
easy to apply in a display. It merely has to be added to already
depicted items. To use mono-chrome formatting aids effectively,
requires experience and, as a rule, it also requires more
programming effort to implement them.
    The second observation is that task support is barely the
starting point for the organization of the information in displays
and the selection of layout tools. Colour is used to mark
conceptual differences between items of information, but these
differences are not necessarily relevant for task performance.
(House keeping information certainly differs from process
information, an engineering unit is not a measured value, etc.).
But why, and how these differences, which are apparently
considered to be important enough to be indicated in two ways in
the layout, affect task performance is never mentioned.

Coding
    Using colour to distinguish between measured values and
engineering units is not only an example of using colour as a
formatting tool, it is also an example of using colour as a visual
code. In a colour code, a colour is used to denote the specific

meaning of a display element. Essential for a true code is that
there is a set of rules for converting one set of data to another.
Display designers use redundant as well as nonredundant coding.
Redundant colour coding is often used as a hybrid formatting tool.
As an example: the house keeping information in a display is
always presented in the top two lines of a display (location code)
and presented in white on yellow (colour code). One does not have
to know the colour code to find the information, if it is known
that this information is located in the top two lines of the
display. And to an experienced user even not to know that should
not be a problem. The content of the information shows what it is.
This is why the presentation of house keeping information is more
a matter of layout than of coding.

Nonredundant colour coding is used in mimic process control
displays in two ways: status and condition coding, and product and
materials coding in flow scheme lines. In the case of nonredundant
colour coding, the user has to learn and remember what it means
when a display item is depicted in a specific colour, otherwise
the information is meaningless (e.g. a pump symbol in pink
denotes a ready-to-use spare pump). For status and condition
coding, display designers rely on familiar colour coding
conventions; red for 'bad', green for 'good', yellow for
'caution'. When more coding levels are needed, colour choices
become more arbitrary. Product coding in flow scheme lines is in
practice fully arbitrary. The way redundant and nonredundant
colour coding are used in CRT-displays bears a strong resemblance
to cartographic coding. This is not surprising. Maps are highly
familiar examples of information displays in which layout and
coding options are intensively used to facilitate the use of the
displayed information. For the general user, maps are road maps
with their prominent colour and shape coding of the different
kinds of roads. Cartographers, however, make different maps for
different purposes, i.e the maps they design are task specific (a
railway map, a demographic map, etc.). On a map, task specific
information is shown prominently, less relevant information is
presented in a detoned way (e.g. unsaturated pastel hues, thin
grey lines), and task irrelevant information is not shown
(municipal borders are not shown on a road map). The usual mimic
CRT-displays are, in contrast, not designed with one main task in
mind. They are best described as intended for three or four tasks
at the same time. One of them is the task of the process operator,
i.e. the displays reflect the ideas of the system designer about
how best to support the task of the process operator. They also
show signal lines and equipment internals. This is the information
considered important by the design engineer and the instrument
engineer themselves. Providing all the information somebody might
need at sometime results for the main intended user of the
displays i.e. the process operator, in an overload of information.
The fact that engineers will never have the thorough knowledge of
the plant and the high familiarity with the displays, the

experienced operators has acquired in the course of time,
strengthens the 'natural' inclination of design engineers to
colour code materials and products in the flowscheme lines. This
certainly supports their own understanding of the displays. For
the experienced operator, however, it is again an increase in the
amount of information he does not need. As a rule it does not help
much to take on the team people with operational experience, even
operators. They too design for a plant they do not know and they
too like colour. The knowledge gained from task analysis has
certainly learned that those who can perform a complex task, do
not have to know how they perform their task and what the task
requirements are. So the rationale that operators know best what
they need is often misleading. All signs point to the conclusion,
that it is the ergonomist to turn to for the expertise required.

DESIGN ENGINEERING AND ERGONOMICS
    Design engineers are reluctant to involve ergonomists in DCS
implementation and interface design. On the one hand their
involvement is an additional activity usually not planned for in
the tight time schedule of the project. On the other hand they
appear not to be convinced of the need to do so. The majority of
the engineers know of ergonomics as the science of 'tables and
chairs and knobs and dials'. How an operator runs his process, and
what he needs to do so, is in their view a technical matter and
therefore outside the scope of ergonomics The help cognitive
ergonomics can provide exactly in this area, is only known to a
few them. This view of the majority makes it difficult to convince
engineers that an ergonomist should have a prominent consulting
role in decisions about the operator interface.
    Circumstances are different when an ergonomist is consulted
because there are serious operator complaints about the interface.
In those cases it helps to understand why engineers design the
operator interface and the displays in the way analyzed above. It
is useful information because it can convince them that a proper
task oriented approach is necessary.

APPLICATION SPECIFIC DESIGN RULES
    Accepting the fact that it is often not possible to directly
involve an ergonomist in a project, a perhaps less desirable way
to affect the design of the interface would be to provide the
engineers with ergonomics guidelines and standards. However,
design engineers find these documents difficult to use. The
guidance provided is neither application nor task specific. To
understand and use the information requires background knowledge.
This complaint is not surprising, because intentionally the
information in guidelines and standards documents is stated in a
general, non-specific way, to make it widely applicable. It is the
task of the ergonomics expert to convert this guidance in context
specific information. One tends to forget that the guidance
documents are firstly intended for use by ergonomists. This point

is discussed in detail by Russell and Galer (1987). This conflict between the intention of guidance documents and their proposed use can be solved by compiling application specific design rules i.e. dedicated to one type of DCS and processes that are comparable. Such a document with detailed design rules for interface and displays can be compiled by an ergonomist, when basing it on a reliable task analysis of the operator(s), the functionality of the DCS, the characteristics of the chosen type of process, and the information contained in relevant guidance documents. This kind of design rules document intended, e.g. for a Honeywell TDC 3000 installed on medium size production platforms, still provides only guidance and no specific rules on some subjects, but the degrees of freedom a designer has, are better defined.

REFERENCES
Hopkins, V.D., 1992, Issues in color application. In <u>Color in electronic displays.</u> eds. H.Widdel and D.L.Post (New York and London: Plenum Press), pp. 191-208.
Russell, A.J. and Galer, M.D., 1987, Designing human factors design aids for designers. In <u>Cognitive engineering in the design of human-computer interaction and expert systems.</u> ed. G.Salvendy (Amsterdam: Elsevier Science publishers), pp. 289-296.
Travis, D., 1991, <u>Effective color displays, theory and practice.</u> (London: Academic Press)

WORKLOAD IN AIR TRAFFIC CONTROL COMMUNICATION

H.B. Nijhuis

National Aerospace Laboratory (NLR)
P.O. Box 90502
1006 BM Amsterdam
The Netherlands

In an experiment the traditional mode of voice communication between air traffic controllers and pilots (a vocal and auditory task) was compared with a data-link mode (a visual and manual task). Eight subjects controlled air traffic, both via voice channels and via data-link in a laboratory simulation. Mental workload was assessed through performance, heart rate variability and subjective judgement.

The results pointed in the direction of a higher mental workload in the data-link mode than in the voice communication mode. The difference in workload could be attributed to a redistribution of the required cognitive resources.

INTRODUCTION

Exchange of information between air traffic controllers (e.g. for the hand over of aircraft from one sector to another) and between air traffic control and aircraft is one of the most fundamental requirements for safe, expeditious and efficient air traffic. Today these tasks are carried out almost entirely by means of radio telephony. Voice messages carry all clearance, advisory and warning information.

Controllers will have to deal with increasing numbers of aircraft. The available radio frequencies are no longer sufficient to accommodate the increasing number of information exchanges that are needed for safe and efficient air traffic. This problem of the so called "frequency congestion" especially exists in air traffic sectors that guide the aircraft directly after departure or just before arrival.

The number of aircraft will keep on growing and some solution for the congestion ought to be found. One such solution is the development of a digital data-link. The proposed data-link solution is a technical solution to a *technical problem* (overload of existing radio channels). The basic *operational problem* in current communication between air traffic control and pilots is the impossibility to exchange the desired amount of information to guarantee a safe, expeditious and efficient handling of air traffic. It is tacitly assumed that data-link will provide a solution for the operational problem.

Data-link

A data-link is a system consisting of a machine and a human operator (a pilot or an air traffic controller). The system must contain an input device to enable the operator to prepare and execute messages and an interface to display received information.

Among its major advantages are the capability of selective addressing. Each intended (and equipped) aircraft or controller workstation can be identified uniquely for reception of particular air traffic control and flight service information, thereby reducing the chance of the wrong aircraft complying to an issued instruction. The possibility of selective addressing lessens the overload problems of the voice radio. The (smaller number of) messages that remain to be communicated by voice, will suffer less from phonetic similarities and overlapping transmissions, thereby reducing ambiguity in controller-pilot communications. The assumption of a reduction in controller time spent on the voice channel after the introduction of a digital data-link was confirmed in a series of studies (for a review see Kerns, 1991).

The technical solution seems therefore appropriate for the technical problem. It is unlikely, however, that the total need for information exchange will decrease after the implementation of a digital data-link. Therefore, a claim for a workload reduction for the air traffic controller is also unlikely. Before going into the so called "multiple resources theory", a better understanding of the term 'workload' is necessary.

Workload

Workload is often described in terms of the demands of a task in proportion to the available mental capacities of the performer. *Mental workload* may be viewed as the extent to which resources of the human information processing system can meet predetermined human performance expectations. The expected performance of air traffic controllers may for example be expressed as their competence in keeping aircraft sufficiently separated (safety).

In the studies so far no effects on pilot or controller workload as a result of data-link have been reported (as measured with subjective questionnaires). However, there are indications of a redistribution of workload in terms of the four dimensions of Wickens' multiple resources theory (Kerns, 1991; Wickens, 1991; 1992). The term resources implies limitations in human mental capacity and has four dichotomous dimensions: input modality (auditory or visually presented information), processing code (spatial or verbal working memory codes), information processing stage (early and central processing versus response selection) and response mode (vocal or manual response). If the demand of a resource exceeds the supply, an overload situation exists and the actual performance will fall short of the expectations. Indices for workload (like heart rate variability) are known to provide contradictory results if they are compared directly with performance measures. This "dissociation" can be understood by considering the strategy of the controllers which will be directed towards maintaining safety levels at all times (Jorna, 1991b). As long as the capacity of one of the resource dimensions is not exceeded and the controller is willing to compensate by trying harder, increasing demands of one task will not decrease the level of performance on an other.

Resource dimensions

Table 1 shows which resource dimensions are hypothesized to be involved when communicating via voice, via a data-link and while monitoring the radar screen (one of the most important air traffic control tasks for the preservation of safety).

Table 1 Nature of used resource dimensions for communication through voice or data-link and radar screen monitoring in an air traffic control environment.

	voice communication	communication via digital data-link	monitoring of radar screen
input modality	auditory (speech)	visual (text)	visual (analogue picture and text)
processing code	verbal	verbal	spatial and verbal
information processing stage	early and late	early and late	early
response mode	vocal	manual	eye movements

Comparison of the communication modes with the monitoring task on the four dimensions shows several possibilities of resource sharing and possible resource interference. For example, the input modality of data-link (visual) is more likely to interfere with the visual input modality of the radar monitoring task than the auditory input modality of voice communication.

To get more insight into the consequences of the introduction of data-link for the workload of the air traffic controller, an (exploratory) experiment was set up to compare the two communication modes.

METHOD

The task the subjects had to perform was controlling air traffic in four scenarios. Two scenarios required voice communication and two used data-link.

The crucial dependent variable was *mental workload*. This variable was measured in three ways. Firstly, a system performance score was obtained, provided by the TRACON simulation facility. Secondly, a physiological measure was used, namely heart rate variability (HRV) in the so called "mid-frequency band". On many occasions in the laboratory as well as in the field, decreases in HRV were observed under "mentally loading" conditions (Aasman, Mulder & Mulder, 1987; Jorna, 1991a, Mulder, 1980). Respiration was measured to control for possible disturbing influences of speech on HRV, not related to workload changes. The third measure of mental workload was the score on subjective scales, the so called "Beoordelings-schaal Subjectieve Mentale Inspanning" (BSMI or in translation "scale for subjective judgement of mental effort"). The BSMI has been validated in a series of experiments (e.g. Zijlstra, 1988). Its validity and sensitivity are well established now. In the current experiment, a questionnaire was designed, which contained -amongst other things- three BSMIs. The first one referred to the task as a whole, the others to the 2 subtasks which comprised the total task (keeping the aircraft safely separated and communicating with pilots and other air traffic controllers).

Eight employees of the National Aerospace laboratory served as subjects. All of them were in their twenties and familiar with air traffic control tasks, although none of them was employed as a controller.

RESULTS

The system performance scores did not differ between conditions; χ^2(N = 7, df = 3, p <.13) = 5.57 and no relation between system performance scores and any of the other measures was found. The lack of any relation between the performance scores and the other two mental workload indices indicates that the tasks did not exceed the individual limits of the resource dimensions of the subjects in any condition. In that case subjects can compensate for increasing task difficulty by a greater mental effort.

The results obtained with heart rate variability were rather straightforward: HRV was significantly lower in the data-link mode than in the voice communication mode; χ^2(N = 5, df = 3, p <.014) = 10.68. This result indicates a higher mental workload in the former condition.

Differences obtained with the BSMI scales were only significant for the communication subtask; χ^2(N = 8, df = 3, p <.07) = 7.05. The communication in the data-link mode was estimated to involve greater effort than the communication via speaking and listening. This result contradicts possible claims of a decrease in mental workload for air traffic controllers with the implementation of a data-link. The BSMIs for the total task and for the separation subtask did not show any significant differences between conditions.

Ideally, the three ways in which the mental workload was measured would indicate exactly the same. As was more or less expected, this appeared not to be the case. In the first place no relationship was found between the system performance score and either of the other two mental workload measures. Apparently, the performance score referred to an other concept than the other measures. Secondly, the variance in HRV (a well validated and reliable measure of mental workload) could be related to the scores on the BSMIs for the separation and the communication subtasks. The relation between the BSMI for the communication subtask (BSMIcom) and HRV were quite clear: a higher subjective effort score on the communication subtask was associated with a lower heart rate variability. Both measures indicate a higher mental workload in the data-link mode as compared with the voice communication mode.

The relation between HRV and BSMI for the separation subtask (BSMIsep) was in the unexpected direction: an high (subjective) effort on the separation subtask (higher mental workload) was correlated with high levels of HRV (lower mental workload). Further analysis revealed that the overall result is a combination of two effects with results were in opposite directions: in the voice communication mode the relation between HRV and BSMIsep was even stronger than in the overall comparison. The contribution of BSMIcom was only marginal in that case. In the data-link communication mode, on the other hand, the situation was completely different: there was absolutely no relation between HRV and the BSMI score on the separation task, but a very strong one between HRV and the subjective effort on the communication task.

It seems as if the two subtasks played a different role in the two communication modes. In the voice communication mode the mental workload mainly originated in the separation task, while there was only a minor role for communication. In the data-link mode on the other hand, the mental workload was almost entirely determined by communication and hardly by the separation subtask. Since the separation subtask was exactly the same for both communication modes (as was confirmed by the scores on the BSMIsep scales), it seems reasonable to suppose that the difference in HRV originated in the different ways of communicating in the two modes.

DISCUSSION

In the introduction we hypothesized about interference within resource dimensions, caused by the concurrent performance of the communication subtask and the separation subtask. The resulting competition between resource dimensions did not seem to cause disturbing interference in the current study, as subjects were rather successful in performing the tasks under both conditions. At the same time, the BSMI for the communication subtask and the HRV results indicated that differences in mental workload existed between the two communication modes. Several explanations for these results are at hand. The first and simplest one is that in the data-link mode more resources had to be mobilised than in the voice communication mode. In other words, the data-link mode was more difficult than the voice communication mode.This explanation seems somewhat counterintuitive though, for the quantitative aspects of the communication task did not differed between the two communication modes (the number of communications and the estimated communication time were the same). In other words: the demands posed by the overall task goals, as defined by the scenarios, were the same in both communication modes.

A second explanation for the difference between the two modes is a redistribution of the use of available resources. The most important hypothesized resource differences (see table 1) can be found in the *input modality* and the *response mode*. Voice communication uses an auditory input (speech), while data-link relies on visual input (text). The input modality of the separation subtask is always visual (monitoring of the radar screen), so that the data-link mode competes for the same resource. Such interference is less likely in the voice communication mode. The difference in interference may have contributed to the lower HRV (= higher mental workload) in the data-link mode.

The response mode dimension in the voice communication mode was vocal, in the data-link mode manual. The subtask of keeping aircraft safely separated did not require such responses. Competition of the communication task and the separation task in terms of the response mode are therefore not likely.

The use of the other two resource dimensions, the processing code and the information processing stage, may also have differed between the two communication modes, thus producing differences in subjective workload (and HRV). Expressing conclusions about these two dimensions on the basis of the available data would be pure speculation. Future research must show exactly to what extent air traffic controllers use the four resource dimensions for their regular tasks and to what extent both communication modes call on the same dimensions. This can be done by setting up experiments in which secondary tasks are performed. See for example Wickens (1991; 1992).

The problem of changes in the mental workload of the controllers (and the pilots) received little attention in the literature on data-link. If the issue was mentioned at all, it was tacitly assumed that the implementation would reduce mental workload. Our experiment does not confirm this assumption. Hopefully, this study forms the start of a systematic investigation into the consequences of the implementation of data-link and other technology for the mental workload of air traffic controllers and pilots.

REFERENCES

Aasman, J., Mulder, G. & Mulder, L.J.M., 1987, Operator Effort and the Measurement of Heart-Rate Variability. Human Factors, 29 (2), 161-170.

Hopkin, V.D., 1991, The Impact of Automation on Air Traffic Control Systems. In Automation and Systems Issues in Air Traffic Control, eds. Wise, J.A., et al. (Berlin: Springer Verlag).

Jorna, P.G.A.M., 1991a, Heart Rate Variability as an Index for Pilot Workload. Proceedings of the sixth International Symposium on Aviation Psychology, 2, 746-751.

Jorna, P.G.A.M., 1991b, Operator Workload as a Limiting Factor in Complex Systems. In Automation and Systems Issues in Air Traffic Control, eds. Wise, J.A., et al. (Berlin: Springer Verlag).

Kerns, K., 1991, Data-link Communication Between Controllers and Pilots: A Review and Synthesis of the Simulation Literature. The International Journal of Aviation Psychology, 1 (3), 181-204.

Mulder, G., 1980, The heart of mental effort. Studies in the cardiovascular psychophysiology of mental work. Doctoral dissertation. University of Groningen, Groningen, The Netherlands.

Wickens, C.D., 1991, Processing resources and attention. In Multiple Task Performance, ed.Damos, D.L. (London: Taylor & Francis).

Wickens, C.D., 1992, Engineering Psychology and Human Performance, second edition (New York: HarperCollins).

Zijlstra, F.R.H., 1988, Meten van mentale inspanning en werken met computers. Tijdschrift voor Ergonomie, 13 (4), 8-16.

Manual handling and posture

A COMPARATIVE EVALUATION OF A HORIZONTALLY MOVABLE AND A STANDARD ADJUSTABLE ARM REST

J.M. PORTER, A.C. SMITH AND D.E. GYI

Department of Human Sciences
Loughborough University of Technology
Leicestershire LE11 3TU

The political and litigious nature of Repetitive Strain Injury (RSI) has created an increasing interest in the design and evaluation of products which are intended to relieve or help prevent such problems. One such product is the 'Relax Arm', a height adjusting, horizontally movable arm rest for keyboard operators. This arm rest was evaluated for its effectiveness in comparison to the traditional height adjustable design. A laboratory study using 12 healthy subjects revealed no statistically significant differences in reported discomfort or EMG activity in the trapezius and extensor carpi ulnaris muscles. However, the use of the 'Relax Arm' led to a significantly smaller wrist extension and ulnar deviation, both of which are advantageous in the prevention of carpal tunnel syndrome. A two week field trial with 9 subjects (4 healthy and 5 RSI sufferers) found that the back, arms and hands were all more comfortable during use of the 'Relax Arm', particularly at the end of the 5 day usage period.

INTRODUCTION

Repetitive Strain Injury (RSI) has been stated to strike at half a million people per year in the UK and one third of all reported injuries at work are disorders of the hand, fingers, wrist, elbow or shoulders (Business Life 1992). The intensive use of a computer keyboard has often been identified as a high risk factor in RSI and journalism has been particularly hard hit with the Los Angeles Times reporting 90 cases since 1983 (Keogh 1989), although the total number of staff was not reported. In the UK one third of journalists at the Financial Times were recently reported to be suffering symptoms (The Times 1992). It has been well documented that diseases of the musculoskeletal system are a major source of health related absenteeism in industry. However, the cost to industry is no longer just in terms of lost working hours as litigation claims of $50-100,00 in Australia and £60-100,00 in the UK are not uncommon (The Times 1992, Kroner Business News 1992).

Boyling (1991) defined RSI as "A collective term for those conditions characterised by discomfort or persistent pain in the muscles, tendons and other soft tissues. This can be a clearly defined clinical condition or an ill defined symptom complex." Many terms other than RSI have been used when describing such injuries including work related upper limb disorders, cumulative trauma disorder, occupational cervicobrachial disorder, and occupational overuse disorder. Diagnostic terms include carpal tunnel syndrome, tendonitis, tenosynovitis, lateral epicondylitis (tennis elbow), medial epicondylitis, compression neurpathy, bursitis, De Quervains disease, thoracic outlet syndrome and adverse neural tension.

There are many causes of musculoskeletal disorders. Any working position places a load on this system which, in time, can lead to discomfort, pain, fatigue and disability as a function of the amount, duration and distribution of the load (Ardnt 1983). With the introduction of the 'computerised office' the trend towards more repetitive and less varied tasks has resulted in a far greater prevalence of 'constrained postures' (Grandjean 1987). Relatively fixed positions of a computer operator's head and arms are required to provide proximal stability whilst operating the keyboard or viewing the screen which results in static loading throughout the body, but mainly in the neck, shoulder, arm and hand (Grandjean et al 1983, Lavaille 1980).

Many employers are now providing 'ergonomic' computer workstations and more attention is being paid to job design in an attempt to reduce the incidence of work related upper limb disorders. This trend will be considerably strengthened now that the CEC Directive 90/270/EEC on the minimum safety and health requirements for work with display screen equipment has recently come into effect. As a consequence, many workstation accessories are now commercially available which are advertised as being ergonomically designed. Unfortunately, most of these have been designed and manufactured without regard to conducting in-depth user trials. Such information (eg. Porter et al 1992) is clearly important in fine tuning the design and also in helping to ensure that there are no contra-indications associated with its use (eg. improves comfort in one body area at the expense of another or of task performance).

This paper describes the evaluation of a recent product called the Relax Arm. It is, to quote the advertising literature, " a new concept in arm support which eliminates static pressure and follows the natural movements of the arm....It is for use whilst typing at a keyboard. It makes finger work easier and alleviates both neck and shoulder pain." The Relax Arm consists of a relatively large and flexible forearm support, mounted on a freely movable arm which can fitted to the same manufacturer's (Rh) chair. This supports the forearm as it moves horizontally in all directions.

The published literature on the value of arm supports and wrist rests has not provided a clear picture. Some studies report decreased muscular loading (Schuldt et al 1987, Weber et al 1984, Granstrom et al 1985) whilst others report increased muscle loading (Bendix and Jessen 1986). However most studies agree that keyboard operators will choose to use an arm support if it is made available.

The evaluation of the 'Relax Arm' was conducted in both a laboratory and a working environment because each method had unique strengths: the laboratory environment being particularly suitable for objective measures whist the working environment enabled a much longer assessment using subjective measures of discomfort.

THE LABORATORY STUDY

12 healthy subjects were selected (6 males, 6 females, age range 25-35 years), all of whom had regularly used keyboards for at least 9 months. 4 of the subjects were touch typists. Subjects undertook 2 half hour typing sessions, in a balanced presentation order, one for each of the 2 computer workstation configurations (ie. the same workstation configuration with either the Relax Arm or Standard arm rest fitted to the chair). The other components of the workstations were positioned as follows: desk height 710mm above floor; screen height 1065mm above floor; screen distance 565mm from desk edge; keyboard height 25-33mm above desk; keyboard tilt 5-10 degrees; footrest height 60-90mm above floor; document holder, adjustable; chair, adjustable.

Measures and Results

The measures that were recorded are given below together with their analysis:

EMG : The descending fibres of Trapezius and the Extensor Carpi Ulnaris were chosen after pilot trials had confirmed their suitability for providing measurable activity whilst typing with an arm support. No significant differences in muscle activity were found using the related t test, although there was a noticeable trend for slightly less overall activity in both muscle groups when using the Relax Arm.

postural angles and distances: Head, neck, trunk inclination, arm flexion, shoulder abduction, elbow, wrist extension, ulnar deviation, thigh inclination, visual distance and acromium to home row distance were all recorded either directly or from photographs on 3 occasions during each 30 minute trial. Mean postural angles and distances were calculated and Table 1 shows the significant differences.

body area discomfort: Discomfort was assessed at the beginning and end of each trial in 25 areas of the body, as well as an overall assessment, using an appropriate body area diagram and a 5 point rating scale (very comfortable, moderately comfortable, neutral, moderately uncomfortable, very uncomfortable). No significant differences were found using the Wilcoxan Signed Ranks test between the 2 arm rests.

self reported stress and arousal: The mood adjective check list (Mackay et al 1978) was administered at the same times as the discomfort rating scales. No significant differences were found.

Table 1: Significant postural differences

Measure	Arm rest	Mean angle in degrees (sd)	p value	Effect of Relax Arm
trunk inclination	Relax Arm Standard	101.4(4) 102.9(4)	<0.05	more upright
wrist extension	Relax Arm Standard	3.2(9) 11.2(9)	<0.0005	decreased extension
right ulnar deviation	Relax Arm Standard	13.5 (3) 16.4(4)	<0.05	decreased deviation
left ulnar deviation	Relax Arm Standard	15.6(5) 20.1(8)	<0.025	decreased deviation

design features assessment for the Relax Arm
A questionnaire covering the ease of adjustment, ease of use when typing, size and padding, comfort and preference was completed at the end of the 2 trials.
adjustability- 92% of subjects felt that the Relax Arm was easy to adjust.
freedom of movement- 58% considered this was improved, 33% felt it was worse.
arm rest size- 75% felt it was about right, 25% felt It was too large.
hardness/softness- 66% felt it was about right, 33% felt it was too hard.
perspiration- 33% felt that their forearm perspired more than usual.
overall comfort- 50% felt more comfortable, 42% less comfortable.
preference- 58% would choose the Relax Arm, 33% the standard rest, 8% no rest.
Discussion of Results.
The lack of any differences in muscle activity may have been due to the finding that the Relax Arm requires considerable 'getting used to' as the body needs to learn how to relax onto the arm rest rather than fight it. A longer adaptation period would be recommended in any future trials. It was observed that many of the subjects supported their forearms on the desk whilst placing their elbows on the standard arm rest. This would effectively provide support for the whole of the lower arm although at the expense of having to adopt an extreme wrist posture.

The Relax Arm's potential benefit was highlighted by the the finding that it encouraged a more neutral wrist posture which would be associated with less pressure of the median nerve in the carpal tunnel thereby decreasing the risk of injury (Ardnt 1983, Sauter et al 1987, Duncan & Ferguson 1974). Grieve (1992) suggests that resting the weight of the arm on the bony front of the extended wrist on the table edge can alter the joint mechanics of the wrist leading to secondary dysfunctions in the tendons and nerves. This problem is effectively prevented by use of the Relax Arm.

The absence of any significant differences in reported discomfort may have been due to the short evaluation time and the relative unfamiliarity of the Relax Arm. This was recognised as a shortcoming at the beginning of the project and this was one of the main reasons for deciding to conduct field trials.

THE FIELD TRIALS

It was initially planned to conduct these trials using solely RSI sufferers because the Relax Arm is marketed at this group. Unfortunately, it transpired that the political sensitivity of the RSI issue prevented medical professionals from disclosing patients' details and many employers would not permit their staff to take part, even if healthy, because they were worried that they might incurr pressure to purchase such equipment if their staff found it to be beneficial.

9 subjects (5 healthy, 5 RSI sufferers, age range 23 to 55 years) were selected, all of whom were in full time employment and were regular VDT users spending a minimum of 50% of the day at the keyboard. These subjects completed a baseline questionnaire and a body chart modified from Sauter & Schlieffer (1991) and Slovak & Trevers (1983) which recorded details concerning any existing musculoskeletal problems and details of any medication or rehabilitation treatment that might affect the comfort assessments during the trial.

The trials lasted for 2 consecutive weeks with each arm rest being used full time for 1 week in a balanced order. Each subject was supplied with an experimental chair (Rh) to which both the arm rests could be fitted. They remained at their own workstation and undertook their normal day-to-day work. Subjects were allowed to adjust any features of the arm rests during the trials to provide optimum comfort.

A body area discomfort chart and a stress arousal inventory (as used in the laboratory study) were completed twice daily, once on arrival at work and once immediately prior to leaving. A design features assessment for the Relax Arm was completed at the end of the 2 week trial.

Results

The stress and arousal scores showed no consistent changes between weeks 1 and 2 or during the use of the 2 arm rests. Unfortunately, only 7 out of the 9 subjects completed the trial. 2 of the RSI subjects, who normally worked without using an arm rest of any description, experienced an exacerbation of their symptoms and declined to finish the study. To avoid losing their data it was decided to keep their existing data and to supplement it for the days omitted. For the purposes of analysis, any body areas that were reported verbally by these 2 subjects to have been aggravated by either arm rest were allocated a score of 5 on the rating scale (very uncomfortable) for that particular arm rest on all the remaining days for that week. All other body areas were given a score of 3 (neutral).

The Wilcoxan Signed Ranks test was used to determine whether any differences between the arm rests were evident based upon their comfort scores on the individual days 1-5. Significant differences were found almost exclusively on day 5 and these are shown in Table 2. Clear trends were also evident for the week as a whole showing that less discomfort was reported with the Relax Arm in the upper back, low back, left shoulder, left & right upper arm, left & right lower arm, right wrist, right fingers and in the overall assessment of body comfort.

The design features assessment for the Relax Arm gave the following results for the field trial subjects:

adjustability- 89% felt that the Relax Arm was easy to adjust.
freedom of movement- 78% felt that this was improved, 22% felt it was worse.
arm rest size- 89% felt it was about right.
hardness/softness- 67% felt it was about right, 33% felt it was too hard.

perspiration- 22% felt they perspired more than usual.
overall comfort- 67% felt more comfortable, 33% more uncomfortable.
preference- 66% would choose the Relax Arm, 11% the standard rest, 22% no rest.

Table 2: Significant differences in reported discomfort on day 5

Body area	Time of day	Arm rest	Median score	p value	Effect of Relax Arm
low back	morning	Relax Arm Standard	2 3	<0.025	less discomfort
low back	afternoon	Relax Arm Standard	2 3	<0.025	less discomfort
mid back	afternoon	Relax Arm Standard	2 3	<0.025	less discomfort
left upper arm	morning	Relax Arm Standard	2 3	<0.05	less discomfort
right lower arm	afternoon	Relax Arm Standard	3 4	<0.05	less discomfort
left fingers	morning	Relax Arm Standard	2 3	<0.05	less discomfort
right fingers	morning	Relax Arm Standard	2 3	<0.05	less discomfort
right thumb	afternoon	Relax Arm Standard	2 3	<0.05	less discomfort

Discussion of results

The field trials revealed a significantly lower extent of reported discomfort in spinal and upper limb areas when using the Relax Arm. Reduced discomfort in the low and mid back may be an indication of reduced spinal loading as studies by Andersson (1980,1987) and Occipinti et al (1985) have shown that the use of arm supports reduces discal and muscular loading of the spine. The lower incidence of discomfort in various areas of the upper limbs may indicate a reduction in muscle and joint loading and the adoption of more neutral joint positions. The finding that some body areas demonstrated a significant difference in discomfort between the arm rests before commencing work, but not at the end of work , and vice versa, is not easily explained.

It is of concern that 2 of the RSI subjects had to discontinue their use of the Relax Arm as this product is marketed at such sufferers. However, the standard rest was perceived as being even worse by these subjects. Comparisons between the healthy and RSI subjects have not been attempted due to the small sample size and varying jobs.

CONCLUSIONS

92% and 78% of laboratory and field subjects, respectively, prefer to use an arm rest rather than no support at all when using a keyboard. Of these subjects 63% and 86% would choose to use the Relax Arm in preference to the standard arm rest.

The Relax Arm significantly reduced wrist extension and ulnar deviation compared to the standard height adjustable arm rest, this more neutral posture being expected to help prevent joint and soft tissue dysfunction.

At the end of 5 days use for each of the arm rests in a working environment it was found that the Relax Arm offered improved comfort in the mid and low back, left upper arm, right lower arm, fingers in both hands and the right thumb. Trends were evident for other body areas.

The Relax Arm may offer benifits to many RSI sufferers, but it should not be distributed to such people without careful consideration of each individual's condition.

REFERENCES

Andersson, G., 1980, The Load on the Lumbar Spine in Sitting Postures, In: Human Factors in Transport Research, eds. D. Oborne and J. Levis, Academic Press.

Andersson, G., 1987, Biomechanical Aspests of Sitting: An Application to VDT Terminals, Behaviour and Information Technology, 6(3), pp257-269.

Ardnt, A., 1983,, Working Posture and Musculoskeletal Problems of VDT Operators - Review and Appraisal, Am. Ind. Hyg. Assoc. J., 44(6), pp437-446.

Bendix, T.,and Jessen, F., 1986, Wrist Support during Typing - A Controlled Electromyographic Study, Applied Ergonomics, 17(3) pp162-168.

Boyling J., 1991, Upper Limb Disorders in the Workplace, OCPPP 'In Touch', No.61

Business Life Magazine, 1992, Taking the Sprain, April Issue.

Duncan, J. and Ferguson, D., 1974, Keyboard Operating Posture and Symptoms in Operating, Ergonomics, 17, 5, pp651-662.

Grandjean, E., Hunting, W. and Piederman, M., 1983, VDT Workstation Design: Preferred Settings and their Effects, Human Factors, 25(2), pp 161-175.

Grandjean, E., 1987, Ergonomics in Computerized Offices, Taylor and Francis Ltd.

Granstrom, B., Kvarnstrom, S., Tiefenbacher, F, 1985, Electromyography as an Aid in the Prevention of Excessive Shoulder Strain, Applied Ergonomics, 16(1), 49-54.

Keogh, G., 1989, Injury Time, Practical Computing, April Issue, pp 64-66.

Kroner Company Administration Briefing, 1992, Latest RSI Damages Award, Issue 7.

Lavaille, A., 1980, Postural Reactions Related to Activities on VDUs. In: Ergonomic Aspects of Visual Display Terminals, Ed. E. Grandjean and E. Vigliani, Taylor and Francis.

Mackay, C., Cox, T., Burrows, G. and Lazzerini, T., 1978, An Inventory for the Measurement of Self Reported Stress and Arousal, Br. J. Soc. Clin. Psychol. 17, 283-284.

Occipinti, E., Columbini, D., Grieco, A., Frigo, C. and Pedotti, A., 1985, Sitting Posture: Analysis of Lumbar Stresses with Upper Limbs Supported, Ergonomics, 28(9), pp1333-1346.

Porter, J.M., Gyi, D.E. and Robertson, J., 1992, An Evaluation of a Tilting Computer Desk, In: Contemporary Ergonomics 1992, 'Ergonomics for Industry', Ed. E.J. Lovesey, Taylor & Francis Ltd, pp. 54-59.

Sauter, S., Schleifer, L. and Knutson, S., 1991, Work Posture, Workstation Design and Musculoskeletal Discomfort in a VDT Data Entry Task, Human Factors, 33(2), pp151-167.

Schuldt, K., Ekholm, J., Harms-Ringdahl, K., Arborelius, U., Nemeth, G.,1987, Neck and Shoulder Muscle Activity During Arm Movements at Work in Various Sitting Postures, with and without Ergonomic Aids. In: International Series on Biomechanics, Vol 6a, Ed. B. Jonsson.

Slovak, A. and Trevers, C., 1988, Solving Workplace Problems Associated with VDTs, Applied Ergonomics, 19(2), pp99-102.

Weber, A., Sancin, E., Grandjean, E., 1984, The Effects of Various Keyboard Heights on EMG and Physical Discomfort. In: Ergonomics and Health in Modern Offices, Ed. E. Grandjean, Taylor & Francis, pp477-483.

The Times Newspaper, 1992, Strain that's a Pain, March 5th Issue.

MUSCULOSKELETAL DISORDERS IN OPERATORS OF WEAVING MACHINERY

N.J. WILKINSON and C.M. HASLEGRAVE

Institute for Occupational Ergonomics,
Dept. of Manufacturing and Operations Management,
University of Nottingham,
University Park,
Nottingham,
NG7 2RD.

The operation of two types of weaving machinery was
studied in a factory manufacturing synthetic material
for use in industrial packaging. The operators were
suffering from a number of musculoskeletal problems,
thought to be caused by the postures and forces
required. The potential for redesign of such machinery
is discussed.

INTRODUCTION

Musculoskeletal disorders at work have been estimated to cost UK
Industry in the region of £1 billion per annum (1988 figures)
(Thompson 1989). The occurrence of such problems could, in many
cases, be prevented by incorporating ergonomic principles in the
design and layout of plant and equipment.

The present investigation came about as a result of concern over
levels of musculoskeletal disorders being experienced in a company
manufacturing synthetic material for use in packaging industrial raw
materials. The machinery used was typical of much equipment used in
the weaving industry and the case study can help to identify
ergonomic improvements for the design of weaving machines.

In the process observed, synthetic thread was produced and
wound as "tape" onto spools which had a maximum diameter of 95
mm when full. These weighed 1 kg. The tape from the spools was then
woven into cloth on circular looms.

The operators in the two departments were largely involved in
unloading spools from the tapeline machine and in loading and
threading these onto the circular weaving machines. Table 1 shows
the incidence rate of musculoskeletal problems which were found
among operators of the tapeline and weaving machines - expressed
as cases observed per 200,000 hours worked - and related sickness
absence, for a 2 year period. Tapeline injuries were non-specific;
weaving injuries were dominated by cases of tenosynovitis.

Table 1. Incidence rate of and time lost through musculoskeletal
disorders over a 2 year period prior to the investigation.

Department	Incidence Rate (No./200,000 hrs)	Time lost (Weeks)
Tapeline	19.08	67
Weaving	3.14	113

The problems are therefore comparable with those in other
industries known to be concerned about the incidence of
musculoskeletal disorders. Armstrong (1986) observed 6.6 cases per
200,000 hours worked at an electronics firm, and 7.0 cases per
200,000 hours at a film processing plant. Armstrong *et al* (1982)
recorded 12.8 cases per 200,000 hours worked at a poultry
processing plant.

METHODS
The following methods were used during the course of the
investigation.

Task Analysis and Job Design
In both departments, a hierarchical task analysis was performed
on the main tasks, in order to identify the main components of the
work and to provide a framework for subsequent posture and force
analyses. This was done by recording operators working at these tasks
using a video camera, then playing back the recording frame by frame.
Depth of task analysis was dependent on the likelihood of a task
element presenting a risk of musculoskeletal problems to the
operator.
Random occurrence sampling (Kanawaty, 1979) was carried out in
the weaving department to establish what proportion of the working
day was taken up by each of the various tasks performed by the
weavers.

Posture and Force Analysis
Using the video tapes mentioned above, typical task cycles were
analysed frame by frame, in a manner similar to that described in
Armstrong *et al*, (1982). Operators' joint angles/positions and hand
actions were recorded with respect to time, task element and
significant force application. A criterion for assessing hand / arm
postures was chosen from the acceptable limits for joint angles
(defined as 75% of 5th percentile maximum voluntary movement)
suggested by Pethick *et al*, (1987), using data given in Chaffin &
Andersson (1984). Significant forces involved in the task were
measured using a simple spring balance and the spools were also
weighed.

Subjective Discomfort Analysis
The injury data was amplified by investigating pain and discomfort
experienced by the operators. A questionnaire was used to obtain
information on history over the previous year, and they were asked

about areas of body affected, symptoms and their severity. It was also interesting to investigate the pattern of development of fatigue or discomfort over a working day by using a discomfort mapping technique (Corlett & Bishop, 1976). Operators were asked, every 2 hours throughout the working day, to rate discomfort felt in various parts of their body.

Physical Workplace Assessment

Key workplace dimensions were measured and compared with anthropometric data. Noise and illuminance measurements were also taken in key areas.

STUDY OF TAPEWINDING MACHINE - RESULTS
Task Analysis and Job Design

The tapeline machine had 64 winding spindles on each side, arranged in 4 equi-spaced rows. Heights of top and bottom rows were 1,58 m and 0,41 m respectively. On the side observed in the study, 2 operators were required to unload all spools, once they had reached a given diameter, and replace these with empty spools. Typically, this took the team 8 minutes, and had to be done every 20 minutes. The timing of this task was very predictable, but was dictated by the machine.

Operators were working a 8 hour shift pattern, 5 days per week.

Posture and Force Analysis

Analysis of posture from the video showed that unloading the tapeline involved high degrees of shoulder medial deviation, wrist pronation and ulnar deviation, and back flexion. Ulnar wrist deviation was noted during force applications when removing spools, which were removed with a 4-finger hook grip, because of constraints imposed by the layout of each winding spindle.

Removal and replacement of a spool required a pushing/pulling force of between 118 and 195 N (+/- 20 N) force over 0.25 seconds, which had to be done once to remove the full spindle, and again to replace the empty one. Spools weighed 1 kg when full.

Subjective Discomfort Analysis

All tapeline operators suffered from discomfort in their neck, shoulders, back and wrists. 66% indicated discomfort in upper arms and lower back. These symptoms were variously described as "pain", "stiffness" and "aches". Significant upward increases in discomfort occurred throughout the working day, in operators' shoulders and legs (both Page's L-trend test, $p<0.01$) and lower arms and lower back (both Page's L-trend test, $p<0.05$).

This significant increase in discomfort shows that, as the task stands, rest periods are too short to allow recovery from periods of exertion.

Discussion

The forceful tasks on the tapeline were assessed on the basis of the classification defined in Silverstein *et al* (1986). Forces required to load and unload the tapeline machine are "high", and the high injury rate and number of complaints of pain bear this out. Clearly, there is a need to reduce the forces involved in the task, but defining an appropriate force limit, based on current literature, is not easy. Many reported guidelines concentrate on manual handling situations, often simulated under laboratory conditions, and cannot safely be translated into a repetitive work situation. However, for this type of task, the limiting factor would appear to be the 4 finger hook grip that operators are forced to adopt to remove spools. Pheasant (1987) quotes 30 N and 15 N as "safe" values for male and female operators respectively. Even so, limits such as these should be applied with caution.

Operators were also forced to bend down, with back flexion > 45 degrees, to unload the bottom two rows of spools. This, in combination with force application, has been linked to low back disorders (Punnett *et al*, 1987). This is supported by the high prevalence of back pain complaints, and the significant upward trends observed throughout a working shift in this area. As is often the case, the requirement for bending was dictated by the layout of the machine. Data included in Pheasant (1987) suggest that a range in working height between 1 - 1,08 m for a mixed population, but this would only accommodate 1 row of winding spindles; a compromise solution of 2 rows could be employed to eliminate bending for the majority of operators.

STUDY OF WEAVING MACHINES - RESULTS
Task Analysis and Job Design
Weaving operators were responsible for keeping all looms running to produce fabric, and this involved a variety of tasks. The task suspected of causing most musculoskeletal problems was that of loading spools into loom shuttles, which involved removing empty or almost empty spools once the machine nad stopped, and replacing them with fresh ones. The machines were arranged so that operators had to lean over part of the machinery to load shuttles. This effectively presented a barrier 0,97 m high and 0,45 m wide: the shuttle was located 0,22 m below the top surface of this obstacle. As the machines stopped automatically once a shuttle spool had run out, or for other reasons such as tape breakage, timing of these tasks was random and operators had some discretion over when they carried out tasks other than the loading. Shuttle loading accounted for 13% (+/- 5%) of the working shift.

This department was operated on a 8 hour shift system, to maintain production for 24 hours/day, 5 days/week.

Posture and Force Analysis
Loading shuttles in the weaving area involved high degrees of shoulder medial deviation and wrist pronation. Wrist rotation from extreme pronation to extreme supination occurred concurrently with ulnar deviation and application of high force with the base of the palm.

Removal and replacement of spools required a pushing/pulling force of 90 N (+/- 20 N).

Subjective Discomfort Analysis
Weaving operators mainly complained of discomfort in the wrists (63% reporting pain in the right wrist, 50% in the left), neck (38%), shoulders and back (both 25%).

Although a small degree of discomfort occurred for various operators, no significant upward trends in discomfort were noted throughout the days of the study.

Discussion
    Loading shuttles involved operators rotating their right wrists from one extreme to the other under ulnar deviation, while applying a high degree of force with the base of the palm. These factors have all been linked, in other studies (Putz-Anderson, 1988; Ayoub & Wittels, 1989; Armstrong et al, 1982), to tendonitis of the wrist, tenosynovitis, and carpal tunnel syndrome. In fact, symptoms of discomfort in the corresponding areas, and especially in the right wrist, were reported by many of the weaving operators. Tenosynovitis accounted for the majority of injury cases. Forces needed to load spools were "high", according to the classification in Silverstein et al (1986). Arriving at a safe limit is hard for this task, due to the very specific posture that the operator is forced to adopt due to the constraints of machine layout. Less than 10 N has been defined as "low" (Silverstein et al , 1986), but application of this figure as a safe limit should be made with caution.
    Although clearances are adequate around each loom, the layout makes access to shuttles awkward. To load fresh spools, operators are compelled to lean over and reach downwards, which leads to the extreme upper limb postures observed. Access to the interior of a circular weaving machine cannot be improved without a fundamental redesign.
    Noise levels in the area necessitated the use of ear protection, the main source of noise being the looms. Noise stress has been cited as a contributory factor to musculoskeletal disorders (Ayoub & Wittels, 1989).

THE DESIGN OF WEAVING MACHINERY
    Tapeline and weaving equipment contain common elements of design, which can be associated with the majority of the musculoskeletal problems observed in the study. Both machines incorporated components that needed to be loaded and unloaded on a regular basis and as rapidly as possible. Spools had to be held securely during machine operation. Those on the winding machine also had to be rotated at speed, and a degree of torque was required to overcome the resistance of the incoming tape. Herein lies a fundamental conflict in the design of the fixtures, because the easier a spool is to remove or replace, the more likely it is to be released during machine operation, with consequent loss of production and risk to safety. Where such components must be removed or replaced frequently, the forces required of an operator should be kept to an absolute minimum. However in most cases, push fits or snap fits, although appropriate at first glance, cannot satisfy the two requirements of retaining the element during rotation and allowing easy repeated removal and replacement over a working day. Automatic locking / release devices are preferable.
    Furthermore, machine layouts should not force operators to adopt inappropriate postures. The structure and layout of the two machines investigated, however, necessitated poor work postures, and

this aspect of their operation had obviously never been considered by the machinery designers.

The timing of the tasks of the operators was also dictated by the machines, and meant that all spools had to be changed very rapidly at the same time. The possibility of phased changes could be considered in redesign of such machinery, spreading the task over a longer period.

For machinery employed in high production environments, ergonomic principles should not be subordinate to design for high rates of production. The resulting injuries or musculoskeletal stress are not only serious for the operators, but have economic consequences in absence, turnover and compensation costs.

REFERENCES

Armstrong, T.J., 1986, "Ergonomics and Cumulative Trauma Disorders", *Hand Clinics*, **2**:3, pp. 553-565.

Armstrong, T.J., Foulke, J.A., Joseph, B.S. & Goldstein, S.A., 1982, "Investigation of Cumulative Trauma Disorders in a Poultry Processing Plant", *American Industrial Hygiene Association Journal*, **43**, pp. 103-116.

Ayoub, M.A. & Wittels, N.E., 1989, "Cumulative Trauma Disorders", *International Reviews of Ergonomics*, **3**, pp. 217-272.

Chaffin, D.B & Andersson, G., 1984, Occupational Biomechanics, New York: John Wiley & Sons.

Corlett, E.N. & Bishop, R.B., 1976, "A Technique for Assessing Postural Discomfort", *Ergonomics*, **19**, pp. 175-182.

Kanawaty, G. (Ed.), 1979, Introduction to Work Study, Geneva: International Labour Office.

Pethick, A.J., Mabey, M.H. & Graves, R.J., 1987, "Development of a Practical Method for Workplace Redesign to Reduce Upper Limb Strain Injury", In: Buckle, P. (Ed.), Musculoskeletal Disorders at Work, London: Taylor & Francis, pp. 239-246.

Pheasant, S., 1987, Ergonomics: Standards and Guidelines for Designers, Milton Keynes: British Standards Institution.

Punnett, L., Fine, L. & Keyserling, W., 1987, "An Epidemiological Study of Postural Risk Factors for Back Disorders in Industry", In: Buckle, P. (Ed.), Musculoskeletal Disorders at Work, London: Taylor & Francis, p. 74.

Putz-Anderson, V., 1988, Cumulative Trauma Disorders, London: Taylor and Francis.

Silverstein, B., Fine, L. & Armstrong, T., 1986, "Hand Wrist Cumulative Trauma Disorders in Industry", *British Journal of Industrial Medicine*, **43**, pp. 779-784.

Thompson, D., 1989, "Identification of Causes and Prevention of Work Related Upper Limb Musculoskeletal Disorders", In: Megaw, E. (Ed.), Contemporary Ergonomics 1989, London: Taylor and Francis, pp. 394 - 400.

WEIGHT GAIN AND LIFTING DURING PREGNANCY

S. SINNERTON, K. BIRCH, T. REILLY AND I.R. MCFADYEN*

Centre for Sport and Exercise Sciences,
Liverpool John Moores University,
Byrom Street,
Liverpool L3 3AF
* Royal Liverpool Hospital,
Prescot Street,
Liverpool L7

Maternal body weight and body size change
throughout the course of pregnancy. The effects
of these changes on manual handling and lifting
were studied in 9 pregnant subjects: 9 non-pregnant
subjects acted as controls. Body mass was recorded
and skinfolds were measured at four sites. An
isometric endurance lift and a vertical and
asymmetric dynamic lift was performed at 12-14 and
18-20 weeks gestation. Despite the increase in body
mass and skinfolds in the pregnant group, there
were no significant changes in the performance
variables.

INTRODUCTION

The increase in body weight and the change in body shape are
the most obvious changes which occur during pregnancy. This may
affect how a woman performs everyday tasks such as lifting and
manual handling. It may also affect the perception of exertion of
the tasks. Nicholls and Grieve (1992) reported an increased
perception of difficulty in the performance of physical tasks as
pregnancy progresses. Weight gain is greatest in the later stages
of pregnancy but the early stages are important in the adaptation
to the changes. Ahlborg et al. (1990) and Peters et al. (1984)
indicated the need to examine specific manual tasks and determine
how their performance is affected by pregnancy. Physiological
and psychological responses to tasks should be monitored, in an
attempt to determine how performance is affected.

METHOD

Nine pregnant women of mean age 32 years ($\pm$ 2.0), height 165.8
cm ($\pm$ 2.7) and body mass (at Stage 1) 64.2 kg ($\pm$ 4.6)
participated in the study. Three of the pregnant subjects were
prima-gravidae and six were multi-gravidae. The control subjects
were of mean age 28.2 years ($\pm$ 5.5), height 164.1 cm ($\pm$ 8.6) and

body mass 65.5 kg ($\pm$ 12.2). The pregnant subjects were tested at
12-14 (Stage 1) and 18-20 (Stage 2) weeks gestation. The testing
was performed on three separate days within one week. On the
first day of testing, measurements were taken of knee height
(measured at the tibial tuberosity) and waist height (measured at
the olecranon with forearms perpendicular to the body). Body mass
was recorded and skinfolds were measured at four sites. Percent
body fat and fat free mass were estimated (Durnin and Womersley,
1974) for the control group. The total of the skinfolds was used
as a measure in the pregnant group.

All subjects signed informed voluntary consent forms. The
study was approved by the Ethics Committees of the Royal
Liverpool Hospital and the Liverpool John Moores University.

Isometric lifting

This was performed on Day 1, using an isometric lifting
dynamometer (Birch et al., 1991). The control and pregnant group
performed the lifts at both knee and waist height. The control
group performed three maximum lifts and the best of these was
used as a reference point. This was described as maximum
isometric lifting strength (MILS). An endurance hold was then
performed. The threshold of the endurance hold was 35-45% of the
maximum lift. The pregnant subjects did not perform the maximum
lifts but, having been matched for height and age, used the
reference points from the control group to enable them to perform
the endurance lift. All subjects were asked to pay particular
attention to their lifting posture. The lift at knee height was a
bent knee lift with the back straight. Heart rate was monitored
throughout the lifts using short-wave radio telemetry (Sport-
Tester). Following the lifts, perceived exertion (RPE) was noted
(Borg, 1962). This provided information about the subjects'
perception of difficulty of the task, both generalised and
localised to the working muscles. It also gave insight into
correct and incorrect working postures.

Dynamic lifting

In order to assess dynamic lifting strength, the
psychophysical method of Snook (1978) was employed. This was
performed over a set distance which was approximately knee to
waist height. Control subjects performed one maximum lift (1RM).
All subjects lifted a tote box (down and up) six times a minutes
for 10 min. Subjects were asked to predict the amount of weight
they could lift, for the specified time and frequency, prior to
the lift. This was described as the rating of acceptable lift
(RAL). The subjects then performed the lift. A set of standard
instructions was used to inform the subjects of the selection
procedure and they were comfortable in the knowledge that they
could adjust the weight at any stage throughout the lift. The
final weight of the box, following the lift was recorded as the
maximal acceptable lift (MAL). This lift was performed vertically
and asymmetrically (through an angle of ninety degrees). The
vertical lift was performed on Day 2 and the asymmetric lift on
Day 3. Heart rate was monitored throughout all lifts and RPEs
were noted as for the isometric lift.

Results were analysed using the Students T-Test. A P value of

0.05 or less was taken to indicate significance.

RESULTS
    The control group showed no significant change in body mass
from Stage 1 to Stage 2; neither was there any significant
difference in skinfolds. The pregnant subjects showed a
significant increase in body mass (P<0.01) from 64.2 kg (± 4.6)
at Stage 1 to 66.8 kg (± 5.3) at Stage 2. The pregant subjects
also showed a significant increase (P<0.01) in the total
skinfolds, from 46.4mm (± 10.4) at Stage 1 to 51.3 mm (± 13.2) at
Stage 2.
    There was no significant difference, between Stages 1 and 2,
in the performance of the maximal isometric lift at knee and
waist height by the control subjects. At both stages the maximal
isometric lifting strength at knee height was significantly
greater than at waist height (P<0.01) in this group. The
endurance time for the control subjects did not vary
significantly between stages, but at Stage 1 the endurance time
was significantly greater at waist height than at knee height
(P<0.05). These data are shown in Table 1. The pregnant subjects
showed no significant difference in the isometric endurance time
between Stages 1 and 2. Within each stage the endurance time at
waist height was significantly greater than that at knee height
(P<0.05). These data are presented in Table 2.

Table 1. Performance data (mean ± s.d.) for control subjects.

Variable	Stage 1	Stage 2
MILS (N)(Knee height)	578 ± 267	573 ± 250
MILS (N)(Waist height)	287 ± 95	346 ± 101
ISOM. END. (s) (Knee)	24 ± 12	31 ± 18
ISOM. END. (s) (Waist)	43 ± 20	43 ± 29
1 RM (kg) (Vertical)	18 ± 3.6	16.1 ± 4.1
1 RM (kg)(Asymmetric)	15.4 ± 2.7	15.3 ± 4.2
MAL (kg) (Vertical)	8.6 ± 1.6	9.2 ± 1.9
MAL (kg) (Asymmetric)	9.1 ± 1.6	8.7 ± 1.7

    With regard to the maximal (1RM) dynamic lifts, by the control
group, these did not change significantly between Stages 1 and 2.
At Stage 1 the weight lifted vertically was significantly greater
(P<0.05) than that lifted asymmetrically. Between Stages 1 and 2,
for the control subjects, there was no significant difference in
the weight of the dynamic lift (MAL), neither was there any
difference between the weight lifted vertically and that lifted
asymmetrically. Mean and standard deviations for these data are
shown in Table 1. With regard to the pregnant subjects, there was
no significant change between Stages 1 and 2 in the weight of the

dynamic lifts (MAL). At both stages there was no significant
difference between the weight lifted vertically and
asymmetrically. These values are shown in Table 2.

The general RPEs did not show any significant variation
between Stage 1 and Stage 2 in the control subjects. The ratings
of general perceived exertion of the pregnant subjects between
Stages 1 and 2 remained unchanged for all performance variables.

Table 2. Performance data (mean ± s.d.) for pregnant subjects.

Variable	Stage 1	Stage 2
ISOM. END. (s) (Knee)	20 ± 10	18 ± 9
ISOM. END. (s) (Waist)	48 ± 27	43 ± 29
MAL (kg) (Vertical)	6.2 ± 0.8	6.1 ± 1.1
MAL (kg) (Asymmetric)	6.5 ± 1.5	6.7 ± 1.4

DISCUSSION

Both the control subjects and the pregnant subjects showed no
significant difference in isometric endurance strength between
Stages 1 and 2. It seems that subjects were able to maintain the
isometric endurance performance in this task as pregnancy
progresses. The greater endurance time at waist height compared
to knee height, for both groups, suggested that the subjects
found the lift at knee height the more difficult. This is in
agreement with Seals et al. (1983) who concluded that the
magnitude of the cardiovascular response to isometric exercise is
directly influenced by the size of the contracting muscle mass.
Williams et al. (1982) reported that more oxygen is consumed when
the lift is from a bent knee position than from a stooped
position, highlighting the increased difficulty of the task.
Despite warnings many women prefer to stoop rather than bend
their knees and the position required for the knee lift in this
study may have induced relatively new demands on subjects.

The results from the isometric lifts, suggesting an increased
difficulty in the task at knee height, were not reflected in the
general RPEs. The maximum strength (for the control subjects) at
knee height was greater than at waist height, but the relative
load (expressed as a percentage of the maximum) was the same at
both knee and waist height. Using the maximum loads from the
control subjects, the relative thresholds for the pregnant
subjects should also be the same. For both groups the endurance
task at knee height was with a greater absolute load at knee
height and this was reflected in the diminished endurance time.

The control group and the pregnant groups did not show any
significant differences in the dynamic lifting performance
between Stages 1 and 2. The weight lifted by the control group in
the 1RM lift was greater for the vertical lift than for the
asymmetric lift. This supports the findings of Mital and Fard
(1986), who found a decrease of approximately 8.5% in the
asymmetric lift.

Although the pregnant women had significantly gained weight and showed an increase in skinfolds, their performance of manual handling and lifting tasks remained unchanged. It is expected that the increase in body mass would increase the demands placed on the women. This was not observed in the present results, probably because the results were taken from the earlier stages of pregnancy when weight gain is least. Later stages of pregnancy might show significant changes in manual handling and lifting performance, initiated by a further increase in body mass and change in body shape. Despite the increase in weight and skinfolds, women seem to adapt to changes in body size and composition in the early stages of pregnancy without compromising the capability for manual handling and lifting.

REFERENCES

Ahlborg(JR), G., Bodin, L. and Hogstedt, C., 1990, Heavy lifting during pregnancy–A hazard to the fetus? A prospective study, International Journal of Epidmiology, 19, 90–97.

Birch, K., Sinnerton, S., Reilly, T. and Lees, A., 1991, The relation between isometric lifting strength and muscular fitness measures, Journal of Sports Sciences, 9, 328–329.

Borg, G., 1962, Physical performance and perceived exertion. (Gleerup: Lund, Sweden).

Durnin, J.V.G.A. and Womersley, J., 1974, Body fat assessed from total body density and its estimation from skinfold thicknesses. Measurement on 481 men and women aged 16–72 yrs. British Journal of Nutririon, 32, 77–79.

Mital, A. and Fard, H.F., 1986, Psychophysical and physiological responses to lifting symmetrical and asymmetrical loads symmetrically and asymmetrically, Ergonomics, 29, 1263–1272.

Nicholls, J.A. and Grieve, D.W., 1992, Performance of physical tasks in pregnancy, Ergonomics, 35, 301–311.

Peters, T.J., Adelstein, P., Golding, J. and Butler, N.R., 1984, Effects of work in pregnancy: short and long-term associations. In Women at work in pregnancy, ed. G. Chamberlain (Royal Society of Medicine, Macmillan: London).

Seals, D., Washburn, R., Hanson, P., Pashter, P. and Nagle, F., 1983, Increased cardiovascular response to static contraction of larger muscle groups, Journal of Applied Physiology, 54, 434–437.

Snook, S.H., 1978, The design of manual handling tasks, Ergonomics, 21, 963–985.

Williams, C.A., Petrofsky, J.S. and Lind, A.R., 1982, Physiological responses of women during lifting exercise. European Journal of Applied Physiology, 50, 133–144.

Acknowledgement: This project was funded by the Health and Safety Executive. This is gratefly acknowledged.

MENSTRUAL CYCLE EFFECTS ON ISOMETRIC AND DYNAMIC LIFTING

K.Birch, I.R.McFadyen, T.Reilly and S.Sinnerton

Centre for Sport and Exercise Sciences,
School of Human Sciences,
Liverpool John Moores University
Byrom St.
Liverpool
L3 3AF

The female menstrual cycle may affect manual handling capabilities. Twelve females performed isometric and dynamic lifting tasks at five points in their menstrual cycle. There was no significant effect of menstrual cycle phase on maximal isometric lifting strength, isometric endurance capacity, maximal dynamic lifting, self-chosen lifting capacity or the ratings of perceived exertion to any of these tasks. The heart rate response to the isometric tasks was elevated in those phases following ovulation (P < 0.05) concurrent with a rise in body temperature. This response seems to have no serious impact on manual handling capacities.

INTRODUCTION
The female reproductive cycle is characterized by regular and predictable alterations in neuroendocrinological processes, which could affect physical performance. There is, however, a lack of research concerning circa-mensal variations in the performance of strenuous physical work such as that required in manual handling. The present purpose was to investigate the effect of menstrual cycle phase on isometric and dynamic lifting capability.

METHOD
Twelve females aged 18-36 years (60.4 ± 8.4 kg), inexperienced in manual handling, participated in the study. All subjects had a regular and healthy (eumenorrheic) menstrual cycle and none was taking any form of oral contraceptive. Each subject's menstrual cycle was divided into five phases corresponding to (i) follicular phase preceding ovulation, (ii) ovulation, (iii)

luteal phase following ovulation and including premenses, (iv) premenses, 72 hours preceding menses and (v) menses or menstrual bleeding. Ovulation was detected by a mid-cycle rise in body temperature of 0.4 degrees centigrade or above (Marshall, 1963). This was determined prior to the experimental period by one month's daily measurement of oral temperature.

Two experiments were conducted on separate days in each phase of the cycle. All experiments were undertaken at the same time of day and the order of testing randomized to eliminate any circadian or order effects.

Experiment 1

The first experiment in each phase utilized an isometric lift dynamometer allowing for the simulation of lifting techniques. The dynamometer consisted of two force bars onto which were mounted strain gauges. On lifting the bars, distortion of the strain gauges was converted to a measure of force via an analogue to digital converter (Unilab, Blackburn, England).

Each subject performed a maximal isometric lift at both knee (the height of the tibial tuberosity of the straight leg) and waist (the height of the elbows with forearms held perpendicular to the body) height. The knee-height lift involved a squat posture and was primarily a measure of leg and back lifting strength (Birch et al., 1993). The lift at waist height was from a standing position and engaged the musculature of the upper body. Each lift was over a 3 s duration and the peak of three attempts was recorded as the Maximal Isometric Lifting Strength (MILS).

At both heights subjects were required to lift and maintain 45% of their maximal lift for as long as possible. Time to fatigue was recorded as isometric endurance capacity.

Experiment 2

At each phase of their menstrual cycle, subjects were presented with a series of visually identical bags weighing randomly between 0.25–2.5 kg, and a box with handles on the side (30 cm x 30 cm x 30 cm). Each subject was instructed to fill the box with as much weight as possible to lift once only from knee (tibial tuberosity) to shoulder (acromial) height; this was the distance through which the handles of the box moved. The weight selected was recorded as the "one repetition maximum" (1RM). The subject was then required, using the bags and the box, to select the maximum weight that she considered comfortable to lift, over the same distance, 6 times per minute for 10 minutes. The instructions to subjects were standardized. Each lift started and finished at shoulder height, the frequency of lift being timed by the researcher. The weight in the box before selection was randomized at each phase. The weight remaining in the box after 10 min was recorded as the Maximal Acceptable Lift (MAL).

The heart rate response to each of the isometric and dynamic lifts was recorded using short-range radio telemetry (Sports tester PE 3000, POLAR ELECTRO, Finland). Perceived exertion (RPE) was rated using the Borg (1962) scale.

Results were subjected to a one-way ANOVA with repeated measures. Differences between phases of the menstrual cycle were examined by post-hoc analysis using Tukey tests. Level of significance was accepted as P < 0.05.

RESULTS
Isometric Lifting

Performance of MILS, endurance time (Table 1) and perceived exertion were unaffected by menstrual cycle phase at either height. The RPE averaged between 11 - 13, or "Fairly Light" to "Somewhat Hard".

Table 1. Performance data for isometric lifting tasks at knee (k) and waist (w) height (Mean ± S.D. in N and s).

	Menses	Follic.	Ovulat.	Luteal	Premens.
MILS(k)	913±264	947±255	884±282	948±251	908±252
Time(k)	35 ±23	36 ±23	43 ±36	35 ±17	39 ±26
MILS(w)	444±105	469±140	457±122	455±128	459±111
Time(w)	34 ±20	38 ±21	40 ±24	34 ±17	37 ±16

The pattern of heart rate response to the isometric endurance task was identical in each of the subjects. On commencing the 45% maximal lift there was an immediate increase in heart rate which continued until the bars were released at fatigue. At this point there was a rapid return to reach resting levels within minutes.

In order to evaluate the effect of menstrual cycle phase on heart rate, each of the subject's heart rate response throughout the endurance task was reported at fatigue, and for 20%, 50% and 80% of the time it took to reach fatigue (total endurance time). The heart rate recorded at fatigue, and at 50% and 80% of the total endurance time for the knee height lift, was greater during the premenstrual phase than during the follicular phase of the cycle (P < 0.5). This phase difference was also noted at fatigue for the endurance lift at waist height. The heart rate attained at fatigue in each phase of the cycle for the knee and waist height lifts are displayed in Figure 1.

When the heart rate data for both lifts were examined at the equivalent time points, 15 s and 30 s of endurance time, all menstrual cycle phase effects were eliminated (P > 0.05).

Dynamic Lifting

Both the 1RM and MAL (Table 2) were unaffected by phase of menstrual cycle (P > 0.05); subjects did choose to lift more during the ovulatory phase, but this difference was not significant. The RPE attained at 5, 8 and 10 minutes of lifting also displayed no significant phase effects.

Table 2. Performance data for dynamic lifting tasks (Mean ± S.D.).

	Menses	Follic.	Ovulat.	Luteal	Premens.
1RM(kg)	15.7±2.6	16.4±2.5	15.7±2.3	15.1±2.8	15.2±2.4
MAL(kg)	8.3±1.6	8.0±1.2	8.5±1.7	8.0±1.4	8.4±1.6

The heart rate response to lifting was again consistent within the sample. Heart rate increased gradually throughout exercise to a steady state at about 8 min. Pre-exercise levels were attained within 5 min of the cessation of lifting. The heart rate response to lifting was greater in those phases following ovulation (Fig. 2). This difference was non-significant.

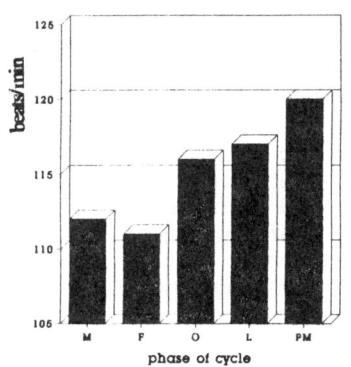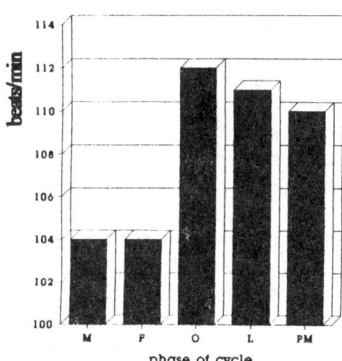

Figure 1. Mean heart rate at fatigue following the isometric endurance lift at knee (left) and waist (right) heights.

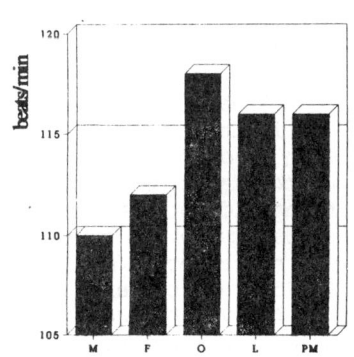

Phase of cycle: M: Menses
F: Follicular
O: Ovulatory
L: Luteal
PM: Premenses

Figure 2. Mean heart rate following 10 min of dynamic lifting.

DISCUSSION

No differences in muscular strength or endurance, either isometric or dynamic, were found throughout the menstrual cycle in this study. Previous studies on handgrip strength have shown both decreased (Wearing et al., 1972) and increased strength (Davies et al., 1991) during menses. In both studies the number of phases within the cycle were few and were established by counting the days following menses. Subsequently the occurrence and time of ovulation in each sample could not be confirmed.

Effects of the menstrual cycle on isokinetic strength and endurance (Dibrezzo et al., 1988) and on more dynamic weight lifting (Quadagno et al., 1991) have also been examined. No phase effects on peak torque or endurance were found for knee flexion and extension at three angular velocities on a Cybex II machine. Quadagno et al. confirmed this, and the present findings, by failing to find an effect of cycle phase on maximal lifting or endurance lifting at 70% maximum in either bench or leg pressing.

Studies of both aerobic and anaerobic exercise have provided no evidence of significant fluctuations in heart rate with menstrual phase (Garlick and Bernauer, 1968; Nicklas, 1989). In the present study, there appeared to be an increase in the heart rate response to both isometric and dynamic exercise during the luteal ($P > 0.05$) and premenstrual ($P < 0.05$) phases of the cycle. This increase could not be explained by changes in exercise intensity and was on average 9 beats/min greater than during those phases preceding the luteal phase.

When the increase in heart rate above resting levels was computed, menstrual cycle fluctuations were present only during those tasks requiring strenuous isometric muscular effort, i.e. during the lift at knee height and towards fatigue at waist height. It appears that the tachycardia produced by sustained isometric contraction is further elevated during the luteal phase. This agrees with Hessemer and Bruck (1985) and Pivarnik et al. (1992) who also recorded elevated heart rate response to strenuous muscular endurance in the luteal phase. These elevated heart rate values may be related to decreased plasma volume (Garlick and Bernauer, 1967), increased capillary permeability and/or reactivity (Jones et al., 1966) or an inability to reach thermal equilibrium (Pivarnik et al., 1992) in the luteal phase. It may also be that subjects were exercising at different percents of their aerobic power throughout the menstrual cycle.

The subjective responses (RPE) to lifting did not mirror the changes in heart rate throughout the menstrual cycle. The ratings for both isometric and dynamic lifting were not affected by menstrual cycle phase. Gamberale et al. (1975) reported an increased perceived exertion to exercise of 40, 70 and 100% maximal oxygen uptake during the premenstrual and menstrual phase. They suggested that this was due to "subjective feelings of discomfort, depression and anxiety associated with premenstrual tension." Higgs and Robertson (1981) reported that RPE was elevated at 100% maximal oxygen uptake only, which would indicate a methodological problem in defining maximal exertion. Furthermore, Gamberale and co-workers' subjects were all recognized to suffer from premenstrual discomfort. The present results substantiate the findings of later studies (Nicklas, 1989) which failed to find differences in perceived exertion associated with menstrual phase in eumenorrheic subjects.

REFERENCES

Birch, K., Sinnerton, S., Reilly, T. and Lees, A., 1993, The relation between isometric lifting strength and muscular fitness measures. Ergonomics, In press.

Borg, G., 1962, Physical performance and perceived exertion, (Lund, Sweden: Gleerup).

Davies, B., Elford, J. and Jannieson, K., 1991, Variations in performance in simple muscle tests at different phases of the menstrual cycle, Journal of Sports Medicine and Physical Fitness, 31, 532-537.

Dibrezzo, R., Fort, I. and Brown, B., 1988, Dynamic strength and work variations during three stages of the menstrual cycle, Journal of Orthopaedic Sports Physical Therapy, 10, 113-116.

Gamberale, F., Strindberg, L. and Wahlberg, I., 1975, Female work capacity during the menstrual cycle - physiological and psychological reactions, Journal of Work, Environment and Health, 1, 120-127.

Garlick, M. and Bernauer, E., 1968, Exercise during the menstrual cycle, variations in physiological baselines, Research Quarterly, 39, 533-542.

Hessemer, V. and Bruck, K., 1985, Influence of menstrual cycle on thermoregulatory, metabolic and heart rate responses to exercise at night, Journal of Applied Physiology, 59, 1911-1917

Higgs, S. and Robertson, L., 1981, Cyclic variations in perceived exertion and physical work capacity in females, Canadian Journal of Sport Science, 2, 145-153.

Jones, E., Fox, R., Verow, P. and Asscher, A., 1966, Variations in capillary permeability to plasma proteins during the menstrual cycle, Journal of Obstetrics and Gynaecology of the British Commonwealth, 43, 666-669.

Marshall, J., 1963, Thermal changes in the normal menstrual cycle, British Medical Journal, 1, 102-104.

Nicklas, B., 1989, The menstrual cycle and exercise: performance, muscle glycogen and substrate responses, International Journal of Sports Medicine, 10, 264-269.

Pivarnik, J., Marichal, C., Spillman, T. and Morrow, J., 1992, Menstrual cycle phase affects temperature regulation during endurance exercise, Journal of Applied Physiology, 72, 543-548.

Quadagno, D., Faquin, L., Lim, G., Kuminka, W. and Moffat, R., 1991, The menstrual cycle: Does it affect athletic performance? Physician and Sports Medicine, 19, 121-124.

Wearing, M., Yuhosz, M., Campbell, R. and Love, E., 1972, The effects of the menstrual cycle on tests of physical fitness, Journal of Sports Medicine and Physical Fitness, 12, 38-41.

This work was funded by a grant from the Health and Safety Executive. This is gratefully acknowledged.

Sport ergonomics

A PHYSIOLOGICAL EVALUATION OF SKI SIMULATORS

T. REILLY, E. KIRTON, E. McGRATH and S. COULTHARD

Centre for Sport and Exercise Sciences
School of Human Sciences
Liverpool John Moores University
Mountford Building
Byrom Street
Liverpool
L3 3AF
ENGLAND

Ski simulators are increasingly being promoted as
training devices, particularly in preparation for
recreational skiing holidays. A series of studies was
conducted to examine the acute and long-term responses
to training on an alpine and a cross-country ski
simulator. The results offer some support for the use
of the simulators in preparation for skiing on snow.

INTRODUCTION

Recreational alpine skiing is nowadays a major leisure and
tourist activity. Participants may engage in skiing for 4 hours
a day during winter holidays and for longer each day of weekend
trips to ski resorts in the mountains. Many tourists may lack
the aerobic fitness and the skiing skills necessary to benefit
physiologically from the skiing activities.

In recent years, the use of ski simulators has been promoted
as a means of aerobic training and skills enhancement prior to
skiing on snow. Skiers have been reported to be capable of
operating at up to 95% $\dot{V}O_2$max whilst exercising on a ski
simulator (Ski-Tone, Oasis Shipping Co.), although no
conclusions were drawn about the fidelity of the skiing actions
(Reilly and McCann, 1990). Skiing specificity has been improved
in subsequent designs, most notably in the Skier's Edge (Redwood
City, CA) used by U.S. skiers. It has not been established
whether this design compromises its use for physical training.

Cross-country skiing is more demanding than alpine skiing in
terms of prolonged loading on the oxygen transport system
(Reilly et al., 1990). As a consequence of this, cross-country
ski simulators are incorporated into the range of exercise
machines used in fitness training centres. Invariably the
validity of the simulators has not been questioned.

This series of studies was designed in order to investigate:-
i) the physiological stress experienced whilst exercising on an
alpine ski simulator; ii) its value in skill acquisition; iii)
the physiological stress induced by a cross-country ski
simulator.

METHODS
Study 1 ; alpine ski simulator: physiological stress
    Ten subjects (6 male, 4 female) volunteered to participate in
the study. This included 5 beginners (1-2 weeks on snow; age
21.6 $\pm$ 3.2 years) and 5 advanced skiers (7-8 weeks on snow; age
20.8 $\pm$ 1.5 years). All gave written voluntary informed consent
to participate in the study.
    Subjects visited the laboratory twice for a test to
exhaustion either on a motor-driven treadmill (Quinton) or a ski
simulator (Skier's Edge). The treadmill test was an incremental
test for measurement of $\dot{V}O_2$max. The $\dot{V}O_2$ was measured
continuously using an on-line system for respiratory gas
analysis (Sensorimedics, Salford) and heart rate was recorded
using short range radio telemetry (Sport-Tester). Perceived
exertion (Borg, 1982) was rated post-exercise for whole-body and
for legs.
    The ski simulator consisted of an aluminium frame, on top of
which were two plates on which the skier's feet were placed.
The plates tilted to each side in order to simulate edging of
the skis. The plates could be moved side to side on a track on
the frame, the resistance to this motion being provided by
rubber strips attached to the plates. The strips could be
adjusted to alter the resistance whilst the exercise intensity
at a given resistance was changeable by varying the frequency of
side-to-side cycles.
    The exercise intensity on the ski simulator commenced at a
rate of 88 cycles per minute. This rate was dictated by a
metronome and was increased every 3 min by 4 cycles/min. The
highest rate attainable was 122 cycles/min; subjects not
exhausted at this point continued at this rate until they were
fatigued.
    Expired air was measured continuously throughout the test.
The highest $\dot{V}O_2$ value reached was attributed $\dot{V}O_2$peak. Heart
rate was monitored by means of radio telemetry and perceived
exertion was rated as for the treadmill test. A resting blood
sample was obtained pre-exercise from a finger tip. Immediately
after exercise and 3 min post-exercise, further blood samples
were collected. The samples were analysed for lactate using the
Analox LM3 machine.
Study 2 ; alpine ski simulator: skills acquisition
    The second study utilised 20 males, mean age ($\pm$SD) = 21.9$\pm$1.8
years. They were equally divided into four groups: expert
skiers, non-skiers who undertook a basic one-month course of ski
instruction on an artificial slope, non-skiers whose instruction
was on the Skier's Edge simulator, and a control group of
non-skiers who had no instruction.
    Electromyography (EMG) of six lower limb muscles was
monitored by means of surface electrodes during a bout of 2 min

rhythmic activity on the ski simulator. The muscles examined
were M. Rectus Femoris, M. Vastus Lateralis, M. Biceps Femoris,
M. Tibialis Anterior, M. Gastrocnemius Lateralis and M.
Gastrocnemius Medialis. Silver - silver chloride surface
electrodes were connected by cable to a multi-channel recorder
(NEC Sanei Instruments, Tokyo) for registering the raw EMG. The
recorder had bioelectrical amplifiers with filters set to record
a bandwidth of 10 to 1000 Hz. The electrode arrangement
provided a balanced input for the amplifier, minimising 50 Hz
artefacts and maximising the signal's common mode rejection
ratio. The 2 min trial was completed by all the groups before
and after two of them undertook the programme of instruction. A
maximum voluntary isometric contraction was performed by each
subject and for each of the muscles prior to the test on the
simulator so that the integrated EMG data could be normalised.

Study 3 ; cross-country ski simulator
    Nine subjects were recruited for a one-month training
programme. These included 6 females and 3 males, mean age
($\pm$SD) = 21.4$\pm$0.9 years. The subjects were divided into three
groups for training on either the cross-country ski simulator
(Slenderworld International, Colne, Lancs), a rowing machine
(Concept II, Nottingham) or a motorised treadmill (Quinton).
Training was conducted 3 times each week for 30 min at a time.
The heart rate was monitored during exercise using short range
radio telemetry (Sport-Tester).
    A range of fitness measures was obtained before and after
completion of the training programme. These included $\dot{V}O_2$max (or
$\dot{V}O_2$ peak) on each of the ergometers (running, rowing and ski
simulator); blood lactate, perceived exertion and heart rate at
maximal exercise, back strength; knee extension and knee flexion
strength, each at four velocities of concentric contraction
(Lido Active, Davis CA); balance (wobble-board) and overall
flexibility (Leighton goniometer)
    The changes in fitness measures over the period of training
were compared between training modes. In view of the small
numbers in each group, no formal statistical analysis was
conducted in comparisons between groups. Training effects among
the subjects as a whole were examined using a combination of
t-tests for individual variables and analysis of variance for
specific ergometry effects.

RESULTS
Study 1 ; alpine ski simulator: physiological stress
    The $\dot{V}O_2$max of subjects attained during the treadmill test was
higher for the advanced group (49.0$\pm$9.1) compared to the
beginners (43.0$\pm$3.6 ml/kg/min). The corresponding heart rate
values were 196$\pm$2 and 202$\pm$9 beats/min. The $\dot{V}O_2$ peak achieved on
the ski simulator was 32.0$\pm$6.6 and 31.3$\pm$3.7 ml/kg/min for the
beginners and the advanced group, respectively. These
represented 65% and 73% of the respective $\dot{V}O_2$max values. This
difference between groups occurred despite the similar heart
rate values of 153 and 156 beats/min, representing 78 and 77% of
maximal heart rates.

The rating of perceived exertion was higher for legs than for whole-body effort on the ski simulator, 16.4 ± 1.7 and 15.0 ± 1.9 (legs) and 12.4 ± 1.5 and 10.2 ± 3.0 (whole-body) for the beginners and the advanced skiers. The time to reach exhaustion on the ski simulator was 29.40 ($\pm$ 4.44) and 34.48 ($\pm$ 3.67) min for beginners and advanced skiers. This reversed performance on the treadmill test where the beginners had the longer times (15.52 $\pm$ 0.5 vs 11.24 $\pm$ 3.66 min).

Significantly higher blood lactate levels were noted after exercise on the treadmill compared to the ski test. Values increased 5-6 fold over pre-exercise levels during the treadmill run whereas increases during the ski test were limited to three-fold. The lactate levels post-exercise were similar for the advanced and novice skiers on the treadmill test. The advanced skiers produced higher lactate concentrations (mean difference 0.7 mM) than the beginners, this difference being noted immediately post-exercise and 3 min following. The lactate levels post-exercise on the ski simulator were 43% and 59% of the peak values following the treadmill test for the beginners and advanced skiers, respectively. This indicates that exercise on the ski simulator was maintained for the most part within the aerobic zone.

Study 2 ; alpine ski simulator: skills acquisition

The experienced skiers demonstrated significantly different EMG profiles to the other groups for all upper leg muscles and for M. Tibialis Anterior (P<0.01). No significant differences were noted for the gastrocnemius muscles (P>0.05).

The EMG profiles were reproducible over the one-month of the study for both the control subjects and the experienced skiers. The two training groups demonstrated alterations compatible with skills enhancement. The duration of muscle activation was decreased in both groups, demonstrating a more economical use of muscular activity. The direction of change in the EMG was towards the profile of the experienced skiers but the extent of this change differed between the two training groups and between the muscles examined. The subjects on the ski simulator showed the more positive changes in the mean normalised amplitudes of the thigh muscles' activity, whilst those learning on the ski slope showed the more positive changes in the phasing of the different muscles.

Study 3 ; cross-country ski simulator

The subjects as a whole showed no improvement in muscle strength. This applied to all the velocities and muscle groups examined (P>0.05). Balance improved significantly with training (P<0.05) but it was not possible to attribute the improvement exclusively to the ski simulator.

The highest physiological responses were observed on the running treadmill, the lowest on the ski simulator with the rowing ergometer being intermediate (Table 1). There was a significant improvement in the highest $\dot{V}O_2$ reached for the rowing ergometer (P<0.01) and ski (P<0.05) tests but the improvement on the treadmill run did not reach significance (P>0.05). There was evidence of a specificity effect in that the greatest improvements on the rowing and ski tests were noted in the subjects using those devices during their training regimen. The lactate levels post-exercise were unaffected by the training programme. The heart rate showed a significant decrease on the ski test only (P<0.05)

Table 1. Physiological responses to maximal effort on an incremental test on each of the ergometers.

	Pre-training Treadmill	Row	Ski
$\dot{V}O_2$ peak(l/min)	2.65±0.25	2.42±0.23	2.04±0.23
Lactate (mM)	8.5±1.8	7.4±1.5	4.0±0.5
Heart rate (beats/min)	193±20	183±10	170±13

	Post-training Treadmill	Row	Ski
$\dot{V}O_2$ peak(l/min)	2.75±0.20	2.56±0.26	2.20±0.24
Lactate (mM)	8.1±2.8	7.3±1.3	3.9±0.4
Heart rate (beats/min)	192±8	187±11	182±13

At the beginning of the training programme, the heart rate during exercise averaged 147, 161 and 141 beats/min for the treadmill, rowing and ski simulator groups, respectively. Corresponding values during the final week of training were 137, 151 and 125 beats/min.

DISCUSSION
The peak values for simulated alpine skiing were lower than those observed during investigation of an alternative ski-trainer by McCann and Reilly (1990). Their subjects were able to obtain 94% and 95% of maximal figures for HR and $\dot{V}O_2$. In the earlier study the peak $\dot{V}O_2$ on the simulator was significantly correlated to the $\dot{V}O_2$max determined on the treadmill (n=14; r=0.72). This contrasted with the present study where the advanced skiers were able to obtain a higher fractional utilisation of $\dot{V}O_2$max. The discrepancies are likely to have reflected the superiority of the previous device (Ski-Tone) for aerobic training and the present machine (Skier's Edge) for skills training.

Nevertheless, present observations support the use of the Skier's Edge simulator for purposes of aerobic training, particularly for skiers of advanced performance. Recreational skiers tend to exercise sub-maximally during ski-classes, with periodic rest pauses for instruction. Their alternation of work-rest cycles can be duplicated on the simulator and the edging of the platforms represent turning actions. Once the basic ski-turns are learned, it is likely that beginners will be able to elevate the exercise intensity on the simulator, thereby enhancing the potential for aerobic training.

The longitudinal study of training on the simulator provided evidence of EMG changes compatible with skills acquisition. The results could not conclusively discriminate between the two

training modes in terms of which was the better. The changes as
a result of using the Skier's Edge were in the directions to be
expected based on EMG profiles from realistic skiing (Clarys et
al., 1988). They provide some support for the use of the
simulator for skills acquisition as well as physical training.

The longitudinal study using the cross-country ski simulator
furnished evidence of improvements in balance and in oxygen
transport in those subjects using the device. Subjects were
able to attain 77% of their $\dot{V}O_2$max whilst working as hard as
they could on the simulator. The exercise intensity whilst
training on the simulator was less than this, the training heart
rate being 83% of the peak at the start of training and 77% in
the final week. The fall in both peak heart rate and training
heart rate over the duration of the study may reflect an
inherent limitation of the simulator. It seems the subjects
were not able to raise the work-rate progressively as their
fitness improved so that a high training stimulus might be
maintained.

Overall, the series of studies offer support for the use of
the simulators in preparation for skiing on snow. Whilst the
simulators did succeed in provoking positive physiological
responses, their limitations as substitutes for snow-skiing
should be acknowledged.

REFERENCES

Borg, G., 1982, Psychophysical bases of perceived exertion.
   Medicine and Science in Sports, 10, 151-154.
Clarys, J.P., Cabri, J., De Witte, B., Touissant, H., de Groot,
   G., Huying, P. and Hollander, P., 1988, Electromyography
   applied to sport ergonomics. Ergonomics, 31, 1605-1620.
McCann, R. and Reilly, T., 1990, Physiological responses to
   exercise on a ski simulator. In: Contemporary Ergonomics
   1990 ed. E.J. Lovesey (London: Taylor and Francis), pp.
   469-472.
Reilly, T., Secher, N., Snell, P. and Williams, C. 1990,
   Physiology of Sports (London: E. and F.N. Spon).

ANGLE SPECIFIC ISOKINETIC TALOCRURAL TORQUE RATIOS

C.M. Graham and G. Garbutt.

School of Health Sciences,
University of Sunderland,
Sunderland,
Tyne and Wear,
SR1 3SD,
ENGLAND.

Maximal voluntary torque (at $30°s^{-1}$) for plantar
flexion (PF), dorsiflexion (DF), inversion (INV) and
eversion (EV) was measured in 9 professional soccer
players (20.8 ± 3.5yr, 178.1 ± 7.8cm, 75.6 ± 8.9kg)
using an isokinetic dynamometer (Lido, Loredan, USA).
Peak torque in PF>DF at each angle, EV>INV from -15 to
10°, INV>EV from 10 to 15°. Torque ratios, PF/DF and
INV/EV were linearly related to joint angle (JA),
PF/DF = -0.53(JA)+2.893, (r=0.967, p<0.001), INV/EV =
0.02(JA)+0.841 (r=0.957, p<0.001). Contralateral
differences (p<0.05) occurred only at extreme JA,
suggesting lower leg function is mainly postural.

INTRODUCTION

An isokinetic evaluation of limb contralateral and
antagonistic muscle group strength can be used to assess muscle
function for injury prevention and rehabilitation. Little
research has reported isokinetic, ankle muscular function,
especially the movements of INV and EV. Need for information on
ankle function becomes apparent when consulting the literature.
Between 17% and 27% of all injuries in sport involve the ankle
(Brady 1990), usually inversion injuries. The nature of soccer
is such that it renders players particularly prone to this kind
of injury. Bearing this observation in mind, it was decided that
professional soccer players would act as subjects for this study.
    The aims of the study were: to determine whether contralateral
differences exist in the lower leg muscle strength between
preferred (P) and non-preferred (NP) kicking legs; and to
determine relationships between antagonistic muscle strength at
specific joint angles (JA) of the ankle. The normative data will
be useful in injury prevention screening processes and to those
involved in rehabilitation work including ankle injury. Muscular
strength is represented by torque production about the joint.

METHOD
Nine male subjects participated in this study, all were
professional soccer players in Division II of the English league.
Table 1 shows some of their physical attributes.

Table 1. Body mass, stature and age of the subjects.

	Body mass (kg)	Stature (cm)	Age (yrs)
Mean ± SD	75.61 ± 8.9	178.1 ± 7.77	20.8 ± 3.53

Subjects were tested bilaterally for maximum voluntary
strength using a Lidoactive dynamometer (Loredan Biomechanical
Inc.,1632 Da Vinci Court, PO Box 115A, Davis, CA95617,USA). To
measure maximal torque a lever arm speed of $30°s^{-1}$ was decided
upon after completing a pilot study and consulting the literature
(Wickiewicz et al., 1984; Wong et al., 1984; Alexander, 1990 and
Gravel et al., 1990). Reciprocal contractions of the antagonistic muscle groups were employed for 6 maximal contractions in
each direction (Baltzopolous and Brodie, 1989), the first being
ignored and the latter 5 used to calculate mean maximal torque
values at 5° intervals. Test ordering was as follows: Subjects
were tested in groups of three, alternate groups performed either
the PF/DF or INV/EV movements first. The leg tested first was
alternated from one subject to the next. These procedures helped
minimise the significance of learning or fatigue effects.

After weighing and measuring height, subjects warmed up on a
cycle ergometer (Monark ergomedic 814E) for 5 mins at a cadence
of 70 r.p.m. and with a work load of 90 Watts. Toe clips were
supplied to enable DF movements, subjects then stretched,
concentrating upon the muscles involved in PF, DF, INV, EV.

For PF/DF movements the subject lay supine with one foot
secured to the dynamometer foot plate, the ankle axis of rotation
concentric with that of the lever arm. The buttock/leg rest was
elevated 10°. The knee JA was set to 70° of flexion (0°= full
extension). The neutral ankle position was designated 0° (foot
sole 90° to shank). Plantar flexion gave positive values.

For the INV/EV movements the subjects lay upon their sides and
the buttock/leg rest was at 10°. The foot plate was adjusted so
the ankle axis of rotation coincided with that of the lever arm.
The foot plate was fixed at 20° of PF, consistent with the findings of Cawthorn et al. (1991) that maximal INV/EV torque occurs
when the ankle is in between 10-25° of PF. The neutral position,
designated 0° was when the foot plate was at 90° to the shank.

To set the ranges of motion (ROM), the lever arm was slowly
rotated until the subjects limits were established. Gravity and
inertia corrections were performed using Lidoactive software.

Subjects carried out 3 sub maximal and 3 maximal reciprocal
movements at $60°s^{-1}$, then at $30°s^{-1}$ as a means of familiarisation.
After 90 seconds of recovery time, 6 maximal reciprocal repetitions were performed as the test procedure. This data was then
saved to disk.

A two sample t-test was used to determine differences between
torque data sets, it was assumed that the variances of the mean
torque's of the 2 samples (the P and NP ankles) were not equal.

RESULTS

For PF no significant differences (p>0.05) were found between the mean torque values for the P and NP ankles except at -15°, torque at the P ankle = 32.8 ± 19.6Nm, NP ankle = 69.2 ± 25.1Nm (p<0.05). Similarly for EV there were no significant differences (p>0.05) between contralateral ankle torque's except at -20°, torque at the P ankle = 17.2 ± 6.2Nm, NP ankle 25.7 ± 4.9Nm (p<0.05). Both of these differences occurred at the extremities of the ROM. For DF and INV there were no significant contralateral differences between torque's (p<0.05)

There were no significant torque differences between contralateral muscle groups in the range -10 to 30° for PF and DF and -15 to 15° for INV and EV, so the contralateral data sets were combined. Mean bilateral torque's are shown in tables 2 and 3.

Table 2. Bilateral mean torque at 5° intervals for plantar flexion, and dorsiflexion (Nm).

Joint angle (°)	Bilateral mean PF torque (Nm)	Bilateral mean DF torque (Nm)
-10	64.38	20.19
-5	86.11	26.40
0	88.18	29.22
5	83.28	30.76
10	67.60	28.58
15	59.93	31.13
20	59.93	27.59
25	30.11	22.97
30	25.38	19.83
Mean	62.77	26.30
SD	22.69	4.31

Table 3. Bilateral mean torque at 5° intervals for inversion and eversion (Nm).

	Bilateral mean INV torque (Nm)	Bilateral mean EV torque (Nm)
-15	12.98	21.81
-10	14.41	21.58
-5	16.09	22.67
0	16.88	21.34
5	17.03	19.17
10	15.83	14.14
15	15.67	13.13
Mean	15.51	19.16
SD	1.38	3.85

Significant differences (p<0.05) were found between the PF and DF and the INV and EV mean torque's.

The antagonistic bilateral mean torque scores were then used for calculating the torque ratios PF/DF and INV/EV. Figures 1 and 2 show the linear relationship between JA and the PF/DF and INV/EV torque ratios. For figure 1. the regression equation for the relationship is

PF/DF ratio= -0.053(JA)+2.893, (r= -0.967, p<0.001).

For figure 2. the regression equation for the relationship is

INV/EV ratio= 0.020(JA) +0.841, (r=0.957, p<0.001)

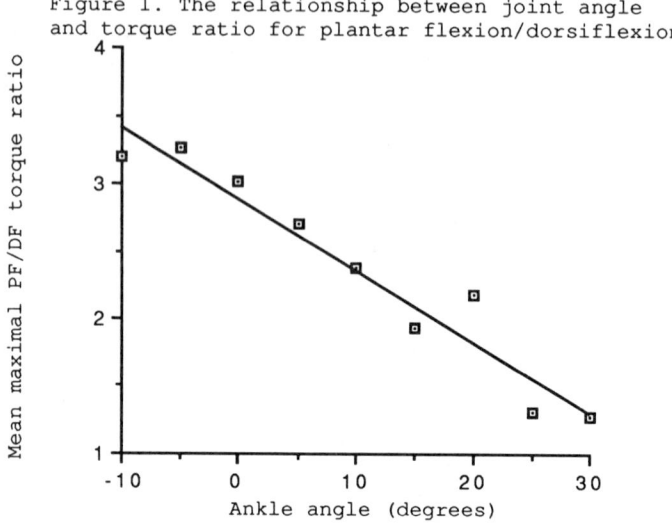

Figure 1. The relationship between joint angle
and torque ratio for plantar flexion/dorsiflexion

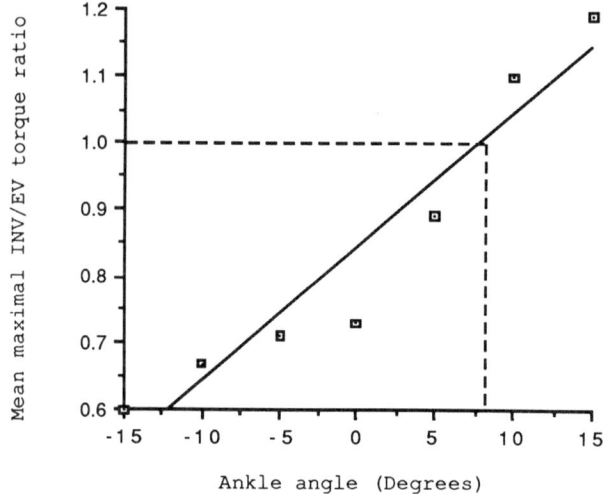

Figure 2. the relationship beteeen the angle of the ankle
and the torque ratio for inversion/eversion

DISCUSSION

The aims of this study were: to determine whether contralateral differences exist in the lower leg strength between non-
preferred (NP) and preferred (P) kicking legs; and to determine
relations between antagonistic muscle strength at specific (JA).

No significant contralateral differences were found for PF and
DF, except at $-15°$ for PF ($p<0.05$). For INV and EV no signifi-

cant contralateral differences were found except at -20° for EV (p<0.05). The significant differences were at JA at the extremes of the ROM. Insufficient subject motivation to exert maximal force at extreme ROM must be considered as an important factor when measuring voluntary maximal torque (Scranton et al.1985).

The finding of no significant contralateral differences for PF and DF supports the work of Scranton et al., (1985) who tested 20 male adults at angular velocities of 0, 30, 90 and $180°s^{-1}$. Agre and Baxter (1987) tested isometric strength of 25 collegiate soccer players, finding no significant differences in PF and DF.

The finding that no contralateral differences existed for INV and EV supports the work of Wong et al., (1984) who found no significant differences for INV and EV (except in women at $30°s^{-1}$) when testing untrained subjects at 30, 60 and $120°s^{-1}$. Findings of the present study support the hypothesis proposed by Scranton et al.(1985) that lower limb musculature is predominantly postural, used equally in maintaining an upright stance, that is, there is no contralateral ankle preference.

To date, all data discussed have considered peak torque without reference to the JA at which the torque was applied. The present study highlights the importance of stating JA, as linear relationships between JA and torque ratio were shown to exist.

Peak torque data from the present study provided peak torque ratios that agree well with the literature, for PF/DF - 3.2. Scranton et al (1985) quoted values of 3 and 2 (low and high angular velocities) for active males. Alexander (1990), found PF/DF≈3 over a range of velocities $(30-270°s^{-1})$, in sprinters.

The mean peak torque ratio was calculated for INV/EV = 0.71. Cawthorn et al (1991) suggested INV>EV quoting values of 1.16-1.17 in active adults. Wong et al (1984) also found INV>EV, 1.11 at $60°s^{-1}$ and 1.25 at $120°s^{-1}$ respectively in untrained athletes.

However, these ratios are to be of little use as no indication of the JA is given. The ratios imply that at any JA, torque production by an agonist muscle group is a fixed percentage of the antagonist torque. This supposition has been shown to be invalid in the present study. Both PF/DF and INV/EV ratios are linearly related to JA as shown in figures 1 and 2. For the PF/DF ratio there was a rapid change from a value of 3.4 at -10°, to 2.6 at 5°, to 1.6 at 25°. The regression equations allowing calculation of torque ratio from JA are: PF/DF ratio = 0.053(JA) + 2.839 Similarly, the INV/EV ratio = 0.020(JA) + 0.841. As shown in figure 2, between -15 and 8° the EV>INV, but between 8 and 15 degrees INV>EV, that is the ratio inverted at 8°.

These findings show quite clearly that marked changes occur in antagonistic strength ratios throughout the ROM. At a given JA the strength of the two muscle groups are not in the same proportion as their strength at another JA.

Observations made on torque-time curves may have implications upon the linear relations proposed due to shifts in the size and shapes of torque curves. When the velocity is increased, peak torque shifts to more distal JA (Prietto and Ciazzo, 1989, Bobbert and Van Ingen Schenau, 1990). The magnitude of peak torque decreases with increasing velocity (Oberg et al, 1986,

Patten Wyatt and Edwards, 1981, Prietto and Ciazzo, 1984, Wickie-wicz et al, 1984, and Wong et al, 1984). These observations suggest that when quoting any torque value or ratio, both the angular velocity of testing and the JA must be stated.

In summary, the results show that in the lower limb, there were no significant contralateral differences in muscle strength. Linear relationships exist between antagonistic muscle group torque ratios and JA. Further work is required to test whether this kind of relation is applicable at different velocities and with different samples. This work will be of particular use if the relationships can be applied to rehabilitation , as muscular deficit over a particular joint ROM will lead to a deviation from the established linear relation. The process may also have applications in screening athletes, to prevent possible injury.

REFERENCES

Agre, J.C., Baxter, T.L., 1987, Musculoskeletal profile of male collegiate soccer players. Arch of Phys Med Rehab 68:147-150

Alexander, M.J.L., 1990, Peak torque values for antagonist muscle groups and concentric and eccentric contraction types for elite sprinters. Arch of Phys Med and Rehab 71:334-9.

Baltzopolous, V., Brodie, D.A., 1989, Isokinetic Dynamometry - applications and limitations. Sports Medicine 8,2:101-116.

Brady, E., 1990, The incidence and factors involved in soft tissue injuries in 82 Dublin soccer players. In unpublished PhD-University of Dublin Trinity college, Dublin 2.

Bobbert, M.F., Van Ingen Schenau, G.J., 1990, Isokinetic plantar flexion: Experimental results and model calculations. Journal of Biomechanics 23,2: 105-119.

Cawthorn, M., Cummings, G., Walker, J.R., Donatelli, R., 1991, Isokinetic measurement of foot invertor and evertor forces in three positions of plantar flexion and dorsiflexion. Journal of Orthopaedic and Sports Physical Therapy. 14,2:75-81.

Gravel, D., Richards, C.L., Fillion, M., 1990, Angle dependancy in strength measurements of ankle plantar flexion. European Journal of Applied Physiology 61:182-187.

Oberg, B., Bergmann, T., Tropp, H., 1986, Testing of isokinetic muscle strength in the ankle. Med Sci Sports Ex 19,3:318-322.

Patten-Wyatt, M., Edwards, A.M., 1981, Comparison of quadriceps and hamstring torque values during isokinetic exercise. Journal of Orthopaedic and Sport Physical Therapy. 3,2:48-56.

Prietto, C.A., Ciazzo, V.J., 1989, The in vivo force-velocity relationship of the knee flexors and extensors. American Journal of Sports Medicine. 17,5:607-611.

Scranton, P.E., Whitesel, J.P., Farewell, V., 1985, Cybex evaluation of the relationship between anterior and posterior compartment lower leg muscles. Foot and Ankle 6,2:85-89.

Wickiewicz, T.L., Roland, R.J., Powell, P.L., Perrine, J.J., Edgerton, V.R., 1984, Muscle architecture and force velocity relationships in humans. J. of Applied Physiology. 61:182-7

Wong, D.L.K., Glasheen-Wray, M., Andrews, L.F., 1984, Isokinetic evaluation of the ankle invertors and evertors. Journal of Orthopedic and Sports Physical Therapy. 5,5:246-252.

ASSESSMENT OF MUSCLE STRENGTH ASYMMETRY IN SOCCER PLAYERS.

N.E. FOWLER AND T. REILLY

Centre for Sport and Exercise Sciences,
Liverpool John Moores University,
Byrom Street,
Liverpool,
L3-3AF.

Isokinetic dynamometry may be used to assess muscle imbalance. The ratio between the strengths of the hamstrings and quadriceps is of particular interest, a low ratio being associated with risk of injury. Four professional soccer players were assessed after musculoskeletal injury. All displayed a lower hamstrings to quadriceps ratio in the injured compared to the uninjured limb. Training on the dynamometer during rehabilitation was found to increase the ratio and contribute to a return to playing.

INTRODUCTION

The role of isokinetic dynamometry in the assessment of muscle function has been well documented (Cabri, 1991). Much interest has been directed towards the determination of muscle imbalances (left/right or agonist/antagonist), as these are believed to be related to injury (Cabri and Clarys, 1991). Greatest attention has been paid to the ratio of the hamstrings and quadriceps muscle groups, as these play an important role in knee joint stability (Westing and Seger, 1989).

Gravity-corrected peak torque has been shown to be the most reliable value for the assessment of muscle function (Kannus and Yasuda, 1992). This provides a measurement of the maximal torque produced by the muscles acting across a joint. Such an assessment does not allow for the identification of particular areas of weakness which may exist in a joint's range of movement. This information can be found from the production of a torque-angle curve. Modern dynamometers with on-line microcomputers allow the accommodating resistance to be constantly monitored and thus the torque-angle curve can be produced.

Studies of the effect of joint angular velocity on muscle function have shown that as the velocity increases so the peak torque decreases, in accordance with the force-velocity characteristics of muscle (Westing and Seger, 1989). In knee flexion and extension exercises, the ratio between the hamstrings and quadriceps is also affected by a change in joint angular velocity (Read and Bellamy, 1990; Cabri, 1991). The hamstrings-quadriceps ratio increases as the velocity increases. It is felt that for assessment of the ratio, slow joint angular velocities should be used. Reported values for the ratio vary between 41 and 81 % depending upon the angular velocity used and the subject's fitness. Read and Bellamy (1990) reported ratios between 73 and 77 % for elite athletes; Westing and Seger (1989) tested ice hockey players and reported a mean ratio of 61 %. These tests were performed at 1.57 and 1.05 rad s^{-1} respectively.

Isokinetic dynamometers are used not only for the assessment of muscle function but also may be used in training and rehabilitation. The accommodating resistance allows a near optimal training load to be present throughout the range of movement. Thus the stimulus is not limited by the weakest point in the range of movement. The controlled nature of the exercises and the ability to achieve a near optimal stimulus make the exercises attractive during rehabilitation. No guidelines exist for the structuring of training or rehabilitation regimens using isokinetic exercises.

In this paper some observations on the assessment of muscle strength imbalance in injured professional soccer players are reported. The use of isokinetic dynamometry during rehabilitation is also illustrated.

METHOD

Four professional soccer players from two top teams on Merseyside were assessed following musculoskeletal injury. Knee flexion and extension were assessed in the sitting position using an isokinetic dynamometer, Lido Active (Loridan, Davis, CA). The seat angle was set at 15° to the horizontal and the back rest at 15° to the vertical. Restraints were placed around the hips and across the knee of the subject.

Subjects were allowed a few minutes to familiarize themselves with the equipment before measurements were made. Two easy repetitions preceded five maximal trials which were recorded. Subjects were instructed to perform maximally for both extension and flexion of

the joint and given verbal encouragement. Readings
were taken for joint angular velocities of 1.05 rad s^{-1}
and 2.09 rad s^{-1}. Isometric flexion and extension were
assessed at seven joint angles for a 5 s maximal
contraction. To avoid fatigue, measures were taken on
alternate legs for each test condition; the uninjured
leg was always tested first.

One player used the dynamometer as part of his
rehabilitation programme. A nine week training period
was used with assessment of muscle function at the
beginning, in the middle and at the end of the
training period.

RESULTS

Peak torque data are presented in Table 1 for each
of the subjects. Imbalances between the injured (Inj)
and uninjured (Good) leg are expressed as a percentage
in Table 2.

Table 1.   Peak torque (Nm) and hamstring-quadriceps
percentage ratio (H/Q %) data for knee extension (Q)
and flexion (H) at 1.05 (Slow) and 2.09 (Fast) rad s^{-1}.

Player	Q Slow Nm	H Slow Nm	H/Q% Slow	Q Fast Nm	H Fast Nm	H/Q% Fast
1. Inj	198	102	51	164	96	58
Good	141	89	63	124	83	66
2. Inj	178	106	59	157	90	57
Good	118	81	69	123	75	61
3. Inj	270	141	52	218	123	56
Good	278	156	56	240	142	59
4. Inj	193	72	37	150	65	43
Good	212	134	63	160	113	71

Table 2.   Mean and range of percentage ratios between
injured (Inj) and uninjured (Good) legs for peak
torque and hamstring-quadriceps ratio.

	Mean	Range
Q Slow (Inj/Good %)	99.3	151-91
H Slow (Inj/Good %)	97.4	131-54
Q Fast (Inj/Good %)	111.0	132-59
H Fast (Inj/Good %)	94.9	120-58
H/Q % Slow (Good-Inj)	77.2	93-59

Comparing the injured and uninjured legs reveals no consistent patterns in the absolute strength values, the injury occurring in the stronger limb in two of the four cases. Examining the hamstring to quadriceps ratio does show a difference between the injured and uninjured limbs. The mean ratio at 1.05 rad s^{-1}, differed by 13 % between the injured (49.8 ± 9.2 %) and uninjured (62.8 ± 5.3 %) legs. A difference of 10.8 % was found at 2.09 rad s^{-1} (Inj 53.5 ± 7.0; Good 64.3 ± 5.4). Increasing the speed from 1.05 rad s^{-1} to 2.09 rad s^{-1} resulted in an increase in the ratio of 3.8 % in the injured leg and 1.5 % in the uninjured leg; this can be accounted for mainly by the drop in the quadriceps torque.

Differences were shown between the preferred and non-preferred legs. The mean percentage difference between preferred and non-preferred legs was 18.5 ± 14.9 % for quadriceps and 24.9 ± 20.1 % for the hamstrings at slow speeds.

DISCUSSION

A consistent imbalance was found between the quadriceps and hamstrings in injured soccer players. The difference in the ratio between injured and uninjured legs implicates this imbalance in injury. The absolute strength values did not differ between the injured and uninjured limbs, as two of the injuries occurred in the stronger limb.

The greater strength in the preferred compared with the non-preferred limb may be explained in functional terms, the greatest concentration of load being placed on the preferred leg during training and match play. This would lead to an imbalance between the two limbs. It appears such an imbalance did not contribute to the occurrence of injury in all the cases examined. The literature suggests that an imbalance between limbs may play a significant role in injury. Knapik et al. (1991) suggested that subjects with an imbalance of greater than 15% were 2.6 times more likely to suffer injury in the weaker leg. In the present study differences were in the order of 20 % and so may have been a factor in the two cases where the injured limb was also the weakest.

The roles of the hamstring muscles in stabilising the knee joint have been well documented (Westing and Seger, 1989; Read and Bellamy, 1990). In soccer the hamstring muscles are used concentrically to flex the knee and extend the hips in preparation for the kick and eccentrically to actively resist knee extension and hip flexion towards the end of the kicking action (Clarys et al.,1988). If the strength of the quadriceps greatly exceeds that of the hamstrings, then the ability to resist knee extension is reduced.

This may result in a forced stretch of the hamstrings and consequent muscle damage. This relationship, between hamstrings to quadriceps imbalance and injury, is supported by the present finding that the injured players all displayed a lower ratio in the injured leg than in the uninjured leg. It is more likely that the imbalance contributed to rather than resulted from the injury. The ratio in the uninjured leg compares with the average ratio in the literature of approximately 60 % (Westing and Seger 1989), the injured leg lying some way below this figure. The use of isokinetic dynamometry to routinely screen players may allow such imbalances to be detected and corrected before they lead to injury and thus reduce the amount of time for which players are out of the game.

Player 4 displayed the lowest ratio between hamstrings and quadriceps (37 %). Following a series of surgical interventions to remove a cyst from the tendon of M.Biceps Femoris, this player trained three times per week for nine weeks using the dynamometer. The training regimen initially included 5 sets of 5 repetitions at 1.05 rad s^{-1} and 2.09 rad s^{-1}. During two sessions each week the emphasis was placed upon the flexion aspect only, while the third session included both flexion and extension. The injured limb was trained during all sessions; the uninjured limb was trained in alternate sessions. Isometric flexion exercises were also performed at 15° intervals between 0 and 90°. Faster joint speeds were introduced up to 5.24 rad s^{-1}, and eccentric hamstring exercises at 2.09 rad s^{-1} as the training progressed.

Table 3. Strength test results for player 4 before (T1), during (T2) and after (T3) a nine week isokinetic training regimen.

	Q Slow Nm	H Slow Nm	H/Q % Slow	Q Fast Nm	H Fast Nm	H/Q % Fast
T1 Inj	193	72	37.3	150	65	43.3
Good	212	134	63.2	160	113	70.6
I/G %	91.0	53.7		93.8	57.5	
T2 Inj	203	92	45.3	161	98	60.9
Good	256	175	68.3	178	145	81.4
I/G %	79.3	52.6		90.4	67.6	
T3 Inj	239	145	60.6	199	134	67.3
Good	245	177	72.2	210	167	79.5
I/G %	97.6	81.9		94.7	80.2	

Muscle function was assessed after five and nine weeks of the training regimen (Table 3). The data demonstrate that the imbalance had been reduced with

the final ratio of hamstrings to quadriceps of 60.7 %
in the injured leg. Training caused an increase in all
strength parameters over the nine weeks. The balance
between preferred and non-preferred leg had been
reduced for both flexion and extension. After this
training period the player returned progressively to
training and to competing in Premier League football.

   The study shows how isokinetic dynamometry can be
used to assess the functional strength of soccer
players. Imbalance between preferred and non-preferred
legs were shown to be large, although no link was
shown between this imbalance and the occurrence of
injury in the players tested. The hamstrings to
quadriceps ratio in the injured leg of 49.8 % was 13%
lower than that in the uninjured leg; this supports
the belief that a ratio lower than 60 % may predispose
a player to injury. A training regimen was implemented
which demonstrates how isokinetic dynamometry can be
successfully used during rehabilitation from muscle
damage in soccer players.

REFERENCES
Cabri, J.M.H., 1991, Isokinetic strength aspects of
human joints and muscles, Critical Reviews in
Biomedical Engineering, 19, 231-259.
Cabri, J.M.H. and Clarys, J.P., 1991, Isokinetic
exercise in rehabilitation, Applied Ergonomics, 22,
295-298.
Clarys, J.P., Bollens, E., Cabri, J. and Dufour, W.,
1988, Muscle activity in the soccer kick. In Science
and Football, ed. T. Reilly, A. Lees, K. Davids and W.
Murphy (London: E and F.N. Spon), pp. 434-440.
Kannus, P. and Yasuda, K., 1992, Value of isokinetic
angle-specific torque measurements in normal and
injured knees, Medicine and Science in Sports and
Exercise, 24, 292-297.
Knapik, J.J., Baumann, C.L., Jones, B.H., Harris, J.M.
and Vaughan, L., 1991, Pre-season strength and
flexibility imbalances associated with athletic
injuries in female collegiate athletes, American
Journal of Sports Medicine, 19, 76-81.
Read, M.T.F. and Bellamy, M.J., 1990, Comparison of
hamstring/ quadriceps isokinetic strength ratios and
power in tennis, squash and track athletes, British
Journal of Sports Medicine, 24, 178-182.
Westing, S.H. and Seger, J.Y., 1989, Eccentric and
concentric torque-velocity characteristics, torque
output˙ comparisons, and gravity effect torque
corrections for the quadriceps and hamstring muscles
in females, International Journal of Sports Medicine,
10, 175-180.

Physiological stress

THE EFFECT ON WORKING POSTURE OF A REDESIGNED INDUSTRIAL SEWING MACHINE

Li Guangyan and Christine M. Haslegrave

Department of Manufacturing Engineering
and Operations Management
University of Nottingham
Nottingham NG7 2RD U.K.

Sewing machinists suffer from musculoskeletal problems which have been attributed to poor working postures as well as to the repetitive hand and arm movements. The potential for improvement of industrial sewing machines was investigated in a study of the effects of varying both table inclination and needle angle, which can change the operator's view of the work task. It was shown that both trunk posture and head/neck flexion could be improved significantly by altering these design parameters.

INTRODUCTION

Studies have shown that there are widespread ergonomic problems in the operation of industrial sewing machines which are used in the garment, shoe and textile industries. Although the machines have been modernised in some aspects, the basic layout of the sewing workplace has been the same for many years.

The sewing operation can be characterised by a sitting posture with the operator's head and trunk flexed forward, the performance of simultaneous but different motions with the two hands, awkward and extreme joint postures and the continuous operation of foot pedals. This main work activity is maintained throughout the whole work shift except perhaps for a few ancillary tasks which permit a certain change in the working position.

As the result of the highly repetitive movements and the poor posture of the trunk and upper extremities, there is a high prevalence of musculoskeletal complaints affecting both back and upper limb among sewing-machine operators (Vihma et al., 1982; Wick and Drury, 1985). Keyserling et al. (1982) for instance, found in a survey of 397 female workers in the garment industry that about 25% suffered persistent musculoskeletal pain in at least one part of their body. The most frequent location of the pain was the hand, followed by the neck and back.

Various solutions for the ergonomic problems have been recommended. Keyserling et al. (1982) suggested reducing the coefficient of friction between the fabric being sewn and the working surface of the machine in order to reduce the resistance of pushing the fabric through the machine. They also suggested inclining the sewing table in a manner similar to a drafting table in the hope of reducing tendency of the operator to lean forwards. Haslegrave and Gregg (1988) made some detailed recommendations for changes needed in the layout of the sewing machine and sewing table, also noting the effects of work organisation.

Two studies have been made previously to design improved sewing machines. Wick and Drury (1985) designed a prototype sewing workstation which included an upholstered swivel chair(adjustable by the user for height and for backrest location), an adjustable foot control and bench-mounted armrests. The sewing machine was tilted 11 degrees towards the operator. The changes were shown to give some improvement in working posture. Delleman and Dul (1989) investigated design parameters for the sewing table and recommended that the table height be adjusted to at least 5 cm above elbow height, and that it be inclined at least 10 degrees towards the operator. They found that the pedal position had no effect on the head inclination.

However, despite such changes to the conventional sewing workstations, Wick and Drury (1985) still found that sewing operators leaned forwards away from the lumbar support of the chair and inclined their head/neck forwards more than 30 degrees from the vertical, so that the magnitude of the postural improvement was not as large as that which might be expected from the change in the table slope (Delleman and Dul, 1989). Thus, the redesigns of sewing machines which have been reported have had some success in improving the operator's posture, but there still appear to be problems with the static muscle load on the operator's neck and back.

EXPERIMENTAL STUDY

The sewing task is highly dependent on vision and the effects of this on posture deserve further attention. The present study was undertaken to explore the possibility of improving the operator's view of the sewing needle, and to measure the effects of this on the posture adopted.

Two experimental sewing machines were adapted from high-speed industrial lockstitch sewing machines (PFAFF-463). These provided adjustment for table slope between 0 and 20 degrees from the horizontal, a sewing head which could be tilted away from the operator so that needle angle could be changed within 0-20 degrees relative to the axis of the sewing table, adjustable pedal distance positions in 5.5 cm increments, and adjustable pedal height.

Nine experimental conditions consisting of three different table slopes and three different needle angles were tested, as shown in Figure 2.

The table height when flat was fixed at 80 cm, (front pivot height for inclined tables). The pedal height was adjusted in accordance with the table slope changes so that the distance between the needle plate and the centre of the pedal remained constant during the whole experiment, and the same adjustment was made in seat height when the table slope was changed for different experimental conditions so that the distance between subject's eye position and the needle plate was constant. A free-swivel chair with height-adjustable seat was used for the experiment.

Eight female subjects participated in the experiment. Their mean age was 34.0 years (SD=7.19, Range=22-46); their mean stature was 166.1 cm (SD=7.75, Range=158.0-182.0). All subjects were professional industrial sewing-machine operators with average experience of 13.1 years (SD=8.92, Range=2-25) and all had normal eye sight. Most subjects were paid an industrial hourly rate, although one was a volunteer, and one was assigned to work as a subject by her employer who paid her normal wages for the time.

Prior to the test session, each subject selected a preferred seat height at which they felt most comfortable for working at a horizontal table. Then, the pedal position (relative to the sewing point) and the pedal height were adjusted to the subject's preference. The seat height and pedal position were kept the same for all experimental conditions, but the operators could choose the horizontal distance from the seat to the table.

Three video cameras were used to record the operator's head/neck flexion, trunk flexion, left/right upper arm extension and abduction. Posture was measured from the video records of the tests using markers which were placed on the shoulder joint, the lateral side of the elbow, the hip joint and near the lobe of the ear on both left and right side of the operator. Two cameras were set on the subject's left and right side with the lens positioned in the frontal plane and at a height slightly below the shoulder of the subject. The third camera was set at the back of the subject with the lens in the sagittal plane and at the same height as the other two.

All subjects wore safety glasses for the eye protection, but the subjects did not experience any effects on task performance from wearing the glasses.

The sewing task was to sew along the four folded edges of a dark rectangle of cotton cloth with the average size of 38x30 cm, measured between the stitching lines.

The subject performed the sewing task for 8-10 minutes under each experimental condition before the video cameras were switched on, then the subject worked for a further 5-10 minutes during which the working postures were recorded. Completing each piece of cloth as the task required was counted as one work cycle. Each subject started the experiment with condition 1 which represented traditional sewing workplace. The other experimental conditions were then presented randomly. It took four to five hours for each subject to complete all nine experimental conditions, and the experiment was divided into two sessions with approximately two hours in each session. The subject had a twenty minute break between two sessions, so that the complete test period was similar to a normal work period.

ANALYSIS OF POSTURE DATA

Videotapes were played back at the same speed as they were recorded and the operators' head/neck and trunk positions were classified by using a transparent template marked with standard posture ranges. Based on the literature (Keyserling, 1986), the trunk flexion/extension was defined (as shown in Figure 1) as the angle (A) between the line through the hip joint and the shoulder joint and a vertical line, measured in the sagittal plane. The trunk posture was classified into four categories: A<0 degrees, A=0-20 degrees, A=21-30 degrees and A>30 degrees.

The head/neck flexion (B) was defined from a line drawn through the marker on the head and the marker on the shoulder joint with the line through the shoulder joint and the hip joint (Figure 1). The postures for the head/neck were classified as Neutral (B=0-20 degrees); Mild Flexion (B=21-45 degrees); Severe Flexion (B>45 degrees), consistent with previous studies (for example, Leonard and Keyserling, 1989).

A stop watch was used to record the amount of time the subject spent in each of posture ranges during the work cycles, and the percentages of work-cycle time spent in these postures were calculated. Three work cycles were analysed for each experimental condition, and the average posture values during the three work cycles characterise the subject's posture in each experimental condition.

Figure 1. Measurement of posture

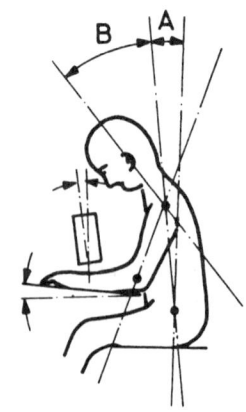

The operator's maximal head and trunk flexion were also recorded for each work cycle, and indicate the extreme positions adopted while the subject was performing 'natural' and 'real' sewing tasks.

RESULTS

Posture measurements
    The tests showed that both table angle and needle angle had a significant effect on the posture of the operator.
    The results showed that the operators tended to adopt a more upright trunk posture as the table slope was increased from 0 to 10 or 15 degrees. However, as the needle was tilted under any table condition, the operators tended to spend more time with the trunk bent forward, which was caused by visual condition changes and is discussed later.
    Figure 2 shows the average effect of different experimental conditions on the subjects' head/neck inclinations, measured as percentage of work-cycle time the head was in each of the posture ranges. It can be seen that, as the table slope was changed from 0 degrees to 10 degrees or 15 degrees while the needle was kept vertical to the table (needle at 0 degrees), the operators' head tended to have more severe flexion, but that if the table angle was kept same and needle angle was increased from 0 to 10 or 20 degrees, the time spent in neutral posture increased and time in severe flexion decreased. This reflected the positive effect of the visual improvement on the operator's head/neck posture.

Figure 2. Effect of table slope and needle angle on head/neck flexion

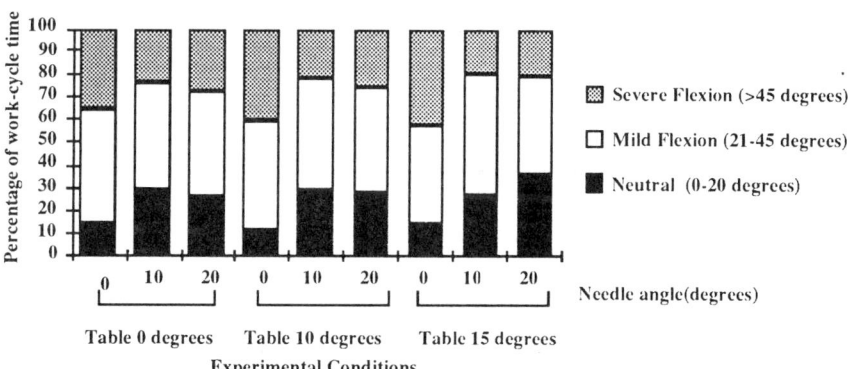

    The needle angle change also had a positive influence on the maximal head/neck inclination, but table slope had no obvious effect on the maximal head/neck flexion.
    On the other hand, the maximal trunk flexion decreased by 8-9 degrees as the table slope changed from 0 to 10 or 15 degrees, but the maximal trunk flexion angle did not change much for different needle conditions when the table slope was constant.

Figure 3. Relationship between maximal head/neck flexion and needle angle

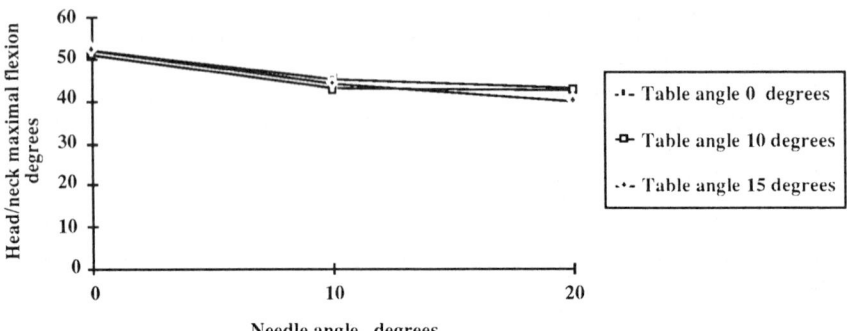

Figure 4. Relationship between maximal trunk flexion and table angle

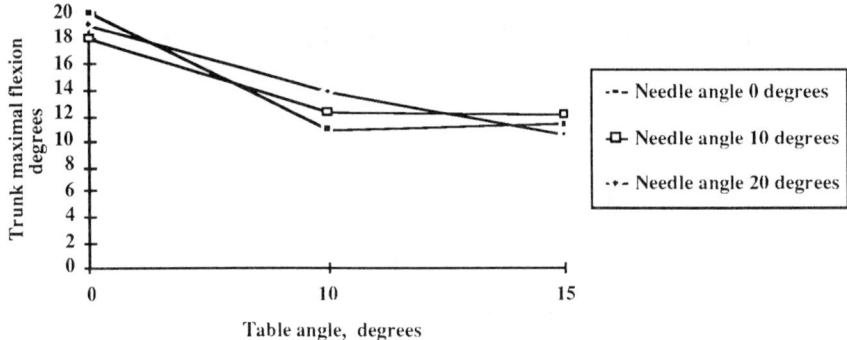

The significance of effects

The effects of table slope and needle angle on the subjects' head/neck and trunk postures were tested by TWO-WAY ANOVA (Related) and the significant relationships for maximal flexion angles are shown in Figures 3 and 4. The results showed that the effect of needle angle on sewing-machine operators' head/neck postures were significant for time spent in neutral flexion and severe flexion at the $p<0.05$ level, and on the maximal head flexion at the $p<0.001$ level, but that the needle angle had no significant effect on the head/neck mild flexion. The table slope had no significant effect on any measures of head/neck posture.

The table slope was shown to have a significant effect on the trunk flexion ranges at $p<0.05$, and on trunk maximal flexion at the $p<0.01$ level. The needle angle changes did not have a significant effect on the trunk posture.

Subjective opinions

During the trials, the subjects were asked for their opinions of the sewing table layouts, rating table slope and view of task on 5-point bipolar scales (too steep-too flat, very good-very poor). Their opinions substantially ally confirmed the posture results.

DISCUSSION

The sewing table slopes and needle angles were both found to affect the sewing-machine operator's working posture. A greater inclination of needle angle reduced head/neck flexion and the maximal head inclination reduced by about 12 degrees as the needle was tilted from 0 to 20 degrees. This result was relatively independent of table slope, so that it may be possible to choose a more suitable needle angle for a better head/neck posture. In this study, the 20 degree needle appeared to be the best choice, but further research is needed on this.

The study showed that as the table slope was adjusted from 0 to 10 degrees, there was a considerable reduction in trunk maximal flexion (for all three needle conditions), but as the table slope was further increased from 10 to 15 degrees, the trunk maximal flexion was not reduced as much as it had been expected, perhaps because the tilted machine head obstructed the operator's view of the work site, and the operator had to bend the trunk forward from time to time in order to relax the severe head inclination. On the basis of these trials, a sewing table inclined 10 degrees towards the operator is recommended, which agrees with the former studies of Delleman et al. (1989) and Wick and Drury (1985).

The effects of adjustments to table and needle were combined in this study, so that these factors can not be viewed separately. However, they do appear to have different effects. Further work is needed to study the sewing operator's visual requirements and their relationship with working posture.

CONCLUSIONS

An industrial sewing machine with adjustable table slope and adjustable needle inclination has potential for improving the machine operator's head/neck and trunk posture while keeping the same sewing quality and speed. A sewing machine with the table at 10 degree inclination and the needle at 10-20 degree inclination is suggested as a preliminary recommendation on the basis of these short trials.

REFERENCES

Delleman, N.J. and Dul, J., 1989, Ergonomic guidelines for adjustment and redesign of sewing machine workplaces. Work Design in Practice, ed. C.M. Haslegrave et al. (London: Taylor & Francis). 155-160.

Haslegrave, C.M. and Gregg, H., 1988, Improving the ergonomics of production sewing tasks. Advances in Manufacturing Technology III, ed. B. Worthington (London: Kogan Page Ltd). 284-288.

Keyserling, W.M. et al., 1982, Repetitive trauma disorders in the garment industry. National Institute for Occupational Safety and Health, NIOSH-81-3220.

Keyserling, W.M., 1986, Postural analysis of the trunk and shoulders in simulated real time. Ergonomics, 29 (4), 569-583.

Leonard, J. and Keyserling, W.M., 1989, A method to evaluate neck and lower extremity postures using simulated real time analysis. Advances in Industrial Ergonomics and Safety 1, ed. A. Mital (London: Taylor & Francis), 593-599.

Vihma, T. et al., 1982, Sewing-machine operators' work and musculo-skeletal complaints. Ergonomics, 25 (4), 295-298.

Wick, J. and Drury, C.G., 1985, Postural change due to adaptations of a sewing workstation. Ergonomics of Working Postures, ed. N. Corlett and J. Wilson (London: Taylor & Francis), 375-379.

PHYSIOLOGICAL MEASUREMENT OF STRESS IN COMPUTER BASED WORK: THE PSI

Margaret Toms and Alfred Masih

University of Wolverhampton
School of Health Sciences
62-68 Lichfield Street
Wolverhampton
WV1 1DJ

Whilst stress associated with computer work is a major concern within HCI, non-invasive, objective methods of measuring stress in a computer based environment are not readily forthcoming. One potentially useful measure, the Palmar Sweat Index (PSI), is considered in this paper. Palmar sweat prints were taken from 16 subjects at several stages during a computerised database administration task. Subjects showed elevated palmar sweating during tasks involving a high memory load as compared with tasks imposing lower loads. The PSI measure was not, however, sensitive to differences between an unstructured command language interface and a structured interface, although clear performance differences were evident between the two interfaces. A post-task verbal report measure of state anxiety was also insensitive to differences between the two types of interface, but was positively correlated with PSI score. It is concluded that the PSI may be a useful tool within HCI research.

INTRODUCTION

The use of computers in the workplace is frequently cited as a source of increased occupational stress. Such stress may arise, for example, from physical strain or discomfort due to poor workstation design and layout, organisational factors such as constraints upon communication and social interaction or lack or training and user support, changes in the nature of the work, perhaps resulting in additional cognitive load, or poor design of software giving rise to an interface which is difficult to use (Briner and Hockey, 1988). This paper is primarily concerned with measurement of stress associated with the last of these.

Ergonomic design of software has been seen as a factor of increased importance in recent years. Indeed recent EC directives explicitly identify software usability as an important consideration in the health and safety of those who use computers in their daily work. Yet evaluation of software usability tends to rely heavily upon performance measures, such as time taken to perform specific tasks and/or the number and types of errors made by typical users, rather than the strain incurred in achieving a given level of performance using a particular system. Measures of affective response to using a system tend to be based almost exclusively around self report measures of user satisfaction. While such self report measures may be a valuable source of information to the software evaluator, they may be insensitive to specific sources of strain within a given system since users may not be consciously aware of the degree of strain imposed.

An alternative is to employ physiological measures in order to study computing stress. For instance, electrodermal activity, heart rate and blood pressure have been used as indices of stress induced by differing system response times (Kuhman, 1989; Schleifer and Okogbas, 1990). However, such physiological measures may themselves involve invasive, stress inducing procedures or may require very expensive equipment.

Consequently, there would seem to be a need for a noninvasive physiological indicator of stress which could be used with a minimum of equipment, yet which would be sensitive to psychological strain associated with computer based work. Recent evidence has suggested that the palmar sweat index (PSI) might be a suitable candidate measure, which could be of particular use in field studies.

The PSI consists of a plastic impression of a fingertip on which the number of active sweat glands can be counted. The impression is taken by applying to the finger a plasticising solution containing graphite and a substance (Formvar) that withdraws from moisture. The resultant print can be removed by means of sellotape. Active sweat glands appear as white holes upon the print and can be counted directly under microscopic magnification. Originally developed by Sutarman and Thomson (1952), the PSI has received relatively little research interest, but recent studies have suggested that, despite its simplicity, the PSI is an objective and reliable measure of physiological changes (Köhler, Weber and Vögele, 1990), and may be a valid alternative to measurement of electrodermal activity (Clements, 1990).

Hockey et al (1989) suggest that one of the main sources of stress associated with computer based work may be the working memory load imposed, particularly in use of new, unfamiliar systems. Using a combination of typical performance measures, the PSI and a self-report measure of state anxiety, this notion is explored here using novice users in a laboratory setting.

METHOD
Subjects
Eight male and eight female undergraduates from the University of Wolverhampton participated as volunteers. As part of their degree course, each had completed successfully at least one Information Technology module which involved use of IBM-compatible personal computers. Thus all subjects had had prior experience of at least one command language based system, although none were familiar with the database administration task used in the experiment.

Database Administration Task
A simulation of a video database administration system was implemented on an Apricot XEN-S 386 IBM-compatible PC. This system represented a database of information pertaining to a home video library. Users of this database could query the system for existing information, and also make changes to information held in the system using a command language interface. Each command required the input of an action, the entity upon which such an action could be performed, and one or more additional qualifying parameters, e.g. 'delete video 12'.

Two versions of a command language interface were used. In the **structured** version, subjects entered the action command and the entity name, and were prompted for subsequent parameters. Although subjects had to provide the appropriate parameter information, they did not have to memorise the order in which such information was required. In the **unstructured** version, subjects entered the action command and entity name, but also had to include all the parameters on the same line before pressing the enter key. Thus, although identical information was required by both versions of the command language, the unstructured version required recall of the sequence in which parameters should be input, as well as the identity of such parameters.

An online hierarchical help system was available to users of the system. At the top level of the system, details of the action commands available were listed (add, delete, modify, list, summarise, find). At the next level, the entities upon which a given action could be performed were detailed (e.g. the delete command could be used with a video or a given item upon a video). At the lowest level, a syntax screen for a given action-entity combination could be examined. The help system could be accessed at any time from a command entry screen by pressing the F1 key. This would clear the command entry screen and display the top level help screen. Once the relevant information was found, subjects could leave the help system and return to command entry by pressing the Escape key. This would return them to the command entry screen which would maintain the same status as that prior to entry into the help system. This enabled subjects to enter part of a command, then make use of the help facility if they were unsure of the remaining command syntax.

Successful completion of a given action upon the database required that the subject input all components of the command sequence correctly. If any errors were made, no part of the command would be executed, and an error message was displayed at the bottom of the screen. If multiple errors were made, the error message referred only to the first error found. After any error, the system required new input of the entire command sequence. After successful input of a command, the command was executed, and a feedback screen displayed indicating the results of that action.

Four female and four male subjects were randomly assigned to work on structured and unstructured versions of the video database system. Each subject was required to carry out a series of administrative tasks using the database simulation. These tasks were blocked according to the memory load imposed by the command necessary to complete the task. Low memory load tasks entailed entry of the required action, entity, and one additional parameter: for instance 'delete video 12'. High memory load tasks entailed entry of the action, entity, and four or five additional parameters: for instance, 'add video 13 vhs blank 180'. Tasks were blocked in sequences of three, such that each subject performed two blocks of low memory load tasks and two blocks of high memory load tasks. Order of block presentation was counterbalanced using an ABBA type design, with half of the subjects using each interface type receiving first a low memory load block, then two high load blocks, then a final low load block, and the remaining half receiving a high memory load block, then two low load blocks, and a final high load block.

Procedure

Subjects were tested individually. Since the main experimental manipulations involved imposition of loads upon working memory, firstly a digit span measure was taken for each subject. Upon completion of the digit span procedure, subjects were instructed to wash their hands. This was necessary to remove any residual sweat, and standardise conditions, prior to PSI measurement.

Palmar sweat prints were taken using the following procedure. The middle finger of the subject's non-preferred hand was blotted with a tissue. Solution was applied to the finger, allowed to dry for 30 seconds, then removed using a piece of sellotape. The resultant print was mounted onto a slide.

Three prints were taken to familiarise the subject with the procedure. These prints were not subsequently analysed, and were taken only to habituate subjects to the measurement procedure itself (Köhler, Zander and Duncker, 1991). A fourth print was then taken. This was used to provide a baseline measure of active sweat glands.

Subjects were given instructions on how to use the database simulation, and completed two practice tasks. These tasks involved the use of commands which were not required in performance of the experimental tasks. Subjects then worked through four blocks of experimental tasks. After completing the second task in each block, a PSI print was taken. An additional print was taken after completion of the last task block.

Finally the Spielberger, Gorsuch and Lushene (1970) State-Trait Anxiety Inventory was administered. Each subject completed both state and trait forms of the inventory, completing the state anxiety form first.

Scoring of the PSI prints
Active sweat glands in each print were counted under 30-fold magnification, using a transparent acetate with a 4mm x 4mm grid square template placed over the print. The grid was centred on the central whorl of the finger print, ensuring that the same area was evaluated across all subjects. Following Köhler, Weber and Vögele (1990), glands situated in the grooves of the print were not counted, since this is an atypical location for eccrine sweat glands.

In addition to the baseline and post-task PSI measures, a single PSI count for each of the low and high memory load task conditions was obtained by calculating the mean of the active gland counts across the two task blocks.

RESULTS
Subjects assigned to the structured and unstructured interface groups did not differ on measure of digit span, $t(14) = 1.2$, *n.s.* (means of 6.9 and 8.4 digits respectively) or trait anxiety, $t(14) = 0.89$, *n.s.* (means of 37.3 and 41.1 respectively). Thus the two groups were comparable on both cognitive and affective measures.

PSI and Self-report data
The PSI scores from prints taken during task performance were subjected to a two factor mixed design analysis of covariance, with interface type and task load as factors, and baseline PSI as a covariate. Adjusted means are shown in Table 1.

There was no effect of interface type, $F<1$, nor interaction between interface type and task load $F<1$. However, there was an overall effect of task load, $F(1,12) = 5.25$, $p<0.05$. High memory load tasks produced elevated palmar sweating, regardless of the type of interface used.

The self report data also indicated no difference in post-task state anxiety between the structured and unstructured interface groups, $t(14) = 0,19$, *n.s.* (respective means of 37.24 and 38.25). State anxiety was positively correlated with post-task PSI score, $r(14) = 0.36$, $p<0.05$, indicating that the PSI and self report measures did have a moderate degree of overlap.

Table 1. Adjusted mean number of active sweat glands per $16mm^2$ for each condition

| | Task load | |
	Low	High
Structured interface	32.8	35.4
Unstructured interface	42.3	47.7

Performance data
Performance data were collected on three measures: mean time to task completion, mean time in the help system per task and total number of errors per task. Each of these measures was subjected to separate two way ANOVAs. (Note: the data from one of the subjects in the structured interface group were omitted from these analyses, since the data file accidentally became corrupted prior to analysis).

Time to task completion showed main effects of interface type, $F(1,13) = 8.34$, $p<0.025$, and task load, $F(1,13) = 65.8$, $p<0.001$, with both of these factors interacting, $F(1,13) = 8.75$, $p<0.025$. This interaction is shown in Figure 1a. Simple effects analysis of the interaction indicated that tasks requiring a high number of command parameters took longer in both structured and unstructured interface conditions, but a difference between interface types only appeared when a high number of parameters was required.

The two interface groups did not differ overall in terms of time spent accessing help, $F(1,13) = 3.47$, *n.s.*, although again high load tasks gave rise to more time spent in the help system, $F(1,13) = 43.68$, $p<0.001$, and there was an interaction between the interface type and task load factors, $F(1,13) = 4.8$, $p< 0.05$. This interaction is shown in Figure 1b. Simple effects analysis indicated that subjects using the unstructured interface required more time in help for tasks requiring a high number of command parameters, although the two interface types were not differentiated on low load tasks.

In terms of errors, however, there were no effects of interface type, $F(1,13) = 1.29$, *n.s.*, task load, $F(1,13) = 1.2$, *n.s.*, nor interaction between the two factors, $F<1$.

(a)   Time to task completion

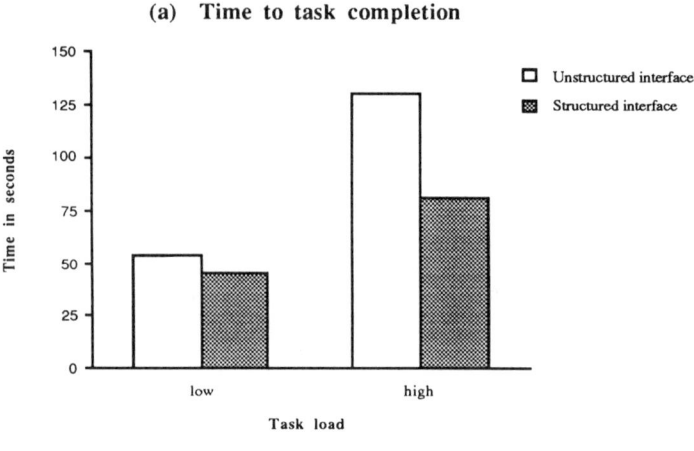

(b)   Time spent in help system

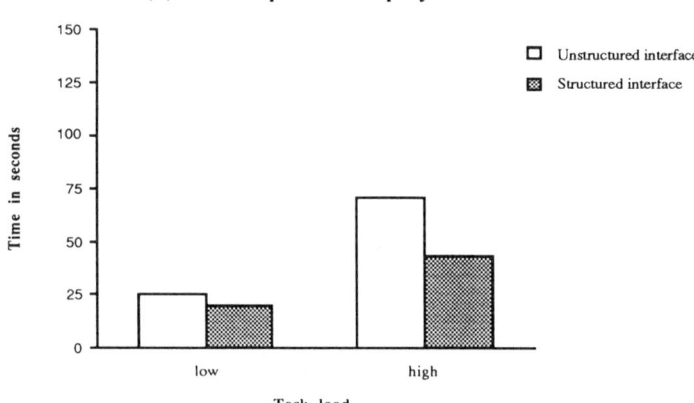

Figure 1. Effects of interface type and task load on (a) overall task time and (b) time spent accessing the help system.

DISCUSSION

The performance data clearly showed that users were affected by an increased working memory load. Subjects did not make more errors when using an unstructured command interface, or in performance of higher load tasks, but they took more time to complete tasks, and spent more time determining the correct command sequence. Although, intuitively, the unstructured interface should be more difficult for such novice users, since less guidance was provided by the system, performance difficulties were only evident on high load tasks, where users required substantially more time to understand and complete the task. This result cannot be attributed simply to the additional number of keystrokes required for high load tasks, since this remained constant between the unstructured and structured interfaces. Rather, the increase in time required seemed to be dependent specifically on the additional memory load imposed.

The stress measures showed a somewhat different pattern. Although increased task load did appear to be genuinely stressful, leading to elevated palmar sweating, the magnitude of this effect was not increased when subjects used the more difficult, unstructured interface. There was no overall difference between the two interface types either on the physiological PSI measure or on self-reported state anxiety. Even with the small sample size used here, the PSI was correlated positively, however, with the state anxiety measure, indicating that it does have construct validity as an affective measure.

Given the size of the sample used in this study, it is probably best regarded as only a pilot study. Further investigation of the use of the PSI within this context is ongoing. Nonetheless the findings reported here suggest that the PSI may be a useful tool for the ergonomist concerned with evaluating software usability. It is relatively simple to use, and does seem to provide an indicator which is sensitive to psychological strain imposed by computer based work. Given the discrepancy here between the patterning of results found using performance measures, and those found using the PSI and state anxiety measures, it would certainly seem unwise to rely on performance measures alone when making inferences about potential strain incurred.

REFERENCES

Briner, R. and Hockey, G.R.J., 1988, Operator stress and computer-based work. In C.L Cooper and R.Payne (eds.), *Causes, coping and consequences of stress at work*, pp115-140. Chichester: Wiley.

Clements, K., 1990, The validity of the Palmar Sweat Index as an alternative to electrodermal activity. Paper presented to the annual meeting of the British Psychophysiology Society.

Hockey, G.R.J., Briner, R., Tattersall, A.J. and Weithoff,M., 1989, Assessing the impact of computer workload on operator stress: the role of system controllability. *Ergonomics, 32*, 1401-1418.

Köhler, T., Weber, D. and Vögele, C., 1990, The behaviour of the PSI (palmar sweat index) during two stressful laboratory situations. *Journal of Psychophysiology, 4*, 281-287.

Köhler, T., Zander, O. and Dunker, J., 1991, Interpreting palmar sweat prints. *Journal of Psychosomatic Research, 35*, 75-81.

Kuhman, W., 1989. Experimental investigation of stress inducing properties of system response times. *Ergonomics, 32*, 271-280.

Schleifer, L.M. and Okogbas, O.G., 1990, System response time and method of pay: cardiovascular stress effects in computer-based tasks. *Ergonomics, 33*, 1495-1509.

Spielberger, C.D., Gorsuch, R.L. and Lushene, R.E., 1970, *The State-Trait Anxiety Inventory (STAB) Test Manual*. Palo Alto: Consulting Psychologists Press.

Sutarman and Thomson, M.L., 1952, A new technique for enumerating active sweat glands in man. *Journal of Physiology, 117*, 51-52.

A POSTURAL STRESS ANALYSIS SYSTEM FOR EVALUATING BODY MOVEMENTS AND POSITIONS IN INDUSTRY

ASH GENAIDY, LIN GUO and ROY ECKART

Department of Mechanical Industrial
and Nuclear Engineering
University of Cincinnati
Cincinnati, Ohio 45221-0116, U.S.A.

DON TISCHBEIN
FERMCO
P.O. Box 398704
Cincinnati, Ohio 45239-8704, U.S.A.

The main objective of this study was to
describe a system that was developed to
analyze the postural stresses around
the wrist, elbow, shoulder, neck, lower
back and lower extremities in the
industrial environment. The system
relies on visual perception to classify
working postures recorded on a videotape.

INTRODUCTION

Working postures are a major risk factor in industry.
Several studies reported the association of working
postures with musculoskeletal disorders in the workplace.
The results of Aaras et al. (1988) indicate that an average
range of motion of 10.1% across the back and shoulder
motions is associated with an average prevalence rate of
0.165. The findings of Ryan (1989) reveal that there is a
threshold of about 45-50% of the time spent standing
creating lower extremities, ankle and feet problems, and
about 25% of the time spent standing creating lower back
symptoms. Punnett et al. (1991) found the following odds
ratios between back disorders and trunk postures: (1) 4.9
for mild flexion; (2) 5.7 for severe flexion; and (3) 5.9
for rotation or lateral bending.

In order to address the above mentioned problems, two
methods have evolved to analyze and to determine the extent
of postural stresses in industry. These are: observational
and instrumentation-based. In observational methods, the
work analyst is involved in the following steps: (1)
observe the operator directly on the job or review a
videotape of the operation; (2) classify the grade of each
working posture; (3) determine the percent of time spent in
that posture. In instrumentation-based methods, the work
analyst attaches a simple device or markers/reflectors
(used in conjunction with an expensive camera system) to
the operator in order to continuously monitor the postural
deviations from neutral positions. Then, the postural
deviations are charted over the working time, in order to
study the relationship between postural deviations and
working time.

Observational methods are simple to use and more

widespread in industry. However, the accuracy of estimating postural stresses depends on the judgment of the work analyst. Thus, the work analyst should be well trained in these methods to be able to correctly identify working postures.

Instrumentation-based methods are more accurate and less widespread in industry. Their use may be limited in industry because of factors such as (1) discomfort felt by the worker due to attaching a device to his/her body, (2) high cost required to invest in a system designed to monitor several joints of the musculoskeletal system, and (3) no need to precisely measure postural deviations for practical purposes.

This study is concerned with the analysis of postural stresses using observational methods. A comprehensive review of the ergonomic literature revealed the following: (1) most observational methods were developed to monitor, at most, three joints for specific industries; (2) some methods did not define enough grades for postural deviations from the neutral position; (3) there is no knowledge of the exact errors committed by the work analyst in analyzing working postures so as to develop structured training programs to address these problems; (4) equal weights were given to the various grades of a given working posture. Thus, research is warranted to study these issues in order to improve the analysis of working postures using observational methods.

The main objective of this study is to describe a system that was developed to analyze the postural stresses around the wrist, elbow, shoulder, neck, lower back and lower extremities in the industrial environment. The system relies on visual perception to classify working postures recorded on a videotape.

METHODS AND PROCEDURES
Postural Classification

Many observational systems were developed to serve specific types of industries. As a result, only the body movements around specific joints were considered in these systems. Similarly, other systems were based on the concept of body positions and work activities. These systems were intended for postural analysis relevant to specific occupations.

It is believed that the practitioner of safety and health should have a postural classification and recording system which would enable him/her to analyze any body movements and positions around all major joints of the body. Thus an attempt has been made in this study to develop a framework for an integrated postural classification and recording system on the basis of review of the published literature. The system describes the common postures adopted around the wrist, elbow, shoulder, neck, low back and lower extremities.

Ranking System

In this study, the effects of the proposed postural classification system on the perceived joint discomfort under similar conditions were examined. Based on the results obtained, a ranking system was developed to give the various body movements the appropriate weight when performing postural stress analysis.

Sixteen male subjects participated in this study. The physical characteristics of the subjects are: (1) age: 29.8 ± 6.5 years; (2) body weight: 70.7 ± 12.0 kg; and (3) stature: 176.9 ± 7.8 cm.

A total of 30 treatments was conducted. All treatments were randomized for each subject. Ten treatments were presented on a daily basis for each subject, thus, all treatments were conducted on a period of three days for each subject.

The subject was asked to hold each postural classification for 60 seconds and a pause of 60 seconds followed each experimental treatment. The subject rated the joint discomfort ratings following the 60-second period for each experimental session using a 0-10 discomfort rating scale where '0' denotes 'no discomfort', '5' 'moderate discomfort', and '10' 'extreme discomfort'.

The subject maintained the static posture in the mid range of the angular deviation of each postural classification. The average range of motion of the U.S. population was considered as the upper limit of the angular deviations for the severe postures (e.g., severe back flexion) and movements comprised of only one non-neutral posture (e.g., radial deviation) (Webb Associates, 1978). All experiments were conducted in a well-ventilated laboratory.

The ranking system was developed on the basis of the following procedure:
(1)   The joint discomfort rating corresponding to each non-neutral postural classification was computed for 95% of the population.
(2)   The lowest joint discomfort ratings value for the non-neutral postures was considered as the baseline and was assigned a value of 1.00.
(3)   The joint discomfort ratings corresponding to the other non-neutral postures were divided by the lowest joint discomfort ratings value and was considered as the weighing factor for the non-neutral posture in question.

All statistical analyses were performed using the Statistical Analysis System software (SAS, 1985a, 1985b).

RESULTS AND DISCUSSION
Postural Classification System
    Table 1 shows the details of the proposed postural classification system. Body movements describe the motion of body parts around the wrist, elbow, shoulder, neck and lower back. Body positions are defined according to the posture of the lower extremities around the hip, knee, and ankle. Additional body positions could be added to the basic list shown in Table 1 in order to fully represent the various body positions encountered in industry.

    The neutral postures are defined within an angular range of 15° for each body movement except for ulnar and radial deviations and elbow flexion/extension. Grandjean (1988) reported that if the postural angle for neck flexion or extension is greater than 15°, the neck muscles will be fatigued rather quickly. Also, Foreman and Troup (1987) defined leaning postures as forward inclination of the trunk of at least 15°. The neutral position for the radial and ulnar deviations is limited to 0° because of their small range of motion. The neutral position is defined for elbow

Table 1. Proposed postural classification
for major body parts

Joint	Postural Classification
Wrist	Neutral $(-15°{\leq}\alpha{\leq}15°; \beta=0°)$ Moderate Extension $(-45°{\leq}\alpha<-15°)$ Severe Extension $(\alpha<-45°)$ Moderate Flexion $(15<\alpha{\leq}45°)$ Severe Flexion $(\alpha>45°)$ Radial Deviation $(\beta>0°)$ Ulnar Deviation $(\beta>0°)$
Elbow	Extension $(0°{\leq}\alpha<60°)$ Neutral $(60°{\leq}\alpha{\leq}120°; \tau{\leq}15°)$ Flexion $(\alpha>120°)$ Pronation $(\tau>15°)$ Supination $(\tau<-15°)$
Shoulder	Neutral $(-15°{\leq}\alpha{\leq}15°; \beta{\leq}15°; \tau{\leq}15°)$ Mild Elevation $(15<\alpha{\leq}45°)$ Moderate Elevation $(45<\alpha{\leq}90°)$ Severe Elevation $(\alpha>90°)$ Extension $(\beta<-15°)$ Adduction $(\tau>15°)$
Neck	Neutral $(-15°{\leq}\alpha{\leq}15°; \beta{\leq}15°; \tau{\leq}15°)$ Moderate flexion $(15<\alpha{\leq}45°)$ Severe Flexion $(\alpha>45°)$ Extension $(\alpha<-15°)$ Lateral Bending $(\beta>15°)$ Rotation $(\tau>15°)$
Low Back	Neutral $(-15°{\leq}\alpha{\leq}15°; \beta{\leq}15°; \tau{\leq}15°)$ Moderate Flexion $(15<\alpha{\leq}45°)$ Severe Flexion $(\alpha>45°)$ Extension $(\alpha<-15°)$ Lateral Bending $(\beta>15°)$ Rotation $(\tau>15°)$
Hip, Knee, Ankle	Standing Sitting Kneeling (either one or two knees contacting the floor) Moderate Squat $(90°{\leq}\alpha<180°)$ Severe Squat $(\alpha<90°)$

flexion/extension within a 60° angular range since only
extreme flexion and extension are considered as potential
risk factors (Armstrong, 1986).

Ranking of Working Postures

The ranking system of multipliers for the non-neutral
postures is given in Table 2. These multipliers will be
used to weigh the different body movements for postural
stress analysis.

It should be pointed out that, based on Table 2, there
is no difference between the moderate and severe flexion
for the neck and low back in terms of joint discomfort
ratings. Also, similar findings are obtained for the
moderate and severe shoulder elevation. Thus, this matter
should be further researched in order to determine if the
proposed postural classifications need to be altered by
combining the moderate and severe flexion of the neck and
lower back, and the moderate and severe shoulder elevation.
It is also recommended to duplicate the experiments
conducted in this study on a large scale industrial
workforce including male and female employees, as well as
workers of various age groups.

CONCLUSIONS

A postural classification was proposed to analyze
movements around the wrist, elbow, shoulder, neck, lower
back, and lower extremities. Furthermore, this study
showed that the various postural classifications produced
different effects on the joints in terms of perceived
discomfort. Thus, a ranking system of multipliers for the
non-neutral postures was developed in order to give the
appropriate weight for each postural classification during
the analysis of working postures.

REFERENCES

Aaras, A., Westgaard, R.H., and Stranden, E., 1988,
    Postural angles as an indicator of postural load and
    muscular injury in occupational work situations.
    Ergonomics, 31 (6), 915-933.
Armstrong, T.J., 1986, Upper-extremity posture: definition,
    measurement and control. In The Ergonomics of Working
    Postures. Models, Methods and Cases, ed. N. Corlett,
    J. Wilson and I. Manenica (London: Taylor and Francis),
    pp. 59-73.
Foreman, T.K. and Troup, J.D.G., 1987, Diurnal variations
    in spinal loading and the effects on stature: A
    preliminary study of nursing activities, Clinical
    Biomechanics, 2, 48-54.
Grandjean, E., 1988, Fitting the Task to the Man. A
    Textbook of Occupational Ergonomics (London: Taylor and
    Francis).
Punnett, L., Fine, L.J., Keyserling, W.M., Herrin, G.D.,
    and Chaffin, D.B., 1991, Back disorders and nonneutral
    trunk postures of automobile assembly workers, Scand. J.
    Work Environ. Health, 17, 337-346.
Ryan, G.A., 1989, Musculo-skeletal symptoms in supermarket
    workers, Ergonomics, 32 (4), 359-371.
Statistical Analysis System, 1985a, SAS User's Guide:
    Statistics, Version 5 Edition (Cary, North Carolina: SAS
    Institute Inc.).

Table 2. Multipliers for the ranking
          system of non-neutral postures

Joint	Body movement	Multiplier
Wrist	Moderate extension	1.53
	Severe extension	2.13
	Moderate flexion	1.28
	Severe flexion	1.85
	Radial deviation	1.28
	Ulnar deviation	1.17
Elbow	Extension	1.06
	Flexion	1.13
	Pronation	1.00
	Supination	1.55
Shoulder	Light elevation	1.45
	Moderate elevation	2.00
	Severe elevation	1.96
	Extension	1.91
	Adduction	1.77
Neck	Moderate flexion	1.62
	Severe flexion	1.64
	Extension	1.47
	Lateral bending	1.89
	Rotation	1.55
Low Back	Moderate flexion	1.98
	Severe flexion	1.98
	Extension	1.74
	Lateral bending	2.32
	Rotation	1.51

Statistical Analysis System, 1985b, SAS User's Guide:
    Basics, Version 5 Edition (Cary, North Carolina: SAS
    Institute Inc.).
Webb Associates, Anthropometric Source Book, Vol. I, NASA
    Ref. 1024, National Aeronautical Space Administration,
    Chapter VI.

MANUALLY DRIVEN FLYWHEEL MOTOR OPERATES
WOOD TURNING PROCESS

J.P.MODAK
PROFESSOR
& DEAN (ACADEMIC)

DEPT OF MECHANICAL
ENGINEERING V.R.C.E.
NAGPUR. (INDIA)

A.R. BAPAT
ASTT. PROFESSOR

DEPT. OF INDUSTRIAL ENGG.
R.K.N.E.C.
NAGPUR. (INDIA)

The present research reports on comparision
of operational characteristics of two systems
to energise wood turning process. These
systems are (1) Manually Driven Flywheel
Motor and (2) on load feeding. The findings
indicate that although flywheel motor is less
efficient it is also 50% less taxing human
energy wise. Provision of analogue
measurement of parameters is the improvement
over the previous research in this area.

INTRODUCTION

Manually driven brick making machine was developed
by Modak (1982) in the absence of any design data
mostly on the basis of intution. This machine consisted
of three units Viz (1) manually driven flywheel with a
speed rising transmission conceptualised in this and
previously published papers as flywheel motor (2)
mechanical transmission consisting of a spiral jaw
clutch and torque amplification unit between flywheel
shaft and process unit input shaft and (3) the process
unit consisting of an auger, cone and die. Inspite of
intution based design of this system it proved to be
economically viable. Hence it was thought to develope
completely and scientifically the entire system.
Modak and Bapat (1987, in press) established
generalised experimental model for the manually driven
flywheel motor. Askhedkar and Modak (1990) establisted
generalised experimental model for the process unit of
this system.

Modak and Bapat (in press) recommend that production processes having horse power requirements far in exceses of average h.p. of a human being but which could be operated intermittently can be energised by this manually driven flywheel motor.

Accordingly the concept was tried for the application to a wood turning process. The findings of this research is the subject matter of the present paper.

EXPERIMENTAL SETUP
General mechanical details
The schematic line sketch of the set-up is described in Fig.1 along with it's photograph in Fig.2. The rider pedals the bicycle mechanism there by rotating the big sproket BP which in turn rotates small sprocket SP through chain CH. Motion of SP is transmitted to the flywheel shaft FS on which the flywheel I is mounted through a speed increasing pair of spur gears G. Carpenter's lathe W gets the drive from FS through belt transmission P.

Speed measurement of flywheel
The X-T plotter XTP is arranged to get the plot of revolutions per minute (rpm) of flywheel I versus time of peddling. DC tachogenerator DCT is connected at the end of FS to facilitate this. Voltage signals developed by DCT are fed to XTP. A digital contact type tachometer is also used to measure the rpm of flywheel I reached at the end of peddling time of 1 minute. Comparision of these two observations serves as dependability of XTP plot.

Input energy measurement
Fig.3 describes the measuring system specially designed and fabricated for this purpose. This enables the measurement of exhaled air of the rider at every instant during peddling duration and it's recording. This can be considered as a modified spirometer. The exhaled air gathered by the mouth piece of the rider is communicated to the float lifting it vertically. Travel of the float is recorded on a stationery scale A and on the graph paper on drum recorder which is driven by a constant speed moter.

Selection of riders
Four riders namely R1,R2,R3,R4 were chosen from student community with identical family back-ground, culture, aptitude etc. This minimizes the extraneous variables associated with riders. Age group is 19-20 years, weight 650 N, height 165 to 168 cm and had slim stature.

EXPERIMENTATION
First of all the Resting Minute Ventilation (R.M.V)

FIG.-1   FLYWHEEL MOTOR & LATHE

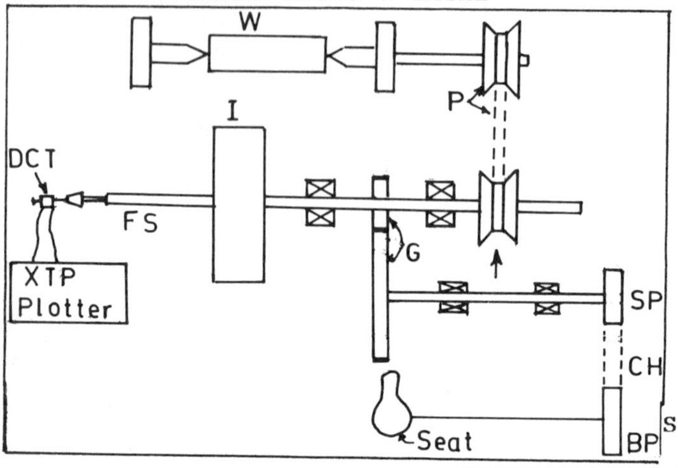

FIG.-3
SPIROMETER

FIG.-2   EXPERIMENTAL SETUP

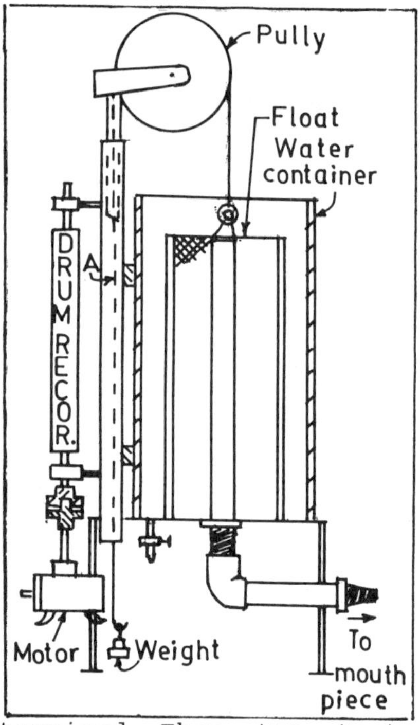

of all the four riders is determined. Then two tests detailed below are performed. Observations and various parameters deduced are presented in Table - 1 and Fig.4.

Flywheel motor as energy source (Test A)

In this test rider energises the flywheel for 1 minute during which the wood turning operation is not performed. After this 1 minute's of peddling it is

FIG.-4 ENERGY INPUT & FLYWHEEL ENERGY VERSUS TIME

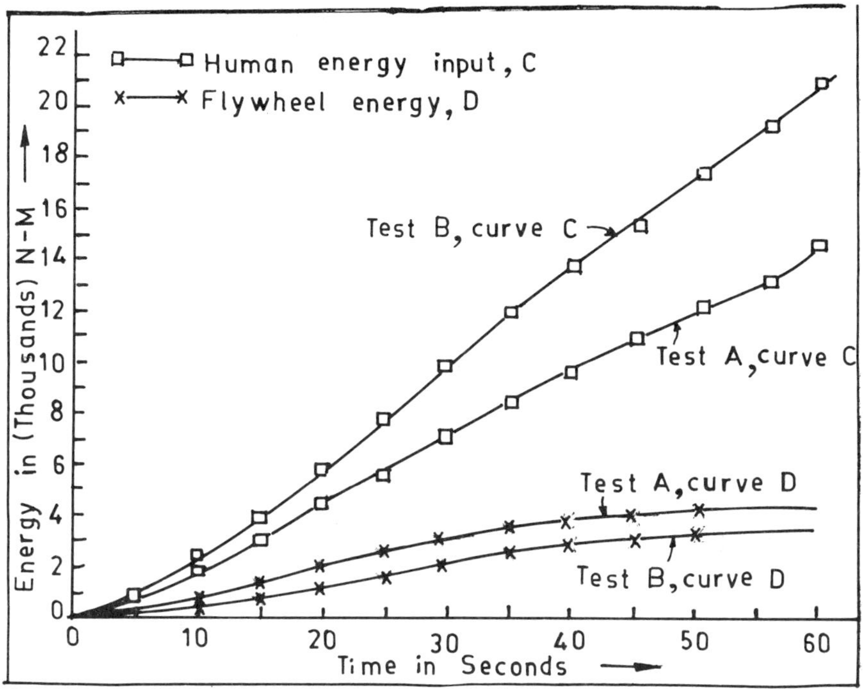

Table 1 EXPERIMENTAL DATA

TEST TYPE	RIDER	VOLUME REMOVED cu.cm	ENERGY INPUT N-m	PROCESS TIME secs	VOLUME PER INPUT cu.cm/N-m
A	R1	16.956	10873.5	84	0.0015593
	R2	17.383	14583.1	76	0.0011919
	R3	19.150	13349.3	59	0.0014344
	R4	27.431	18986.3	81	0.0014447
	AVG.	20.230	14500.0	75	0.0014076
B	R1	47.1000	13760.60	137	0.0034228
	R2	56.7900	20632.30	128	0.0027524
	R3	37.7500	24485.90	130	0.0015417
	R4	62.4100	25175.00	145	0.0024790
	AVG.	51.0125	21013.45	135	0.0025490

stopped and imidiately wood turning is commenced which continues till whole kinetic energy of the system is exhausted. Time of operation of the lathe is recorded.

On load test (Test B)
In this test as the rider starts peddling wood turning is simultaneously commenced. The rider pedals

only for 1 minute but the wood turning continues till
the kinetic energy of the system is exhausted.
    During both these tests for every observation XT
plotter plots rpm of main shaft FS versus time and drum
recorder plots float travel versus time. From these
plots the plots of energy stored in the flywheel versus
time and of energy input by the rider versus time
averaged for all the riders are deduced. These have
been presented in Fig. 4 for tests A & B.

ANALYSIS OF RESULTS
Overall averaged performance of Test A
    Total energy input per cycle of operation of 135
seconds is 14500 N-m out of which 4200 N-m is left in
the flywheel before commencing wood turning process.
The difference 10300 N-m energy can be attributed
towards (1) error in estimation of energy input from
exhaled air (2) energy required for accelerated
activities of body physiological functions and (3)
energy required to over come windage & friction losses.
It is rather difficult to apportion this 10300 N-m
energy in to it's above stated components. The
efficiency of this test is 28.96 %.
Overall averaged performance of Test B
    Total input energy per cycle of 135 seconds is 21000
N-m. Out of this 3500 N-m is left in the flywheel after
stopping peddling. 3500 N-m energy is available for
remaining part of wood processing which continues for
allmost same period of 75 seconds as in Test A after
peddling. Hence it may be predicted that in this phase
material removed may be 16.85 cu cms. Hence during
peddling the material removed may be about 34.15 cu cm
which might need 7090 N-m energy. Hence the energy
needed to overcome losses during peddling is 10410 N-m
in this test as against 10300 N-m in test A. Infact as
this is on load test this loss should have been much
greater. But the fact that it is not so may be because
the wood turning process may not be appreciably loading
the system. Effectively 10590 N-m of 21000 N-m input
energy is utilised in this test for wood turning giving
50.42 % efficiency.
Salient operational characteristics
    Fig.4 indicates that input energy rate is higher for
test B than for test A infers that human energy input
is higher, higher is the load. There is no significant
rise and difference in the terminal speed attended by
flywheel after 40 seconds in both the tests. This
infers that (1) rider's maximum frequency of thigh
oscillations decides the terminal speed and (2) it is
independent of load.
Detailed analysis of losses
    In both the tests frictional loss due to

considerably increased bearing loads on account of system unbalance at high speed is expected to be appreciable. This frictional loss can be proved to be proportional to the area under curves D in fig. 4. Let this loss for test A & B be F1 & F2 respectively. If x and y are considered as fractions of total energy input attributing to the error in input energy estimation from exhaled air & energy requirements for enhanced body physiological functions respectively then following relations defining the breakup of total energy loss hold true.

For test A, $10300 = x*14500 + y*14500 + F1$ ..N-m...(1)

For test B, $10410 = x*21000 + y*21000 + F2$ ..N-m...(2)

If areas under D curves are assumed to be equal, from equation (1) & (2); it can be deduced that

$$110 = 6500*(x + y) \quad ..N\text{-}m............(3)$$

Equation (3) gives $x+y = 0.0169$. This proves that the energy loss 10300 N-m in test A and 10410 N-m in test B is essentially on account of enhanced frictional forces at the bearings due to appreciably increased bearing loads at high speed due to system unbalance.

CONCLUSION

Cycle time of both the tests is 135 seconds. With the present system, test B is more efficient than test A by 74 % where as B is more taxing than A by 50% from the human energy input point of view. If mechanical system is fabricated and assembled with greater accuracy the overall system efficiency is likely to be increased considerably. Experimental verification of this may be considered as suggested further work. Peddling time should be 40 seconds only for any test.

REFRENCES

Askhedkar R.D. " Optimisation of manufacture of bricks through development of manually operated brick making machine." Ph. D. thesis of Nagpur University, India,1990. Under the supervision of Dr. J. P. Modak.

Modak J.P. "Manufacture of lime-flyash sand bricks using a manually driven brick making machine." Project report - project sponsored by MAHAD Bombay India,1982.

Modak J.P. and Bapat A.R. "Design of experimentation for establishing generalised experimental model for a manually driven flywheel motor." Proceedings International AMSE, France conference Modelling and Simulation New Delhi 1987, Volume B, pp 127-140, Oct' 87.

Modak J.P. and Bapat A.R. "Formulation of generalised experimental model for a maually driven flywheel motor and its optimisation." Accepted for publication in Applied Ergonomics.

OXYGEN COST OF TREADMILL AND ROAD RUNNING

ANDREW DUGGAN and JOHN F PATTON*

Army Personnel Research Establishment
Ministry of Defence
Farnborough, Hampshire, GU14 6TD.

*U S Army Research Institute of
Environmental Medicine
Natick, MA 01760-5007, USA.

This study assessed the validity of using oxygen cost (VO2) data obtained during treadmill running to estimate VO2 of road running. Twelve male soldiers performed a best-effort road run and their run speeds were recorded. VO2 was then measured during treadmill and road running at these speeds. The relationship of VO2 to run speed did not differ significantly between treadmill and road. Treadmill and road running VO2 values were highly correlated (r=0.97) and the means were almost identical. It is concluded that VO2 of best-effort road running can be accurately determined from measurements made on the treadmill.

INTRODUCTION

Determination of the oxygen cost (VO2) of running can be desirable for a number of reasons, including:

1) evaluation of the physical workload imposed by certain tasks;

2) assessment of the degradatory effects of items of clothing (eg footwear) and personal equipment (eg webbing) on physical work capacity;

3) quantification of training intensity;

4) assessment of energy balance;

5) investigation of the efficiency of human locomotion.

Direct determination of VO2 requires the collection and analysis of respired air samples. This is more readily performed during treadmill running than overground running since the respired air can then be sampled at a static location. Further advantages of using the treadmill are that the exercise is performed within the controlled environment of the laboratory, run speed can easily be set and held constant, the subject can be closely monitored, medical support can be kept close to hand, and the analysis equipment can be kept alongside the test area allowing rapid determination of results.

For these reasons, the treadmill has often been used in investigations into the VO2 of running (eg. Bransford and Howley, 1977; Duggan, 1989; Epstein et al., 1987). The direct applicability of data from such investigations to overground running is supported by the studies of McMiken and Daniels (1976) and Bassett et al. (1985) in which no difference was found between VO2 of overground and treadmill running at the same speed. On the other hand, the validity of generalising from treadmill running to overground running is brought into question by studies which have reported energy expenditure on a treadmill to be less than on a track during running (Pugh, 1970) and less than on a road during walking (Daniels et al., 1953).

The present study assessed the validity of using treadmill running to determine the VO2 of best-effort road running in boots.

METHOD

Twelve healthy male soldiers acted as subjects for the study. They had all had previous experience of running on a treadmill. Subject characteristics are summarised in Table 1.

Table 1. Subject characteristics (n=12)

	mean	sd
age, y	27.2	6.2
height, cm	172.1	4.4
weight, kg	71.9	8.9
fat, % body weight*	18.1	4.8
VO2max, ml/(kg.min)#	53.3	5.6

*Body fat was estimated from four skinfold thicknesses according to the method of Durnin and Womersley (1974).
#Maximal oxygen uptake (VO2max) was measured directly during uphill treadmill running using a discontinuous protocol based upon that described by Taylor et al. (1955).

Subjects performed a best-effort 2.4 km run on smoothly paved and tarmacked footpaths and roads. The clothing worn comprised underwear, denim combat trousers, PT vest, and military combat boots which extended to the calf region of the lower leg (mean weight 1.7 kg per pair). The run was timed and the mean run speed was calculated for each subject.

One to two days after the 2.4 km timed run, subjects ran on a treadmill (Woodway ELG2) at zero gradient wearing the same clothing as above, including the boots. After a 3 minute warm-up, treadmill speed was set as close as possible to the best-effort 2.4 km timed run speed of the subject. The subject ran at this speed for 6 minutes, with expired air being collected into Douglas bags in the last minute.

Inspired and expired air samples were analysed for both O_2 and CO_2 concentrations by paramagnetic (Taylor Servomex OA186) and infra-red (Beckman LB2) gas analysers, respectively. Expired air volume was measured with a chain-compensated spirometer (Warren E. Collins). VO2 was calculated according to standard methods.

Two days after the treadmill assessment, each subject ran alongside a motor vehicle (Land Rover) around a level 2 km circuit of tarmacked roads. Subjects wore the same clothing as above. The vehicle was driven as close as possible to the best-effort 2.4 km run speed of the subject, the speed being checked by measuring the time to cover a marked distance and adjusted as appropriate. Run speed (Road Speed) was calculated from the time to cover the last 500 m of the circuit. Towards the end of the run, the subject began breathing through a valvebox, the expired side of which was connected by wide-bore plastic tubing to Douglas bags in the back of the vehicle. Two timed collections of expired air were taken into plastic Douglas bags over the final 500 m of the course. These were analysed as described above. All subjects had been running for at least 5.5 minutes before the expired air collections commenced.

RESULTS

As intended, 2.4 km best-effort timed run speed, treadmill speed and road speed were all very similar. These variables were highly correlated with one another (r>0.97) and did not differ significantly (ANOVA, p>0.05), the mean (sd) values being 14.1 (1.4), 14.0 (1.3) and 14.0 (1.5) km/h, respectively.

Treadmill and road running VO2 values are plotted against each other in Figure 1. There was a very high correlation between these variables (r=0.97) and it is apparent in Figure 1 that the points all lie close to the line of identity. VO2 of running did

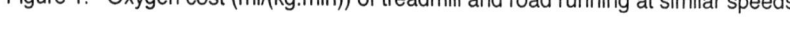

Figure 1. Oxygen cost (ml/(kg.min)) of treadmill and road running at similar speeds

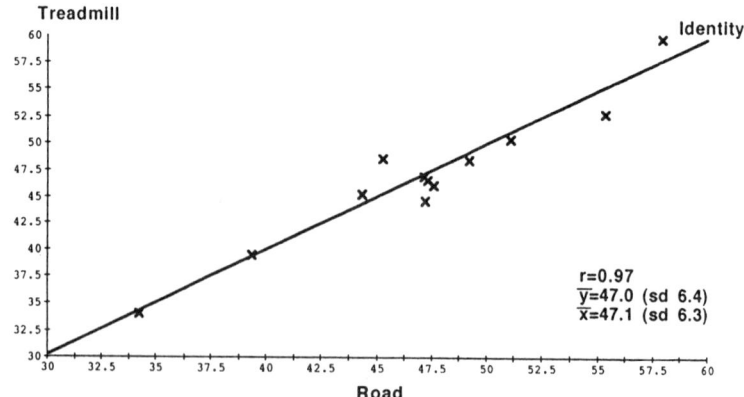

not differ significantly between treadmill and road (t-test,
p>0.05), the mean (sd) values being 47.0 (6.4) and 47.1 (6.3)
ml/(kg.min), respectively.

The relationship between VO2 and run speed is shown for both
the treadmill and road in Figure 2. It can be seen that the VO2-
run speed regression lines for treadmill and road running were
very close to one another over the range of speeds tested.
Statistical analysis by Reeves Method for the examination of
several linear regressions to see if they are from the same
population revealed that the two regression equations did not
differ significantly in either slope or position.

Figure 2. Oxygen cost - run speed relationship on treadmill and road

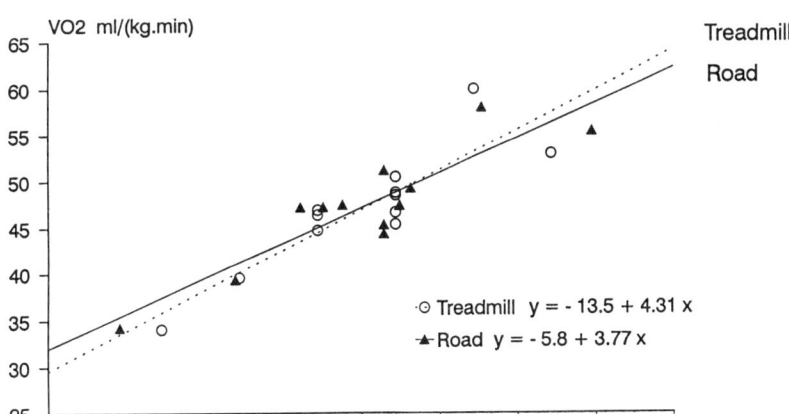

DISCUSSION

The results of the present study support the findings of
McMiken and Daniels (1976) and Bassett et al. (1985) that there
is no difference between the oxygen cost of treadmill and
overground running. The overground running was carried out on a
running track by McMiken and Daniels (1976), on both a running
track and road by Bassett et al. (1985) and on a road in the
present study. From the common finding that VO2 did not differ
between treadmill and overground running it is inferred that any
difference in the effects of road and running track surfaces on
VO2 is small. Other surfaces may have highly significant
effects, however. Thus, Soule and Goldman (1972) reported large
differences between the energy costs of walking on widely varying
surfaces and Zamparo et al. (1992), for example, found the energy
cost of running on sand to be 23 % greater than on firm surfaces.
Caution should be observed, therefore, in applying VO2 data
obtained during treadmill running to surfaces other than road or

running track.

Wearing heavy, high combat boots has previously been reported to markedly increase the VO2 of running (Duggan, 1989). Such boots were worn in the present study but not in the studies of McMiken and Daniels (1976) or Bassett et al. (1985). The similarity of the results from these studies suggests that the effects on VO2 arising from the biomechanical limitations imposed by these boots applies equally during treadmill and road running.

McMiken and Daniels (1976), Bassett et al. (1985) and the present study all compared VO2 of treadmill and overground running over a very similar range of run speeds. At higher run speeds, differences between the energy costs of treadmill and overground running might occur. Thus, in overground compared with treadmill running there is the extra energy cost of overcoming air resistance and this will become significant at high run speeds (Davies, 1980). This extra energy cost is proportional to run velocity raised to the second power, rising from 2 % of the energy cost of running at the pace of an elite marathon runner (18 km/h) to 7.8 % during a sprint (36 km/h) (Davies, 1980). Furthermore, Elliott and Blanksby (1976) and Nelson et al. (1972) found that there were certain gait differences between treadmill and track running that occurred at high run speeds but not at the speeds that fell within the range covered in the present study. The impact of these biomechanical changes on VO2 was not assessed but it seems reasonable to postulate that there may be an effect. It should be noted, however, that for periods in excess of a few minutes, only very fit individuals could sustain run speeds above the range used in the present study.

Wind velocity during overground running was very low in the studies of McMiken and Daniels (1976) and Bassett et al. (1985). During the present study, mean wind velocity on an adjacent airfield was 5.4 (sd 1.1) m/s but the subjects were partially shielded from the wind by the vehicle. It is important to note though, that the energy cost of running against the wind can increase sharply with wind velocity and even running with the wind on another part of the circuit does not fully compensate for this (Davies, 1980). Energy cost may also be increased by crosswinds and running the bends with the wind at right angles to the runner (Davies, 1980). VO2 data from treadmill running should, therefore, only be applied to road running performed under relatively calm conditions.

It is concluded that, over a wide range of run speeds, oxygen cost data obtained during treadmill running may be directly applied to best-effort road running providing wind speeds are low.

REFERENCES

Bassett, D.R., Giese, M.D., Nagle, F.J., Ward, A., Raab, D.M. and Balke, B., 1985, Aerobic requirements of overground versus treadmill running, Medicine and Science in Sports and Exercise, 17, 477-481.

Bransford, D.R. and Howley, E.T., 1977, Oxygen cost of running in trained and untrained men and women, Medicine and Science in Sports, 9, 41-44.

Daniels, F., Vanderbie, J.H. and Winsmann, F.R., 1953, Energy cost of treadmill walking compared to road walking. US Army Quartermaster Center, EPD Report 220, Natick, Mass., cited by Givoni, B. and Goldman, R.F., 1971, Journal of Applied Physiology, 30, 429-433.

Davies, C.T.M., 1980, Effects of wind assistance and resistance on the forward motion of a runner, Journal of Applied Physiology, 48, 702-709.

Duggan, A., 1989, Run test performance in boots and running shoes. In Contemporary Ergonomics 1989, Proceedings of the Ergonomics Society's 1989 Annual Conference, ed. E.D.Megaw (London: Taylor and Francis), pp. 386-391.

Durnin, J.V.G.A. and Womersley, J., 1974, Body fat assessed from total body density and its estimation from skinfold thickness: measurements on 481 men and women aged from 16 to 72 years, British Journal of Nutrition, 32, 77-97.

Elliott, B.C. and Blanksby, B.A., 1976, A cinematographic analysis of overground and treadmill running by males and females, Medicine and Science in Sports, 8, 84-87.

Epstein, Y., Stroschein, L.A. and Pandolf, K.B., 1987, Predicting metabolic cost of running with and without backpack loads, European Journal of Applied Physiology, 56, 495-500.

McMiken, D.F. and Daniels, J.T., 1976, Aerobic requirements and maximum aerobic power in treadmill and track running, Medicine and Science in Sports, 8, 14-17.

Nelson, R.C., Dillman, C.J., Lagasse, P. and Bickett, P., 1972, Biomechanics of overground versus treadmill running, Medicine and Science in Sports, 4, 233-240

Pugh, L.G.C.E., 1970, Oxygen intake in track and treadmill running with observations on the effect of air resistance, Journal of Physiology (London), 207, 823-835.

Soule, R.G. and Goldman, R.F., 1972, Terrain coefficients for energy cost prediction, Journal of Applied Physiology, 32, 706-708.

Taylor, H.L., Buskirk, E. and Henschel, A., 1955, Maximal oxygen intake as an objective measure of cardio-respiratory performance, Journal of Applied Physiology, 8, 73-80.

Zamparo, P., Perini, R., Orizio, C., Sacher, M. and Ferretti, G., 1992, The energy cost of walking or running on sand, European Journal of Applied Physiology, 65: 183-187.

Any views expressed are those of the authors and do not necessarily represent those of the Army Personnel Research Establishment, the Ministry of Defence or Her Majesty's Government.

Sitting, standing and workspace

DEVELOPMENT OF A THREE-DIMENSIONAL APPROACH TO INVESTIGATE THE ERGONOMICS OF STANDING.

R.S. Whistance[1], L.P. Adams[1/2] and R.S. Bridger[1]:

1. Department of Biomedical Engineering, University of Cape Town Medical School, Observatory, 7925 Cape Town, South Africa.

2. Medical Biophysics, South African Medical Research Council, C/o Department of Biomedical Engineering, University of Cape Town Medical School, Observatory 7925, Cape Town, South Africa.

This paper discusses the use of a low-cost PC-based near real-time photogrammetric system for studying postural changes in standing subjects working under various constraints. Stereophotogrammetry has had limited use in ergonomics research although it has several major advantages. The method is relatively non-contact, cost effective, portable, is accurate to within 2 - 3mm and allows the subject to be examined rapidly. Preliminary results of the study are reported which confirm these advantages.

INTRODUCTION

In many work situations, operators must adapt themselves posturally to the constraints imposed on them by their work environment. These adaptations are often relatively fixed and are often repeated throughout the working day over a period of many years, resulting in postural misalignments which may increase the risk of discomfort and both short and long-term disorders. The study of postural adaptations to external environment is therefore an essential part of the evaluation of industrial workspaces.

The problem of how to translate three-dimensional information about the human form into a two-dimensional form was faced five hundred years ago by Renaissance artists who adopted the principle that art should faithfully represent nature. They overcame the problem by studying anatomy and skillfully applying this knowledge to their work. However, it was Sir Charles Wheatstone who realized in 1832 that it was possible to reproduce the three-dimensional perspective we get with binocular vision by making two drawings from slightly

different view points and looking at these stereoscopi-
cally.

Biostereometrics began to develop with the appearance of the first double image stereocameras in the
1850's. It is based on the principle that the surface
of any biological structure can be regarded as an in-
finite number of points. If the three-dimensional co-
ordinates of enough of these points are known for a
particular application, a comprehensive measurement can
be made for the part under investigation. The most
versatile stereometric technique is stereophotogrammetry which provides a sound methodology for collecting
information describing the shape of the human body in
three dimensions.

In the past, stereophotogrammetric methods,
although very successful, have been labour intensive
and time-consuming which severely limited the scope and
number of studies which could be performed. Developments in solid state cameras, in computer hardware and
in image processing technology have lead to real- or
near real-time photogrammetry.

According to Sheffer and Herron (1989), the method
has several major advantages. Among these are: (1) the
method is for the most part non-contact and does not
distort the true shape of the human subject; (2) it
allows the subject to be examined relatively quickly as
measurements are made afterwards from the photographic
record; (3) it provides a permanent record which is
easily stored for retrospective examination using the
same or an entirely different set of measurements;
(4) it has redundancy, i.e. it is possible to decide at
a later stage which data is really needed for a particular study and the data can be used for more than one
study; (5) the apparatus is portable and can be used in
a variety of experimental conditions; (6) it is
accurate to within 2mm in a measurement of 2 - 3 metres
and the accuracy is independent of movement and (7) it
is cost-effective.

The current investigation uses a low-cost PC-based
near real-time photogrammetric system for studying the
posture of standing workers. Although postures and
back shapes have been investigated biostereometrically
by other disciplines, (e.g. Turner-Smith and Thomas,
1989)(Pineau et al, 1986), the methodology has been
under-exploited by ergonomists.

Standing posture in this study was examined under
three degrees of foot constraint and at two task distances. The hypotheses that foot position and task
distance will influence standing posture were tested.

METHOD
a)   Equipment
     The computerised system consisted of the following

equipment: Sanyo video cassette recorder with still
picture playback VHR-D500SA (digital), 10 Phillips
video tapes, two channel image mixer (Video Wiper PVW-
1), 2 CCD monochrome cameras (Burle TC652EX Series
Cameras), 2 zoom lenses (Computâr 8.5mm 1:1.3),
Phillips monitor CM8833, control points, vertically
mounted mirror 1mx1m, 2 lamps with 100 watt globes and
dimmer, MBM 32 personal computer, Matrox PIP-512 image
processing card, fiducial markers mounted on stand.
Other equipment: Black-painted workbench with adjustable height, removable foot constraint and foot-rail.
b) Procedure
Semi-nude subjects who had already been standing
for one hour were prepared by sticking spherical reflective markers mounted on 15mm stalks on the skin
over vertebra prominens, the tips of 7 spinous processes, the dimples over the posterior superior iliac
spines and the right anterior superior iliac spine.
Flat reflective markers were also stuck on the right
side over the skull behind the ear, over the greater
trochanter, the knee joint, the ankle joint and the
little toe.
  Nine male and nine female subjects were videotaped
as they worked on a jigsaw puzzle at the workbench in
a laboratory with black walls. Subjects worked at a
task distance that was either constrained or unconstrained. In addition to this, foot position was constrained by a board, unconstrained or employed the use
of a footrail on the bench. This gave six different
experimental conditions under which the subjects
worked, in random order, for ten minutes each. A
mirror was positioned so that it reflected the markers
on the subject's back when they were obscured by the
body.
c) Observation and mathematical procedures
The images to be measured were selected by viewing the split screen video and were captured on the
personal computer equipped with the image processing
card. Only images of working postures were selected
and selections were made at approximately sixty second
intervals.
  Each image was then digitized so that the two dimensional co-ordinates of image points on both the
left and the right side were known. A computer program was then run using two-dimensional co-ordinates of
the fiducial marks and a two-dimensional linear transformation to correct for shift of the video frames due
to the pause function of the video machine. This same
software then derived the three-dimensional co-ordinates of the points on the body by a method of projec-
tive transformation as discussed by Adams (1981).

Where three-dimensional co-ordinates corresponded
with points beyond the plane of the mirror, i.e. where
mirror image points were selected and digitised, a
further transformation was performed in order to translate these points from the mirror image space to the
body space.

d) Presentation of results

A further computer program was run to derive postural angles using the relevant spatial co-ordinates. The
program calculated seven average postural angles for
each subject under each of the six experimental conditions. The seven postural angles were: Trunk Flexion
(TF), Hip Flexion (HF), Pelvic Inclination (PI), Trunk
Inclination (TI), Neck Flexion (NF), Knee Flexion (KF)
and Plantar Flexion (PF). With the exception of Plantar
Flexion, which is defined as the angle between the
heel/toe line and the heel/knee line, these angles are
defined in Bridger (1988).

Additional software found the average orientation
in space for the spinal and pelvic markers for all subjects for each experimental condition. These were
plotted out as stereopairs to be viewed stereoscopi-
cally. The perspective from which the stereopairs were
viewed could be pre-defined. Stereopairs could also be
printed out for individuals.

e) Experimental results

Statistical analysis of the resulting postural
variables supports the hypotheses that both task distance and foot positon influence standing posture.

Task distance was found to have a significant
effect on posture. Subjects had more neck and less trunk
flexion when working at an unconstrained task distance
than they did when working at a constrained distance.
They also had less flexion at the hips, less backward
tilting of the pelvis and had larger angles of trunk
inclination, i.e. were more upright, working at the unconstrained distance. Space constraints prevent detailed discussion of these findings.

Foot position was also found to affect posture significantly. In particular, use of the footrail resulted in more backward tilting of the pelvis, more plantar
flexion in the supporting foot, less knee flexion in
the supporting leg and less flexion of the trunk than
other foot positions. The constrained foot position
resulted in more hip flexion than the unconstrained
foot position.

CONCLUSIONS

Although the theory of steereophotogrammetry is
complex and sophisticated mathematics must be applied
in order to produce the final results in this system,
the software, once developed, presents no obstacle to

researchers without a photogrammetric or advanced mathematical background. It can be run through a friendly, menu-driven system.

The only drawback to the current methodology is that the digitization process is manual and therefore time-consuming. Although it is possible to search images automatically for object points beyond a given brightness threshold, it was not possible to do it in this case because the investigator had to decide when to select "real" points and when to select "mirror" points.

Biostereometric analysis has been used successfully in many disciplines including medicine, dentistry, biomechanics, anthropology and biology and its accuracy and reliability have been established. The results of the current experiment are well within expectations based on previous findings using more traditional methods. The method appears well suited for testing hypotheses about workstation design and is considered by the authors to present the basis for further work.

REFERENCES

Adams, L.P., 1981, X-ray stereo photogrammetry locating the precise, three dimensional position of image points, Med. and Biol. Eng. and Comput., 19, 569-578.

Bridger, R.S., 1988, Postural adaptations to a sloping chair and work surface, Human Factors, 30(2), 237-247.

Gutschow, B.A., 1990, A near real-time photogrammetric study regional body surface motions of human beings during respiration. MSc. Thesis, University of Cape Town.

Herron, R.E., 1972, Biostereometric measurement of body form, Yearbook of Physical Anthropology, 16, 80-121.

Hertzberg, H.T.E., Dupertuis, C.W. and Emmanuel, I., 1957, Stereophotogrammetry as an anthropometric tool, Photogrammetric Engineering, 23, 942 - 947.

Pineau, J.C., Ignazi G. and Prudent, J., 1986, Biostereometric analysis of spine curvatures on living body. Biostereometrics 1985, in Proc. SPIE 602, 252 - 256. Eds. Coblentz, A.M. and Herron, R.E.

Sheffer, D.B., and Herron, R.E., 1989, Biostereometrics. In Non-topographic Photogrammetry, ed. Karara (American Society for Photogrammetry and Remote Sensing, Falls Church, Virginia), 359 - 366.

Turner-Smith, A.R. and Thomas, D.C., 1989, Evaluation of back shape using the ISIS scanner. Biostereometrics 1988 - Fifth International Meeting. In Proc. SPIE 1C30, 16 - 25. Eds. Baumann, J.U. and Herron, R.E.

A FUNCTIONAL SIT-STAND SEAT.

H. GREGG and E.N. CORLETT

Institute for Occupational Ergonomics
University of Nottingham
Nottingham
NG7 2RD

It is still customary to define seating using the
principles identified by Akerblom (1954). To a great
extent this is because the proposals for chairs which
can exploit an open angle at the hip have required a
forward sloping seat surface, from which the sitter
slides off, and the measures to combat this have not
been widely accepted by users.
A reconsideration of the requirements for sitting,
using as a starting point the secure support of the
sitter at such a height as will maintain an open
angle at the hip, suggested a new form of seat
surface which was shown to have advantages in
providing a wider range of postures for the sitter.

INTRODUCTION
The opportunities which are available for the human body to
transmit its weight to the ground range from standing upright on
one foot, as with an Australian aboriginal at rest, or lying on
the ground. Within this range much attention has been focussed on
sitting. However this term covers a wide range, from squatting to
leaning on a surface which scarcely supports the buttocks.
Conventionally we can see sitting as a process which reduces
the amount of body weight which is transmitted through the feet.
This definition would include leaning on a wall as part of
sitting, so we must qualify our definition to say that some
component of this body weight reduction passes through the
buttocks. Given such a broad definition can we view sitting so
that we come to new conclusions about it?
The purpose of proposing this attempt at a new view is to break
the perspective which links sitting with the commonly perceived
view of the "right angled sitting posture", which usually comes to
mind when sitting is mentioned. For many years we have started
from this posture and modified our view of a seat from it. In
consequence seat design has come to view the right angled posture
as appropriate unless something prevents it.

Sitting requirements

Mandal (1981) has been one of the most forceful authors in his presentation of an alternative sitting posture, demonstrating that an open angle at the trunk improves the opportunities to maintain a lumbar curve, and that this can be achieved by using a sloping forward seat surface. His proposals arose from observation of seated people at their jobs. If we are prepared to resist the temptation to provide a solution to an immediate problem, can we make a further improvement on Mandal's work?

The contribution of the backrest to the redistribution of bodyweight on a seat was discussed by Corlett and Eklund (1984) independently of the role of the seat surface. Strictly this is not correct, since it assumes a horizontal seat for the stability of the sitter, but it does allow the contribution of the backrest to be considered on its own. A later study, (Eklund and Corlett 1986) proposed that seat design should start with an analysis of requirements independently of any concept of a seat. A table presented the factors influencing the requirements for sitting arising from the sitter's circumstances, with relevant means of measurement and the consequences of the various demands on the sitter's body. Table 1 illustrates the functional factors in sitting.

Table 1. Functional factors in sitting.

The *task*	The *sitter*	The *seat*
Seeing	Support weight	Seat height
Reaching	Resist accelerations	Seat shape
Exerting forces	Under-thigh clearance	Backrest shape
	Trunk-thigh angle	Stability
	Leg loading	Lumbar support
	Spinal loading	Adjustment range
	Neck/arms loading	Ingress/egress
	Abdominal discomforts	
	Stability	
	Postural changes	
	Long-term use	
	Acceptability	
	Comfort	

Yet there is still a gap; how do we arrive at a chair design from such a table and the subsequent measurements? Any design will be affected by the work to be done, we cannot use, say link and task analyses to arrange a workspace without some assumptions about how the users will behave. For instance if we arrange things in close proximity one of our assumptions is that the users will be relatively static in the workplace, probably seated. But we know that being seated all the time is not necessarily a good thing. Winkel (1985) for example has shown the adverse effects on the legs of prolonged sitting. So our workspace design must arrange for movement, as must the job design.

The design consequences

This movement requirement is not just getting up and down from the seat, but changing position on it. The sitter must be adequately supported but also be able to reach, get up or down and lean back with equal facility, otherwise some of these activities will be neglected. The facility to change posture readily is an important factor in the design.

Can we see some general concepts appearing?
    (i)The workplace must, as far as possible, allow for the users
to sit or stand at will. We should not be designing for a
standing, or a sitting workplace unless there are working
requirements which make this essential.
(ii) To move from sitting to standing and vice versa must be easy,
hence the moment of body weight about the feet should be kept low.
Otherwise assistance from the arms will be needed or leg muscle
loads will be high.
(iii) A major proportion of body weight should not be on the seat
surface, but be distributed between the seat, the backrest and the
feet. This reduces sub-surface pressures and helps to allow
changes in weight distribution between the various possible weight
bearing surfaces.
(iv) The resultant position of the sitter should permit effective
vision, and allow ergonomically acceptable positions of the limbs
to exert the necessary actions and forces.
    If we are to adopt these ideas, there are some interesting
consequences. For the first concept, it suggests that workplaces
will in general be higher than is now conventional practice. When
we consider the second concept, the seat will also be higher, with
major impact on its design which we will come to later. The third
concept implies that the users' posture will permit this weight
redistribution, and in particular that the backrest will be in use
more than is often seen in practice. For our fourth concept the
usual ergonomic procedures for design, ie putting the users in
appropriate positions and building the work requirements around
them, will be the appropriate procedure, rather than laying out
the workplace and fitting the users in afterwards.

THE RESULTS
    Adopting this channel of thought produced an altered perception
of a work seat. It was necessary to devise a seat which would let
the rest of the concepts be adopted and at the same time be
practical and comfortable. It had been observed (Eklund et al.
1982) that there was a limit to the amount of body weight which
could be comfortably borne on the feet, which in turn indicated a
proportion to be borne by the ischii. At the same time it was
recognised that under-thigh support was desirable, for stability
and a feeling of not sitting on a bar. It was also observed that
unless the major body weight was carried on a horizontal surface,
the use of the backrest would push the sitter off the seat.
    As a result of these and other studies a seat, curved in the
sagittal plane and with the facility to rotate forwards as the
seat height was increased, was required. The study by Eklund et
al. (op cit) showed that the rate of forward tilt with increase in
height was similar for all users, providing that the starting
point at the lowest position was set for the individual user.
Furthermore the backrest position would not need altering with
height change, since the relative position between the lumbar
curve and the ischii did not alter significantly.
    Tests with seats of this design have been done at supermarket
checkouts, airport booking counters and at sewing stations. In
particular, tests were done in offices in relation to keyboard
operation and for assessing whether a fixed height desk could be
used by people with a wide range of stature. In this last
situation it was demonstrated that footrests were not necessary,
even very short users could sit with body weight supported, both

feet comfortably on the ground and with full access to the whole desk space whilst using the backrest.

We propose that the seat development described here presents opportunities for the provision of seated workplaces where the users may be comfortably seated, but stand at will. This choice will require using a desk which is higher than the conventional desk. It will be recalled that up to the early part of this century the tall desk was the norm, the 30" work table is a relatively new introduction. What is more, the old high desk had a sloping surface, another aid to avoiding a crouche-forward posture (Mandal, op cit).

Given that the provision of a chair of the form described allows a greater range of postures, it also allows for more variety in the arrangement of working equipment. Add to this the fact that people of different heights may use the same seat without the need for a footrest, and the economies of the design are evident.

It must be noted that the special features of this design are covered by world patents.

REFERENCES

Akerblom, B. 1954, Chairs and sitting. In: Human factors in equipment design. W.F., Floyd and A.T., Welford (Eds) (H K Lewis and Co. Ltd. London).

Corlett, E.N., and Eklund J.A.E., 1984 How does a backrest work? Applied Ergonomics, 15 111-114

Eklund, J.A.E., and Corlett E.N. 1986 Experimental and biomechanical analysis of seating. In: The ergonomics of working postures: models, methods and cases. J.R. Wilson, E.N. Corlett and I. Manenica (Eds.) (Taylor and Francis Ltd. London).

Eklund, J.A.E., Houghton C.S., and Corlett E.N., Industrial seating. Report on some pilot studies. Unpublished internal report, Dept. of Production Engineering and Production Management, University of Nottingham.

Mandal, A.C., 1981 The seated man, the seated work position, theory and practice. Applied Ergonomics 12 19-26.

Winkel, J.,1985 On foot swelling during prolonged sedentary work and the significance of leg activity. Arbete och Halsa no.35 Arbetarskyddsstyrelsen. Stockholm

FISHING VESSEL WHEELHOUSE DESIGN - A CASE STUDY

STELLA MILLS

Department of Information Technology
Cheltenham & Gloucester College of Higher Education
Oxstalls Lane
Gloucester
GL2 9HW

Small in-shore fishing vessels have not
enjoyed the ergonomic attention that has
been afforded to larger vessels,
particularly in the area of wheelhouse
design. This paper uses, as a case study,
a modern 65 feet OAL side trawler and
examines ergonomic aspects of its
wheelhouse design

INTRODUCTION

Although the fishing industry is one of the
oldest in the world, there have been few studies
made of the ergonomic aspects of fishing vessels,
especially relating to specific areas such as the
wheelhouse. Some attention has been given (e.g.
Hadley, 1988) to naval vessels but these studies
are generally too specialised to give more than a
cursory aid to the improvement of the design of the
in-shore fishing vessel.

There are a number of factors which effect the
design of a particular fishing vessel (Mills, 1992)
the most obvious being the positioning of the
fishing gear on the deck. For example, side
trawling requires the wheelhouse to be placed aft
of amidships, while stern trawling necessitates the
wheelhouse being placed amidships or forward. The
engines (propulsion, winch and tank emptying) are
usually placed under the wheelhouse to maximise
storage tank area for the catch. It is possible to
construct different designs on the same hull
(Fyson, 1985) but this invariably means that only
the hull 'skin' remains unchanged.

CASE STUDY

Figure 1: Side Trawler under Construction

The fishing vessel under study is a 63 foot overall length (OAL) side trawler constructed in steel and purpose designed for trawling mussels. She has four engines which are located in the engine room underneath the wheelhouse, two Cummins for propulsion, one for winch operations and one for emptying the storage tanks. Some modifications were made in design during construction; in particular, a hatch allowing access directly from the wheelhouse to the engine room had to be removed to comply with a DTI inspectors' ruling that it was too great a fire-hazard. In addition a stanchion post was removed and

the window (originally non-opening) above the wheel
was replaced with an opening one (but without a wet
weather wiper) so that the skipper could shout orders
directly to the crew on deck. The boat generally
makes round trips of 24 hours each and carries a crew
of two men, although provision is made for up to four
crew members, plus skipper.

The bridge is placed aft (Figure 1) and of the
16 windows in the wheelhouse (excluding the half-
door windows) six open to allow adequate
ventilation, although with the engines underneath,
it may be necessary at times to open the doors as
well. The doors open inwards through necessity due
to the elevated position of the wheelhouse and
these could be an accident hazard when open due to
further restricting the limited space of the
wheelhouse. Also, in case of fire, doors should
not open inwards. The alternative would be to fit
sliding doors which would increase vision and space
in the wheelhouse, remove the accident/fire hazard
and allow for fully adequate ventilation.
Ventilation inlets (Mercel, 1971) are not used. As
heating is by electric hot-air fan, placed near the
ceiling instead of by the floor since hot air
rises, some means of constant fresh air intake
should be provided (eg a window constantly open
during working operations).

Noise and vibration were not measured as the
wheelhouse was not observed fully operational at
sea. However, that there must be some noise and
vibration is obvious due to the engines being
directly beneath the wheelhouse but these may only
be a problem when the skipper is already subject to
mental overload such as when he is working the
winch, judging the catch, manoeuvring the vessel
and giving orders to the crew all at the same time
(Stoop, 1990). The vibration could contribute to
motion-sickness, although given the space
limitations for tanks, engine-room and crew
accommodation, it is difficult to see how the naval
architect could have changed the engines' position
in order to reduce vibration in the wheelhouse.

Lighting in wheelhouses has been investigated
by Waters and Ivergard (1983) who carried out some
experiments in order to find an alternative night
illumination to red. Their results were
inconclusive and consequently red has continued to
be used because it does not destroy the skipper's
dark adaptation as much as other colours. However,
red light is an unnatural environment and some
skippers prefer the selected use of white light as
well as red. It is not surprising, therefore, to

Figure 2: View of Skipper's Controls

find that the wheelhouse under study here has white
light and red light for the paper-chart and
propulsion controls.   Another problem with red
light is that some electronic charts need the user to
be able to distinguish between red lines and figures,
but since the Regulations still require the use of a
paper-chart this problem can, of course, be overcome
simply by using the paper-chart at night - which is
what happens on the vessel studied here.   However,
the warning 'on' lights for the light panel (deck and
vessel lights) are red possibly causing error in a
red night environment.

The skipper's seat is fully adjustable and conforms
to the requirements of the EEC Directive 90/270,
although the equipment, being on board a means of

transport, is actually exempt from this Directive.
The seat is fully adjustable for height, slope and
forward/backward on an adequate runner. The back is
also fully adjustable and the arms have full rotation
horizontally. The head-rest, which extends beyond
the top of the back of the seat, has a large
upward/downward range of movement.

Turning now to the instrumentation, the engine and
winch dials are all the accepted white calibration on
black background with additional associated colouring
where necessary. For example, the rudder angle dial
has green colouring to indicate starboard movement
and red for port movement. The positioning of the
propulsion engines dials to the left of the skipper
at waist height could be a potential hazard since it
requires a turn of the head to view these and thus
forward and deck vision is totally lost. A possible
improvement would be to position these either side of
the main compass in front of the wheel, thus
interchanging them with the light switches. The
winch instrumentation is in the centre of the forward
side of the wheelhouse and contains a joystick for
fine manoeuvring of the wheel (with the left hand)
while controlling the winch engine throttle with the
right hand. Outside, for on-deck use, there is an
emergency shut-off for the propulsion engines, but
otherwise it is necessary to move to control the
winch engine and the propulsion engine at the same
time. While for an experienced skipper this would
only be necessary in the case of an emergency, it is
just such a time when accidents are most likely to
happen.

The vessel carries four electronic aids: an echo-
sounder, a raster scan radar, an electronic chart,
and an auto-pilot. There is also a small auto-pilot
navigator which is used as a back-up. The vessel
does not have a fish-finder as she trawls only for
mussels and the skipper prefers to use water-depth
and area knowledge for locating the best beds. The
visual display units are mounted on fully rotational
and tilting stands to the right of the wheel (Figure
2) with the electronic chart above the master scan
radar (just visible on the left in Figure 2) and the
echo-sounder above the autopilot (on the right in
Figure 2). These are obviously positioned for
maximum ease of use since when the skipper is at the
winch controls it is more important for him to be
able to see the echo-sounder (giving the depth of
water) than the electronic chart or radar. However,
as can be seen in Figure 2, reflection is an obvious
problem.

In addition, the vessel carries VHF radio (ship to
shore), full distress signals, circuit breakers,

bilge alarm, smoke/heat alarms, security alarms for when the ship is in harbour, binoculars and a portable television for use by the crew on the outward and return journeys to and from the mussel beds.

CONCLUSION
Overall, the wheelhouse is well-equipped and generally the equipment is well positioned. However, there being no engine control within easy reach of the winch controls could be a hazard in an emergency since the skipper would have to move to regulate the propulsion engines. The doors opening inwards could prove hazardous in case of emergency evacuation caused for example, through fire; sliding doors would be a solution but may block vision when open. Views vary as to whether it is an advantage to have the electronic chart above the paper chart so that they can be viewed simultaneously; this writer suggests it depends on the usage of each and the area knowledge of the skipper. A major problem is reflected glare from the visual display units which may be helped by anti-glare screens and delicate adjustments of position. It is interesting that the legal suggestions for reducing human error and accidents given by Stoop (1990), are all adopted in the positioning of the visual display units under study here except for their integration into 'one display' - an advanced feature yet to appear in small fishing vessels. Thus this vessel has in general a well designed wheelhouse which gives an efficient work area in keeping with modern trends.

REFERENCES
Fyson, John, 1985, Design of Small Fishing Vessels, (Farnham: Fishing News Books Ltd.).
Hadley, M. A. , 1988, Present Trends in Naval Bridge Design and Integrated Navigation, J. Navigation, 41(2), 276-287.
Mercel, D. I., 1971, Ergonomic Considerations in the Design of a Distant Water Stern Trawler Bridge Layout, M.Sc. Project, Work Design and Ergonomics, The University of Birmingham.
Mills, S., 1992, Ergonomic Aspects of Wheel-house Design, The Business School Research Journal, Cheltenham and Gloucester College of Higher Education, Autumn, 1992, 32-40.
Stoop, J., 1990, Redesign of Bridge Layout and Equipment for Fishing Vessels, J. Navigation, 43(2), 215-228.
Waters, T. L., and Ivergard, T., 1983, A study of red and white light on the chart table for navigation at sea, Ergonomics, 26(4), 349-358.

Drivers and driving

DRIVERS AT TRAFFIC SIGNALS: A QUALITATIVE ANALYSIS

Sandy Robertson

University of London Centre for Transport Studies,
University College London,
Gower Street,
London
WC1E 6BT

A matched sample of 193 redrunning vehicles and 193
which encountered the amber signal without redrunning
was obtained from video recordings of three signal
controlled junctions. Various driving situations were
identified and an analysis of the association between
redrunning and these situations was undertaken.
Investigation of some possible mechanisms relating to
red running the situation at traffic signals was
undertaken using graphical methods. The situations
associated with redrunning were dithering, complex
traffic situations, and being in a platoon but not
leading it.

INTRODUCTION

There are about 20,000 injury accidents at traffic signals
every year and many of these are related to violation of the
traffic signals. The study of driver behaviour at traffic signals
may provide insights into the reasons for violation of traffic
signals and ways of reducing the incidence of violation. This
study builds on earlier, numerical analysis by the author (Allsop
et al 1991, Robertson 1990). Although the numerical analysis of
the physical situation at traffic signals provides much useful
data, there are a number of behaviours and/or driving situations
which cannot be identified using such methods. A qualitative
approach using observational techniques can identify many of these
behaviours, but on its own cannot put these into the context of
the physical situation. This study integrated both aspects by
using a qualitative analysis of the data which had been analysed
numerically in the earlier study. A statistical analysis of the
results of this study, together with more detailed descriptions of
the methodology can be found in Robertson (1992).

At traffic signals drivers approaching a set of signals which
have just turned from green to amber may be in one of four

physical situations a)can clear (go), b) can stop (stop), c) can
stop and can clear (option) and d) can neither stop nor clear and
will run the red no matter what their action (no option). The four
situations are discussed in detail in Allsop et al (1991_. The
drivers must make a choice of whether to attempt to stop at or
before the stopline or whether to attempt to clear the stopline
before the signals turn to red.

Current methods of analysis generally make assumptions
regarding factors which influence the physical situation, for
example fixed values of response time and levels of acceleration
and braking. They tend to be centred around traffic
characteristics such as speed and distance at amber onset (for
example the speed/distance diagram (Baguley and Ray 1989)) or a
probabilistic model (Chang et al 1985). While such approaches are
appropriate for some types of investigation they do not easily
lend themselves to detailed investigation of behaviour. The author
has proposed an alternative, user-centred representation based on
the acceleration and braking required by drivers. This paper
demonstrates the use of this representation to investigate one of
the factors which was identified as being associated with red
running during the first phase of this study.

METHOD
   Video recordings of three signal controlled junctions at
Norwich, Takeley and Chelmsford were used. The initial data
extraction and numerical analysis of these is described in Allsop
et al (1991). A sample of 193 redrunning and 193 non redrunning
vehicles matched for site, vehicle type, direction of movement,
speed and distance at amber onset were extracted from the
numerical data. Video tapes of these vehicles were then viewed and
each vehicle was categorised in terms of a number of driving
situations which are detailed below. A pilot study using the 17
pairs of vehicles from the Norwich site was undertaken to assess
the appropriateness of the categories. There was some modification
to the existing categories and 4 additional categories were
included. These are indicated in the descriptions by an
asterisk.The categories are briefly described below. A more
detailed discussion of the categories can be found in Robertson
(1992).
   1)Obscuration of the traffic signals. This included vehicles
where the driver's view of the signals was obscured at amber onset
(for example being preceded by an HGV).
   2) The presence of a vehicle ahead of the subject vehicle. A
vehicle was placed in this classification if there was another
vehicle between the vehicle in question and the stopline at amber
onset.
   3)The presence of a vehicle following the subject vehicle. A
vehicle was put in this classification if there was another
vehicle upstream judged to be visible from vehicle under
consideration at amber onset.
   4)The presence of a vehicle stationary at the junction.

5)The presence of a vehicle stationary on the approach to the junction.

*6)The driver of the vehicle could observe signals turning to green at beginning of signal cycle.

7)The presence of pedestrians in the vicinity of junction.

*8)Complex road and traffic conditions. This was a subjective judgement of the degree of complexity of the road environment and traffic conditions. This included vehicle interaction (such as overtaking) which meant that there could be active components in the driving situation/behaviours.

*9)Wet or damp road conditions.

10)Dithering by drivers. This was, in general, only able to be observed in the case of vehicles which attempted to clear. A driver was regarded as dithering if the motion of the vehicle suggested a change of mind with regard to the appropriate decision.

*11)Vehicle in a platoon, but not leading the platoon.

It was hypothesised that there would be an association between redrunning and being in categories 1,2,3,6,8,10 and 11.

The association between redrunning and being in the above categories was initially analysed using χ^2 tests for single categories and log-linear modelling using the GLIM system (Aitken et al 1990) for multiple categories.

A graphical analysis developed by the author (Robertson 1990, Allsop et al 1991) was then used to investigate the effects of an increased response time resulting from being in the categories identified by the log-linear modelling as having an association with redrunning upon the situation (go, stop, option, no option) facing the drivers at amber onset. Only the category dithering will be considered in this paper.

RESULTS

The results of the χ^2 tests and the log-linear modelling are detailed in Robertson (1992),but a brief summary is included here for completeness. The χ^2 tests indicated that in at least some circumstances obscuration of the traffic signals, the presence of other traffic ahead of and/or behind the vehicle, being in a complex traffic situation, being in a platoon but not first, and dithering were individually associated with redrunning (p<0.05). The log-linear modelling indicated that when all the categories were considered together those which had a significant association with red running were being in a complex traffic situation, being in a platoon but not first, and dithering. The last factor bears some investigation as it occurred in about 20 per cent of the redrunning vehicles and in less than 1 per cent of non redrunners. The log-linear modelling indicated that there was about 58 times more redrunner who were judged to dither than non redrunners.

GRAPHICAL ANALYSIS

The graphical analysis used the what the author has called the a,b representation, which is a plot of the calculated levels of braking versus acceleration required for an individual vehicle to

stop or clear respectively. The calculated levels of acceleration
(a) and braking (b) take speed (v) and distance at amber onset
(d), duration of the amber signal (T), and response time (t) into
consideration. T is assumed to be 3s, which is the standard
duration in the UK, and is assumed to be one second.

The plot may be divided into sectors which indicate the
situation which the drivers faces in terms of the levels of
acceleration and braking required to clear and stop successfully.
The boundaries of the sectors are specified by the maximum levels
of a and b which are acceptable to the drivers. As the sample is
matched for speed and distance at amber onset (and therefore
acceleration and braking required) a plot of these shows pairs of
near coincident points. This is shown in Figure 1. which indicates
the division of the a,b plane into the four zones corresponding to
a maximum braking of about $5.5ms^{-2}$ and maximum acceleration of
about $1ms^{-2}$. In Figure 1 it can be seen that all the drivers
should be able to negotiate the signals successfully. It could be
argued that there was some deliberate violation of the signals by
redrunners who were in the stop situation, but this does not
explain the number of redrunners in the go and option zones who
would not need to violate the signals in order to negotiate the
signals successfully.

Figure 1. a,b representation: No effect of dithering

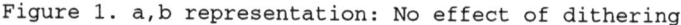
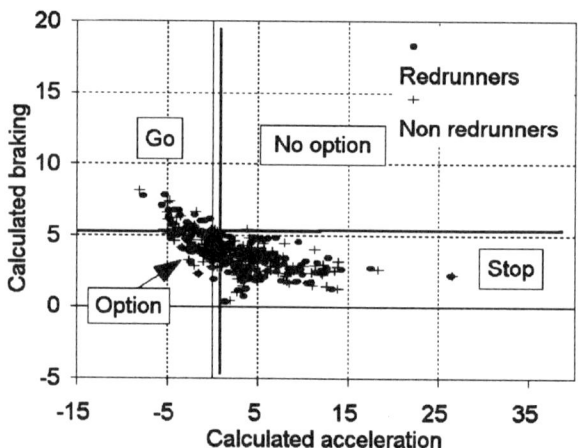

Consider the case of a dithering driver. In general the
observed dithering involves the vehicle slowing initially then
accelerating or carrying on at the same speed. The implication of
this observed behaviour is that the driver has initially made a
decision to stop, has acted on that decision, and then has
reversed the decision and then acted again . In terms of the
physical situation this has two implications a) effective t is
increased and b) effective v is reduced.

This can be incorporated into the calculations of a and b.
First consider the case of increased t. Assume that the effect of

dithering is to increase t from 1s to 2s. A plot with drivers who
dithered having t set at this value produces the plot shown in
Figure 2. As can be seen, a number of the red runners have been
shifted into the no option zone; however some redrunners are still
in go or option zones.

Figure 2. a,b representation: t=2s for dithering drivers.

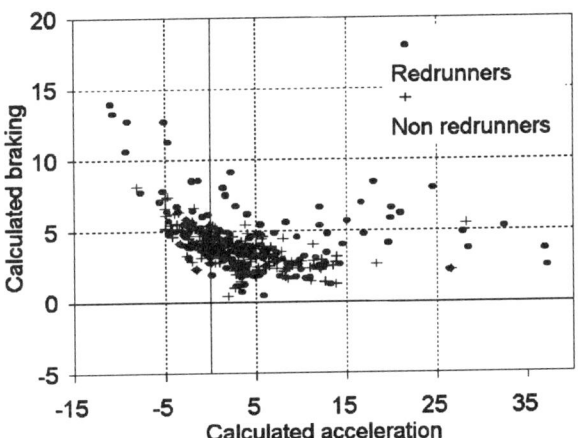

If dithering affected only the assumption regarding t then,
although some of the red runners who could have cleared under the
original assumptions were shifted to the no option situation,
there are many which were not. The plot in Figure 3 shows the
effect of reducing the effective v of drivers who dither by 10 per

Figure 3. a,b representation: t=2s, v=0.9v for dithering drivers.

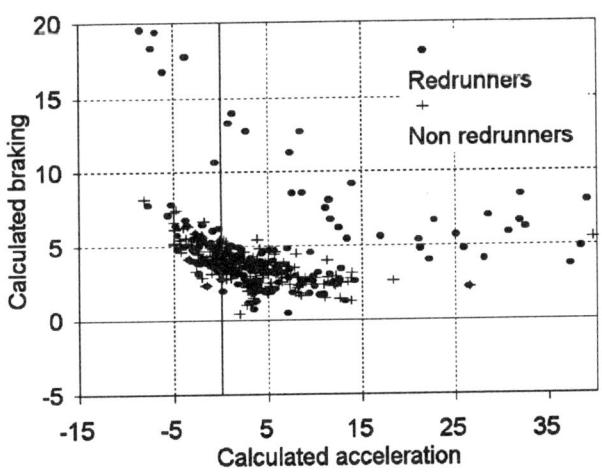

cent. In this plot it can be seen that a greater number of (but still not all) redrunners have been shifted to the no option situation.

It can be seen that there a number of redrunners who dithered which should have been able to stop by braking at less than $4ms^{-2}$. These drivers are likely to have either severe constraints on the level of braking which they could use or are deliberately violating the signals.

The theoretical influence of factors which effect the situation with which a driver is faced has thus been used to demonstrate that apparent inconsistencies between predicted and observed behaviour may be explained by situation specific factors, for example increased t and decreased effective v as result of dithering. This suggests that difficulties in decision making on the part of the driver contribute to redrunning, and thus identifies an area in which research could make a positive contribution to reducing red running.

ACKNOWLEDGEMENTS

The author gratefully acknowledges the Rees Jeffreys Road Fund for funding this work and Richard Allsop for advice and guidance.

REFERENCES

Aitken, M., Anderson, D., Frances, B. and Hinde, J., (1990) Statistical Modelling in GLIM Oxford Science Publications, Clarendon Press, Oxford.

Allsop, R.E., Brown, I.D., Groeger, J.A. and Robertson, S.A., (1991). Approaches to modelling driver behaviour at actual and simulated traffic signals. Contractor Report CR-264. Transport and Road Research Laboratory, Crowthorne, England.

Baguley, C.J. and Ray, S.D., (1989) Behavioural assessment of speed discrimination at traffic signals. Research Report RR-177. Transport and Road Research Laboratory, Crowthorne, England.

Chang, M.S., Messer, C.J. and Santiago, A.J., (1985) Timing traffic signal change intervals based on driver behaviour. Transportation Research Record, 1027, 20-29.

Robertson S.A. (1990) Dynamic framework for modelling driver behaviour at traffic signals. Transport Studies Group Note. University College London. (Unpublished).

Robertson S.A. (1992) Driver behaviour at traffic signals: a qualitative analysis. In G.B.Grayson and J.F.Lester (Eds.) Behavioural Research in Road Safety. Transport and Road Research Laboratory, Crowthorne. (In press).

A STUDY OF COOPERATION EXTENDED TO TRAPPED MERGING DRIVERS

TAY WILSON
MARIE GODIN

Laurentian University
Ramsey Lake Road
Sudbury, Ontario, Canada P3E 2C6

Undesirable, counter-effective driving practices in
merging situations, where the number of lanes suddenly
decreases, can lead to increase delays and driver
annoyances which may ripple into other costly
undesirable driving activities. This study of co-
operation extended to seven hundred trapped merging
drivers revealed that assistance was greatest during
evening rush hour; but this cooperation often took the
form of vigorous braking. Cooperation extended to
trapped merging drivers was greater on Wednesdays than
on Mondays or Fridays. Finally, the probability of
assisting a trapped merging driver was, perhaps
counter-intuitively, lower when an opportunity existed
for the assisting driver to change lanes; thus
possibly indicating some degree of inertial driving.
Recommendations for locale specific driver improvement
programs based upon the findings of this study with
the aim of developing desirable driving practices
which increase traffic flow and safety are made.

Wilson (1983, 1991) has noted that in most endeavours to
intervene successfully in changing undesirable practices, a major
component is the identification and assessment of complex of
activities in which individuals are presently engaged, in a detail
that is as rich in the appropriate context as possible. A major or
perhaps the major part of the context in which driving occurs is
social. It is thus surprising that relatively little effort has
been devoted to the social description of what drivers are doing
in particular locales. This approach to might, in today's parlance
be called studying the ecology of Driving has been applied to the
study of pedestrian crossing incidents by Wilson and McArdle
(1992). Undesirable, counter-effective driving practices in
merging situations, where the number of lanes suddenly decreases,
can lead to increased delays and driver annoyances which may
ripple into other costly undesirable driving activities. In order
to begin to provide a data base upon which desirable driving
practice countermeasures might be developed, there is here
reported a study of the circumstances under which assistance by
passing drivers is provided or not provided to trapped merging
drivers upon the occasion of the disappearance of one of the
traffic lanes. The location chosen was a busy intersection in a

moderate sized Canadian city.

METHOD

Figure 1 depicts the two locations in which merging activities were observed, at one of the busiest intersections of Sudbury, a Canadian city of 100,000. After a pilot study to choose the hours most representative of morning and evening rush hour, a suitable period of low traffic density and to develop scoring protocols, 719 trapped would be merging drivers over 54 observation periods (half in each location) each lasting 25 minutes during a three week period were made. At location 1, subjects were westbound drivers becoming trapped when attempting to enter the traffic stream from the merge lane because of traffic from Highway 17 East during one green light or from traffic from Second Avenue during the next green light. Each 25-minute period of observation contained approximately 30 green lights. At location 2, subjects were eastbound drivers becoming trapped after crossing the intersection on the green light. This could only occur in the instances when there were vehicles in both lanes or if one or other lane was actively used for passing. There were approximately 19 green lights per observation period. The number of approaching vehicles who were in a position to assist a trapped merging driver but who failed to so do was scored in each case.

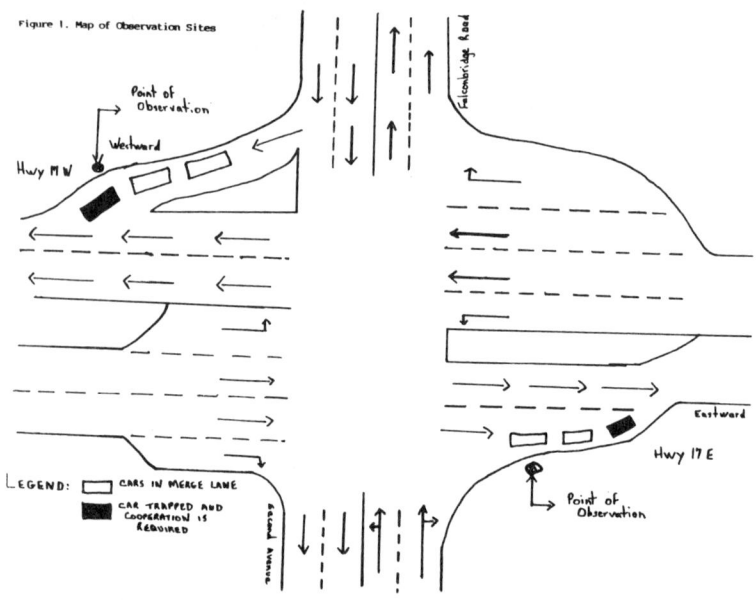

Figure 1   Map of Road Geometry Where Merging Lanes Disappear at Two Observation Sites in Sudbury, Canada

Observations were carried on for 25 minutes at each site, in an alternating sequence, on Mondays, Wednesdays and Fridays during the periods 8-9 AM, 9:45-10:45 AM and 4:30-5:30 PM. Traffic volume during the observation periods was estimated by averaging the count of vehicles for the first and last five minutes of each observation period. The overall average traffic volume for an observation period was 633 for the AM rush hour, 249 for the mid-morning period and 513 for the PM rush hour. Each case was scored on the following variables: site location, day and time, number of trapped driver in merge lane, whether the potential assisting driver had the opportunity of changing lanes and method of assistance provided to the trapped driver (slow down, lane change, full stop or forced braking). The reliability of the rating scheme was estimated by having a second observer scoring drivers during the second week of observation. No instances of disagreement in scoring between the two raters occurred during this time.

RESULTS

Of 719 trapped merging drivers over three weeks, 433 received assistance which allowed them to enter the traffic stream and 286 were forced to wait until the end of traffic. Subjects were passed by a significantly greater number ($F(1,717)=162$, $p < 0.01$) of drivers when cooperation was not occurring (M=2.18, s=1.65, N=286) than when it was occurring (M=0.77, s=1.31, N=433). Drivers receiving assistance were evenly distributed between the two locations (217 westbound and 216 eastbound) but 96% of drivers forced to wait for traffic were from the westbound merge lane (Chi-square $p < 0.01$). There was a significant time of day difference in the numbers of vehicles passing trapped drivers without providing assistance ($F(2,716)=4.99$, $p < 0.01$,) over the three observation times: non-rush hour mid-morning time (M=1.68, s=1.68, N=156), the morning rush hour (M=1.29, s=1.56, N=236) and evening rush hour (M=1.19, s=1.59, N=327). Post hoc Scheffe tests revealed that the number of vehicles assisting trapped drivers was significantly higher ($p < 0.05$) during the PM rush hour than during the non-rush hour morning period. In the evening rush hour about twice as many vehicles were assisted as were kept waiting (223 versus 104) while in the morning rush hour (129 versus 107) and morning non-rush hour (81 versus 75), the numbers of each category were approximately equal. This difference was chi-square significant ($p < 0.001$). No significant overall week-day differences in proportions of assistance to non-assistance for trapped merging drivers were found.

When there was opportunity for an assisting lane change by a potentially co-operating vehicle, more than twice as many trapped vehicles were forced to wait until the end of traffic as were helped (186 versus 89) while, when there was no opportunity for lane change, the number of forced-to-wait vehicles compared with assisted vehicles was 89 versus 128. This difference was chi-square significant ($p < 0.001$).

Table 1 shows the chi square significant ($p < 0.01$) distribution of assistance to trapped driver by type of assistance (slowing down, changing lane, sudden strong braking and coming to full stop). As can be seen from the table, during evening rush hour, two to three times the proportion of trapped drivers were assisted by slowing down compared with the proportion assisted during either morning rush hour or mid-morning periods. Furthermore, during the evening rush hour, two to three times the proportion of assisting drivers braked suddenly and hard as did in the other time periods.

Because of the great difference in proportion of offered assistance between west bound and east bound sites, the data was now analyzed separately. A significant time of day differences in mean numbers of drivers per session forced to wait until end of traffic was found in the eastbound lane ($F(2.224)=9.23$, $p < 0.01$); with the non-rush hour morning session (M=1.18) having significantly (Scheffe test, $p < 0.05$) more drivers forced to wait than in either the morning rush hour (M=0.20) or in the evening rush hour (M=0.43). Moreover, for the westbound site, the proportions of drivers forced to wait compared to those assisted on Monday (89 versus 58), Wednesday (78 versus 89) and Friday (108 versus 70) differed significantly (Chi-square, $p < 0.02$) with Wednesday having the lowest proportion of drivers forced to wait. No such differences were observed in the eastbound site.

Table 1. Type of Assistance Provided to Trapped Drivers by Time of Day
(Chi-square 22.36, df=6, p less .01)

Assistance Method	Morning Rush	Mid Morning	Evening Rush
Slowing Down	69	54	159
Changing Lane	18	8	10
Full Stop	19	5	12
Brake Vigorously	23	14	42

DISCUSSION

The results indicate the following tentative conclusions at the location studied. The probability of a trapped driver being assisted is highest at the evening rush hour. The probability of assisting trapped drivers is, counter intuitively, lower when the potential assisting driver has an opportunity to move into a free lane to assist the trapped driver. This result supports the notion of a state of relatively inertial driving, particularly when traffic volume is lower. In evening rush hour, there is a relatively higher amount of assistance to trapped drivers by means of slowing down and, more dramatically, by vigourous braking. This latter result suggests relatively increased aggression on the part of trapped drivers at this time of day. Finally when the two observation sites were separated, one site showed a day of the week effect with Wednesday being the day of greatest cooperation.

A locale based driver improvement initiative based upon those results (see Wilson, 1991) would presumably alert drivers to the danger of sudden braking during evening rush hour at merging sites, encourage drivers to drive on Monday and Friday more like they drive on Wednesdays with regard to merging assistance, and alert drivers to the dangers of inertial driving (failure to lane change cooperatively) during relatively low traffic times.

REFERENCES

Wilson, Tay, 1983. Undesirable aspects of British driving. In
    Road Safety, Praeger, London, pp. 206-9.
Wilson, Tay, 1991. Locale Driving Assessment - A Neglected Base

of Driver Improvement Interventions. In <u>Contemporary Ergonomics</u>, Praeger, London, pp 388–393.

Wilson, Tay, and McArdle, G., 1992. Driving Style Caused Pedestrian Incidents at Corner and zebra Crossings. In <u>Contemporary Ergonomics</u>, Praeger, London, pp 388–393.

DRIVING RESEARCH AND THE IOWA DRIVING SIMULATOR

by James Buck*, James Stoner**, John Bloomfield# and Tom Plocher#

*Industrial Engineering and **Civil-Environmental Engineering Departments, The
    University of Iowa, Iowa City, IA 52242, USA
#Honeywell Systems Inc., Minneapolis, MN USA

Public automotive transportation is receiving renewed emphasis
worldwide. Both the public and Government officials are aware that
ergonomics poses several critical problem. This situation creates a
challenge to ergonomic researchers because of the problem magnitude
and inherent dangers. The University of Iowa focused on this
challenge by going after and winning the National Advanced Driving
Simulator (due in 1996), building the Iowa Driving Simulator in
1992, accepting Honeywell System's invitation to go after the
Federal Highway Administrations "Automated Highway Project"
and winning it. This paper presents that story.

INTRODUCTION

Work on the Iowa Driving Simulator IDS started around the beginning of 1991.
Several University of Iowa researchers were working on the National Advanced
Driving Simulator NADS concept and mechanical designs for some military
vehicles at the time. Consequently, these people had some design problems
associated with building NADS and some operational testing problems with the
other vehicles in clearly mind. Then a sudden opportunity arose whereby the
University could acquire a computer driven visual image generator system from the
US Dept. of Defense. With this promise in hand and the University's commitment
to build a simulation bay on a current building, IDS was launched. It seems like
irony but in the movie called, Field of Dreams," the leading character determined
that he would build a lighted baseball park on a location out in the middle of farm
country and "They would come!" These researchers and the University of Iowa
took on an analogous attitude. Oddly, that fictional baseball park, a tourist
attraction, is not 50 kilometers from the University.
   The story below tells:
      1. How IDS has progressed since this start.
      2.Why Driver-in-the-Loop-Simulation (DILS) is needed for ergonomic
         research.
      3.Some of the implications and capabilities on research planning and
         experimentations with a moderate level DILS. and
      4.Challenges posed in the Automatic Highway Research Program.
Two conflicting statements summarize this story: The first is an ancient Asian
saying that, "The man who rides the back of a tiger is afraid to get off." and the
second is that the ride and view are fascinating.

THE IOWA DRIVING SIMULATOR

The Iowa Driving Simulator IDS has several major subsystems consisting of:
      1.The Automobile Cab which serves as part of the immediate driver
         interface,
      2.A Graphical Database which provides visual images for the driver outside
         of those in the cab,
      3.An Audio-Vibratory System to provide other sensory driving cues,
      4.A Motion Base for the cab to give movement and orientation cues to
         drivers, and

5.System Control and Integration which interconnects to and controls these subsystems.
Various aspects of the automobile-driver models of IDS are embedded in those various subsystems. Some of those models are part of the driver-to-system communications. Drivers steer the vehicle, apply braking, select direction and gearing. Accordingly, there are steering models, braking models, and drive-train models which couple elements of the cab subsystem to Systems Control and Integration. There are also models for communications from the system to the driver. Those models include tire-suspension models, aerodynamic models, and weather models, in addition to all of the models needed to display sensory input to drivers.

While the acquisition of the graphical database triggered the IDS project, it should be stated that this group of researchers had experiences in automotive suspensions, highway systems, computer controllers, and simulation for ergonomics research. However, automotive driving simulation posed some extreme challenges because: 1.Events impose much higher frequencies than aircraft simulation, 2.Visual scenes are much closer than with aircraft, and 3.Ergonomic requirements are far more severe technologically. Some breakthroughs in the proposed NADS development helped greatly and the high quality graphics of the graphic database provided much of the needed visual improvement. One breakthrough was in parallel computing efficiency. That technology provided the means to achieve super-fast computations for these models without having to go to approximations in those computations. Another breakthrough was in discovering how to breakup model computations for rapid and accurate updating.

The Cab Subsystem

Consists of an instrumented Ford Taurus less engine and part of the trunk. Four people fit comfortably in this principal vehicle of DILS. Potentiometers were added to the brake and accelerator pedals, as well as the gear-selector. Signals from these potentiometers went through an A/D converter to the system integrator. Also, the steering wheel was fitted to a digital encoder for direct hookup to the system integrator. That integrator in turn controlled all of the cab instruments. This subsystem provided part of the communications between the driver and the car system.

Graphic Database

Consists of an Evans and Sutherland CT6 model with texture and streaming effects. Those two effects add further cues of motion as perceived by the driver. There is a domed screen around much of the cab that reflects projected images from this visual system. Each projector generates about 50 degrees horizontally and 40 degrees vertically. Three projectors provide the forward images while one projector focuses for rearview scenes. The visual update of this system is every 16 2/3 ms; yielding a 60 Hz refresh rate.

This visual database now has about 50 miles of urban expressway and rural roadway. Both roadways have full highway markings and the country roadway additionally has crossroad intersections with stop signs and in one place a traffic light. There are other signs and various farm building along those roadways. A few cloverleaf interchanges connect the rural and urban roadways. In addition, there are other vehicles in this graphics database which move about in intelligent ways; such as gradually slowing down as they approach stop signs and avoiding collisions with each other or the principal car. Turn-signals and brakelights work on these other cars all the time unless specifically turned off. Head- and taillights appear on these cars.

The Audio-Vibratory System

Consists of 8 high-amplitude and high-frequency speakers located on the cab roof. Digital-analogue tapes were made at various wind speeds on the road, on various roadways, both with and without passing automobiles. These tapes were made at similar locations to the speaker locations so as to maintain stereo-directional effects in the principal vehicle. In playback, these sound are dithered, blended, and amplified before being added to the speakers as directed by the system integrator. This sound system as a 20 to 20K Hz frequency spectrum and a dynamic range of 95 db. Accordingly, this system covers most of the population. It is even capable of getting us in trouble with OSHA. While there are plans for creating random feedback vibrations into the steering wheel, that feature did not exist at the writing of this paper.

The Motion Base System.

A CAE hexapod base has been acquired and installed in the simulation bay for motion and car orientation control. This instrument previously operated a B52 trainer which weight about 6000 lbs. Since the current cab is only about 1/3 of that load, acceleration capability has been almost doubled to about 1.5 g and the frequency response is abut 10 Hz. The movement of this hexapod is about $\pm$ 60 inches. Classical washout algorithms are currently in use with this motion base but newer adaptive algorithms are being investigated. These washout algorithms attempt to provide a similar feel in the simulated vehicle to the actual automobile in a similar situation while preventing the motion base from exceeding its limitations. Also these algorithms provide tilt-control which helps fool the inner ear in sensing acceleration changes.

The System Controller and Integrator

Consists of a variety of computers and software. One computer is a Harris host computer that coordinates all of the other subsystems. Another is a Gould computer that performs all of the graphical database calculations. There is a third Alliant computer with over twenty internal parallel computers that perform all of the rapid computations for the motion control and washout algorithms. These computers are interconnected through ethernet and to all of the peripheral controls, sensors, A/D converters, and visual displays inside of the cab.

Another aspect of the system controller and integrator is the IDS scenario control and data collecting features. Scenario control is handled in software from the host computer. There is control of simulated traffic, dynamic control of traffic lights, environmental conditions control for both weather and external illumination, and all special effects. External traffic which is in view by the driver is limited to 14 displayed vehicles; each preprogrammed with behavioral characteristics such as gap acceptance tendencies or preferred headway. However, additional vehicle can be displayed but without independent action.

Tracking and data collection are also part of this system controller and integrator. Highway segments are classified by the type of roadway, its degree of curvature in each direction, and any superelevation (i.e. banking). There is also a series of available data which can be directly recorded as an experiment unfolds. Users have a variety of standard options of data including: root-mean-square from the lane centerline, lateral within-lane movements, line edge crossings, locations of nearby traffic, speeds, and other conditions of the car.

WHY DRIVER-IN-THE-LOOP SIMULATION IS NEEDED IN ERGONOMICS!

While there is considerable controversy about the degree of fidelity that is needed for DILS in various applications or experimental conditions, practically no ergonomists argue that DILS of some level of fidelity is not needed (Buck, Alford,

Deisenroth, 1978). Many design principles of aircraft design and training came from DILS. Some of the safety procedures of nuclear power control is from DILS. With automobiles, one of the clearest needs for DILS is experimental safety. Experimental situations where people can easily lose lives or sustain serious injuries are clear cases when DILS must be used. However, there are other reasons for using a DILS for ergonomic research besides safety. One is that DILS provides consistency that can not be obtained with road research. Driving tracks can not assure consist roadways, weather, traffic, nor can any of these three important features be controlled without DILS. Another reason is speed in replicating conditions with DILS which can not be accomplished with test tracks or regular roadways. Driving in situations involving traffic control lights can be perfectly controlled even with minor variations in speeds or the vehicles or slight variations of timing and sequencing using DILS but not otherwise (Buck,1977).

SOME IMPLICATIONS OF DILS IN DRIVING RESEARCH!

Simulators of even moderate levels of fidelity are expensive to own, startup, and operate. Most ergonomists are used to low-cost experiments with task abstractions. In such cases subjects are brought in one at a time at the subject's convenience and tested. The high cost of DILS requires system operators to keep high utilizations of the simulators with little or no lost time between successive subjects. Accordingly, simulator are often scheduled in time blocks of 2 or 4 hours. Because of those high costs per hour, experimenters try to pack as much experimental inquiry into each time block as possible. Since the cost per hour of simulator time is 1 to 2 orders of magnitude greater than the costs of subjects' times, experimenter are enticed to play the airline game of overbooking subjects. Therefore, no-show subjects or subjects who get sicks, or have other problems, there are always additional subjects available to keep the simulator experiments ongoing. With more subjects available than are likely to be needed by the simulator, there is a temptations to run other experiments on those subjects who are serving in standby status. In any case, DILS of moderate fidelity or greater requires a different subject acquisition, use, and payment strategy than the ones most familiar to ergonomists.

SOME CHALLENGES IN THE AUTOMATED HIGHWAY RESEARCH PROJECT.

Several concepts of driving have been proposed which is believed to increase highway safety and trip expectancy simultaneously with automobile throughput (Alicandri and Moyer, 1992; Saxton, 1980). Thus, more cars can be carried to urban centers at peak densities without losses in safety and without building more highways. One of the concepts to achieve this result is to have groups of cars traveling along together at speeds of 1103 to 153 kilometers per hour (70 to 95 miles per hour) separated by forward distances of 1 to 3 meters and lateral distances a fraction of a meter and to do that while all are under automatic control. If that group of cars is at a uniform speed and separation and it is free from nonautomated vehicles, then that concept has improved feasibility. Major problems with this concept are getting vehicles into and out of the automated states; especially getting cars to disperse out of those groups under emergency conditions. A little reflection makes it obvious that special highway facilities need to be dedicated to cars that are capable of these automated actions while there are other facilities and cars which are not. Therefore, experiments need to be made to see if a majority of a population of drivers can manually perform needed vehicle control actions in joining into automated groups and dispersing from those groups under a very wide combinations of circumstances with varying kinds of special highway facilities. It is clear that this form of research would be difficult if possible to do without some form of DILS.

Some additional problems are indicated in the following test scenario. -- Consider a group of three or more vehicles and the test vehicle is the first, last, or middle of the group. The group is expected to travel in automated mode on a dedicated lane next to one or more lanes of cars travelling in manual control mode. Additional groups of cars may be as close as 2 sec. travelling time ahead or behind. If the driver in the automated car is instructed to take control and manually direct the vehicle into the adjacent manual-vehicle lane. It is clear that difficulty in performing that task depends upon: 1.Automated lane speed, 2.Location of adjacent automated car groups ahead and behind. 3.Position of the automated car within its group. 4.Mode of giving forewarning and timing before takeover. 5.Speed and speed variability of adjacent nonautomated cars at the time of automated car takeover. 6.Location of the nearest nonautomated vehicle in the nonautomated lane. and 5.Maneuvers of the other cars in the automated group following takeover. Even without considering some of the other variables indicated earlier (e.g. various highway facilities), this list clearly indicates a major experimental difficulty is the vast numbers of experimental conditions. This situation brings about another challenge of finding exploratory experimental designs which can efficiently define the range of the values of principal variables, criteria sensitivity within that range, and the identification of those first-order interactions with either very weak or very strong sensitivity.

REFERENCES

Alicandri, E. and Moyer, M. J., 1992, "Human Factors and the Automated Highway System," Proceedings of the Human Factors Society, 36th. Annual Meeting, Atlanta, GA, USA

Buck, J.R., 1977, "New Strategies in Ergonomic Research and Design," Proceedings of the American Institute of Industrial Engineers, Spring Conference, Dallas, Texas

Buck, J.R., Alford, E.C., and Deisenroth, M.P., 1978, "Man-in-the-Loop Simulation," SIMULATION, Vol. 30, No. 5, May 1978, pp. 137-144.

Saxton, L., 1980, "Automated Highway System - Considerations for Success," Vehicular Technology Society, IEEE

THE BENEFITS OF "PRE-INFORMATION" IN ROUTE GUIDANCE SYSTEMS DESIGN FOR VEHICLES

G.E BURNETT and A.M PARKES

HUSAT Research Institute
The Elms, Elms Grove
Loughborough
Leics. LE11 1RG.

This work compares methods of presenting route guidance
information to drivers. Field trials were conducted,
with each of sixteen subjects undertaking two conditions,
each on a different route, one with visual information
only and the other with both visual and auditory
information. Half the subjects were given pre-
information messages warning them of the distance to the
next turn. Dependent variables measured were perceived
mental workload, frequency and duration of glances to the
display, and subjective views. Presentation of visual
and auditory information was the preferred mode, reducing
both mental workload and time spent looking at the
display. Pre-information helped in the navigation task.

INTRODUCTION

There is currently extensive world wide research and
development being carried out regarding in-car navigation or
route guidance systems, such that these systems will be widely
available in the near future. In the worst case these systems
could overload the driver with additional information and
seriously distract the driver from the primary task of driving
safely. Professional groups of automotive workers and
prospective manufacturers are therefore looking to the human
factors community for recommendations and guidelines on usable
and safe interfaces.

It is clear that systems should provide the simplest possible
display in terms of information extraction. If distraction is to
be minimised, it is necessary that the mode of information
presentation permits ease of perception, identification and
interpretation of navigating information. Various media (text,
symbols, maps, speech, or a combination of these) can
potentially be used to present such information to drivers.
Streeter (1986) concluded that speech was the 'best' mode in
terms of reduced navigational errors and distraction, although
she made her comparison with maps and not alphanumeric or

symbolic displays. Parkes and Coleman (1990) compared the
effectiveness of directional symbols, printed text and voice
simulation in a simulated route guidance task, and found the
last was the easiest method to follow. However, as pointed out
by Southall and Twiss (1988), although such voice orientated
displays offer no visual distraction, this does not necessarily
mean that they are less demanding of attention or perception.
        This experiment has sought to compare the effects of two
different modes of information presentation (visual information
only and visual and auditory information) on driver behaviour.
In addition, the effects on driver behaviour of giving drivers
advance warning or "pre-information" regarding the next turning
manoeuvre, have been investigated.
    The field was chosen as the environment in which to undertake
this experiment. It could then be confidently predicted that
the captured data would correspond to real road situations.
However, as Parkes (1991) discusses, road trials are inevitably
associated with reduced control of variables. A rigorous
experimental design and consistent procedure were necessary to
reduce these effects.

METHOD
    Sixteen volunteer subjects took part in the experiment; eight
male and eight female. 75% of subjects were under 30 years old.
Subjects drove an instrumented Saab 9000i test vehicle, on two
matched routes on suburban and rural roads, of around 20 minutes
duration each. Two video cameras were mounted in the vehicle,
one to capture drivers head and eye movements, one to record the
external road scene.
    Route information was presented to the driver by either
visual display alone, or in combination with an auditory
display. Visual information took the form of symbols presented
on a ten inch Macintosh LCD computer screen, mounted on the
vehicle dashboard centreline at a height of the steering wheel
centre. The orientation could be adjusted to the subjects'
preference. Hypercard software was used to generate the visual
and auditory displays under the direction of an experimenter in
the rear passenger seat.
    A two-way repeated measures design enabled each subject to
experience two conditions, each on a separate route, one with
visual guidance only, and one with visual and auditory guidance.
Half the subjects received pre-information of the forthcoming
manoeuvre during the route, half did not.
    Symbols presented to the driver were of three types:
    straight ahead - the driver should continue to follow the
current road.
    preparation - first given around 100 yards before a
required manoeuvre and following an alerting auditory beep. The
preparation symbols incorporated count down information to the
next turning which could be at a turn off road, T-junction,
crossroads or roundabout.
    pre-information - presented following an auditory beep,
and immediately following a turning manoeuvre. This showed the
form of the next manoeuvre and the approximate distance to it.
The symbol remained visible for five seconds before reverting to
the straight ahead symbol.   Examples of the three different
symbol types follow:-

Figure 1. Examples of the three symbol types

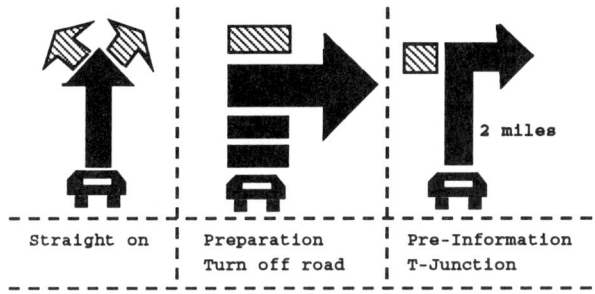

| Straight on | Preparation Turn off road | Pre-Information T-Junction |

Auditory messages were recorded digitised speech prompts, presented following an auditory beep, and simultaneous with the start of the visual presentation. Only simple messages were used.

The following dependent measures were taken: time to complete the route, number and type of navigational errors, visual glance frequency and duration, and subjective indices of mental workload; NASA-RTLX (Hart and Staveland 1988), and Modified Cooper-Harper Scale (Wierwille and Casali 1983), and a subjective questionnaire addressing usability and acceptability. Glance durations were measured to an accuracy of 1/25th. of a second, and were defined as the time between the eyes leaving the road ahead to view the display screen, and returning to the road ahead.

Prior to the experiment, subjects were told of the nature of the trials, and were instructed to drive as normally as possible. Familiarisation involved the subjects driving three routes, each of approximately ten minutes duration. The first familiarised the driver with the cars' controls, the other two were driven prior to each experimental route, and familiarised the subjects with the forthcoming route guidance messages. Subjects were free to ask any questions during the familiarisation stage.

RESULTS

Table 1 shows that during a journey subjects spent significantly more time looking, and made significantly more glances towards the route guidance display when a.) Visual Information only was presented, compared to when both visual and auditory information was given, and when b.) Pre-Information was given, compared to when it was not.

Table 1. Visual behaviour of subjects

```
-------------------------------------------------------------------
                                        Frequency    % time spent
                                        of Glances   looking at
                                                     the display
-------------------------------------------------------------------
                        Visual Information
Pre-Information              Only            66           6.5
   given              Visual & Auditory
                         Information         49           4.0
-------------------------------------------------------------------
                        Visual Information
No Pre-Information           Only            55           3.6
   given              Visual & Auditory
                         Information         19           1.5
-------------------------------------------------------------------
```

The average duration of a glance made towards pre-information symbols (equal to 0.93 sec) was significantly higher than for any other symbol type. The average duration of a glance for all symbol types was 0.78 sec.

Table 2 reveals that subjects made significantly more glances towards the preparation symbols when they were given pre-information compared to when they were not.

Table 2. The average frequency of glances made towards preparation symbols as a function of pre/no pre-information

```
          -------------------------------------------------------
                                          Frequency of Glances
          -------------------------------------------------------
          Pre-Information messages given            2.1
          No Pre-Information messages given         1.7
          -------------------------------------------------------
```

Table 3 shows that subjects rated the driving and navigating tasks as significantly less demanding mentally, for both the NASA-RTLX and Modified Cooper-Harper (MCH) scales, when visual and auditory information was presented compared to visual information only. The NASA-RTLX scale varies from 0 (low perceived mental workload) to 100 (high), whereas the MCH scale values in table 3 correspond to minimal driver effort (0.5), and mild difficulty in the driving and navigating task (2.0).

Table 3. Overall perceived mental workload

```
          -------------------------------------------------------
                                Mean NASA-RTLX    Median MCH
                                   values           values
          -------------------------------------------------------
          Visual Information Only        29            2.0
          Visual & Auditory Information  22            0.5
          -------------------------------------------------------
```

The questionnaire given to subjects revealed that the visual and auditory information was perceived to be significantly easier and less stressful to navigate with, compared to the visual information only. Furthermore, a clear majority of subjects (81%) had either a preference or strong preference for the visual and auditory mode of information presentation. All subjects who were presented with pre-information messages found them 'helpful' or 'very helpful' in the navigation task.

DISCUSSION

When both visual and auditory information was presented, fewer single glances were made and less time was spent glancing towards the route guidance device, compared with the display of visual information only. Overall mental workload values were reduced, and drivers rated the combined mode as easier and less stressful with which to navigate. Furthermore, they overwhelmingly preferred it to the visual information only conditions.

These findings suggest that a combined mode of information presentation should be adopted for the route guidance systems of the future. However, whilst maintaining that this may well be the case, it is necessary to discuss some important design issues that arose during the course of these experiments.

The auditory components were deliberately designed to be simple. However, despite the simple nature of the auditory commands, many subjects still commented on the temporal and transient quality of auditory information, thus requiring sustained concentration. This point could be resolved to a certain degree with a system that incorporated a repeat function. The responses suggest that auditory messages should be as short as possible if concentration is to be minimised. An example of a suitable message would be "Turn left at the end of the road". Any further contextual information required to make a turn successfully and be reassured that the correct manoeuvre had been made could then be contained in suitably designed visual symbols.

The presentation of pre-information messages increased the frequency of glances, and the time spent glancing towards the route guidance device, an obvious result of extra information being presented to drivers. The question then arises as to whether the extra information is a safe amount when it leads to visual attention being drawn from the road ahead. Subjective ratings indicate that drivers did feel both safe and comfortable when using the pre-information compared to when they were not. However, despite rating the information as being at least helpful in the navigation task, subjects did not find it led to fewer errors. Nor was pre-information found to lessen the neutral demand of the navigation task. This point raises the question of whether the extra information is actually useful, or indeed a hindrance in navigating. Some subjects did, at least initially, confuse the preparation and pre-information symbols.

When pre-information messages were presented, more glances were made towards the preparation symbols than when they were not. This result is somewhat surprising, since we would perhaps predict that glance frequency would decrease as a result of subjects being pre-warned of the form of the turn they were about to make. This result may have arisen because subjects still felt some benefit from regular confirmation glances to the display when the primary task demands allowed them to do so.

CONCLUSIONS

According to a number of criteria, the combined mode of visual and auditory information presentation was found to perform better than the visual information only. However, several subjects commented on its temporal nature, even for such simple messages used here. It is therefore recommended that auditory commands are kept short and the majority of the contextual and conformational information is contained in suitably designed symbols.

Pre-information was overwhelmingly considered to be useful by those who were presented with it. However, some subjects confused the pre-information and preparation symbols. A more appropriate display might contain pre-information symbols that were permanently present, and presented in a different format (in terms of size, shape etc.) from the preparation or immediate contextual symbols

Presentation of pre-information messages was found to increase the number of glances made towards preparation symbols. If there is a comfort factor associated with glancing at an in-car display then this effectively questions the use of glance frequency as a suitable measure of screen complexity.

REFERENCES

Hart, S.G. and Staveland, L.E., 1988. Development of NASA-TLX (Task Load Index) : Results of Empirical and Theoretical Research. Human Mental Workload. eds. P.A. Hancock and N. Meshkati (North Holland: Elsevier Science publishers), pp. 139-183.

Parkes, A.M. and Coleman, N., 1990. Route Guidance Systems: A comparison of methods of presenting directional information to the driver. In Proceedings of the Ergonomics Society's 1990 Annual Conference, ed.E.J. Lovesey (London: Taylor & Francis), pp 480-488.

Parkes, A.M., 1991. Data Capture Techniques for RTI Usability Evaluation. In Proceedings of the 1991 DRIVE conference pp. 1440-1456.

Southall, D. and Twiss, M.K., 1988. Human Factor considerations in the design of in-car navigation systems, In Proceedings of the 1988 ISATA conference, Florence, Italy.

Streeter, L.A., Vitello, D., and Wonsiewicz, S., 1986. How to tell people where to go: Comparing navigational aids. International Journal of Man/Machine Interaction, 22, pp. 549-562.

Wierwille, W.W., and Casali, J.G., 1983. A validated scale for global mental workload measurement applications, Proceedings of the Human Factors Society Annual Meeting, pp 129-133.

CARPHONE USE AND MOTORWAY DRIVING

A.M.PARKES, S.H.FAIRCLOUGH AND M.C.ASHBY

HUSAT Research Institute
The Elms, Elms Grove
Loughborough
Leics. LE11 1RG.

This study sought to identify changes in driver behaviour
due to handsfree telephone conversations carried out
during motorway driving. 18 volunteer subjects either
drove in silence or whilst completing verbal tasks on a
carphone. No evidence for a change in driving behaviour
in terms of speed choice, lane occupancy, accelerator use
or overtaking manoeuvres was found. However mental
workload did increase. The results are presented in
relation to other studies, and safety implications are
discussed.

INTRODUCTION

The aim of this study was to identify changes in driver
behaviour due to a handsfree telephone conversation carried out
during a motorway journey. Previous research has addressed the
problem from a variety of viewpoints, using different
independent and dependent variables within a range of test
environments. Four studies (Quenault 1968, Wetherell 1981,
Mikkonen and Backman 1988, Brookhuis et al 1991) indicate no
driving performance change whilst using a carphone, and five
(Brown and Poulton 1961, Brown et al 1969, Fairclough et al
1990, Alm and Nilsson 1990 and 1991) indicate change; most
typically a reduction in driving speed, but also a decrease in
ability to control the lateral path of the vehicle, and an
increase in driver reaction time. Of the real road studies,
changes in driving behaviour were associated with a driving task
with high task complexity i.e. complex manoeuvres or urban
traffic conditions.

Previous work at HUSAT has concentrated on handsfree
carphones in terms of conversational ability and driving ability
on urban roads. Parkes (1991), in a study investigating decision
making ability in four communication conditions (face to face,
telephone to telephone, driver to passenger and carphone to
office), showed consistent drop in structured mental task scores
when talking on a carphone. In the range of tests used, it was
found that when the driving task interacted with the carphone

task it resulted in difficulty in remembering verbal or
numerical data, and in making correct interpretations from
background information. Fairclough et al (1990) found that
drivers not only found speaking (to a passenger or into a
carphone) and driving harder than driving alone; but they also
made strategic reductions in speed in order to cope with the
increased mental workload. The increase in workload also
manifested itself in an increase in physiological activity
(heart rate) associated with the carphone condition.

Alm and Nilsson (1990) used an advanced moving-base driving
simulator to assess drivers' behavioural changes in response to
use of a handsfree carphone. It was found that a carphone
conversation (in the guise of a Working Memory Span Test) had a
negative effect on drivers' reaction times. Speed also decreased
with the onset of a call. It had a negative effect on drivers'
lane position, worse so when the tracking component of the
driving task was hard. Mental workload was rated higher when
driving whilst using the carphone.

If a consistent pattern can be gleaned from the studies
above, it would seem that priority can be given to the primary
task (driving) whilst talking on a carphone, without observable
decrements in performance, so long as some threshold point is
not reached. The reports of Mikkonen and Backman (1988)
indicates that drivers may increase their level of activation to
cope with the increased workload of doing two things at once.
However the studies of Alm and Nilsson (1990 and 1991) are
interesting because they indicate that even in low workload
driving conditions in a simulator, changes in safety related
performance could be detected.

It is impossible to control all intervening variables in a
real road experiment, nor is it possible to look at performance
at the edges of normal safety margins. Therefore it was decided
to investigate low complexity driving in the relatively
constrained environment of a three lane motorway, with moderate
traffic flow.

Previous studies would suggest that in such circumstances,
gross measures of driver behaviour such as average speed and
lane choice would be relatively unaffected by the addition of a
carphone conversation. We would also expect there to be an
increase in reports of mental workload and mental effort, and a
short term dip in speed at the onset of the carphone call.

METHOD
18 subjects were recruited from research and secretarial
staff of the research institute. The 10 female and 8 male
subjects ranged in age from 21 to 44.

Two experimental conditions were employed in the study. This
allowed for comparison between two driving situations; driving
on the motorway in silence (control condition), and driving
completing one of two tasks via a handsfree carphone (carphone
condition).

Within the carphone condition, two verbal tasks were
presented to subjects in the course of the experimental
investigation. Both involved components of numerical memory and
were quantifiable. The first type of task took the form of a
Sternberg memory test. The subject was presented with a list of
five numbers and then asked to confirm the presence or absence

of an individual digit. This task was called the 'NUMBER LIST' task.

The second type of verbal task involved the subjects listening to a list of three-digit numbers. After each number had been read the subjects had to repeat the number and state if it was higher or lower than the preceding number. This task was called the 'NUMBER JUDGEMENT' task.

The subject received six verbal tasks (three of each) in the course of the experimental trial. Two of these tasks were carried out in the stationary vehicle as a practise trial. An experimenter based at the Institute played the role of 'speaker' to administer the verbal task. Each verbal task lasted approximately two minutes in duration.

The vehicle used in the study was an instrumented Vauxhall Cavalier 2.0 GLi fitted with a Motorola 6800X handsfree cellular telephone. Video recordings were made from inside the vehicle of the external scene. As part of the 'handsfree' package, the carphone uses a small microphone and loudspeaker. The microphone was attached to the sun visor in front of the driver and the loudspeaker fitted into a recessed area above the glove box in front of the passenger. This set-up requires no extra head movement on behalf of the driver in order to hear and be heard properly whilst making a call.

The experimental trials took place on the three lane southbound M1 motorway, through Nottinghamshire and Leicestershire from Junction 26 to Junction 20. Trials took place at off-peak times in the mid-morning and early afternoon. On three occasions during the planning stage, traffic flow measurements were taken from different points along the intended route. Average traffic flows per hour were: Inside lane 515 cars, 352 HGV; Middle lane 940 cars 100 HGV; Outside lane 476 cars 0 HGV.

The experimental measures used during the study may be grouped into three categories.

a) *Vehicle parameters*: Captured electronically at a rate of 2Hz, these measures were speed before/during each telephone call and accelerator position.

b) *Observation data*: This data was scored from the video recording taken by the in-vehicle cameras. Frequency counts were taken for the number of vehicles overtaken and number of lane changes. The amount of time spent in each of the three motorway lanes was also scored in seconds.

c) *Subjective workload measures*: Two subjective questionnaires were used in the course of the investigation. The first was a version of the Modified Cooper-Harper Scale (MCH) (Wierwille et al, 1985) which had been specially adapted for the experiment and a modified version of the NASA-Task Load Index questionnaire (TLX) (Hart and Staveland, 1988).

The experiment took the form of a repeated measures design. The subjects driving behaviour was monitored over a single journey during which they received four telephone calls. Each call lasted for a duration of two minutes and was followed by an interval of three minutes. The intervals between the telephone calls were captured as the experimental control condition.

The order of presentation of the two types of verbal task (JUDGEMENT and LIST) were counterbalanced across subjects. Each task type contained three groups of experimental material, this

was also counterbalanced across subjects. The order of
presentation of the two subjective questionnaires (the MCH and
the TLX) was also counterbalanced across subjects.

The subjects were familiarised with the dashboard controls
and the handsfree carphone whilst seated in the stationary
vehicle. They then received a practise trial during which they
performed the two verbal tasks in the stationary vehicle. The
subjects then drove to a motorway junction in order to
familiarise themselves with the handling characteristics of the
experimental vehicle. This was followed by a fifteen minute
practise drive on the motorway during which they received two
shortened versions of the two verbal tasks as an additional
practise trial.

The subjects stopped at a service station in order to perform
the first half of the NASA-TLX questionnaire before performing
the experimental drive. This journey lasted approximately
twenty-five minutes, during which time the subjects received
four telephone calls. The operation of the carphone was
completely handsfree except for a single button press needed to
'end' the call after the verbal task had been completed.

On completion of the experimental journey, the subjects
stopped at a service station in order to complete the TLX and
MCH questionnaires.

RESULTS
The vehicle parameters captured were analysed for
significance ($p<0.05$) using Wilcoxon signed-rank tests.

Vehicle Parameters
Vehicle speed
No significant change in speed was observed between the
control and carphone conditions. Also, no drop in speed
immediately after onset of the call was observed. There was no
difference between the two experimental tasks.
Accelerator position
No significant changes in accelerator position resulting from
a carphone conversation were apparent.

Driver Behaviour Measures
Subjects' strategic driving performance on the motorway,
recorded on videotape, was analysed post hoc for the two minute
periods during and prior to the carphone conversation. Wilcoxon
signed-rank testing was used to test for significance ($p<0.05$).
Number of vehicles overtaken
There was no statistical significance in the difference in
the number of vehicles overtaken between the control and
carphone conditions.
Number of lane changes
There was no statistical significance in the difference in
the number of lane changes between the control and carphone
conditions.
Time spent in each motorway lane
There was no statistical significance in the difference in
the time spent in each motorway lane between the control and
carphone conditions.

Modified Cooper-Harper Scale

Wilcoxon signed-rank testing showed significant increases in both of the speaking conditions over the control condition ($p<0.01$). Mean ratings were 'Control'= 2.7, 'Judgement'= 4.1, 'List'= 3.8

Task Load Index (TLX) Questionnaire

The TLX score represents an index of subjective mental workload for driving and the two verbal tasks. Significant differences ($p<0.05$) were found between the control condition and the carphone conditions (Wilcoxon signed-rank tests). There was no difference between the carphone condition tasks. Mean scores were 'Control'= 36.93, 'Judgement'= 48.01, 'List'= 54.38.

DISCUSSION

No evidence for a change in driving behaviour was found in this study. Strategic level choices of speed; tactical level choices of lane occupancy; and operational level activity such as accelerator pedal depression, appear consistent across the experimental conditions.

Analysis of the subjective responses of the drivers revealed an increase in perceived workload when the carphone tasks were introduced. This was demonstrated by a MCH score of under 3 for the control condition (suggesting acceptable driver effort for adequate safe driving) and around 4 for the carphone conditions (suggesting moderately high effort to attain adequately safe driving). This semantic categorical difference is not large, but is a clear indication of the additional load imposed by the relatively straightforward verbal tasks involved. It might be expected that a secondary task of more complex open ended negotiation style dialogues would reveal greater subjective differences. Analysis revealed a clear difference in RTLX scores between the control and carphone tasks. The combination of MCH and RTLX results indicate that mental workload is increased by the introduction of a carphone task to levels where workload is judged to be high but not approaching the point where maximal levels of driver effort are required to maintain safe driving. It must be remembered however, that call duration's in this study were strictly limited to periods of two minutes. Sustaining the subjectively experienced high mental workload for more protracted periods, may have led to greater demands on resources.

This study did not attempt to replicate previous research (Alm and Nilsson 1991) that had shown a decrease in choice reaction time to events presented undertaking a simulated driving and carphone task; due to the difficulties of performing such tests on real roads.

CONCLUSION

The presence of a carphone call in moderate traffic motorway conditions did not result in a level of workload that produced behavioural adaptation by the drivers, as measured in terms of speed, lane choice, overtaking strategy or accelerator position. Nor was there any indication of an immediate reduction in speed as an alerting response to the onset of an incoming call.

Measures of subjective mental workload revealed that the combination of low difficulty driving, with simple verbal tasks, raised overall workload to the point where drivers began to feel unsafe whilst driving. Other studies of driving behaviour in simulators have linked a similar rise in mental workload to increased lateral deviation, and increased choice reaction time (Alm and Nilsson 1991).

The findings from this study support the current concern that the act of using a carphone whilst driving may for many drivers, in many traffic scenarios pose an increase in the risk of an accident due to a reduction in safety related driving performance.

REFERENCES

Alm, H. and Nilsson, L. 1990. Changes in driver behaviour as a function of handsfree mobile telephones: A simulator study. Report No. 47. DRIVE Project V1017. Swedish Road and Traffic Research Institute, Linkoping, Sweden

Alm, H. and Nilsson, L. 1991. The effects of a mobile telephone conversation on driver behaviour in a car following situation. Report No. 73. DRIVE Project V1017. Swedish Road and Traffic Research Institute, Linkoping, Sweden

Brookhuis, K.A., DE Vries, G. & DE Waard D. 1991. The Effects of Mobile Telephoning on Driving Performance. Accid. Anal. & Prev.. Vol 23, No4, pp. 309-316.

Brown, I.D. and Poulton, E.C., 1961. Measuring the spare 'mental capacity' of car drivers by a subsidiary task. Ergonomics, 4, 35-40.

Brown, I.D., Tickner, A.H. and Simmonds, D.C.V.,1969. Interference between concurrent tasks of driving and telephone. Journal of Applied Psychology, 53, 419-424.

Drory, A., 1985. Effects of rest versus secondary task on simulated truck driving performance. Human Factors, 27 (2), 201-207.

Fairclough, S.H., Ashby, M.C., Ross, T. and Parkes, A.M. 1990. Effects on driving behaviour of handsfree telephone use. Report No. 48. DRIVE Project V1017. HUSAT Research Institute, Loughborough University, U.K.

Hart, S.G. and Staveland, L.E. 1988. Development of the NASA-TLX (Task Load Index): Results of empirical and theoretical research. In P.A. Hancock and N. Meshkati (Eds.) Human Mental Workload. North Holland, Amsterdam, pp 139-183.

Mikkonen, V. and Backman, M., 1988. Use of car telephone while driving. Technical report No. A39. Department of Psychology, University of Helsinki.

Parkes, A.M. 1991. Drivers business decision making ability whilst using carphones. In Lovesey, E.J. (Ed.) Contemporary Ergonomics 1991. London: Taylor and Francis.

Quenault, S.W., 1968. Task capability whilst driving. TRRL Report No. LR166, Transport Research Laboratory, Crowthorne, Berkshire.

Wetherell, A., 1981. The efficacy of some auditory - vocal subsidiary tasks as measures of the mental load on male and female drivers. Ergonomics, 24 (3), 197-214.

Wierwille, W.W., Rahimi, M. and Casali, J.G., 1985. Evaluation of 16 measures of mental workload using a simulated flight task emphasizing mediational ability. Human Factors, 27 (5), 489-502.

Telecommunications

HUMAN FACTORS GUIDELINES FOR IBC* DESIGNERS

A M CLARKE & G ALLISON

HUSAT Research Institute
Loughborough University of Technology
The Elms, Elms Grove
Loughborough
Leicestershire LE11 1RG.

This paper describes the human factors guidelines that have been produced as part of the RACE ISSUE* project. The guidelines cover four main topics; Videoconferencing, Videotelephony, Multimedia and Usability Evaluation. Details of the coverage of each guidelines document is given along with details of how these may be obtained.

INTRODUCTION

Future telecommunications will allow people to communicate not only by the spoken word but visually as well. Emerging digital telecommunications, particularly broadband networks, will enable groups of people to work co-operatively while being located miles apart. The range of benefits is enormous but in order to reach their full potential, telecommunications services must be designed in a way that they are easy to use and offer users the functionality they want. Human factors has a vital role to play in ensuring that these requirements are met.

Designers of future broadband systems need prescriptive information in order to design systems from a good human factors basis, following the best advice that can be given. However, the best designed system still needs testing and designers need help in how to conduct usability evaluations. In order to help designers both to design good systems and to conduct usability evaluations with system prototypes, a number of guidelines documents have been produced on the RACE ISSUE project:

---

* IBC : Integrated Broadband Communications

* The ISSUE (IBC Systems & Services Usability Engineering) project 1065 was part of the CEC funded RACE programme - Research and Development in Advanced Communications Technologies in Europe - and acknowledgement must be given to other partners in this work, Fondazione Ugo Bordoni, Italy; Telefonica Investigacion y Dessarollo, Spain; Saritel, Italy; PTT Research, Netherlands; Telesystemes, France and Technical Research Centre, Finland.

Videoconferencing

Guidelines to support both user organisations considering videoconferencing or wishing to change the way in which they use it at present, and service providers and suppliers committed to improving the usability of videocommunications services.

Videotelephony

Human factors guidelines that cover aspects of videotelephony design such as the equipment design, video switching and user procedures both for standalone systems and those that integrate advanced videotelephony with the workstation.

Multimedia Display Characteristics

The user interface to these advanced services has been made richer but potentially more complex by the introduction of new media such as sound and moving images. These guidelines support the designer by providing advice about how the many display media can be used to best effect.

Usability Evaluation

A short introduction to human factors techniques for designers. Sections on designing user trials and data gathering offer basic advice to designers who wish to evaluate their products.

The guidelines that have been produced are based on the results of experimental trials and surveys that were conducted in the ISSUE project along with information extracted from an extensive literature search. Since all the guidelines offered need to be interpreted for the context in which they will be applied, wherever possible the guidelines are accompanied by a rationale.

In order to exploit the advantages of different presentation media, ISSUE has developed its human factors guidelines in modular form as a four volume series supported by short videos.

ISSUE sees the videos as having a dual purpose: dissemination of information and persuasion. Obviously, they contain examples of guidelines or how they are applied in a realistic design scenario, but they should also persuade the viewer of the benefits to be gained by adopting human factors techniques. It is hoped that after watching a short video, the viewer will want to read the more detailed guidelines.

Research has shown that there are many problems to overcome when presenting guidelines in text form (Mosier and Smith, 1986). To avoid many of these problems, ISSUE devised a presentation style to enhance the guidelines' readability. This uses a facing pages layout as shown in the figure overleaf.

The actual guidelines are kept as brief as possible and shown on the right-hand page. Supporting text which may be a description of the research from which the guideline has been derived, further explanation, caveats etc. is shown on the left-hand page. To help differentiate between the two different typestyles are used i.e. *Futura* 11 point for the guidelines and *New Century Schoolbook* 9 point for the explanatory text. Each information package is self-contained, and there is extensive cross referencing to information covered in other packages.

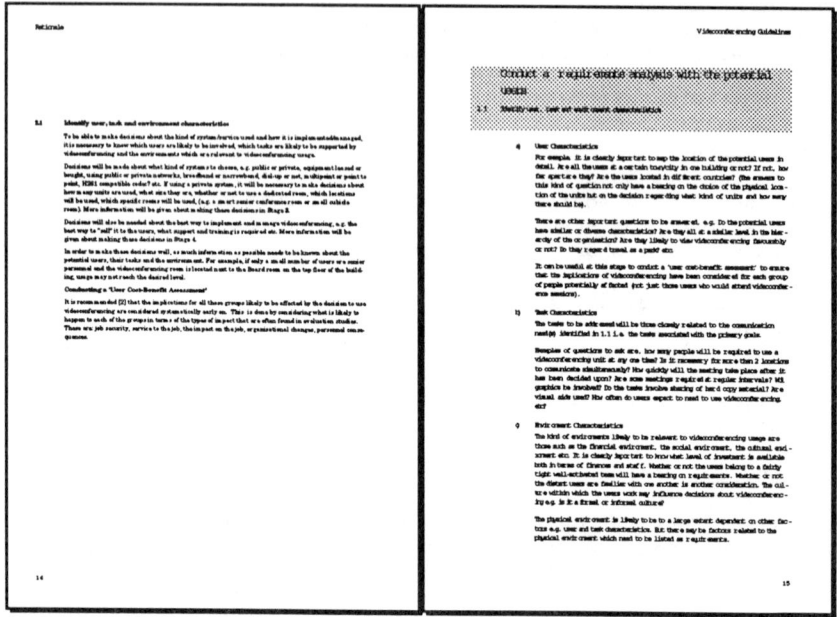

Figure 1   Facing pages layout for guidelines

The guidelines have been desktop published, printed on A4 paper, and have been glue bound. The videos are available in VHS format only.

VIDEOCONFERENCING GUIDELINES
    Videoconferencing is currently the most widespread form of videocommunications. A number of the larger multinational organisations have already invested in videoconferencing suites (i.e. rooms that typically accommodate 6 people round a table), costs are shrinking all the time and therefore many more small to medium organisations may be considering investing in videoconferencing and the earlier purchasers may be considering updating their existing equipment. Organisations can be faced with a bewildering array of sales information. How best to decide whether the organisation should use videoconferencing? Then, if it is decided to proceed, how best to select the appropriate equipment for the organisation's needs and how best to manage the service internally?
    Between the years 1990 to 1993 a survey of videoconferencing was conducted in order to establish the needs of organisations. It was conducted in 3 parts:
1 A survey of Telecommunication Managers addressing the organisation's needs.
2 A survey of actual day to day users addressing end users' needs.
3 A longitudinal survey to investigate whether opinions and needs had changed over a time.

As a result of the survey (and in consultation with other relevant literature) written guidelines have been produced to help an organisation make the following decisions:

1 Whether an organisation should invest in videoconferencing.
2 What type of system/service to have. Having decided to use videoconferencing the guidelines then aid the organisation in implementing, managing and evaluating the system.

The ISSUE Guidelines describe 6 stages for successful videoconferencing. These aid an organisation in selecting an appropriate system to match its users' needs and then how best to implement and manage the system.

The six stages identified in the guidelines are;

1 Establishing the Need for Videoconferencing
2 Conducting a User Requirements Analysis
3 Selecting the appropriate systems/services
4 Implementing and managing the chosen systems/services
5 Evaluation
6 Acting on the Evaluation

The cycle of stages is designed to be iterative, continuing for as long as the technology is used. Although the stages are presented as being linear in sequence, there will often be times when it is more appropriate for them to be carried out in parallel with each other. The word organisation, in the guidelines, refers to an autonomous organisational unit. This will normally be an entire organisation but may in the case of large organisations be only one part of the organisation.

A video has been produced to be used in conjunction with the written guidelines. It shows a case study of a typical organisation with the problem of deciding whether to use videoconferencing, illustrating the stages that need to be undertaken to ensure successful videoconferencing and suggesting that viewers need to consult the written documentation for further detailed information.

VIDEOTELEPHONY

Over the first three years of the ISSUE project a considerable amount of experimentation was conducted looking at various aspects of videotelephony from operating procedures to how the technology is best used to promote effective multi-party communication. This has led, amongst other things, to the development of a family of pictograms for videotelephony functions that achieved the highest user preference scores in a European study conducted under the auspices of ETSI (Böcker 1992).

The videotelephony guidelines adopt a facing pages layout as described previously. The human factors guidelines for videotelephony covers aspects of videotelephony design such as the equipment design, video switching and user procedures both for standalone systems and those that integrate advanced videotelephony with the workstation.

There are sections describing basic video and audio functions, discussion of image and audio quality, user control and system feedback, convening multipoint videoconferencing sessions and teleworking.

Selected guidelines are illustrated in the video which has three main parts: standalone systems, multipoint systems, and PC integrated systems.

MULTIMEDIA

The human factors guidelines document produced presents basic advice on the various media e.g. text, still image, full motion video, speech etc., discussing their respective strengths and weaknesses, and how they should be used to best effect on service interfaces. This main section is preceded by sections on general display characteristics such as screen size, resolution, lay-out. Other sections consider the use of colour, windows, design of icons and pictograms.

Another section deals with the interaction with the system or service, which is split into input devices and dialogue styles. It is followed by a section dealing with the manipulation of information, the most important aspects of which are user control over information and information retrieval/exploration. Video agents, feedback, help functions and messages are covered in a section on user support. Finally the last two sections deal with physical properties of human-machine interaction such as screen position, ambient illumination and noise.

There is also an accompanying video illustrating some of the guidelines. The aim of the video that has been produced is to present those aspects of multimedia display characteristics that are best explained by using the powerful visual medium of video. The video comprises three parts. The first gives an overview of the most relevant aspects pertaining to screen layout e.g. windows, colour, screen position and orientation, etc. The second part considers the different media: text, graphics, still image and video. The final part deals with the integration of the media into a coherent end product.

USABILITY EVALUATION

This volume differs from the rest in that it focuses on how designers can make use of human factors techniques during product design and evaluation. Its aim is to persuade designers to adopt a user centred approach to system and service design, and particularly to make use of prototype based evaluation to generate user requirements and to diagnose product failings before they are too costly to rectify.

The authors accept that there is no way that referring to a handbook can convert a designer into a skilled human factors practitioner. Nevertheless, there is some basic guidance and advice that can be offered to designers such that they might avoid some of the more obvious mistakes often made when employing people in product trials e.g. the inadvertent selection of a biased sample of users.

This volume contains a guidelines document, illustrative case study, in both written and video formats, and three technical supplements. The technical supplements cover key skills in greater detail, namely: experimental design and procedure, data interpretation and analysis, and constructing and using interviews and questionnaires. The main guidelines document provides a short introduction to human factors techniques for

designers. The document proposes a seven stage model of the
evaluation process: stage 1 defining the problem; stage 2
considering the context; stage 3 developing the prototype
stage 4 planning the evaluation; stage 5 carrying out a pilot
study; stage 6 running the evaluation; stage 7 making
recommendations.

Although the written case study is a hypothetical one, it
draws heavily on the experience of the designers that were
working on an experimental multimedia retrieval system prototype
used in the ISSUE project. It illustrates the application of
the guidelines and will be particularly valuable for those
designers who do not have access to the accompanying video. The
brief video presents a case history of a company introducing an
electronic multimedia travel brochure.

The technical supplements' aim was to provide an appreciation
of the technical background to evaluation studies so that
evaluators, designers and project managers can realistically
judge the need for expert help when planning a detailed
usability evaluation.

CONCLUSIONS
    The ISSUE project in its first three years conducted
laboratory research and surveys into key human factors aspects
related to videocommunications and multimedia retrieval. In the
final year of the project this material along with information
from the literature has been converted into guidelines for IBC
service designers. The four volumes of the guidelines
containing the written information along with the videos are
available from the authors of this paper.

REFERENCES

Böcker, M., (1992), Report about the results of an evaluation
    study of pictograms for point-to-point videotelephony. ETSI
    STC-HF1 document. Status: Approved at ETSI HF meeting,
    Vienna, June 1992.
Mosier J. N. & Smith S. L. (1986) "Application of guidelines
    for designing user interface software." Behaviour and
    Information Technology,Vol 5 No 1, 39-46.
RACE Deliverable 65/HUS/-/DS/A/014/B2 (1992) "Videoconferencing
    Guidelines for User Organisations and Service Providers".
RACE Deliverable 65/FUB/-/DS/A/007/B2 (1992) "Human Factors
    Guidelines for Videotelephony".
RACE Deliverable 65/HUS/-/DS/A/015/B2 (1992) "Usability
    Evaluation - Guidelines for IBC Service Designers".
RACE Deliverable 65/HUS/-/DS/A/016/B1 (1992) "Planning to
    Videoconference" Video.
RACE Deliverable 65/TID/HFG/DS/A/006/B1 (1992) "Videotelephony -
    The New Communication Form" Video.
RACE Deliverable 65/TID/-/DS/A/007/B1 (1992) "Multimedia Video"
    Video.
RACE Deliverable 65/HUS/-/DS/A/017/B1 (1992) "Usability
    Evaluation - Guidelines for IBC Service Designers" Video.
RACE Deliverable 65/PTT/-/DS/A/010/B1 (1992) "Human Factors
    Guidelines for Multimedia".

USABILITY TESTING IN THE "REAL WORLD": EVALUATING A MULTI-FUNCTION TELEPHONE SYSTEM.

PATRICK W. JORDAN and KEVIN C. KERR

UIPM Group
Glasgow Interactive Systems cenTre (GIST)
University of Glasgow
Glasgow G12 8QQ
Scotland

A collaborative study involved a university based evaluation group, and a multinational electronics company. The aim was to evaluate the usability of a high-end (multi-function, state-of-the-art) telephone system.

Because the evaluation group could only advise from a distance, and were operating under tight time constraints, the project was a particularly challenging one. This paper reports on the approach to the evaluation, the field work done, and the way in which the data was analysed.

INTRODUCTION

The subject of this study was a state-of-the-art telephone system produced by a multinational electronics firm. As well as allowing the user to make and receive calls, the phone incorporated a host of additional features, including (amongst others): conference calls, following calls, voice mail box, call interception, memory, re-dial, and a built in answer-phone.

An employee of the company (from here on referred to as the investigator) approached the User Interface Performance Measurement (UIPM) group at Glasgow University and asked for help in evaluating the system for usability. His boss had given him two weeks in which to conduct the evaluation and report back. A department in his place of work had been involved in the design of the phone which was installed across the site. However, the company decided that the system should be checked for usability bugs before it was made commercially available.

The UIPM group agreed to help in the evaluation. This was a chance to conduct an evaluation away from the "sterile" surroundings of the laboratory. The sharp constraint on time made the evaluation a particularly challenging one. A further constraint was that the group would be involved from a distance rather then going on site.

THE UIPM METHODS

The UIPM group was formed in 1989 to develop and refine the application of seven assessment methods to human - machine interaction. Each method involves taking measurements of user performance at the interface, making them distinct from analytic or predictive approaches to interface assessment, such as task analyses or expert walkthroughs. A brief description of each method is given here, however for a more detailed review see Jordan (1993)[1].

Focus group

A group of users brought together to discuss a particular issue - perhaps experiences of using a machine, or ideas about characteristics which they would like it to have.

Think aloud protocol

This method involves asking users to make comments on what they are doing and thinking whilst using an interface. Meanwhile the investigator observes, listens, and takes notes.

Incident diary

A mini - questionnaire which the user completes each time they get stuck when performing a task. Particularly useful for picking up bugs which cause infrequent errors. However, should be kept short or else the user will not fill them in.

Feature checklist

A list of the systems functionality. Users tick the features that they have heard of or use regularly. The information provided enables the investigator to see whether the features which the designer has provided are actually used in practice.

Questionnaire

A printed list of questions. Depending on the design of the questionnaire the respondent may be given boxes or scales to mark, or make open ended responses.

Semi-structured interview

Here the investigator constructs an agenda of questions about an interface. He or she thens talks around these with the interviewee.

Experiment

The investigator sets users a series of tasks and quantifies performance in terms of, for example, time on task and errors. As the most formal of the seven methods, experiments need to be tightly controlled so that the measurements taken are fairly precise. Probably provide the most unambiguous data.

---

[1]The methods are also illustrated on a video "User interface performance measurement: assessing human-machine interaction", presented by Patrick Jordan, and available from: Media services, University of Glasgow, 64 Southpark Avenue, Glasgow G12 8LB, Scotland.

APPROACH TO THE PROBLEM: CHOOSING THE METHODS

The methods to be used were chosen after considering the constraints on, and the resources available for, the evaluation. The two major constraints were that no-one from the UIPM group could go on sight, and that the investigator had only two weeks in which to complete the evaluation. On the plus side, however, there was an on-site user population, including designers who also acted as trouble-shooters for the rest of the staff. The company were also co-operative and enthusiastic about the study.

This meant that any combination of methods chosen would have to be quick to set up and require little expert knowledge - the investigator was new to interface evaluation - but could probably elicit a fair amount of users' time. These constraints immediately ruled out three of the methods:

> Experiments - require expert knowledge to design, and usually take quite a while to set up (for example, they shall usually require piloting).

> Semi-structured interviews - preparation and validation of questions can take time. Also, they can be time consuming to administer, as they require a 1:1 ratio of investigator's time to interviewee's time.

> Incident diary - most useful for events which occur rarely. By the time the diaries had been prepared there would not be much time left to collect data.

Although it had originally been anticipated that focus groups would be run, it proved difficult to arranging times suitable for groups of staff to meet. Thus, three methods were left for use in the study: checklists, think-aloud protocols, and questionnaires.

DOING THE GROUNDWORK: WHERE MIGHT THE USABILITY PROBLEMS OCCUR

Having decided on the methods to be used in the study, the next stage was to decide which aspects of the design to focus on. Clearly, with so many functions the "problem space" in which to look for bugs was potentially vast. The next step, then, was to narrow this problem space down. This was done by supplying the investigator with written material which would enable him to perform an "expert walkthrough" - an analysis of the machine's design in relation to a set of ergonomic principles for usability. The material supplied included an introductory article outlining ten principles for ergonomic design (Jordan 1992), and a book which contained a detailed checklist of design guidelines (Ravden and Johnson 1989). Having checked the design against these guidelines (the clarity with which the book is written means that this requires little or no ergonomics expertise), the investigator was able to identify areas in which the design might be suspect from a usability point of view.

A further source of useful preliminary information were the designers who worked on the site, particularly in their capacity as trouble-shooters. A questionnaire was drawn up with general open ended questions, asking both for the designers' opinions of the machine (eg. what are its best / worst aspects, how would they improve it), and about the sorts of problems they felt users were having (eg. what are the most common queries / complaints). The investigator issued this to the designers.

From the information supplied from the walkthrough and the designers, ideas were generated about some of the problems that might occur, and potential bugs in the design. This formed a start point for the study. The aim now was to find out if these problems really were there, and if so whether they were of any importance to the users. Furthermore, might there be any problems which had not been anticipated?

THE STUDY: APPLYING THE METHODS
Questionnaire
This study provided an ideal environment for issuing a questionnaire - a large, easily identifiable, user population who were all likely to respond (they could complete the questionnaire in work time, and indeed might even be pressured to do so by their managers). Because of this the questionnaire could simply be distributed and collected later, without the investigator having to hover over respondents to ensure completion. Questionnaire length was not particularly an issue either. Of more concern was the validity and reliability of the questionnaire. To save time, a questionnaire was compiled from a combination of other pre-validated ones (see Wong and Rengger 1990) - including only the questions thought relevant to this machine. Generally speaking questions fell into three categories: questions about how well the interface is designed (eg. "(do) messages from the interface always mean what they say?", or "(is the) position of the messages on the screen (consistent)?"), questions about the feelings that interaction with the machine engenders (eg. "(do) you feel threatened by the machine?", or "(is) working with (this) machine stimulating?"), and questions about how easy, or difficult interaction is (eg. "(is) the machine easy to use?", or "(did you) need to learn a lot of things before get going with this machine?").

Checklists
A simple checklist was drawn up listing the machines entire functionality. Users were then asked to put a tick against the functions which they had used. A second checklist was also drawn up, but this covered only an area of the interface where it was predicted that there might be usability problems. Here, users were asked not only to tick the features which they were familiar with, but also to write down what they thought the function of each feature was - a check on users' understanding of this part of the interface.

Think aloud protocols
The investigator set some of the users specific tasks and asked them to think aloud as they were doing them. The tasks set were chosen, as far as possible, to represent the range of the machines functionality, and to utilise parts of the interface where bugs were anticipated. The investigator also asked some of the users to think aloud whilst going about their usual interactions with the system. Not only were these think alouds useful for bug detection, they also helped reveal as to why users might use some features, yet ignore others.

APPROACH TO ANALYSIS
The application of each of the methods generated a lot of data, mainly of a qualitative nature, which it was then necessary to make sense of. Due to considerations of commercial confidentiality, it would be inappropriate to reveal details of the results However, it is worth reporting the approach taken to data analysis.

The investigator was provided with a framework by which to make sense of the data gathered - a multi-component definition of usability. The definition was based on that given by the International Standards Organisation (ISO) (Brooke, Beven, Brigham, Harker, and Youmans 1990). However, the ISO's broader definition is broken down into five components - guessability, learnability, experienced user performance, system potential, and re-usability - which relate to the learning aspects of usability (see table 1). After reading a paper in which the meaning of each component was explained (Jordan, Draper, MacFarlane, and McNulty 1991), the investigator was able to use his judgement in classifying the data as to which component of usability it related to. Thus, instead of merely producing a long list of usability problems, he was able to make important distinctions, for example, between bugs that caused initial learning difficulties and those which resulted in persistent errors.

Guessability	The effectiveness, efficiency, and satisfaction with which a user can perform a task for the first time.
Learnability	The effectiveness, efficiency, and satisfaction with which a user can reach an asymptotic level of performance on a task, after the first successful task completion.
Experienced user performance	The asymptotic level of effectiveness, efficiency, and satisfaction with which a user can perform the task.
System potential	The limit on user performance imposed by a machine's design.
Re-usability	The effectiveness, efficiency, and satisfaction with which a user can perform a task after a protracted period away from the task.

Table 1. Five component definition of usability (based on Jordan et al 1991).

CONCLUSIONS

The company were delighted with the outcome of the study. Within two weeks they had been presented with a great deal of high quality information, all of which had been obtained from tests on real users. Not only did they now know what the usability problems with the system were, they also knew (because of the structured model of usability employed) what their effects were. In turn, the comprehensiveness of the findings, when coupled with the observations from the expert walkthrough, made the formulation of potential design solutions comparatively straightforward. Furthermore, the company had received all this information without having to hire an "expert", or compromise any of their own confidentiality.

Application of the methods had proved effective in a situation where time constraints were tight, and where no member of the UIPM group had access to the users or the product. This suggests that the methods can succeed in the "real world", not just the laboratory, and that by using the right combination of them constraints on evaluations can be overcome.

Acknowledgement
The work of the UIPM group is supported by DTI / SERC grant
GR/F/39171/IED4/1/1109 - Measurement of User Interface Performance.

REFERENCES
Brooke, J., Beven, N., Brigham, F., Harker, S., and Youmans, D., 1990, Usability
statements and standardisation - work in progress at ISO. In Human-Computer
Interaction - INTERACT' 90, ed. D.Diaper, D. Gilmore, G. Cockton, and B.
Shackel (North Holland: Elsevier)
Jordan, P.W., 1992, Ergonomic design for vehicle users of new technology interfaces.
Engineering Designer (May). pp. 16 - 20.
Jordan, P.W., 1993, Methods for user interface performance measurement. In EJ
Lovesey (ed), Contemporary Ergonomics 1993 (London: Taylor and Francis).
Jordan, P.W., Draper, S.W., MacFarlane K.K., and McNulty, S-A., 1991.
Guessability, learnability, and experienced user performance. In D. Diaper and N.
Hammond (eds), People and Computers VI. (Cambridge: Cambridge University
Press).
Ravden, S.J. and Johnson, G.I. 1989, Evaluating Usability of Human - Computer
Interfaces: a Practical Method. Chichester: Ellis - Horwood.
Wong, G-K and Rengger R, 1990. The validity of questionnaires designed to measure
user satisfaction of computer systems. NPL Report DITC 169/90. London: National
Physical Laboratory

THE ROLE OF EXPERIENCE IN PERFORMANCE ON DIFFERENT TYPES OF TELEPHONE MEMORY RETRIEVAL TASKS.

G.J. Gelderblom
Delft University of Technology,
Faculty of Industrial Design Engineering, Jaffalaan 9,
2628 BX Delft, The Netherlands.

A. Bremner
Loughborough University of Technology, U.K.

Operational difficulties with consumer products can
be due to a mismatch between user expectations
through experience and the 'designed' operation of a
product. By means of an experiment focused upon the
operation of the memory retrieval function of
different telephones the effect of experience on
performance was examined. Both experience acquired
prior to and during the experiment were taken into
account. Moreover, the *way* the respective experiences
were stored was established. The results showed that
experience is stored in a procedural manner and
therefore does not facilitate performance on
telephones requiring different retrieval procedures.
Also, if subjects had prior experience it seemed to
be essentially unaffected by newly acquired
knowledge.[1]

INTRODUCTION

The operation of consumer products can be difficult because
of a mismatch between the users' expectations and product
functioning. Such interactive difficulties are obviously of a
cognitive nature. The users' expectations can be seen as being
based on knowledge derived from prior product interactions (e.g.
experience), however, when a user is confronted with a new
product, the product itself offers 'knowledge' on its operational
aspects. This knowledge is likely to be combined with the users
existing knowledge based on experience. Norman (1989) refers to
these two knowledge types as 'knowledge in the world' and
'knowledge in the head'. Successful cognitive interaction with a

---

[1] The experimental work described in this paper was performed at Delft University of
Technology, The Netherlands under the supervision of Dr. H.H.C.M. Christiaans.

product depends then on the *fit* between these two sources of knowledge. If the two are complementary the user should have sufficient information to operate the product successfully. However, if they are less than complementary or perhaps contradictive, operating problems are likely to occur. Even with very simple products a mismatch between the way the product's been designed to function and a user's expectations can lead to operating failures. This phenomenon is illustrated in a study by Gelderblom and Christiaans (1991) where subjects' difficulties while trying to use, what was to them, an unknown type of can opener are described. Subjects tried assuredly and repeatedly to operate the unknown opener in the same way they would their own. It can be assumed then that the knowledge inherent in the unknown product was not sufficient to convince the subjects of an operating procedure different to the one founded on their own experience. At this point it should be pointed out that a can-opener has only one purpose and the functioning of it can, in principle, be deduced from its appearance.

To gain more insight into the role of experience concerning the operation of more complex consumer products an experiment was carried out in which subjects were asked to (1) give a description of how the retrieval of a telephone number from memory *in general* is executed, (2) specify their expectations towards the operation of four different models separately, (3) retrieve a telephone number from the memory of those telephones and (4) describe their experience regarding telephone memory while specifying products they believe adopt a comparable operating procedure.

The subjects' general descriptions regarding the operation of the telephone memory gave an indication of their mental models (Wilson and Rutherford, 1989); their opinions being based on prior experience with telephone memory functions. Their verbal expectations for each specific telephone gave an indication of the influence of current observation. Knowledge concerning each telephones' operation had to be provided either by their design features or by the subjects' experience.

The experimental aims were to establish the influence of experience gained prior to the experiment, and the influence of experience gained during the experiment. In addition it would be possible to see whether performance differences were related to the four different telephone designs.

METHOD

Twenty subjects, both sexes equally represented, participated in the experiment for which they received a modest financial reward. Four different telephones were used which embodied four different design solutions for number retrieval from memory. They were considered to be a reasonable representation of the memory retrieval functions available on modern domestic telephones. Below are illustrations of the telephones used and a description of the retrieval procedures.

Model 1: BT's Relate 200.

Model 2: PTT's Monza 10

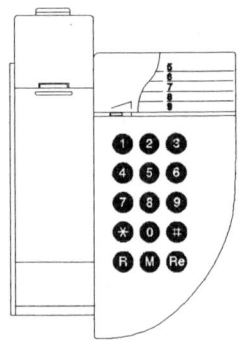

One key to press for direct
selection of the required
telephone number.

First a key to enter the
memory mode and then a keypad
number for selection of the
required telephone number.

Model 3: BT's Converse 200

Model 4: BT's Relate 400

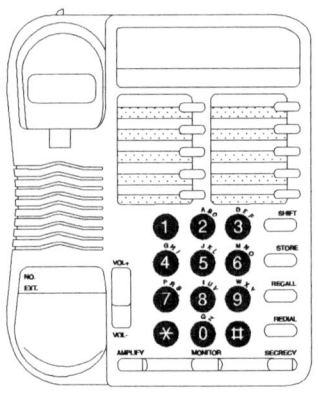

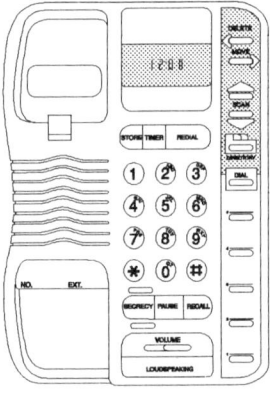

Double-functioned memory
buttons that require the
SHIFT-key to be pressed if
the second telephone number
associated with that button
is required.

Memorized-telephone number-
directory system with display
feedback. Requires DIRECTORY
activation, scanning and
finally selection of the
required telephone number by
pressing the DIAL key.

The subjects were first given a short introduction to the concept of the memory function on modern telephones. In it nothing was mentioned regarding the precise operation of number memory retrieval. Following the introduction subjects were asked to explicate their general conception of the retrieval of a telephone number from the telephone's memory. Then the first of four telephones was presented. For each specific telephone subjects were asked to give a procedural description on how the appropriate telephone number should be activated from memory. Then they were invited to perform the actual retrieval task. After trying individually to operate the telephones subjects were asked once again to expound on their procedural operation. The four telephones were presented in a randomized order over the subjects. During the operation of the telephones the subjects were instructed to think aloud so that their thoughts could be recorded.

Finally, after the operation of the telephones the subjects were asked about their experience with the memory retrieval function on telephones and whether its operation reminded them of any other products. This was done to get an impression of their experience. All the subjects' actions and utterances were recorded on video.

RESULTS

Experience

To establish the subjects' level of experience a distinction was made between general experience with one of the four design solutions and specific experience with any of the *actual* telephones. Eight of the twenty subjects had experience with the telephone memory concept but none were familiar with the particular designs. The experienced subjects were all familiar with the same type of retrieval system.

General descriptions

The content of the subjects' general descriptions depended on their experience; the experienced users mainly expressed verbally their experience *prior* to the experiment - regardless of the telephones they had operated during it. Inexperienced subjects in contrast developed various concepts; some of which were entirely new.

The level of abstraction of these general descriptions was independent of experience; explication of the subjects' general views resulted in procedural descriptions rather than the general descriptions requested.

Procedural descriptions

The experienced subjects gave fewer false procedural descriptions than the inexperienced group. They expected the procedure appropriate for *their* telephone to be applicable to the other types of telephones. The inexperienced group came up with numerous diverse expectations, many of which were not suitable for any telephone.

Performance
    Performance differed greatly over the four telephones. This
difference was not related to the sequence in which the
telephones were operated; ie the presentation order did not
affect the subjects rate of success. The experienced users
however performed best on the type of telephone they were
familiar with.

DISCUSSION
    The experienced subjects in this study were familiar with
one particular concept of memory retrieval within telephones.
However, since none of them were familiar with any of the four
actual products the experiment also dealt with the telephones
'guessability' (Jordan and O'Donnel, 1992). The study revealed a
clear hierarchy in guessability of the four telephones. The
operating procedures for three of the telephones were far from
self-evident which means that these designs do not provide
sufficient information for successful operation.
    Concerning the influence of experience obtained prior to the
experiment; the results showed that experience with one of the
four memory retrieval systems gave those subjects an advantage in
operating the telephone using that same system. However, on the
other systems, the effect of that experience was insignificant.
For those systems the users' knowledge in the head did not
facilitate performance.
    Looking at the influence of experience obtained during the
experiment the results showed that any acquired knowledge did not
evolve into a *general* understanding of memory retrieval. This
could explain the absence of transfer of newly acquired knowledge
over the four telephones throughout the experiment. Also, because
generally subjects' experience was stored as procedural
information and the operation of unfamiliar telephones requires
different procedures the individual memory retrieval tasks were
not facilitated. This accounts too for the knowledge of the
experienced users; their knowledge acquired prior to the
experiment seemed to be essentially unaffected by the experience
obtained during it.

REFERENCES

Gelderblom, G.J., and Christiaans, H. (1991) Mental models and operating consumer products. In <u>Proceedings of the 11th Congress of the International Ergonomics Association</u>, eds. Y. Queinnec & F. Daniellou (London: Taylor & Francis) pp. 613-615.

Jordan, P.W., and O'Donell, P.J. (1992) Quantifying guessability, learning, and experienced user performance. In <u>Proceedings of the Ergonomics Society's 1992 Annual Conference</u>, ed. E.J. Lovesey (London: Taylor & Francis), pp. 404-410.

Norman, D.A. (1989) <u>The Psychology of Everyday Things</u>, (New York: Basic Books).

Wilson, J.R., and Rutherford, A. (1989) Mental Models: Theory and applications in Human Factors. <u>Human Factors</u>, 31, pp. 617-634.

Poster presentations

PRACTICAL EXPERIENCES WITH CONSUMER PRODUCTS, USERS AND PROTOTYPING.

G. I. Johnson[1] and E.P.G. van Vianen

Applied Ergonomics Group,
Philips Corporate Industrial Design,
Philips International B.V., Gebouw SX,
Postbus 218, 5600 MD Eindhoven.
The Netherlands.

This poster describes some of our experiences of user-centred design and evaluation of consumer products, using prototypes. The types of consumer products used as examples here are representative of everyday, contemporary, interactive, 'high-end' (i.e., many featured) products. Our work concerned prototypes of products from the consumer electronics areas of in-car systems and television products. This approach used screen-based, interactive prototypes which were then employed in user trials to assess usability. Involvement of users at an early stage is seen as critical, functional prototypes providing many possibilities. A discussion of the practical issues faced when employing prototypes in usability analysis is presented.

INTRODUCTION

The notion of prototyping has for many years attracted much attention as a means of turning a product concept into something which can be experienced and perhaps evaluated by human factors specialists and potential end-users. The praises of prototypes, particularly those of the rapid variety, have been sung, along with iterative design and user involvement (see e.g. Gould and Lewis, 1985; Wilson and Rosenberg, 1988). However, there are often a number of practical difficulties with this approach (see Lim and Long, 1992).

The goal of this paper is to describe some of our experiences in the use of product prototypes for usability assessment. We do not draw great distinctions between prototypes, simulations and mock-ups: In this case we refer to screen-based functional versions of interactive products, used early in the design process, with the goal of conducting usability investigations with users. In other words, prototypes as graphical representations of products such as radios, that can provide the focus for user trials.

Our experiences and impressions are drawn from a number of consumer electronics projects such as the development of user interfaces for in-car systems and advanced

---

[1] New address: NCR (E & M) Ltd., Cognitive Engineering, Technology Development Group, NCR Self-Service and Financial Systems Division, Kingsway West, Dundee DD2 3XX. Scotland. e-mail: Graham.Johnson@Dundee.NCR.com

televisions, as seen in the wider context of our work in involving the end-user in the design of everyday products (Johnson and Van Vianen, 1992; Van Vianen and Johnson, 1993).

These types of product are seen as offering rather different challenges to the prototype developer and the human factors specialist when compared with specialist computer or software products. Not only do consumer products have the broadest of user groups, they are also usually subjected to highly pressured development cycles with increasingly shorter and shorter lead times. Consumer product production is typically for a mass market, and is frequently heavily influenced by previous versions or families. This can mean that prototypes, other than traditional 'appearance' (industrial design) models, are relatively rare within consumer product development, unless a concerted effort is made to raise usability issues to the top of the product quality agenda.

Clearly, some consumer products, such as in-car systems, can also be characterised by attributes such as providing entertainment (e.g. playing and listening to music), comprising discrete controls and displays, possessing modes, and so on. The context of use is very often non-occupational; but again, that can vary greatly.

The complexity of contemporary consumer products is gradually increasing. Many are complex and highly interactive providing the end-user with many choices, advanced features and functions. A bewildering array of possibilities are currently offered to the user of say, the latest domestic audio equipment. To capture the essence of a potential product as a prototype in the early (concept) stages of a development inevitably involves making a number of compromises. We discuss below some of the practical issues and trade-offs in using prototypes of consumer products for usability analysis.

OUR USE OF PROTOTYPES

The purpose of prototyping can easily be lost when multi-disciplinary teams begin to enthuse about, and attempt to own, a prototype. For instance, one could imagine the opportunities of demonstrating products to management, versus the usability evaluation of specific functional areas within a user-product dialogue as potentially competing goals.

Our use of on-screen prototypes has largely been within the context of development projects dealing with products such as those given in Table 1 (below). In all cases, the prototypes sought to attain a high level of functionality, even where this included the creation of a simulated environment -- for instance, fictitious television and radio channels. In each case a graphical representation of the product was created, one with which a user could interact, either via a mouse or the touch-screen.

Table 1. Types of consumer product prototypes for user-based evaluations and example usability issues investigated.

Product type.	Example issues of interest.
• ADVANCED IN-CAR STEREO SYSTEMS	control configurations, displayed information needs, audio feedback, consistency, etc.
• TELEVISIONS	on-screen menus, menu depths, enhanced remote control designs.
• FUTURE IN-CAR 'TRAFFIC INFORMATION' PROVIDERS	overall interaction design, use of modes, message comprehension, etc.

To use a prototype a user would press (e.g. via the touch screen) the 'on-off' button shown on the screen, and then proceed to operate the 'product' via the simulated controls and displays, carrying out specified tasks or simply exploring the prototype, what it offers and how it works.

For many of the projects, the notion of a throw-away prototype was not always clearly stated, although it was accepted from the outset that the prototypes would have a limited life (i.e. only within the development cycle) and that the tools chosen would not provide 'code' or script of real use in the final product.

Using a prototype within user trials was typically intended to provide concrete feedback on the deficiencies and strengths of the design and parts of the design. Our approach has been to employ a screen-based prototype as the focus for exploratory and task-based investigation of user reactions, video-recording user behaviour wherever possible.

BUILDING PROTOTYPES

Over the past few years we have experiences with several 'off-the-shelf' packages that can be employed in straightforward screen-based prototyping. For example, we have encountered use of Apple Macintosh packages such as Hypercard, Supercard, Authorware Professional, and Macromind Director as well as in-house (Philips) packages such as the MSDOS-based Rapro-T for a number of consumer product prototypes. More recently, we have made full use of Supercard. As to the merits of these packages, that is dependent upon the various requirements of the prototype, some of which are discussed later. It is not the main purpose of this paper to recommend particular packages -- see e.g. Hix and Ryan (1992) for a useful paper on prototyping packages (for user interface development).

In general, the prototypes were made on the basis of early specifications, in various forms, of the product's functionality. Rather than roughly describing a scenario or storyboard of product use, we focused on a functional model that imitates the product-to-be in as face valid a manner as possible. Of course, the nature and detail of the specification is fundamental to the success of the prototype build. The construction of a fully functioning prototype demands that all paths (possible interactions) be defined, and the process of adequately specifying the prototype is critical at this stage.

PROTOTYPES AND USABILITY EVALUATION

The thrust of this paper concerns the use of prototypes, such as those constructed with Authorware or Supercard, in the context of usability analysis. There are a number of issues which arise when carrying out user trials with product prototypes, dealt with below.

Audio feedback

In a great many cases, the use of speech and non-speech audio is an important aspect of the product prototype. For many entertainment products (e.g. a radio) there is a need to simulate the 'normal' feedback of say tuning. This involves the recording of relevant music passages, 'noise', and so on. Using large amounts of audio will, of course, have some implications for the memory and possibly speed of the prototype.

Additionally, auditive feedback is associated with key depressions; for instance, in the cases of function confirmation or as an 'error' tone. The extent to which audio feedback enhances the face validity of a product prototype cannot be underestimated. In our experience, audio characteristics greatly influence users' opinions of the overall validity of the prototype.

Visual representation of the product

Representing the product on the screen is a prime example of an area where trade-offs are to be made. On the one hand, realism is sought via textured graphics (perhaps 'scanned in') to give some sense of depth, and an accurate visual feel for the prototype; on the other hand, the constraints imposed by the hardware and software will often restrict the ability to produce say animation of any complexity. Obviously, the input of cards, cassettes, CDs and the like are typically animated events and often play a major role within the functionality of the product.

Other issues here concern the question of scale and dimensioning, e.g. whether the product prototype should adopt a 1:1 scale as it appears on the screen. Similarly, approximations to the correct colours and textures (of crucial importance in product design) also have to be compromised. Our experiences are mixed with regard to visual representations of the product. As one might expect the 'accuracy' of a visual representation is chiefly dependent upon the purpose of the evaluation: If subjective assessment of appearance is one of the main issues to be investigated then high quality representations are required, if certain aspects of the interaction (e.g. the way in which several fundamental controls relate to one another) then a less complex visual can be used, augmented if necessary with 'off-line' sketches, drawings or renderings.

Finally, there is the issue of the product display. Many consumer products offer Liquid Crystal Displays (LCDs) and have particular font characteristics. Again, the visual representation of these is subject to trade-offs; e.g. development time of bespoke fonts, attempts to 'copy' LCD colour attributes. The difficulties in drawing any tentative conclusions about legibility, etc. should be obvious from such an artificial element. However, it is possible to study the comprehension of displayed text, icon-type feedback, timing of displayed information, etc.

Means of interaction

With a screen-based product prototype there is, from the outset, such problems as tactile feedback, key/button travel, rotary controls, and so on. Users are normally offered a mouse and/or a touch-screen to operate the prototype, having been received a demonstration and some training with a similar prototype. Controlling the amount of training (users typically receive a minimum, and will cease when they are ready) and discerning which button presses are intended and which are slips is often complicated. For the latter, exploratory (i.e. not task-driven, or prompted: a free exploration of the prototype) parts of an evaluation where users are encouraged to comment (think aloud) on what they are doing provides much insight into error situations of the slip and mistake kind. A related issue is that of being able to validly assess control proximity and such like via the artificial media of the screen-based prototype.

Response times

As we might expect, the complexity of a product prototype can have a great effect upon the response times for certain operations. In our experience, the successful approximation of the speed (say of feedback) is critical to the success of the product prototype in conveying a convincing image. The user-product interaction of interest is so closely dependent upon the prototype's ability to react as the product would, that this is the last area in which compromises can be made. However, it is often found that specifying the speed of various reaction or operations is extremely difficult if say a mechanical mechanism has not yet been fully developed and integrated.

Users' views of prototype validity

Part of the questionnaire and interview stage within the user trial should always address the validity of the prototype. Users need to be questioned about the extent to which they found the prototype realistic, which aspects were most artificial, how it could be more realistic, whether they felt their performance (say with simple tasks) was influenced and if so how, and so on. Prior to finalising the prototype, issues such as these can be tackled at pilot stage. The ability of prototypes to be convincing is quite surprising -- for instance, a user who reaches (to the screen) to receive an ejected cassette.

Interestingly, there were no problems with the many users recruited for trials who had no previous computer experience. Although initially unfamiliar with a mouse/touch screen, they were able to interact with the 'product' in the same fashion as did those with extensive computer experience.

Influence upon performance

Data gathered from user interaction with a product prototype of a qualitative kind (e.g. time on task data) must always be treated with care. De Vries (1990) demonstrated, in a control study, that for an in-car stereo system, when a production version a relatively complex Hypercard prototype of that product were compared in terms of users' time on task, task effectiveness, and error data there were no statistical significant differences between the two conditions.

Context and prototypes

The importance of the context as a factor in usability evaluations of product prototypes cannot be ignored. Screen-based prototypes of complex in-car products can rarely be placed in a driving environment. The lack of context possible with some prototypes is clearly a factor to e taken into consideration.

Supporting or dictating role?

The role of the prototype is often a subject of great debate. For some it is a panacea, for others it assists in the process providing supporting evidence, and for others it is a live functional specification to be taken very literally. In our experience, the role of the product prototype and its evaluation must be seen in a supporting role, rather than as a 'live' specification. Complexity of the eventual product, and the development time and effort necessary to constantly edit and modify a prototype can usually be better spent in working towards a basic early (α) version of the product itself.

How functional, how fragile?

Is it necessary to have a fully functional prototype? Well, that seems to depend on the use to which it will be put (i.e. which aspects, and to what depth, will be evaluated, and by whom). On the issue of robustness of prototypes, it is our view that any prototype (from paper and pencil through to basic button rows with related display) is useful within development. Emphasis must always be on the fact that it is the product prototype (and not the user) that is being evaluated.

On the nature of evaluations with product prototypes

Given some of the characteristics of product prototypes mentioned, there are several questions raised about the type of user-based evaluation suited to this situation. In many respects, a pure usability engineering approach (e.g. Brooke. 1986) is not always the most suitable -- criteria concerning time and error data may not be readily gained, dependent upon the 'realism' of the prototype. More importantly, the purpose of an evaluation at this stage in a development is normally to identify the strengths and weaknesses of a design concept, and to play a diagnostic role. With such objectives it is useful to focus on more qualitative data, albeit within a task-oriented evaluation. Typically, users' interactions with prototypes can be the subject of discussions with users, users ought to be allowed to freely explore prototypes, and to think aloud where appropriate.

CONCLUDING REMARKS

On the basis of our experiences with prototypes of consumer products we can draw a few conclusions, that are of potential use to others embarking on similar work.

Prototyping consumer products, despite many practical problems, can deliver a wealth of useful information. This paper has discussed some of the important factors that seem to determine the success of prototype use with users when assessing usability, e.g. visual representation of the product, response times, context, and types of evaluation design.

Many questions, encountered when using prototypes, remain open and require further investigation. Some concern the validity or fidelity of prototypes (an overlap with simulator issues), others are more related to the general product design process and the role played by prototyping activities therein.

ACKNOWLEDGMENTS

We are most grateful to Philips Corporate Industrial Design for the support of this work. Thanks are also due to the many developers within the Philips organisation for their assistance, and to Gordon Allison at HUSAT Consultancy for his technical support of Supercard-based projects. Note that the views expressed within this paper are the authors' and not necessarily those of the companies they represent.

REFERENCES

Brooke, J.B., 1986, Usability engineering in office product development, In People and Computers: Designing for Usability, Proceedings of HCI '86, Harrison, M.D. and Monk, A.F. (eds.), (London: Cambridge University Press), 249-259.

Gould, J.D. and Lewis, C.H., 1985, Designing for usability -- key principles and what designers think. Communications of the ACM, 28, 300-311.

Hix, D. and Ryan, T., 1992, Evaluating user interface development tools, In Proceedings of the Human Factors Society 36th Annual Meeting, (Santa Monica, CA: Human Factors Society) 374-378.

Johnson, G.I. and van Vianen, E.P.G., 1992, Comparative evaluation of basic car radio controls, In Contemporary Ergonomics 1992, Lovesey, E.J. (ed.) (London: Taylor & Francis), 456-462.

Lim, K.Y. and Long, J.B., 1992, Pitfalls of rapid prototyping: Observations on a commercial system development project, In Contemporary Ergonomics 1992, Lovesey, E.J. (ed.), (London: Taylor & Francis), 354-364.

Vianen, van E.P.G. and Johnson, G.I., 1993, 'Gebruikersschetsen: Een aanvullende methode', In preparation.

Vries, de, G., 1990, Audible help and usability of the DC794 Car Stereo, Philips Corporate Industrial Design (CID), Eindhoven. AE Group Report: Confidential.

Wilson, J. and Rosenberg, D., 1988, Rapid prototyping for user interface design, In Handbook of Human-Computer Interaction, Helander, M. (ed.) (North-Holland: Elsevier Science Publishers), 859-875.

SPATIAL OPERATIONAL SEQUENCE DIAGRAMS IN USABILITY INVESTIGATIONS

Graham I. Johnson

NCR (Manufacturing) Ltd.
Cognitive Engineering, Technology Development,
NCR Self-Service & Financial Systems Division,
Kingsway West, Dundee DD2 3XX.

This paper describes the application of Spatial Operational Sequence Diagrams (Spatial OSDs) to the analysis of usability, as it relates to the 'walk-up-and use' product category. Whilst many within the human factors community are, rightly, concerned with highly interactive computer-based systems, and apparently employ the very latest methods, we must not ignore the more traditional techniques such as Spatial OSDs. It is illustrated that Spatial OSDs can provide a most useful contribution to product usability evaluation and that the overall power of Spatial OSDs is greatly enhanced when considered in careful combination with other (task) analysis methods. The goal of this exploratory work is to provide insight into the cognitive aspects of the user-product interaction, where users can be regularly consulted on their actions and the relationships between them.

INTRODUCTION

There is a growing interest in usability evaluation and the development of appropraite methods and techniques, more specifically in software usability. This is probably due, in part, to the increase in interactive products and systems, and the general awareness of ease of use as a major issue in the overall success of interactive products. It seems that everyday products are also gaining more interest in the human factors community (e.g., Norman, 1989; Thimbely, 1991), as are various methods developed to evaluate their usability (e.g., Baber and Stanton, 1992; De Vries and Johnson, 1992).

One such everyday, interactive product is the Automated Teller Machine (ATM), perhaps better known in the UK as the cash dispenser or the 'hole in the wall' machine. ATMs, are prime examples of public, walk-up-and-use products. They are employed by most main banks, serving a wide range of end-users (Bell and Scobie, 1992). ATMs, because of this user range, the tasks to be carried out via the system, and the environment offer a number of challenges to those concerned with usability, design and the cognitive aspects of user-product interaction (see e.g. Payne, 1991; McLean, Young, Bellotti and Moran, 1991).

Clearly, there are many useful approaches to usability evaluation that can be adopted when assessing (different) ATM designs. The scope of applicable methods is great. This short paper explores the potential of only one of these approaches, that of augmented Spatial Operational Sequence Diagrams (Spatial OSDs). Its suitability in the study product usability with regard to everyday products such as ATMs is discussed.

SPATIAL OPERATIONAL SEQUENCE DIAGRAMS (SPATIAL OSDs)

This technique, as its name suggests, attempts to capture the fundamentals of interaction, or operations, in a spatial or geographical manner. The diagram drawn takes the form of a map (of the 'equipment') upon which links between (sequential) operations are plotted. Typically, a Spatial OSD will illustrate relationships between specific areas, controls, and displays, often in the context of many control-display panels or a large workstation. This is conducted for sub-tasks, the accumulation forming an overall picture of the user-system interaction.

Whilst the notion of operation sequence diagrams and their variants is by no means new (see Kurke, 1961), the technique does appear to have found less use within current human factors practice. This is possibly as a result of: (a) the recent domination of human-computer interaction (HCI) within human factors applications, (b) the limited use of the technique when employed solely, and (c) difficulty with combined display-control configurations, i.e. much of HCI is only concerned with screen-keyboard-mouse configurations, and therefore, has little interest in Spatial OSDs. In fact, it is arguable that the majority of HCI attends to that which is 'non-hardware' -- meaning menus, status lines, dialogue boxes, overall system structure, feedback messages, and so forth.

However, people interact with 'everyday' computer-based products in many other ways; for example, 'intelligent' consumer products (such as video recorders and cameras), and public access terminals (such as ATMs and advanced vending machines). In these cases we often find control panels, special function keys, separated displays and distributed controls. It is suggested that such user interface configurations invite the Spatial OSD-type approaches. Furthermore, the nature of the interaction (task) can often be more rigid or sequential than the types of interaction offered by most conventional, office-based computer systems.

INVESTIGATING AN ATM USING SPATIAL OSDS

The goal of this investigation was to ascertain what type and amount of information a relatively simple Spatial OSD could capture when applied to different high street ATM types. The common task of withdrawing an amount of cash with a receipt was chosen as one which is representative of everyday use.

The contemporary ATM offers many features to the user, including deposit facilities, statement and cheque book orders, passbook updates, and balance inquiries, alongside the most common use of obtaining cash. Thus, there are many elements to the 'hardware' of the user interface, most of which concern themselves with the delivery or acceptance of envelopes, money, cards, etc. Typically, these modules are distributed within the overall fascia to the ATM. The usual transaction involves the entry of a card, entry of a PIN (or Personal Identification Number), and menus used in conjunction with function keys.

The Spatial OSDs below (Fig. 1(a) and (b)) illustrate the general cash withdrawal sequences for typical high street ATMs. As can be seen from Figure 1 (a) and (b) the number of steps is perhaps more than we would expect (or remember), and the locations (re-)visited are numerous. Obviously, as the complexity of the diagram increases considerable issues emerge as to the clarity of the representation and thus the amount of information that can be conveyed. Nevertheless, the utility of the Spatial OSD approach was demonstrated. It should be noted that despite the location of the main elements within an ATM -- such as the dispenser for cash -- the whole interaction is heavily influenced (perhaps dictated) by the screen-based dialogue (e.g. the number of steps, confirmations required, menus offered, etc.).

It is perhaps interesting to consider the variety of useful information that could be included or combined with the types of diagram shown in Figure 1. For example, it ought to be possible to show:

- temporal aspects, e.g. those stages at which hesitation occurs, or where the user must, say, wait for the machine's response;

- handling (input, output) aspects, e.g. where cash is withdrawn or envelopes deposited; and,
- error characteristics, e.g. where 'slips' and mistakes occur, derived from detailed behavioural data.

Figure 1. (a).   Spatial OSDs for a typical high street ATM.
Sequence illustrates a cash withdrawal

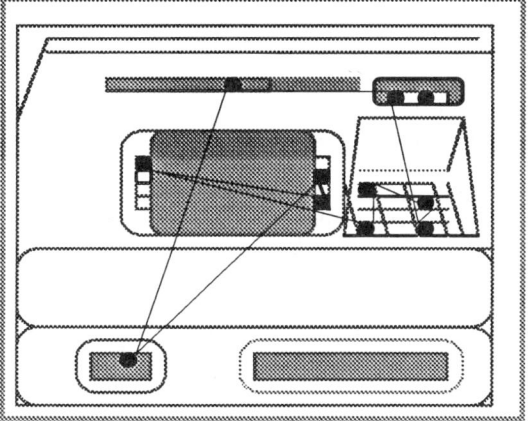

Figure 1. (b).   Spatial OSDs for a typical high street ATM.
Sequence illustrates a cash withdrawal.

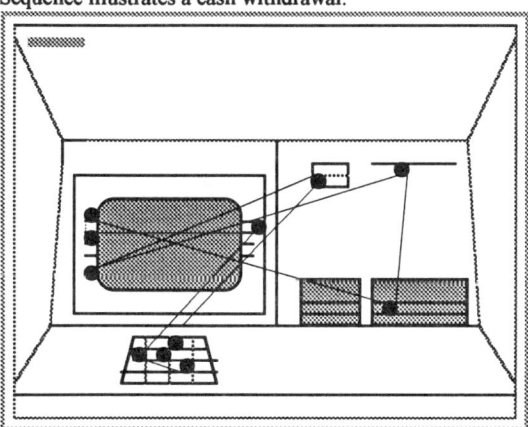

The high face validity of Spatial OSDs offers the opportunity for interaction analysis involving returning to users with preliminary representations of the tasks. The reason for doing this is to check with users from which data was orginally obtained, the validity of the interpretation of the data. It is often found that the off-site explanation and validation of described tasks or interaction with users themselves is a difficult part of the process. Spatial OSDs seem to promote a clear organisation of task data (steps) that is more readily understood by users than many other techniques with relatively abstract representations. Examples of the latter would include formal notations, tables and hierarchies.

There are, of course, also disadvantages in the use of Spatial OSDs and these ought not to be ignored. The way in which a pictorial representations of the sequences become increasingly complex is very often a problem. It is largely dependent upon the size of the task, type of layout, and number of conditionals present within the task.

For those concerned with the types of mental model end-users might hold, it is the cognitive aspects that are of most interest, where the Spatial OSD is regarded as a useful framework upon and through which relevant annotation can be developed. This is considered in the next section.

AUGMENTING SPATIAL OSDs TO CAPTURE COGNITIVE ASPECTS

The idea that Spatial OSDs can be combined with other data sources in order to enhance their effectiveness in providing insight into the user-task-system relationship is not particularly new. However, if we closely examine the information needs (the requirements) of a method which provides an good overview of the interaction, then it is obvious that the all-important cognitive aspects of the task are not captured via Spatial OSDs.

If we look at the various options open for the augmentation of the diagrams, using say a key to represent temporal aspects or task types then the OSD becomes rather complex. The likely degree of complication in representation implied by this approach for anything other than the simplest of tasks is one of the factors that suggested the combination of Spatial OSDs with established cognitive (user) modeling approaches. In this way it ought to be possible to annotate a relatively self-explanatory Spatial OSD with task and knowledge relevant descriptions.

Specifically, GOMS-type (Card, Moran and Newell, 1983), and CCT (Cognitive Complexity Theory)-type analysis methods (Kieras and Polson, 1985; Kieras, 1988) seemed appropriate to these sequential activities (i.e., relatively few conditionals). As an example consider the description of the basic steps within a generic ATM cash transaction, partially shown in Figure 2 (similar to that offered by Johnson (1992)).

Whilst others have examined the problems and possibilities of CCT with regard to system usability (see e.g. Knowles, 1988), it is generally felt that there are some severe limitations to the formal CCT (and related goal structure) approaches in tackling usability in a general sense. Such limitations would include their possible sense of abstractness (from the task) via notation, the overemphasis of knowledge amount (rather than type), their focus on error-free interaction, and an apparent difficulty in handling qualitative data.

It seems highly appropriate that procedural knowledge types be represented in terms of a production system style notation, such as that shown in Figure 2. The ability of this to take form and offer meaning to users themselves is surely increased as a result of organisation into the type of framework offered by a Spatial OSD.

DISCUSSION

Having briefly explored the potential for Spatial OSDs and their possible combination with methods such as CCT as far as self-service product types are concerned, it seems reasonable that such combination ought to offer more in their study than perhaps formal (notation only) approaches. Despite the utility of analytic approaches, the need for behavioural measurement will always be relevant in the full investigation of usability issues.

Figure 2. CCT-type description for a high street ATM cash withdrawal sub-tasks.

```
            If
            (the goal is to withdraw cash)
            Then
            (Do sequence (insert card)
            (read screen message)
            (type PIN number)
            (type enter)
            (read screen menu)
                (choose from:
                    BALANCE
                    DEPOSIT
                    STATEMENT
                    CASH
                    OTHER....)
            (select CASH)
            (read screen menu)
                (choose from:
                    SAVINGS
                    CHECKING
            (select CHECKING)
            (read screen menu)
            (choose from:   $ 20,
                            $ 40, .. ..... etc.
```

Clearly, this work is at an early and exploratory stage. Further work within the context of extensive behavioural data collection and its analysis is required. Yet the potential of these combinations seems promising. One could imagine, for example, the general task background of Spatial OSDs assisting greatly in the overall structuring of say verbal protocol data, for specific tasks or operations, or in the analysis of knowledge requirements in a (product) transfer situation.

Finally, it should be noted that this type of augmented Spatial OSDs are probably applicable to other areas of human factors interest beyond ATMs, such as in-car system (control and display) design. Also, the potential use of eye movement data would also offer many possibilities for detailing and structuring sequence-biased task domains.

ACKNOWLEDGMENTS

The author is grateful to NCR (Manufacturing) Ltd. for supporting this work. The views expressed within this paper are the author's, and not necessarily those of NCR.

REFERENCES

Baber, C. and Stanton, N.A., 1992, Defining 'problem spaccs' in VCR use: The application of task analysis for error identification, In Contemporary Ergonomics 1992, ed. E.J. Lovesey, (London: Taylor & Francis), pp. 418-423.

Bell, K.R. and Scobie, G.E.W., 1992, Multimedia technology:Banks and their customers, International Journal of Bank Marketing, Vol. 10, No. 2, 3-9.

Card, S.K., Moran, T.P. and Newell, A., 1983, The Psychology of Human-Computer Interaction, (Hillsdale, NJ: Lawrence Earlbaum Associates).

Johnson, P., 1992, Human-Computer Interaction; Psychology, Task Analysis and Software
        Engineering, (London: McGraw-Hill).

Kieras, D.E. and Polson, P.G., 1985, An approach to the formal analysis of user complexity,
        International Journal of Man-Machine Studies, Vol. 22, 365-394.

Kieras, D.E., 1988, Towards a practical GOMS model methodology for user interface design.
        In Handbook of Human-Computer Interaction, M. Helander (ed.), (North-Holland;
        Elsevier Science), 135-157.

Knowles, C., 1988, Can cognitive complexity theory (CCT) produce an adequate measure of
        system usability? In People and Computers IV; HCI'88, ed. D.J. Jones & R. Winder,
        (Cambridge: CUP), 291-307.

Kurke, M.I., 1961, Operational Sequence Diagrams in systems design, Human Factors, Vol.
        3, 66-73.

McLean, A., Young, R.M., Bellotti, V.M.E. and Moran, T.P., 1991, Questions, options, and
        design criteria: Elements of design space analysis, Human-Computer Interaction,
        Vol. 6, 201-250.

Norman, D.A., 1989, The Psychology of Everyday Things. (Basic Books).

Payne, S. 1991, A descriptive study of mental models, Behvaiour and Information
        Technology, Vol. 10. No. 1, 3-21.

Thimbeley, H., 1991, Can humans think?, Ergonomics, Vol. 34, No. 10, 1269-1287.

Vries, de and Johnson, G.I., 1992, Het gebruik van GOMS en de 'limted information task' bij
        de evaluatie van een autoradio, Nederlands Tijdschrift Voor Ergonomie, December.

SOME ERGONOMIC CONSEQUENCES OF PLAYING FIELD HOCKEY

THOMAS REILLY AND JANINE TEMPLE

Centre for Sport and Exercise Sciences,
School of Human Sciences,
Liverpool John Moores University
Mountford Building
Byrom Street
Liverpool
L3 3AF
ENGLAND

The purposes of this project were to determine the
physical and physiological responses to dribbling a
hockey ball under different conditions and establish
the relations between back muscle strength and spinal
responses. Dribbling a hockey ball was found to
induce greater physical, physiological and subjective
strain on players over and above normal running. The
strain was accentuated when dribbling was executed in
a crouched posture. Results suggest that back muscle
strength may have a protective function in such
conditions.

INTRODUCTION

Field games, such as the football codes and hockey, impose
unique physiological and physical demands on their participants.
The physiological demands of hockey are determined by work-rate
requirements of the games. These are accentuated by the added
strain associated with performing game skills such as dribbling
and playing the ball with the hockey stick whilst running at
sub-maximal and near-maximal speeds.

The semi-crouched posture in hockey is likely to impose undue
physical strain on the spine, particularly when fast locomotion
is also required (Fox, 1981). This is reflected in injury
statistics emanating from sports medicine clinics in the United
Kingdom. Cannon and James (1984) reported that 16% of patients
over a 4-year period, who were referred to their clinic for
athletes suffering from back pain, were hockey players. This
vulnerability of hockey players was confirmed by Reilly and
Seaton (1990). Altogether, 53% of their respondents to a survey
of male hockey players reported experience of lower back pain.
They also demonstrated that dribbling the ball in particular was
a potential cause of lower back pain syndrome. This is likely
to be accentuated when dribbling is done in a semi-crouched

posture. The compressive force on the spine associated with
dribbling results in a loss of stature which can be measured as
shrinkage in spinal length.

The purposes of this study were to:-
i) determine the amount of shrinkage occurring over a short
period of dribbling a hockey ball;
ii) compare different dribbling techniques in terms of shrinkage
induced and degree of effort perceived;
iii) establish the relation between back strength and spinal
responses.

METHODS

Six female hockey players, aged 19-22 years, acted as
subjects. They gave written informed consent to participate in
the experiment. They first visited the laboratory for
familiarisation with experimental procedures.

The experimental conditions entailed three trials at two
speeds of locomotion on each occasion. These speeds were 5 km
and 9 km h^{-1}, each for 5 min duration. The three sessions
incorporated normal running, dribbling a hockey ball and
dribbling a hockey ball in an enhanced crouched position on a
motor driven treadmill. A three-sided rebound box was erected
towards the front of the treadmill to keep the ball on the belt
whilst dribbling was being performed. The board took the spin
off the ball and returned it to the subject.

A 20 min rest period in Fowler's position (see Leatt et al.,
1985) preceded the 5 min exercise bouts (both 5 km h^{-1} and 9 km
h^{-1}) on each occasion. Back strength was measured pre-exercise
and post-exercise using a spring-loaded dynamometer (Taki Kiki
Kogyo, Tokyo). Perceived exertion was rated after each 5 min
exercise bout according to Borg (1970).

A reference measure of stature was taken at the beginning of
each session using the procedures described by Reilly et al.
(1991). The subjects had previously been trained on the
stadiometer until they could fulfil experimental requirements of
an SD of <0.5 mm over 10 consecutive measurements of stature.
Spinal shrinkage was measured at the end of the 5 min exercise
bout whilst regain in stature was measured 20 min post-exercise.

RESULTS

Shrinkage incurred was greater at 9 km h^{-1} than at 5 km h^{-1}.
This occurred at all the experimental conditions (Table 1).
This complies with previous observations of Garbutt et al.
(1990) that the exercise intensity affects the degree of spinal
loading induced. The effect of dribbling (both orthodox and
crouched) was significant, though only at the higher running
speed ($P < 0.05$).

Table 1. Mean (± SD) shrinkage in stature (mm) for each of the
two speeds and the three experimental conditions (n=6).

speed (km h^{-1})	Running	Dribbling	Crouched
5	1.8±1.3	2.0±0.6	2.0±0.9
9	2.0±1.5	2.6±2.1	2.5±1.6

The rating of perceived exertion was greater after crouched
dribbling than for orthodox dribbling, which was in turn higher
than for running. These differences were significant according
to analysis of variance (P<0.01). This trend was reflected in
the heart rate responses to exercise which were greatest for
crouched dribbling and lowest for normal running (Table 2).

Table 2. Rating of perceived exertion (RPE) and heart rate
(beats min^{-1}) for the two experimental speeds and the three
modes of activity (n=6).

speed (km h^{-1})		Running	Dribbling	Crouched
5	RPE	7.0±1.1	12.3±1.0	12.5±0.6
	HR	81 ±5	103±9	137±5
9	RPE	11.2±1.7	15.3±1.0	18.2±0.8
	HR	139±2	147±8	165±3

Crouched dribbling caused a decrease in back strength
compared to the other conditions (P<0.01). This result was
significant according to analysis of covariance when differences
in back strength at the start of each session were eliminated
within subjects (P<0.05). The mean value for back strength
pre-exercise was 100.4 N (see Table 3). The decrease in back
strength was greater following the higher running intensity of
9 km h^{-1} than that after 5 km h^{-1} (P<0.05).

Table 3. Mean back strength (N) at end of each exercise period
(n = 6).

speed (km h^{-1})	Running	Dribbling	Crouched
5	99.4	96.2	89.9
9	95.1	94.3	81.7

A significant inverse correlation was observed between
perceived exertion and back strength (r = -0.84 and r = -0.87).
The regain in stature was significantly related to back strength
(r = 0.85).

DISCUSSION
    The physiological responses to dribbling demonstrated that
the crouched posture imposed a greater stress than did the
orthodox dribbling condition. This was reflected also in the

perceptual responses to exercise. Reilly and Seaton (1990)
considered that the additional energy cost in dribbling may be
accounted for by postural factors and partly by the arm exercise
in using the hockey stick. Crouched dribbling may accentuate
postural factors, increase thoracic restrictions and further
increase stride rate (and decrease stride length); these might,
in combination, explain the severity of exercise in this
condition.

These trends were not duplicated by the shrinkage data. This
measure was more responsive to the speed of movement than to the
different experimental conditions. The magnitude of the
shrinkage was similar to that previously reported for male
hockey players (Reilly and Seaton, 1990). The shrinkage
incurred was significantly greater when dribbling than when
running normally at 9 km h^{-1}. A longer period of exercise might
have separated the effects of the crouched and normal dribbling
conditions but would not be representative of realistic
hockey-play.

Crouched dribbling did cause a greater deterioration in back
muscle strength post-exercise than the other conditions. The
intensity of exercise was also an influential factor. Whether
this decrease was due to muscle fatigue, back stiffness or a
decrease in motivation for maximal voluntary effect can not be
determined from the present study. The strength of the back
muscles does seem to be a relevant consideration, in view of the
inverse relationship with subjective effort and the regain of
stature post-exercise.

REFERENCES

Borg, G., 1970, Perceived exertion as an indicator of somatic
    stress. Journal of Rehabilitation Medicine, 2, 92-98
Cannon, S.R. and James, S.E., 1984, Back pain in athletes,
    British Journal of Sports Medicine, 78, 159-164.
Garbutt, G., Boocock, M.G., Reilly, T, and Troup, J.D.G., 1990,
    Running speed and spinal shrinkage in runners with and
    without low back pain. Medicine and Science in Sports and
    Exercise, 22, 769-772.
Leatt, P., Reilly, T. and Troup, J.D.G., 1985, Unloading the
    spine. In Contemporary Ergonomics, ed. D. Oborne (London:
    Taylor & Francis), pp. 227-232.
Reilly, T. and Seaton, A., 1990, Physiological strain unique to
    field hockey. Journal of Sports Medicine and Physical
    Fitness, 30, 142-146.
Reilly, T., Boocock, M.G., Garbutt, G., Troup, J.D.G. and Linge,
    K., 1991, Changes in stature during exercise and sports
    training. Applied Ergonomics, 22, 308-311.

AN ERGONOMIC EVALUATION OF BOARDSAILING HARNESSES

T. REILLY, E. BRYMER and M.S. TOWNEND

Centre for Sport and Exercise Sciences
School of Human Sciences
Liverpool John Moores University
Mountford Building
Byrom Street
Liverpool
L3 3AF
ENGLAND

This study was designed to evaluate three different
harnesses used by boardsailors. These were a chest, a
waist and a seat harness. The harnesses were evaluated
using force data, mathematical modelling, EMG and
spinal shrinkage during 30 min of simulated boardsailing. Results indicated an effect of skill level,
body mass, sailing stance and harness type. The better
harnesses were the chest and seat type: their specific
effects reflected individual differences.

INTRODUCTION

The sailboard was first patented in 1968. Boardsailing or
windsurfing received formal international recognition when it
was accepted as an Olympic sport in 1984. It has now grown to
enormous popularity as a water-sport and the design of equipment
for the sport has attracted specialist attention. This applies
not only to the boards and rigs (comprised of sail, mast and
boom) but also to the equipment designed for the boardsailor.

One of the major innovations in apparatus has been the
harness, introduced to increase time to fatigue by supporting
the body and decreasing the muscular effort needed from the arms
and shoulders. The chest harness was designed to decrease
discomfort and it eventually was altered to a waist fitting.
This was converted into a seat harness by attaching webbing to
it and attaching the webbing around the thighs. The migration
of harness design from the chest to the seat meant that the
point of application of force moved from the chest to the waist
and on to the hips. At present all three styles of harness are
available, albeit in a range of design forms.

The muscular effort demanded in boardsailing is mainly
isometric: the lower back, shoulder and arm muscles have been
identified as suffering the greatest strain (Loquet et al.,

1984; Van Gheluwe et al., 1988; Allenand Locke, 1989). The
lower back has also been implicated in investigations of injury
profiles among participants (Ullis and Anno, 1984). There is
still a lack of data on forces involved in this activity, the
loading incurred on the spine and how these might be affected by
harness type. The purpose of this study was to evaluate three
different harnesses – a chest, a waist and a seat harness–used
by boardsailors.

METHODS
    Nine subjects (7 male, 2 female) volunteered for the study
after giving written informed consent. The subjects, aged 19–27
years, were classified according to their level of experience in
boardsailing as low skill (less than 6 months harness use),
intermediate (6–24 months harness use), high skill (more than 24
months harness use). There were three subjects in each class.
    Each subject undertook a 30 min period on a sailboard
simulator designed to induce conditions approximating Force 5 on
the Beaufort scale. The performance was undertaken for 30 min
(at the same time of day and on separate days) using each of the
three harnesses and a fourth session when no harness was worn
acted as a control.
    The simulator consisted of a 2 m tall metal frame constructed
of steel tubing. The frame was connected posteriorly and
anteriorly by 45.7 cm lengths of the same tubing. A 48 cm 'H'
frame was welded to the top of the construction, the cross-piece
being positioned parallel to the sides with a pulley connected
to it. The sailboard was 3.6 m long and had a universal joint
to permit movement in three planes. The rig consisted of a mast
cut down from the top to 3 m, with a notch at its top. The boom
was attached at a height of 1.54 m, corresponding to the
shoulder-line of a sailor 176 cm in stature. Rope was attached
from the clew end of the boom to the back and top of the mast.
The boom was fixed with a 90° angle between it and the mast. A
further Y-shaped web of rope was connected from the top of the
mast to both ends of the boom. The common point of the Y
corresponded to the centre of pressure for a sail 5.6 m^2
(Townend, 1984). The force corresponding to sailing with a
5.6 m^2 sail in the minimum wind speed registered as Force 5 on
the Beaufort scale was 23 kg. This was applied at the centre of
pressure via a line that extended from the centre of pressure
and over the pulley.
    Harness to rig forces were recorded during each session. A
load cell (Sengamo, Sussex) was connected between the harness
line and the harness. The results were recorded using an
analogue meter (Ter voltage recorder). The system was
calibrated using masses of 20, 25, 35 and 40 kg. Photographs
were taken simultaneously with the force measurements for a
two-dimensional mathematical simulation of all harness
conditions (Townend, 1984). The predicted values of rig forces
were compared with forces measured during the tests.
    Electromyography (EMG) of M. Latissimus Dorsi, M. Rectus
Abdominis and M. Vastus Lateralis was obtained using surface

electrodes on both sides of the body. The electrodes were connected by cable to a polygraph (NEC Sanei-Instruments, Tokyo) for recording of the raw EMG. The polygraph had bioelectric amplifiers with filters set to record a bandwidth of 10 to 1000 Hz. The electrode arrangement provided a balanced input for the amplifiers, minimising 50 Hz artefacts and maximising the signals' common mode rejection ratio. Traces were recorded throughout the 30 min session and the EMG was normalised by reference to a 6 s maximum isometric contraction prior to the commencement of the test.

Precision stadiometry was used to evaluate spinal shrinkage (Wilby et al., 1987). Subjects were trained prior to the study for measurements on the stadiometer. They were deemed to be trained when 10 consecutive recordings were produced with an SD <0.5 mm. Measurements were obtained before and after the simulated boardsailing session which was preceded by a 20 min period of standing at rest.

The four conditions were administered in random order on separate days. When not wearing a harness (the control condition), the subjects were unable to maintain their activities continuously and were allowed to rest: the number of short rest periods ranged from 5-9 for this condition.

RESULTS AND DISCUSSION

The harnesses were effective in reducing fatigue. All subjects were able to complete the 30 min sessions without interruption for rest, except in the control condition.

The highest mean forces were observed whilst using the seat harness. The mathematically derived forces overestimated the measured values (Table 1), indicating that the two-dimensional model was inadequate. The sailing action was found to be three-dimensional in nature, consisting of twists to face the direction of motion, lean away from the direction of motion and lean away from the rig. A three-dimensional model should be developed to reflect these observations. The forces recorded varied according to the extent the subjects were able to maintain a relaxed stance on the board, the more relaxed stance being associated with the lower forces.

Table 1. Comparison of means of the mathematically determined and measured forces (N) for the three harness types.

Harness	Measured	Calculated
Chest	289.5	366.7
Waist	273.6	408.1
Seat	335.0	441.7

The greatest shrinkage was found to occur with the waist harness ($P<0.05$). In some subjects this harness had similar effects to the 'no harness' condition. A significant correlation was observed between skill level and shrinkage during use of the chest harness ($r = 0.731$; $P<0.01$), the more

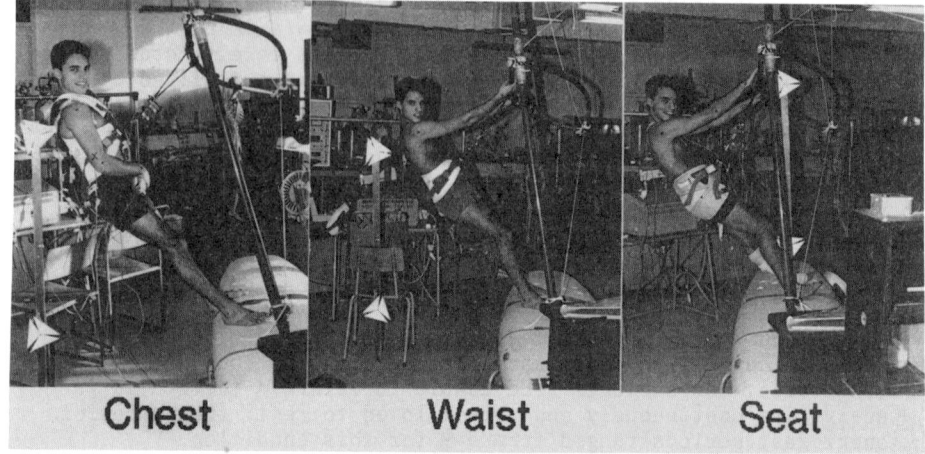

Fig 1.   Use of chest (left), waist (middle) and seat (right)
         harness on the simulator

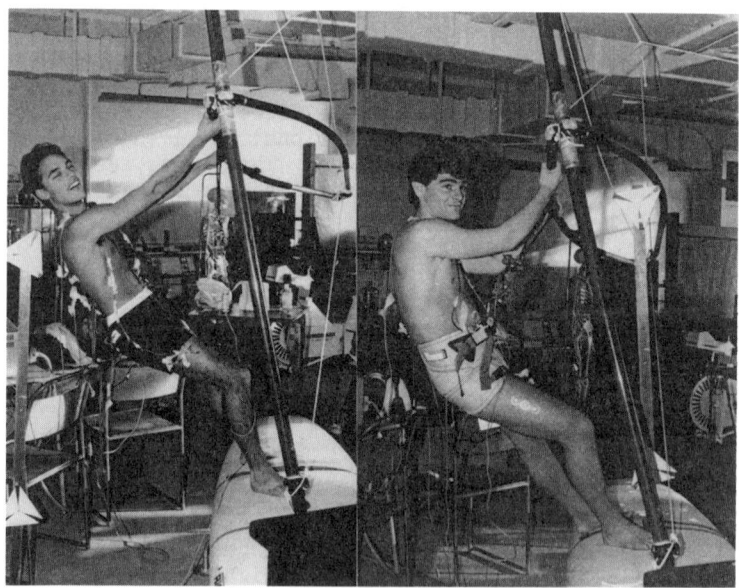

Fig 2.   'Two-dimensional' stance with no harness (left)
         contrasted with a twist observed in the sailing
         stance (right).

experienced subjects incurring the highest losses in stature. An increase in stature was observed in the experienced sailors using the seat harness, the correlation between change in stature and skill level being r = 0.773.

Besides the skill level, body mass also seemed to influence the results. Change in stature and body mass were significantly correlated during two of the conditions. The correlations between shrinkage and body mass were 0.937 for the chest harness and -0.827 for the seat harness. These indicated that the heavier subjects incurred more spinal loading using the chest harness but least using the seat harness. When regression lines for body mass and change in stature were superimposed for the chest and seat harnesses, the point of intersection was at 70kg.

The EMG traces reflected the differences in posture adopted when using the different harnesses. Observations confirmed that boardsailing with a harness tended to decrease muscular effort. The reductions in EMG were evident for M. Rectus Abdominis and M. Latissimus Dorsi but not for M. Vastus Lateralis which was less influenced by harness use. Normalised EMG values (as % MVC) indicated that the highest strain was on M. Latissimus Dorsi for the seat harness. A majority of the recordings exceeded 5% MVC, suggesting that muscular fatigue should affect the boardsailor during a prolonged session (Hagberg, 1981). High correlations were consistently found between the different muscles, indicating fatigue would be manifest among the active muscle groups. For the chest condition, a negative correlation was noted (r = -0.99) between the normalised EMG of M. Vastus Lateralis and body mass, indicating that the lighter subjects relied more on exerting force through the leg.

Overall, the results demonstrated that all harness types were effective in reducing fatigue. The waist harness induced the greatest amount of shrinkage. Individual differences in body mass, sailing stance and level of experience influenced the effects of the chest and seat harnesses. It is likely that environmental conditions will further affect results in realistic boardsailing. Results on the simulator would lead to the recommendation of the seat harness for experienced sailors and those of body mass 70 kg or greater, and the chest harness for less skilled and/or lighter performers.

ACKNOWLEDGEMENT

The authors gratefully acknowledge the help of staff at "Ultra Sport" for providing the harnesses for this research.

REFERENCES

Allen, G.D. and Locke,S., 1989, Training activities, competitive histories and injury profiles of elite boardsailing athletes. Australian Journal of Science and Medicine in Space, 12 (2), 12-14

Hagberg, M., 1981, Muscular endurance and surface electromyogram in isometric and dynamic exercise. The American Physiological Society, 51, 1-7.

Loquet, H., Guilbert, J., Jacquesson, G. and Milbled, G., 1984,
     Etude electromyographique sur simulateur de la pratique de la
     planche a voile. Larc Medical, 5, 306–312.
Townend, M.S., 1984, Mathematics in Sport. Chichester, Ellis
     Horwood.
Ullis, K.C. and Anno, K., 1984, Injuries of competitive board
     sailors. The Physician and Sportsmedicine, 12 (6). 86–93.
Van Gheluwe, B., Huybrechts, P. and Deporte, E., 1988,
     Electromyographic evaluation of the arm and torso muscles for
     different postures in windsurfing. International Journal of
     Sports Biomechanics, 8, 67–78.
Wilby, J., Linge, K., Reilly, T. and Troup, J.D.G., 1987,
     Spinal shrinkage in females: circadian variation and the
     effects of weight-training. Ergonomics, 30, 47–54.

METHODS FOR USER INTERFACE PERFORMANCE MEASUREMENT

PATRICK W. JORDAN

UIPM Group
Glasgow Interactive Systems cenTre
Department of Psychology
University of Glasgow
Glasgow G12 8QQ
Scotland

Seven methods for assessing human-machine interaction
are outlined. Each of the methods involves taking measures
of performance at the interface - making them distinct from
analytic methods, such as task analyses or "expert
walkthroughs". The methods are compared along a number
of dimensions, to help the investigator choose a set of
methods for an evaluation.

INTRODUCTION

There are a number of approaches which may be taken to the
assessment of user interfaces. These include task analyses, expert
walkthroughs, and measurement of user performance (see Jordan 1992
for an overview). Seven methods for the latter approach are reviewed
in this paper.

Information gained from the methods may be used for a variety of
purposes. Examples are comparing the performance of two similar
products, detecting bugs in the interface, or seeing whether an
interface is suitable for a particular task, or a particular group of users.

Everyone concerned with user interfaces should find the methods
useful. There is no substitute for seeing users struggling with tasks
which might otherwise have been thought easy. Even small amounts of
practical measurement with prototypes may bring great benefits to a
design.

The seven methods reviewed are: Focus groups, Think aloud protocols,
Incident diaries, Feature checklists, Questionnaires, Semi-structured
interviews, and Experiments. The order of presentation is loosely based

on the degree of formality of each method - starting with the informal focus group, and moving through to the tightly controlled experiment.

THE METHODS
Focus groups

The focus group is a group of users brought together to discuss a particular issue. Discussion could centre on, say, users' experiences of using an interface, or ideas about the sorts of characteristics and functionality they would like a system to have (requirements capture).

The leader of a focus group will have an agenda of questions, in order to ensure that a range of topics are discussed. The leader's task is to ensure that everyone in the group gets the chance to have a say. However, he or she should try to let the users talk as freely as possible, and avoid "leading" questions.

The leader need not necessarily be the designer of the interface under review. However, it is often useful for the designer to sit in on the group and listen to what is being said. Often his or her participation may involve little more than introducing themselves and then listening to the discussion, however they may decide to join in, or perhaps answer questions if appropriate.

An advantage of the focus group is that it can be used at any stage of the design. Users may discuss an idea, a prototype, or their experiences of using a finished product. Because users may have little idea of, say, technical constraints on design solutions, they won't be able to tell the designer exactly what to do, and some of their suggestions might be impractical. However, data from focus groups can still provide the designer with information about the ways in which users interact with products and the problems that they may have - things which might otherwise be difficult to anticipate. O'Donnell, Scobie and Baxter (1991) report the use of focus groups in evaluating the interface to a central heating controller.

Think aloud protocols

This method involves asking a user to give some form of commentary about that they are doing and thinking whilst using an interface, in response to prompting from an investigator. Meanwhile, the investigator observes, listens, and takes notes.

The think aloud is a particularly useful method as it helps to identify not only what kinds of problems occur, but also why they occur. This means that of lot of useful information can be gained from each user. During the think aloud session the users may be set specific tasks to perform, or may simply be asked to explore the an interface freely. The former approach is useful for identifying specific bugs, whilst the latter may help to explain why users use some parts of an interface yet ignore others.

The investigator may vary the level of prompting that they give the users. The more the user speaks the more information the investigator is likely to gain. However, too much prompting may interfere with user performance, or lead to them making things up in order to answer a question.

Incident diaries

These are mini-questionnaires which users are asked to complete when an "incident" of some kind occurs. This might mean, for example, when a user makes an error or has difficulties on a task, or when they are confused by a system action. Typically a diary might contain questions about what type of problem has occurred, how troublesome a problem was to the user, and if and how it was resolved.

Incident diaries are useful when relatively infrequent problems occur, and the investigator cannot be there to observe them. The method is also cheap in terms of investigator time and effort, as having decided on a set of questions, these can them be copied and sent to many users. However, it s not always clear whether respondents are describing problems accurately - indeed they might sometimes lack the technical vocabulary to do this. It is also important that the diaries are short and clear, so that they can be completed quickly and easily - otherwise users may not bother completing them.

Feature checklists

Basically a feature checklist is a list of an interface's features or functionality. Users are asked to tick the features that they have noticed, or used before. This information enables the investigator to see whether the features that the designer has provided are used in practice. Features and functionality can be listed in a number of ways, for example as a list of command names, or as semantic descriptions of particular tasks. Edgerton (1992) found that checklists offered considerable advantages over open recall, in the context of asking respondents to recall commands used at an interface.

Questionnaires

These are printed lists of questions. Depending on the design of the questionnaire, the respondent may be given scales or boxes to mark, so that responses may be "scored", or they may reply to open ended questions.

The longer a questionnaire is, the less likely a user is to complete it. It can be difficult to get users to complete mailed out questionnaires, and often the return rate is less than 25%. Unfortunately, the 25% who do return the questionnaire may not be typical of the group that you wish to hear from. Indeed, it may be that those motivated to complete and return the questionnaire are people with particularly strong, and thus perhaps unrepresentative, views. It is particularly important that questionnaires sent by mail are completely self explanatory. One way of avoiding low response rates is for the investigator to stay with the respondent while he or she completes the questionnaire. This enables the investigator to help explain any questions which the user might not understand, and to ensure that no questions are missed out..

Questionnaires can be a comparatively cheap way of obtaining information, as once a questionnaire has ben designed it can be issued to any number of users However, if the investigator has to monitor as they complete the questionnaire, then the cost of administering it will increase sharply. A number of pre-validated usability questionnaires are available (for example, the Index of Interactive Difficulty (IID)[1] (Jordan and O'Donnell 1992)).

Semi-structured interviews

With this method the investigator constructs a series of questions to use as a schedule for the interview. The investigator will then sit with the interviewee and talk informally with them around this schedule. Each item on the schedule has an opening question and a number of prompts which the investigator can use to help form follow-up questions.

After the interview, the responses may be sorted into pre-determined categories and quantified. However, a major benefit of this method is that the user may say something unexpected about the interface, which the investigator would not have anticipated otherwise. This can be a source of useful qualitative information.

Experiment

Here the investigator sets subjects a series of tasks, and quantifies performance in terms of measures such as time on task and number of errors made. The objective data obtained from experiments is often useful in addressing specific design questions. For example, if considering whether, say, a word processing package should be menu or command entry driven, two alternative versions of the package could be constructed, and then compared experimentally over a series of tasks.

Experiments need to be tightly controlled, so that measurements taken are fairly precise. This can take time and effort and a degree of specialist knowledge may be required. However, as the most formal of the methods reviewed, experiments probably provide the most unambiguous data.

CHOOSING METHODS FOR AN EVALUATION

Often it is appropriate to use the methods in combinations. For example, after conducting an experiment with an interface it might be appropriate to use, say, a questionnaire at the beginning to check for users' previous experience with an interface. Perhaps a semi-structured

---

[1]The IID can be obtained free of charge from Glasgow Interactive Systems cenTre, University of Glasgow, Glasgow G12 8QQ, Scotland.

interview might also be conducted afterwards, to elicit opinions about the interface used. Subjects might also have been asked to think aloud at some point.

When choosing methods for an evaluation, it may be useful to compare them along a number of dimensions. This enables the investigator to see how the various methods fit in with any constraints on an evaluation One dimension, which has already been mentioned, is cost. Questionnaires, incident diaries, and checklists are relatively cheap for the investigator, as it is the respondent who bears much of the cost, in terms of time, effort, and distraction from other things. At the other end of the scale are experiments, which require a great deal of preparation, and usually the presence of the investigator whilst being run.

Another dimension is the number of users likely to be needed in an investigation. Think alouds and focus groups can start generating useful information with only small amounts of users, whilst, once again, experiments are at the other end of the scale, perhaps requiring around a dozen users before anything can be reliably established.

The methods also differ in that some are used for taking measures at the time of interaction, whist some are retrospective. With experiments and incident diaries, for example, the performance is measured at the time, whereas with, say, checklists or questionnaires, the user may be asked to recall previous interactions. Another issue is the type of data required. For example, experiments are a source of objective performance measures such as time and errors, whilst, say, the responses to a semi-structured interview reflect users' subjective opinions.

Whichever methods are applied, it is important that the users involved in the investigation are really those for whom the interface was designed. In particular, it may be wrong to expect the staff of a design office to be representative of a user population. Expert knowledge, particularly participation in the design process, can make things seem obvious, when to many people they are not obvious at all.

CONCLUSIONS

Seven methods for user interface performance measurement have been reviewed. Hopefully this has given an insight into their properties and how they may be helpful in interface evaluation. For more detailed information about the methods, and some examples of their application, it may be worth following up some of the references given below. Even a small amount of measurement of real users performing tasks with an interface can make the difference between a design that works well and one that users ignore.

Acknowledgement

The work of the UIPM group is supported by DTI / SERC grant GR/F/39171/IED4/1/1109 - Measurement of User Interface Performance.

Video
The application of the methods discussed are illustrated on a video "User interface performance measurement: assessing human-machine interaction", presented by Patrick Jordan, and available from: Media services, University of Glasgow, 64 Southpark Avenue, Glasgow G12 8LB, Scotland.

REFERENCES
Edgerton E.A., 1992, A comparison of the feature checklist and the open response questionnaire in HCI evaluation, Computing Science Research Report GIST - 92 - 2, University of Glasgow.
Jordan P.W., 1992, Making friends with the user, Professional Engineering, October 1992.
Jordan, P.W.., and O'Donnell, P.J., 1992. The Index of Interactive Difficulty. In EJ Lovesey (ed), Contemporary Ergonomics 1992. (London: Taylor and Francis).
O'Donnell P.J., Scobie G., and Baxter I.,1991, The use of focus groups as an evaluation technique in HCI, People and Computers VI, Cambridge University Press, 1991.

A CHECKLIST FOR THE ASSESSMENT OF DISPLAY SCREEN EQUIPMENT

M ANDERSON   and   E D MEGAW

Human Focus Limited   Industrial Ergonomics Group
Mill Green Business Park   The University of Birmingham
Mitcham   Birmingham
Surrey CR4 4HT   Edgbaston B15 2TT

Recent health and safety legislation requires employers to perform a risk assessment of display screen equipment (visual display units). In the present study, a checklist with accompanying guidance was developed to enable assessors with minimal instruction in ergonomics to carry out this risk assessment. This paper describes how the checklist was developed and highlights the problems that were encountered during this process.

INTRODUCTION

The European Community Directive on 'the minimum safety and health requirements for work with display screen equipment' (90/270/EEC) was adopted on the 29th May 1990. The aim of this Directive is to reduce the potential health risks associated with the use of such equipment; namely, physical (musculoskeletal) problems, visual fatigue, and mental stress. Although the Health and Safety Executive consider work with display screen equipment to be a low-risk activity (HSE 1992), the large number of users makes it desirable to attempt to reduce the risks involved.

On the first of January 1993, the "Health and Safety (Display Screen Equipment) Regulations 1992" implemented the terms of this particular Directive in the United Kingdom.

This legislation consists of nine Regulations, and a Schedule of 'minimum requirements' that must be met by equipment covered by the Regulations. The employers' obligations can be summarized as follows -

- Analyse workstations to assess health and safety risks.
- Reduce any risks found to the lowest possible levels.
- Provide training for display screen equipment users.
- Provide eye and eyesight tests, and corrective appliances if necessary.
- Inform users of any health and safety measures taken.
- Design jobs to provide rest breaks or changes of activity.
- Ensure that display screen equipment meets certain minimum requirements as outlined in the Schedule to the Regulations. These cover the equipment, the environment, and the operator-computer interface.

GENERAL PROBLEMS WITH THE DIRECTIVE/REGULATIONS

The emergence of the Directive on display screen equipment caused some surprise among manufacturers of such equipment, as they were not consulted in its production. In addition, when taken together with the other Regulations effective from the first of January 1993 (such as those concerned with manual handling), this legislation places a substantial burden upon the employer.

Hughes (1992) suggests it would have been more appropriate to produce a general directive on office work, and within this 'work with display screen equipment' could have been a subset. As many of the requirements of the Directive 90/270/EEC apply to office work generally, this would appear to be a more logical approach. Therefore, it is felt that unnecessary attention has been directed to people defined as display screen equipment "users", with the needs of the rest of the office population largely ignored.

It is not clear whether the analysis of risk and compliance with the minimum requirements of the Schedule are independent of each other. It is likely that if a risk assessment is conducted and no risk is found, then the display screen equipment will comply with the Schedule to the Regulations. Likewise, if the equipment meets the minimum requirements of the Schedule, then many of the risks of display screen work will not be present.

It must be stressed, however, that a risk assessment should not be solely based on the Schedule; but should take into account other factors covered in the Regulations such as work routine, users' eye-sight, and the provision of training/information for users.

PROBLEMS WITH THE SCHEDULE TO THE REGULATIONS

Several of the minimum requirements are quite vague, making it difficult to ensure compliance with them. Some are difficult to interpret, while others cannot be simplified to a checklist level. Commenting on the Schedule, Hughes (1992) states that:

'The approach is fundamentally flawed. Instead of stating clearly the Ergonomic requirements that a workstation must achieve, for example, 'VDUs shall be easy to read accurately and without discomfort by the expected user population in their working environment for their VDU tasks', it contains a hotch-potch of incomplete and ambiguous screen characteristics' (p. 3).

Stanton and Baber (1992) point out that various important topics are omitted completely, such as interaction styles and the use of different input devices.

In the guidance to the Regulations, the Health and Safety Executive suggest that compliance with the minimum requirements can be achieved by referring to the British Standard BS 7179. This standard - "Ergonomics of design and use of visual display terminals (VDTs)" was designed to "...provide recommendations to promote the health and safety of visual display terminal (VDT) users to ensure that they can operate VDT equipment efficiently and comfortably" (Part 1, p.2). It is intended that the standard should be of use to manufacturers, purchasers and users of VDT equipment.

There are several inconsistencies between the requirements of the Schedule and the recommendations of BS 7179. For example, the Regulations require an absence of flicker on the display screen whilst BS 7179 suggests a criterion of no perceived flicker for 90% of the user population.

This standard also makes no reference to the operator-computer interface for example; making compliance with this section of the Schedule very difficult. There is little guidance in the Regulations to aid employers with the assessment of the interface, and the relevant parts of the International Standard IS 9241 are not yet available.

Therefore, it is not clear how employers are to go about assessing their display screen equipment in this respect. As an example, Stewart (1992) questions how one can determine whether a piece of software is '...adaptable to the users' level of knowledge or experience'. He suggests that the drafters of the Schedule 'have mistakenly assumed that the software interface can be treated at the same level of specificity as the anthropometric requirements for workstation furniture'. (p. 136).

It is also suggested in the guidance to the Regulations that the requirements of the Schedule can be met by reference to the International Standard IS 9241 "Ergonomic requirements for office work with visual display terminals". This standard (and the subsequent European standard EN 29241) represents the best ergonomics consensus; however, at the time of writing it is largely incomplete. It is not expected to be produced in full until 1994 at the earliest, which is way out-of-step with the required date for implementation of the Directive (December 1992).

Ideally, the minimum requirements of the Regulations should have been held back until the relevant parts of IS 9241 are produced. The Schedule could then refer to the appropriate sections of this standard instead of listing its own vague and ambiguous requirements. This would have prevented a great deal of confusion for those who are required to comply with its specifications.

Some employers may decide to postpone the risk assessment of their display screen equipment until the production of IS 9241 is complete. This standard may provide a good basis for an assessment checklist; however, by delaying the risk assessment, employers may not be fulfilling their immediate obligations of the Regulations.

The requirements of the Schedule (and the Regulations as a whole) cover all display technologies (such as liquid crystal displays), whilst the guidelines in BS 7179 (Part 3) and IS 9241-3 only refer to conventional cathode ray tube (CRT) displays. This is a great problem with standards on the whole - the technology is changing rapidly and the standards are often out-of-date when they are eventually produced. Recommendations such as the reduction of ambient lighting to between 300 and 500 lux may be understandable for negative-polarity screens (light text on a dark background); however, with an increase in the number of positive-polarity screens (and positive-polarity software such as 'Microsoft Windows'), higher levels may now be set.

As with BS 7179, there is a mismatch between the requirements of the Schedule and the contents of IS 9241. As Hughes (1992) points out, radiation and noise are not covered at all in this standard.

The Schedule contains no actual criteria for compliance, which means that there can be no certification of products. Manufacturers are able to state 'Complies with BS 7179: Part 3: 1990' for example; however, statements such as 'Complies with the requirements of the EC Directive/Regulations' are meaningless.

This situation is complicated by the fact that several of the requirements, for example regarding flicker, are dependant on an interaction of several aspects of the display screen equipment including the software, hardware and the environment. It is therefore difficult to identify whether the hardware or software manufacturer can state compliance, if any.

A more general problem with the use of standards in the interpretation of the Schedule is that several of the requirements are concerned with how the display screen equipment is used at a workstation, not merely the design specification of such equipment. Stewart (1992) states that '...the analysis requirements ...involve assessing individual workplaces, not just samples or standard configurations' (p. 127).

CONSTRUCTION OF THE CHECKLIST

It is important that the risk assessment checklist produced can:
- identify possible risks to health and safety.
- identify whether equipment conforms to the minimum requirements of the Schedule to the Regulations.
- be used by an assessor who has received minimal instruction in ergonomics.

The Schedule to the Regulations formed the basic structure of the checklist. Each of the minimum requirements was translated into a number of appropriate questions that could, in most cases, be answered with a 'YES' or 'NO' response. Questions were also devised to obtain information on the task performed, the daily work routine and the users' eyesight.

Where possible, BS 7179 and IS 9241 were used in constructing appropriate questions. However, due to the above problems with these standards, several other sources were referred to. These included - ANSI/HFS 100-1988, BS 5940 (Office furniture), the German standard DIN 66234, the Swedish 'Screen Facts' and 'Screen Checker', CIE Lighting Guide (1984) and various Health and Safety Executive documents (for example, 'Visual Display Units' [1983] and 'Seating at Work' [1991]).

A template (overlay) was designed to aid physical measurements, for example the dimensions of screen characters. By placing this template over the character it was possible to determine at a glace whether it complied with the recommendations for character height and width.

As an illustration of the checklist design, one of the minimum requirements is: "The characters on the screen shall be well-defined and clearly formed, of adequate size and with adequate spacing between the characters and lines". This was translated into the following questions:

- Measure the height of a capital 'M' on the screen using the template provided. Is it in the range 3 mm to 4 mm?
- Enter three capital 'H' in a row on the screen. From your viewing position, can you see a clear space either side of the middle letter?
- Enter the following two rows of letters on the screen. Do any of the letters touch?
      ygpjppjqp
      khbkthhkt
- Usually the characters on the screen are made up of dots or segments (a dot-matrix display). Are these individual dots visible from your normal viewing position?
- Some characters may be easily confused on the display screen. Type the following pairs of letters and numbers: QO Z2 XK IL I1 S5 S8 B8 VU O0 (capital O/zero). Can they all be easily distinguished from your normal viewing position?

To aid interpretation of the checklist, some response boxes indicated that the user should provide additional information. For example, if the user felt that their chair was not comfortable, they were requested to state exactly what was wrong with it.

The checklist thus consisted of a combination of objective and subjective questions, to increase the amount of information generated by an assessment.

The assessment supervisor scored the checklist by placing a marking template over the response columns, revealing answers that did not meet the requirements or recommendations.

It is acknowledged that many assessors may experience great problems when interpreting the results of the assessment and implementing remedial action. Therefore, for each of the checklist questions, some guidance was provided in a separate document. This guidance was compiled from sources that included those used in the writing of checklist questions. As an example, if the user states that there are reflections on the screen, the guidance suggests how they could be reduced.

Pilot studies were conducted at several stages in the construction of the checklist to determine whether the wording of the questions was understandable and obtained the required information, and to highlight any possible problems that may arise in a full risk assessment of display screen equipment.

CONCLUSIONS

The checklist produced permits a more objective and structured assessment of display screen equipment than that achieved by reference to the original terms of the Regulations. Although the checklist is time-consuming to administer, the Regulations do require that the risk assessment is comprehensive.

By its very nature, the checklist had to be simple to use. The relevant standards are frequently very technical in nature, and therefore some of the recommendations could not be translated into suitable checklist questions (for example, measurement of modulation transfer function area [MTFA], or screen displacement).

It was particularly difficult to interpret the requirements of the operator-computer interface section of the Schedule. This was partly due to the vagueness of the minimum requirements, and partly due to a lack of any established criteria in this area. This section of the checklist may be considered deficient until the publication of IS 9241 (especially parts 8 to 19).

In the writing of guidance for a risk assessment checklist it is not possible to identify all problems that may occur and provide comprehensive suggestions for remedial action. This is especially the case with the assessment of display screen equipment, as there are many different display screen tasks and technologies, and risk may arise from a variety of interacting factors.

The assessor also needs to bear in mind that any interpretation of the Regulations and the checklist needs to take the individual circumstances into consideration. Should the assessment identify any problems that are not simple to rectify (for example, complex lighting situations), assistance from an expert may be required.

In spite of these problems, the material generated in the study enables an assessor with little prior knowledge of ergonomics to identify and evaluate risks, and to implement some remedial measures.

REFERENCES

ANSI-HFS 100-1988. The American National Standard for Human Factors Engineering of Visual Display Terminal Workstations. The Human Factors Society Inc. (1988).

British Standard BS 5940 (1980). Office furniture. Part 1. British Standards Institution, London.

British Standard BS 7179 (1990). Ergonomics of design and use of visual display terminals (VDTs) in offices. Parts 1 - 6. British Standards Institution, London.

CEC (1990) Minimum Safety and Health Requirements for Work With Display Screen Equipment Directive (90/270/EEC) Official Journal of the European Communities No L 156, 21/6/90.

DIN 66234. (1981). Characteristic values for the adaptation of workstations with fluorescent screens to humans. Parts 1 - 9. German DIN Association.

Health and Safety Executive (1983). Visual display units. HMSO, London.

Health and Safety Executive (1991). Seating at work. HMSO, London.

Health and Safety Executive (1992). Display screen work: Guidance on the regulations. HMSO, London.

Hughes, J. (1992). The Manufacturer's perspective. Paper presented at the VDU and Display Screen Equipment Seminar, Cavendish Conference Centre, London, 16 July 1992.

IS 9241. (1992). Visual display terminals (VDTs) used for office tasks - ergonomic requirements. Parts 1 - 4. International Standards Institution.

Stanton, N. and Baber, C. (1991). Introducing EC directive 90/270/EEC. The Ergonomist, November 1991, No. 257.

Stewart, T F M (1992). The role of HCI standards in relation to the directive. Displays, 13 (3), 125 - 133.

TCO (1986). Screen checker. The Swedish Confederation of Professional Employees, Stockholm, Sweden.

TCO (1991). Screen facts. The Swedish Confederation of Professional Employees, Stockholm, Sweden.

USER STRESS IN AUTOMATIC SPEECH RECOGNITION

R.GRAHAM* and C.BABER**

*HUSAT Research Institute
The Elms, Elms Grove
Loughborough, Leics LE11 1RG

**Industrial Ergonomics Group
School of Manufacturing and Mechanical Engineering
University of Birmingham, Birmingham B15 2TT

The change in users' speech under stress is one factor
which may prevent automatic speech recognition (ASR)
from fulfilling its potential. Subjects' speech during a
dictation task under time pressure was recorded, and
passed through a speech recognizer. Both the dictation
performance of the speaker and the recognition accuracy
of the ASR device were found to be significantly lower
in the stress condition. The results were explained in
terms of vocal and cognitive effects in the stressed
speaker.

INTRODUCTION

The capabilities and use of automatic speech recognition
(ASR) have increased dramatically since the first commercial
devices appeared in the early 1970's. ASR has been successfully
implemented, for example, in industrial inspection, avionics,
postal sorting and computer aided design. It allows increased
mobility around the work area and hands- or eyes-free operation.
For example, in an avionics application, by shifting workload
from the visual modality to the auditory, a pilot can maintain
his gaze out of the cockpit while simultaneously giving commands
to his system. Other advantages of speech over alternative forms
of input include reduced cognitive demands and the relative ease
in training users.

However, ASR continues to fall short of manufacturers claims.
Some of the problems are due to the inability of recognizers to
cope with variations in the user's speech in different
situations or conditions. Stress, fatigue, noise and the
performance of the system itself have all been identified as
factors which may contribute to such vocal inconsistencies.
Since most ASR devices work on the principle of matching speech
to stored templates, any speech variations will have a
detrimental effect on ASR performance. In order to make devices
more robust, it is necessary to know the extent and nature of
these effects. The present study is concerned with one of the
possible causes of speech variation, namely stress.

Three main types of stress have been distinguished: physical stress results from such stressors as noise or vibration whereas cognitive stress arises from excessive workload,which might be caused by a number of concurrent tasks or the speed with which tasks must be performed Finally, affective or emotional stressors are more subjective and long-term in nature, and as such they are less relevant to ASR.

The effects of stress on the speech signal have been studied since the start of the century, one of the first findings being a tendency for subjects to talk more loudly under high levels of ambient noise. In many subsequent studies, stress has also resulted in an increase in the mean fundamental frequency or the pitch variability of the voice. Another common observation is an increase in speaking rate and an associated decrease in word duration (see Scherer, 1981, for a review). As well as these vocal effects, the higher processes of speech planning and execution may be affected by stress, causing an increase in the number of speech errors made.

The consequences of such changes for ASR, however, are less well documented. Studies into ASR in noise have found performance decrements at high levels; Paul et al (1986), for example, observed that a negligibly small recognition error rate in quiet conditions rose to 3% at 95dB, 48% at 105 dB and 62% at 115dB when the noise entering the microphone was filtered. Vibration and acceleration appear to have a less drastic effect, with small performance effects at very high stress levels.

Cognitive stress effects on ASR have been tested by giving subjects a concurrent task, or tasks, to be carried out with the primary speech task. Simpson (1986) compared a simple aircraft warning classification task with a condition where a secondary task of target tracking was added. Although fewer errors were made in the simple task, this difference was not significant due to the large individual differences across subjects. Further investigation found that most of the errors caused by stress were due to an insufficient pause between words. Hence, the problem could be solved by using a connected-word recognizer rather than an isolated-word system. Other researchers have found similar non-significant results with concurrent task loading.

Time pressure, another form of cognitive stress, might be expected to have more significant effects due to an increase in speech rate. However, an extensive literature search has found no work directly relating to time stress and ASR, and the present study attempted to change this.

The main aims of the study were to test the effects of time stress on the human speaker and on the ASR device. These measures have different consequences for the use of ASR. If human performance (in terms of dictation accuracy) is poor under stress, extra training or coping strategies might be necessary. Alternatively, if the speaker performs well but the speech recognizer's ability is affected, additional work into developing more robust recognition algorithms is required. The ultimate aim of such research is to ensure that the overall human-machine recognition system is resilient to stress effects.

METHOD

Fifteen subjects were tested at the Speech Research Unit at DRA, Malvern, all of whom were closely involved in the speech research field. They comprised eleven males and four females, aged 20 to 38, with Southern English or non-extreme regional accents.

Video recordings were made which panned across a sequence of parked cars. These were played on a standard 20-inch colour television set via a VHS video. Subjects wore a headset mounted microphone and were required to speak the number-plates of the cars using the ICAO alphabet (alpha, bravo, charlie etc) and the digits zero to nine (eg. 'bravo one zero six lima lima papa'). Their speech was recorded onto DAT tape and later passed through the AURIX (ASTREC-216) speech recognizer. This continuous-word system used speaker-independent, male or female templates and a 500-word vocabulary.

Those subjects who were not already familiar with the ICAO alphabet were given a period of training until they were able to read characters without hesitation or error. They were then informed of the experimental structure and seated in a low-lit soundproof booth facing the television. A headset was fitted allowing voice recording and communication with the experimenter. They were told that they would be presented with videos of number-plates and were required to read out as many as possible, although they were not expected to be able to read all of the plates. Instructions were also given to keep to the experimental vocabulary only and to speak clearly and consistently. Prior to the experimental phase, subjects were given a practice block of trials similar to the experimental blocks described below.

The experiment consisted of six blocks of video recordings of approximately 90 seconds each, showing parked cars with their number-plates clearly legible. The number-plates were of the standard British type with 6-7 characters made up from the letters A to Z and digits 0 to 9. Three blocks showed a mean of 20 cars being panned across ('slow') and three had a mean of 33 cars ('fast'). The design was within-subjects, each subject carrying out all six blocks, with the order balanced across subjects. After each block, subjects completed a mental workload questionnaire consisting of the six scales from the NASA-TLX (Task Load Index), rating their experiences of mental demand, physical demand, temporal demand, performance, effort and frustration level during the previous block on a 10-point scale.

Transcripts were made of the subjects' speech output and these were compared the number-plates shown in the videos to score human performance. The performance of the recognizer was assessed by comparing its output to the transcripts of subjects' speech. Finally, a measure of the overall ability of the human-computer system to carry out the task of number-plate recording could be calculated from the above scores.

RESULTS

It was first necessary to confirm that the experimental paradigm was successful in putting subjects under stress. Results from the mental workload questionnaire showed that the fast condition was rated as significantly more stressful than the slow condition on all six scales (paired comparison t-tests

with 14 df: mental demand t=7.25,p<0.0001, physical demand
t=2.76,p=0.016, temporal demand t=10.11,p<0.0001, performance
t=10.03,p<0.0001, effort t=7.92,p<0.0001, frustration
t=7.20,p<0.0001)

Human performance was scored in terms of the percentage of
number-plate characters correctly spoken. Table 1 gives these
mean scores across the 15 subjects:

Table 1: Percentage of characters correctly spoken

              Slow condition   Fast condition
Mean   98.4%   71.9%
(Standard dev) (1.35%)   (13.39%)

This represented a highly significant difference between
dictation performance in the slow and fast conditions (paired
t(14)=8.04, p<0.0001).

Table 2 shows the types of error made in dictation
performance. Substitutions involve speaking the wrong utterance
for a character (eg "pee" instead of "papa") whereas additions
involve speaking words where none should have been spoken (eg.
"three six seven" instead of "three seven").

Table 2: Mean number and type of spoken errors

Type of error   Slow condition   Fast condition

Substitution   2.4 (sd 1.60)   11.5 (sd 2.38)
Addition   0.9 (1.46)   3.7 (3.31)

Total   3.3 (2.15)   15.2 (11.1)

Significantly more speech errors of both types were made in
the fast condition compared to the slow.

Table 3 below shows the performance of the speech recognizer
as a percentage of words uttered by the speaker which were
correctly recognised. The results from two of the fifteen
subjects are omitted due to incomplete recognition results.

Table 3: Percentage of words correctly recognised

              Slow condition   Fast condition
Mean   67.8%   62.3%
(Standard dev) (7.71%)   (10.63%)

The performance of the speech recognizer was found to be
significantly worse with stressed speech (paired t(12)=3.39,
p=0.0053).

Finally, the performance of the whole system (ie human
speaker and speech recognizer) was analysed. This showed that
67.6% of words were correctly recorded from video to ASR output
in the slow condition, but only 47.5% were recorded in the fast
condition. Again this represents a highly significant difference
in overall accuracy between the slow and fast conditions (paired
t(12)=8.52,p<0.0001).

DISCUSSION

An initial observation was that the experimental paradigm was successful in inducing stress in subjects; performance was affected by time pressure and subjects rated the fast condition as more stressful on all dimensions. The main findings were significant decrements in the performances of both the speaker and speech recognizer with time stress. Subjects spoke a smaller proportion of words correctly and made more substitution and addition speech errors. Similarly, the ASR device recognized a smaller proportion of words correctly when the speaker was under stress. The findings have strong implications for both the users of ASR and the designers of recognizer systems.

Under conditions of time pressure, speakers' memory for the experimental vocabulary was disrupted. Characters from the ICAO alphabet were briefly forgotten and subsequent performance was affected as subjects attempted to recall these items. It was apparent that those speakers who were more familiar with the alphabet were affected to a lesser extent by the time pressure. Although the vocabulary is likely to be highly practiced in any real ASR application, it would be expected that speech performance could be improved by a longer, more involved training period. A learning effect amongst all subjects was also observed. One explanation for this is that subjects were able to develop better strategies for coping with the stress. Thus, with ASR applications involving stressful situations, it would be useful to expose subjects to this stress during training.

Analysis of subjects' speech errors showed that many of the substitutions under time stress involved the speaker reverting to more familiar responses, eg. "pee" for "papa" or "twenty-five" for "two five". Also subjects tended to have problems remembering certain ICAO words under stress. Both of these findings have implications for the design of ASR vocabulary; it is important to make it natural and memorable for the user, a principle known as habitability.

The first observation of the speech recognizer results was that the overall system performance was quite poor with a mean recognition accuracy of around 60%. It must be noted, however, that fully speaker-independent templates and the complete recognizer vocabulary of 500 words were used. Substantial improvements would be expected with a reduced vocabulary and by incorporating some syntax information into the recognition algorithms.

User stress was found to significantly impair speech recognition performance. Given the speaker results discussed above, this suggests that the quality as well as the content of speech were detrimentally affected by stress. It might be expected that the rate of speech would rise with time stress. However, even this would not fully explain the results, as previous research has found that the main difference between slow and fast speakers is a reduction in the pauses between words, which should not affect the accuracy of a continuous-word speech recognizer. Further analysis of the speech signal under time stress would be necessary to test for possible increases in the intensity and frequency of the voice, although there is quite good evidence from previous studies that such changes do occur.

Various possible solutions have been suggested to improve an ASR system's ability to cope with user stress. Firstly, if poor recognition performance under stress is going to be hazardous to life (in a combat aircraft, for example), or less dramatically, cause ASR to be unproductive, then its use must be limited. One way of doing this is to use ASR in conjunction with other input devices; for example, allowing keyboard entry in emergency situations. More acceptable solutions lie in controlling the user's voice through training or feedback. Instructions to users that their voice is drifting from the templates has proven useful. Alternatively, specific feedback about certain vocal characteristics might be given.

The ASR system itself, in particular the training of the voice templates, can be altered. For example, template training can be carried out in a more 'task-like' stressful environment rather than the benign conditions usually used. Paul et al (1986) have also found success with 'multi-style' training in which templates are based on a combination of five styles of speech (normal, fast, clear, loud and question pitch) to encompass the variability observed in stressed speech. A final solution is to make the algorithm or templates adaptive, such that changes in speech are tracked and templates constantly updated. Early attempts have met with moderate success, although problems arise if templates are inadvertently updated with the wrong word.

To summarise, the experiment found significant performance decrements with user stress, confirming the importance of research in this area. Further work is necessary to analyse the speech changes accompanying stress, and to test the usefulness of the solutions that have been suggested to counteract the problem.

REFERENCES

Paul, D.B., Lippmann, R.P., Chen, Y. & Weinstein, C.J., 1986, Robust HMM-based techniques for recognition of speech produced under stress and in noise. In Proceedings of Speech Tech '86 (New York: Media Dimensions Inc), pp 241-249.

Simpson, C.A., 1986, Speech variability effects on recognition accuracy associated with concurrent task performance by pilots, Ergonomics, 29(11), pp.1343-1357.

Scherer, K.R., 1981, Speech and emotional states/ Vocal indicators of stress. In Speech Evaluation in Psychiatry, ed. J.K. Darby (New York: Grune & Stratton), pp. 171-220

ACKNOWLEDGMENTS

The experiments in this study were carried out at the Speech Research Unit at DRA, Malvern. The authors would like to thank Andrew Varga, Brian Mellor and all the subjects at Malvern for their help in carrying out the study, and Kate Hone at Birmingham for her support.

DRIVER STATUS MONITORING: CAN IN-VEHICLE I.T. DETECT THE IMPAIRED DRIVER?

S. H. FAIRCLOUGH & S. J. HIRST

HUSAT Research Institute
The Elms, Elms Grove
Loughborough
Leics. LE11 1RG.

In-vehicle sensor technology and associated software are currently under development to monitor the psychological status of the driver. Using this technique, psychological impairment resulting from the use of alcohol, drugs etc. or naturally occurring degradation such as fatigue are inferred from associated changes in vehicle control. An experiment was conducted in order to investigate which vehicle sensors were sensitive to the effects of the maximum legal limit of Blood Alcohol Content (BAC). 12 subjects performed a number of driving tasks over two sessions on a closed-course circuit, a placebo and an experimental condition where the subjects received a mixture containing approximately 0.05% BAC. Vehicle parameters, psychophysiological variables and subjective self-assessment data were collected.
The results revealed that the driver's operation of the steering wheel was the most sensitive indicator of alcohol impairment. There was an inter-subject correlation between subjective self-assessment and steering wheel operation, i.e. those subjects who felt themselves to be significantly impaired demonstrated the most erratic steering behaviour. Implications for the development of a driver state monitoring device are discussed.

INTRODUCTION

The concept of driver status monitoring describes a system that collects data from a variety of vehicle sensors and then infers the psychological state of the driver. A number of factors have averse affects on driving performance, most notably, alcohol (Lowerens et al, 1987), certain drugs (Brookhuis et al, 1986) and time-on-task (Lisper et al, 1986). Previous studies have demonstrated the feasibility of monitoring decrements in driving ability unobtrusively (Thomas et al, 1989) via recourse to vehicle parameter data. However the practical utility of developing a monitor system relies heavily on the sensitivity of the vehicle parameter data. If changes in driving behaviour either anticipate or precede

changes in the driver's psychological status, then vehicle
parameter data has limited diagnostic value. The feasibility
of a driver status monitor (DSM) relies on the co-occurrence of
changes in both psychological state and their associated impact
on vehicle parameters.

A number of DSM devices have been developed and introduced
on the commercial market. Thomas et al (1990) report that in
the USA alone, over 200 DSM devices have been patented.
Amongst these systems, two approaches are apparent, those that
rely on vehicle parameters alone, (for example Nissan's Safety
Drive Adviser (Yabuta et al, 1985) that uses steering wheel
movements) and those dependent on direct recording of
physiological parameters such as the British Dormalert. The
latter variety are described by Thomas et al (1990) as
'extremely unreliable and obtrusive'. The problem with non-
obtrusive vehicle parameter systems is that the precise
relationship between driving performance and psychological
state is still to be established.

Previous research funded by the Commission of European
Communities (CEC) under the DRIVE Programme established an
experimental protocol under the auspices of the DREAM project
(Brookhuis et al, 1991). During this project, a number of
demonstration experiments to investigate this relationship were
conducted in laboratory, simulator and real road settings
(Thomas et al, 1989). For example, Thomas et al (1989)
describe changes in the lateral deviation of the vehicle as
correlated with electroencelographic (EEG) changes in cortical
activation as a function of alcohol intake.

This line of investigation has been continued into the DETER
project, also funded by the CEC-funded DRIVE Programme. DETER
is an acronym for Detection, Enforcement & Tutoring for Error
Reduction. The aim of the DETER project is to develop and test
a number of monitoring, tutoring and enforcement prototype
systems.

METHOD

The aim of this study was to investigate how the ingestion of
alcohol affected drivers' behaviour on a closed-course circuit.

Independent variable.

The independent variable was alcohol administered in
sufficient quantity to induce a blood alcohol content (BAC)
level approximately equivalent to 0.05% (the maximum permissible
under U.K. law). The alcohol was presented in a study via an
Alcometer s-m2 (Lion Laboratories, UK).

Test environment and apparatus

The experimental trials took place on a disused airfield
situated near the village of Langar in Nottinghamshire. The
test vehicle used in the trials was developed by the HUSAT
Research Institute. The electrocardiogram signal was captured
via a 8-channel MacLab system and stored on a Powerbook 170
running 'Chart' software. Both laptop computers were situated
on the rear seat of the vehicle.

Experimental design.

The study was implemented as a repeated measures design. The
subjects took part in two experimental sessions, separated by at

least 24 hours. During one experimental session the subjects received alcohol and in the other, a placebo. The latter was achieved by simply presenting the peppermint cordial and water mixture minus the vodka. The order of presentation for the alcohol and placebo trials were counterbalanced across the subject group.

Subjects.
   Twelve subjects took part in the experimental study (six males and six females). All subjects were aged between 20 and 25 years and had held a full driving license from 2 to 8 years. On average the subject group achieved a minimum annual mileage of 2000 miles with an upper limit above 6000 miles. Unfortunately one of the male subjects could only complete one of two experimental sessions and was duly disregarded from the final data set.

Dependent variables.
   Three distinct groups of dependent variable were captured and analysed in the course of the study.
   The first group originate from the in-vehicle sensors and are associated with vehicle dynamics and control. They included brake pedal position, accelerator pedal position, steering wheel movement, mean speed (in mph), rate of deceleration and rate of acceleration (both expressed in ms^{-2}).
   The second group are those electrocardiogram (ECG) measures associated with mental effort and arousal. Both heart rate (expressed in beats per minute, bpm) and the variability of the cardiac inter-beat interval (IBI, expressed in milliseconds, ms).
   A questionnaire was administered to the subjects after each experimental task. The subjects were also asked to rate their perceived level of alertness on the seven-point Stanford Sleepiness Test (1 - active, vital, wide awake, 2 - functioning at a high level, not at peak, 3 - relaxed, not at full alertness, responsive, 4 - a little foggy, let down, 5 - fogginess, slowed down, starting to lose interest, 6 - sleepiness, woozy, fighting sleep, 7 - struggling to stay awake) developed by Hoddes et al (1973).

Procedure and Tasks.
   Following an initial introduction, the subjects familiarised themselves with the test vehicle via a fifteen-minute drive on the real road. The placebo/alcoholic mixture was then administered and the subjects were driven by an experimenter to the test-site (a journey of approximately fifteen minutes). When the subject arrived at the test-site, s/he was fitted with the electrocardiogram (ECG) electrodes and all relevant monitoring software and connections were tested. The subjects were asked to perform three driving tasks, the order of which was counterbalanced across the subject group. Two driving courses were marked on the track using a mixture of traffic bollards and paint. One driving course was arranged in the shape of an 'oval' (hereafter referred to as OVAL), the shape and width of the course was determined by a pair of bollards. The subjects were instructed to drive in an anti-clockwise direction around the oval course at a speed no higher than

thirty mph. The subjects were instructed not to speak and told
that they would perform the task for ten minutes.
In the second course, the curves of the oval were removed
leaving two 'straight' sections, each of which ended in a
painted marking of double, broken lines (e.g. 'give way' lines
on U.K. roads). The subjects were instructed to stop the
vehicle at the white line and to turn to the left. The subjects
completed thirty left turn manoeuvres in the course of this
'junction ' (hereafter referred to as the JUNCTION trial). The
third experimental task involved subjects driving down a
straight approach section and judging a gap represented by two
3' high cardboard boxes. Seven gaps were designated, 1.5, 1.68,
1.86, 2.04, 2.22, 2.4 and 2.58 metres respectively. The first
was impassable, the second and the third just passable and the
others easily passable. Subjects completed 14 trials (each of
the seven gaps twice) and the order of presentation for the gaps
was randomised throughout.
     Alcometer testings were carried out before and after each
trial and questionnaires were administered after each trial.

RESULTS
     The junction trial failed to reveal any statistically
significant differences between driving behaviour in the alcohol
and the placebo trial.
     The data originating from the oval driving trial has been
divided into two groups, straights and bends. The former data
set was captured whilst the vehicle travelled along the straight
sections of the oval and the bends on the respective curved
sections of the track. Unfortunately one set of subject data
has been lost from this group due to inclement weather
conditions. The oval data during the BEND sections revealed a
number of significant differences between the two experimental
conditions.

Table 1. Differences in vehicle parameters over the BEND
sections of the oval track for both alcohol and placebo
conditions (n=10).

PARAMETER	PLACEBO	ALCOHOL	Statistical significance
Steering wheel std. dev.	12.1	13.2	p<0.01
Mean speed (mph)	17.1	16.9	
Deceleration (m/s^{-2})	-0.8	-0.7	p<0.05
Acceleration (m/s^{-2})	1.0	1.0	

     The gap acceptance trials revealed no significant differences
on number of correct/incorrect refusals, speed of approach and
distance travelled between experimental conditions. However
brake position appeared to be higher (though not statistically
significant) during the last 40 metre section of the approach
during the alcohol condition.

A statistical analysis of the ECG records across both experimental conditions failed to reveal any significant difference between the two conditions.

An analysis of the questionnaires revealed no significant differences between subjects' scores on the SSS (Stanford Sleepiness Scale) across the two experimental conditions.

Correlation coefficients were calculated to indicate the relationship between vehicle parameters and psychological indicators of impairment/stress on an inter-subject basis. This data is presented for the data originating from the oval trial during the alcohol condition. The significant correlation's (in *italics* in table 2) were not present in the placebo data.

Table 2. Correlation matrix between selected vehicle parameters and psychological indicators during the oval trial for the alcohol condition only (n=10).

Steer. SD	Steer. SD				
Deceleration	0.2	Decel.			
Heart rate	0.1	*0.7*	HR		
Mean IBI var.	-0.4	-0.5	-0.7	IBI	
Sleepiness	*0.8*	0.4	0.5	*-0.7*	Sleep

DISCUSSION AND CONCLUSIONS

The data originating from the junction trials (table 1) failed to reveal any significant differences between the alcohol and placebo conditions. This may have been due to the relatively short sampling period of ten seconds however, the raw data revealed a number of important inter-subject differences. In essence the driving styles during the approach period varied between those subjects who exhibited caution (i.e. slowed down early, decelerated smoothly) in the alcohol condition and those who made no such provision. Hence in the data we see that subjects are inclined to begin their approach at higher speeds and decelerate at much higher (and intrinsically more 'uncomfortable') levels during the alcohol condition.

The oval data was divided into straight and bend sections of the track. The data originating from the straight sections gave some evidence that subjects tend to produce higher amplitude steering wheel adjustments when under the influence of alcohol. This impairment of perceptual-motor tracking behaviour reflected by the mean standard deviation of the steering wheel movement is much more pronounced when the subjects attempted to negotiate bends on the oval track.

Subjects in the alcohol condition tend to decelerate at a higher level on the straight sections of the oval track but to decelerate at a significantly lower rate on the bend sections. This may be indicative of some problems of motion and distance perception associated with alcohol. Therefore the subjects fail to make optimal speed-distance judgements during their approach to the bend and decelerate at a higher rate (a finding reflected in the junction data); and similarly fail to slow down the vehicle at the appropriate rate when negotiating the bend.

These types of conjecture illustrate how vehicle parameters
might be related to psychological impairment effects within the
context of the road layout.

A majority of subjects failed to note any awareness of the
effects of alcohol during the experimental trials. This finding
was also reflected by the inter-subject differences that are
represented in table 2. The underlying principle of these
findings may be that some of the subjects were more affected by
the influence of alcohol than others. Hence those subjects who
rated themselves as less alert on the SSS scale were also those
who (a) exhibited the highest deviations of steering wheel
adjustments and(b) increased their investment of mental effort
and showed lower levels of IBI variability and (c) showed
increased physiological stress as indicated by an increase in
heart rate.

To summarise, the HUSAT study illustrates how impairment
effects are apparent in vehicle parameters, particularly
steering wheel behaviour. The results show how vehicle
parameters and their interpretation are subject to the influence
of road layout. The study also demonstrates the huge influence
of inter-subject variability.

REFERENCES

Brookhuis K. A., Louwerens J. W. & O'Hanlon J. F., 1986. EEG
energy-spectra and driving performance under the influence of
some antidepressant drugs. In Drugs and Driving, ed.
O'Hanlon J. F. & de Gier J. J. (London: Taylor & Francis),
pp. 213-221.

Hoddes E, Zarcone, V, Smythe, H, Philips, R & Dement, W C, 1973.
Quantification of sleepiness: a new approach.
Psychophysiology, 10, pp. 431-436.

Lisper H-O., Laurell H. & Van Loon J., 1986. Relation between
time to falling asleep behind the wheel on a closed track and
changes in subsidiary reaction time during prolonged driving
on a motorway Ergonomics, 29, 3, pp. 445-53.

Louwerens J. W., Gloerich A. B. M., de Vries G., Brookhuis K. A.
& O'Hanlon J. F. The relationship between drivers' blood
alcohol concentration (BAC) and actual driving performance
during high speed travel. In Alcohol, drugs and traffic
safety, ed. Noordzij, P. C. & Roszbach, R. (Amsterdam:
Excerpta Medica, pp. 183-187

Thomas D. B., Herberg K-W., Brookhuis K. A., Muzet A. G.,
Poilvert C., Tarriere C., Norin F., Wyon D. P., Schrievers G.
& Mutschler H. Demonstration experiments concerning driver
status monitoring. DRIVE Project V1004 DREAM, December 1989.

Thomas D. B., Herberg K-W., Brookhuis K. A., Muzet A. G.,
Poilvert C., Tarriere C., Norin F., Wyon D. P., Schrievers G.
& Mutschler H. The feasibility of developing a device for
monitoring driver status. DRIVE Project V1004 DREAM,
December January 1990.

Yabuta K., Iizuka H., Yanagishima T., Kataika Y. & Seno T. The
development of drowsiness warning devices. Technical Report,
Nissan Motor Company, also presented at the Tenth
International Conference on Experimental Safety Vehicles,
Oxford, 1985. Cited by Thomas et al (1990).

HUMAN FACTORS CHECKLIST FOR SERVICE EVALUATION

A.M. CLARKE, G. ALLISON & T. HEWSON

HUSAT Research Institute
Loughborough University of Technology
The Elms, Elms Grove
Loughborough
Leics. LE11 1RG.

This poster paper describes a Human Factors checklist (QUEST) that has been developed for Human Factors Expert service evaluation. The work reported here has been developed on the LUSI* project. Extracts from the checklist will be shown on the poster.

INTRODUCTION

There has been a trend over the last ten years towards an integration of technology between computer systems and telecommunication systems. As a result of improvements in technology and reduction in costs particularly in terms of the networks, there has been a convergence of telecommunications, computing and video technologies.

It is now possible to network together quite complicated multimedia systems. Linking more than two systems together adds another problem in terms of the user; how to handle multipoint working. Videocommunications are now more widespread. Improved Codecs along with the provision of wider bandwidths have made this possible. As well as traditional videoconferencing rooms, desktop videotelephony will soon be common place along with video windows being incorporated into PC systems. The types of systems that will be possible in the near future will be videomail, videomessaging, multimedia retrieval, PC integrated videotelephony to name but a few.

How can a human factors expert evaluate these systems/services?

* The LUSI project 2092 is part of the CEC funded RACE II programme - Research and Development in Advanced Communications Technologies in Europe and acknowledgement must be given to other partners in this work, Telefonica Investigacion y Dessarollo, Spain; TELES GMBH, Germany; University of Bonn, Germany and Telesystemes, France).

QUEST

The Questionnaire for Usability Evaluation of Services and
Terminals has been developed as a tool for a Human Factors
expert to identify potential usability problems with services
and products. It is hoped that there will be a strong
correlation between the problems elicited from experimental
trials and those identified by QUEST. If this is the case than
it will be possible to use QUEST as a standalone tool for
investigating any new services.

It may be the case that after comparing the results of the
parallel trials amendments will be made to QUEST, before it will
meet our objectives.

DEVELOPMENT OF QUEST

QUEST was built around checklists for evaluating purely the
interface aspects of a system, looking at feedback issues,
abbreviations, wording, navigation through text, and other
common interface problems. There have been a number of these in
existence for a number of years checklists for the evaluation of
standalone computer systems e.g. Ravden & Johnson 1989,,
Sniderman 1986, Scapin et al 1988. In order for the evaluation
tool to be appropriate for service use, other aspects had to be
covered within the Questionnaire e.g. service description,
identification of the user, directories of the users, call up
procedures etc.

Initially the questionnaire came in two parts; part 1 to be
filled in by a human factors expert, and part 2 to be completed
by the system designer. The second part included all technical
questions regarding information rates, as well as compliance
with a variety of standards.

The two part structure of the checklist unfortunately made it
far too long and time consuming to complete. As a result most
of the standards questions were deleted and many of the
remaining technical questions were incorporated into the first
part. There has now been produced a short section towards the
end of QUEST covering the technical aspects of the equipment.

The questionnaire can now be completed by a human factors
expert with limited technical assistance. The expert will "walk
through" the service, completing QUEST in a sequential manner.
A mixture of question types are used, including questions with
simple yes/no answers, questions requiring answers to be made on
rating scales and others requiring more detailed written
explanations. The type of question being dependent on the
information required. The rating scales questions were designed
to simply help identify problem areas and not attempt to
quantify any problems as more or less important than others. As
QUEST is designed to be a diagnostic tool to identify problem
areas with services and not to be a tool for comparing
services.

ISSUES COVERED BY QUEST
QUEST is currently made up of six sections;

The Service Overview: Extracting basic information on the name of the service, what type of service it is, what input devices are used etc.

Hardware Issues: This section asks questions about the keyboard, screen, displays, casework, maintenance, and portability.

Communication Issues: These issues include setting up the call, security features, call handling, signing off procedure.

Service Usability Issues: The degree of user support, training, navigation, interface features, are all covered in this section. As well as the characteristics of the task, dialogue design and the consideration given to the prospective environment.

Service Usability Summary: An overall impression of the system, and the service, rating the impression of usability and learnability.

Service Usage and Technical Questions: This section will be difficult for the Human Factors expert to complete without some help. It covers questions of service distribution, cost of the service, and billing policies. In addition the technical questions of information rate, display resolution etc. are asked.

CONCLUSIONS
 QUEST has been developed as an aid for a human factors expert in the task of performing service evaluation. Its value is that it adopts a structured "walk through approach" in order to ensure that all aspects are covered in the expert evaluation. It is currently being used in the LUSI project, however, it is the authors intention to use it more extensively with other RACE project's services.

REFERENCES

Ravden, S. J. & Johnson, G. I. (1989) Evaluating Usability of Human-Computer Interfaces: A practical method, Ellis Horwood.

Scapin, D. L. (1990) Organising Human Factors knowledge for the evaluation and design of interfaces. International Journal of Human-Computer Interaction, 2, 203-229.

Shneiderman, B. (1986) Designing the User Interfaces: Strategies for Effective Human-Computer Interaction, Addison Wesley.

SUCCESSFUL VIDEOCONFERENCING

A.M. CLARKE & S.M. POMFRETT

HUSAT Research Institute
Loughborough University of Technology
The Elms, Elms Grove
Loughborough
Leics. LE11 1RG.

This poster paper describes a set of written human factors guidelines which aim to support user organisations and service providers in achieving successful videoconferencing. The poster will describe the six stages that have been identified for an organisation to consider when introducing videoconferencing.

INTRODUCTION

Videoconferencing is slowly enjoying an increase in popularity as technology improves and prices fall. It is however, being used with varying degrees of success. As with any tool, which particular one is chosen and how it is implemented will be crucial to the success of its use. Human factors has shown that for IT systems, the decisions made will be highly dependent on the potential task, user and environment characteristics. The same is true for videoconferencing.

The RACE ISSUE project (IBC Systems and Services Usability Engineering) conducted a survey of organisations using videoconferencing in Finland, Italy, Spain and the UK. On the basis of the survey results and an extensive literature search, the project has produced a set of written guidelines which aim to support user organisations and service providers in achieving successful videoconferencing. To illustrate the guidelines there is a video portraying the decision making process of a Company investigating the feasibility of introducing videoconferencing.

The primary message of the guidelines is that videoconferencing technology can be an enormously useful tool, helping an organisation to be efficient, to be flexible in terms of organisational structure, and to improve the quality and speed of communication both internally and externally. This can only be achieved however if:

- there is a real need for videocommunication

- the service matches the needs of the intended user groups

- the service is well-managed providing adequate technical and user support

- the service is evaluated regularly so that it can respond to changing requirements

- the evaluations are acted upon.

STAGES FOR SUCCESSFUL VIDEOCONFERENCING
The following diagram shows the stages involved in successful videoconferencing.

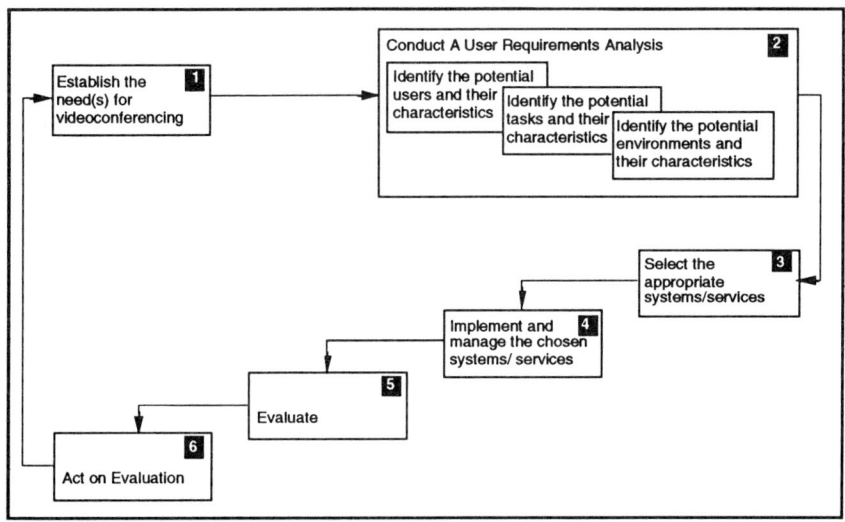

STAGE 1   Establish the need(s) for videoconferencing
This involves identifying at least one target application and clarifying your main objectives for videoconferencing.

STAGE 2   Conduct a User Requirements Analysis
This involves identifying the user, task and environment characteristics and producing a set of user-based requirements for the videoconferencing.

STAGE 3   Select the Appropriate Systems/Services
This involves matching the available services/systems with the requirements (including financial) and recommendations available, then selecting the service provider(s) and supplier(s).

STAGE 4   Implement and Manage the chosen Systems/Services
   This involves selecting the service manager or management
   team, working with the service providers and suppliers,
   working with the users and the potential users as well as with
   other managers in the organisation, to ensure a smooth running
   operation for the service.

STAGE 5   Evaluate
   At appropriate points in the life of the service, it will be
   necessary to carry out an evaluation of the videoconferencing
   in order to plan future strategy/policy. This will involve
   determining the effects of using videoconferencing and
   identifying areas for improvement.

STAGE 6   Act on the Evaluation
   There are 3 possible outcomes from the evaluation; i) continue
   unchanged to the next evaluation point, ii) implement
   appropriate changes to the next evaluation point or iii)
   discontinue videoconferencing.

   Further details of each of these stages can be found in the
guidelines, and will be shown on the poster.

CONCLUSIONS
   The guidelines document in a draft format which was
distributed for feedback, was well received by both user
organisations and service providers. The guidelines were
updated in the light of feedback received and are currently
being printed. It is hoped that these and the accompanying
video will be useful tools for user organisations and service
providers.

REFERENCES

RACE Deliverable 65/HUS/-/DS/A/014/B2 (1992) "Videoconferencing
   Guidelines for User Organisations and Service Providers".

RACE Deliverable 65/HUS/-/DS/A/016/B1 (1992) "Planning to
   Videoconference" Video.

DEVELOPING A DATA COLLATION TOOL FOR MANPRINT IN THE UK

W I HAMILTON and H WALTERS

EPP Human Factors
168 Cheltenham Road
Bristol
BS6 5RE

JS 33/2 MANPRINT
Ministry of Defence
Whitehall
London, SW1A 2HB

In April 1991, the Ministry of Defence introduced the MANPRINT human factors quality assurance initiative for land systems procurement. MANPRINT may be seen as involving both a technical function and a management or data control function. While the technical function is addressed by the human factors tools already in existence, there is a shortage of data control and reporting tools.

By far the major overhead associated with MANPRINT, for both the MOD and UK industry, will be the need to document, collate and report human factors assessments for each of the six MANPRINT domains. There is, therefore, a need for a reporting tool which serves the needs of the UK defence community in responding to MANPRINT. To meet this requirement, the MOD, together with EPP Human Factors have embarked on a research project to develop such a tool. Called Man-Q, the tool supports a range of documentation and reporting tasks which must be performed under MANPRINT. This paper details the work done in the development of Man-Q and describes its functionality. The conclusion includes recommendations for additional tools needed to support the UK's MANPRINT programme.

GENERAL DEFINITION OF REQUIREMENTS
Background

In 1991 the Ministry of Defence (MOD) introduced the MANPRINT Plan for Quality Assurance in Human Factors in the procurement of military land systems (MOD 1990). MANPRINT demands the early analysis and specification of the human elements of any new system with respect to the so called 'domains' of manpower, personnel, training, human engineering, and safety and health hazard assessment.

MANPRINT provides a formal mechanism to ensure that human factors issues are addressed early and often throughout the development of a new system. The basic principles of its application are that the suppliers of equipment to the MOD, will regularly analyse and report the consequences of design decisions in terms of their effects on:-

• Manpower, Personnel and Training requirements

• Human Factors Engineering/Ergonomics

• System Safety and Health Hazards.

These are by no means new issues. On one project or another most suppliers will have been faced with constraints and problems related to some or all of these areas. Up to now, however, most of the topics covered by the MANPRINT domains will have been overshadowed by other design driving issues, such as cost and technological options. MANPRINT changes that by raising the profile of human factors to put it on equal terms with other design drivers.

Even suppliers who regularly apply ergonomic principles to their designs will now find that the demand to do so has increased. Additionally, and perhaps most significantly, all suppliers will now have to declare their commitment to human engineering right at the outset. Technical proposals will have to include a special section on MANPRINT, and a MANPRINT plan for system development, which shows how the supplier intends to monitor and assess the impact of the design on the soldier. When the project is in full flow the supplier will be expected to make regular 'Returns' to the MOD in order to demonstrate that the design is being properly analysed in terms of the requirements of MANPRINT, and that any problems identified are being adequately resolved.

Hamilton (1992) points out that MANPRINT will affect project development efforts in **two main ways**. First of all, MANPRINT demands substantial effort on the human factors analysis and assessment of the system design. To meet this requirement, suppliers will have to apply human engineering expertise to the design development process. Human factors engineers will have to be involved in all aspects of the system design to assess its consequences for the operators and maintainers. A substantial fraction of the overall project resources will have to be allocated to this, the MANPRINT technical effort.

Secondly, the output of this human factors work has to be collated and reported back to the MOD. This represents a new aspect of project management which will require that formal procedures are observed in the handling of MANPRINT data and in the preparation of summary reports on progress made under the six MANPRINT domains. Figure 1 depicts this management and technical distinction and shows how the two functions might be related.

Figure 1: Information flow between the Management and Technical functions of MANPRINT compliance

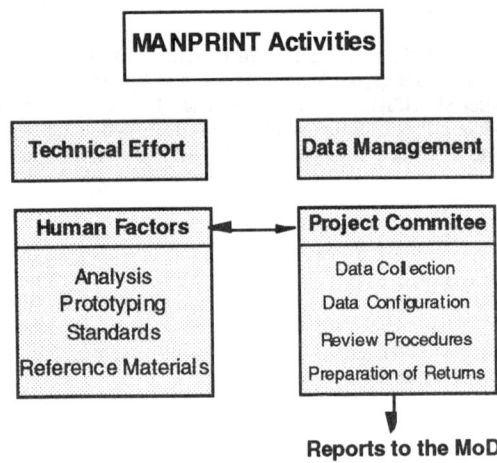

To accommodate MANPRINT, most firms will have to make some modification to their current project management practices. The MANPRINT technical effort will obviously produce vast quantities of data which will have to be carefully recorded and meticulously cross-referenced. It would be the responsibility of a MANPRINT Project Committee to monitor the flow of information and to assimilate new developments - see Hamilton (1992) for a full explanation of this function. At the very least some form of 'hard copy' data control procedure will have to be implemented: ideally, though, a computer data-base is needed. It was the aim of the Man-Q research programme to establish the feasibility of, and the requirements for just such an information management tool.

A MANPRINT Project-Data Management Tool: Man-Q

There are currently available a number of software tools for MANPRINT and human factors work generally (e.g. HARDMAN and CREWCUT[1], Micro SAINT and the Human Operator Simulator[2], and SAMMIE[3]). However, they are all focused on the technical function of MANPRINT.

---

[1]HARDMAN and CREWCUT are both product names of the US Army Reseacrh Laboratory.
[2]The Human Operator Simulator and Micro Saint are both product names of Micro Analysis and Design Inc..

The success of the MANPRINT initiative is likely to depend upon its cost implications for industry: data management and reporting will constitute perhaps the major overhead for a MANPRINT qualifying project. Consequently, there is an urgent need for a software aid to promote efficiency and reduce the cost of the management effort.

EPP Human Factors, together with the MOD, have developed a prototype data management aid for MANPRINT. It is called Man-Q (short for MANPRINT Quality assurance) and is so named because the MANPRINT Project Management function will most probably be located within the design quality management function.

Man-Q has been organised so as to allow project summaries to be entered, as well as analysis data for the six MANPRINT domains. Man-Q is designed to support the MANPRINT data item descriptions developed in the United States (DOD 1989) and modified for use under UK MANPRINT. With Man-Q the Project Manager can:

- Track Key MANPRINT milestones

- Log assessment data

- Cross reference actions

- Conveniently produce MANPRINT reports for the MOD.

In what follows the details of the Man-Q application are set-out together with a specification of the assumptions made in its development.

THE MAN-Q DEVELOPMENT PROGRAMME
Phase 1: Prototype Development
The original concept for Man-Q was simply to create a set of forms on which to record MANPRINT domain data items. As these were developed, however, the need for data cross referencing became more and more obvious. Eventually it was decided that the forms should merely act as the front-end to a more complex data structure in which the data could be manipulated and reorganised as required. At this stage the Man-Q development was focused on the creation of a data base application.

The environment selected for the initial development work was FileMaker Pro[4]. FileMaker is a flat file data management system which offers comprehensive sorting and cross referencing capabilities. It is very simple to use and this convenience allowed many ideas to be quickly developed and tested, until the range of functionality for Man-Q had at last been defined.

Man-Q was to be a tool which would be used primarily by the MOD's suppliers. It was to support the suppliers in assembling and managing the vast quantities of data which would be generated through the technical analyses for each of the MANPRINT domains of any qualifying project. Man-Q would also prompt the contractor whenever a report was due for a given domain, and would be able to provide that report at the touch of a key.

While the work on the prototype had successfully defined what the tool should do, in order to be truly effective it needed to be totally consistent with the MOD's newly emerging requirements for MANPRINT reports. Consequently, at this stage EPP Human Factors sought the support of the MOD in specifying the precise details of Man-Q's operation, such as the exact intervals at which MANPRINT reports are required to be submitted, the nature of health and safety hazard classifications and the optimum formats for domain reports.

In the next phase of the programme EPP Human Factors and the MOD worked closely together to develop Man-Q as a comprehensive stand-alone application.

Man-Q as a Stand-Alone Application
Man-Q needed to be made available on the most popular and common computer platforms. As a prototype it had been developed on the Apple Macintosh[5] and could be transferred to the

---

[3]SAMMIE (System for Aiding Man-Machine Integration Evaluation) is a product name of Compeda Limited.

[4] FileMaker Pro is a trade mark of the Claris Corporation.

[5] Apple and Macintosh are trade marks of Apple Computers Inc..

Windows[6] version of FileMaker. However, the FileMaker environment was seen as insufficiently powerful to support the final version. Man-Q needed to be able to handle many-to-one data relationships and to provide a powerful indexing and sorting facility. For this reason the final development was switched to a full relational data base environment.

FoxPro 2.0[7] was chosen for the final development since it is a powerful language with full query language support, and it also has built-in development tools which make the programming work faster and simpler. In addition versions of FoxPro are, or will shortly, become available for the DOS, Windows and Macintosh operating systems/environments.

THE MAN-Q INTERFACE AND FUNCTIONALITY
Who will be the user ?
It is most likely that there will be more than one class of user for Man-Q.
Super users : *Super users* are the MOD personnel whose job it will be to monitor and evaluate project proposals, and to receive and collate MANPRINT reports while a MANPRINT qualifying project is running. The MOD user will need to be able to operate Man-Q as an aid to the evaluation of technical programmes and system MANPRINT management plans. They will also want to be able to modify the functions built into Man-Q to tailor the program to new projects. This class of user may be regarded as a *super user* because they will have access to functions which will affect other user's interaction with Man-Q. The *super users* will use a version of Man-Q which runs within the FoxPro environment.
Contractors : The next level of user will be the *contractor*. The *contractor* will usually be a user group. Probably the group or department within the supplier organisation which is responsible for MANPRINT. The *contractor* will have access to all of Man-Q's files but will not be able to modify the program in any way. They will operate Man-Q within a run-time version of FoxPro and will be able to enter and edit MANPRINT data (including the importing of data records supplied by sub-contractors) and compile reports. They will not be able to delete records which have been supplied by sub-contractors, nor will they be able to change or override any of the project specific functions added by the *super users*.
Sub-contractors : Sub-contractors are the third class of user. Essentially they are the same as contractors, with access rights only to browse, edit and delete records which they themselves have created.

The interface
The user interface to Man-Q has been designed to be as simple as possible. Many of the intuitive qualities of the prototype version on the Macintosh have been preserved in the final version.
The interface essentially employs a WIMP style dialogue. The user navigates the files and records by pointing and clicking on buttons and enters data either by selecting items from pop-up menus, or by simple keyboard text editing.
The powerful sorting and searching facilities of the FoxPro environment are available to all users and these too are accessed via menu selection and dialogue boxes. However, there is no necessity for all users to have to manage these facilities since the most complex search, sort and report activity sequences can be combined into a custom function which is initiated by the click of a button. At the time of writing user trials on the interface were about to commence.

File structure and function
Man-Q comprises seven related data files (see Figure 2) in which the user records data from the various technical assessments performed for the six MANPRINT domains. The files are related by a project name field so that the user can group together all the information relating to a single project.
Each of the files (and their associated report forms - see later) are accessed from the Main Menu screen. This screen contains buttons by which the user can select to enter a file. The Main Menu is shown in Figure 3. In addition to the main menu, there is a standard button palette with which

---

6 Windows is a trade mark of the Microsoft Corporation.

7 FoxPro is a trade mark of the Microsoft Corporation.

the user can navigate around the records in a specific file. This palette has buttons to move to the previous, first, next and last records in the file or the selected group.

Figure 2: The Man-Q program file structure.

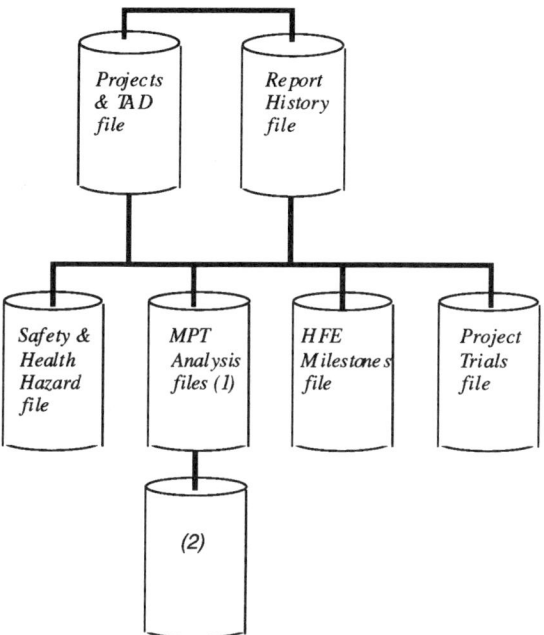

Once the user has accessed a file he/she is presented with a data entry screen with a number of fields which can be filled in either by typing text or by selecting items from predefined value lists. The function of each of the files is as follows:-

Projects & TAD file: The Project & TAD file is the central file for the MANPRINT data management function. There are two input screens for this file. Both screens serve as an output or browse screen as well as the input or edit screen. The user can switch between these screens using buttons. The first contains a detailed description of each MANPRINT qualifying project with which the supplier is currently involved; a breakdown of subcontractors and their responsibilities; and, provides a record and schedule of reports made or due for each of the MANPRINT domains. The dates on which reports are required for each domain are calculated automatically by Man-Q based on the start dates of the Project and the planned in-service date for the equipment. The project summary file will be available to all users but it is anticipated that it would generally be the MANPRINT manager's responsibility to maintain this file.

The second screen holds the target audience description (TAD) for each project. The manpower and personnel requirement is broken-down by Civilians, Regulars and Reservists and by trade type (using the REME trades classification) and skill level/rank. The TAD will normally be entered by the MOD prior to the delivery of the MANPRINT file for a new project and can not be edited by anyone other than a *super user*.

Safety & health hazard file : The safety and health hazard descriptions file is the first of the data bases which act as the collection point for the data generated by the MANPRINT technical function. When a health and safety assessment has been performed on a design, any hazards identified must be recorded on a data item description (DID) form. This form summarises the problem and acts as a record of actions taken to solve it and of the final solution chosen.

Hazards are potentially the most controversial of all of the topics addressed by MANPRINT. For this reason each hazard DID is given a unique log number which can never be reassigned nor deleted. The log number will also identify the supplier or subcontractor who generated the hazard DID, therefore offering complete traceability and security.

Figure 3: The MANPRINT menu in Man-Q.

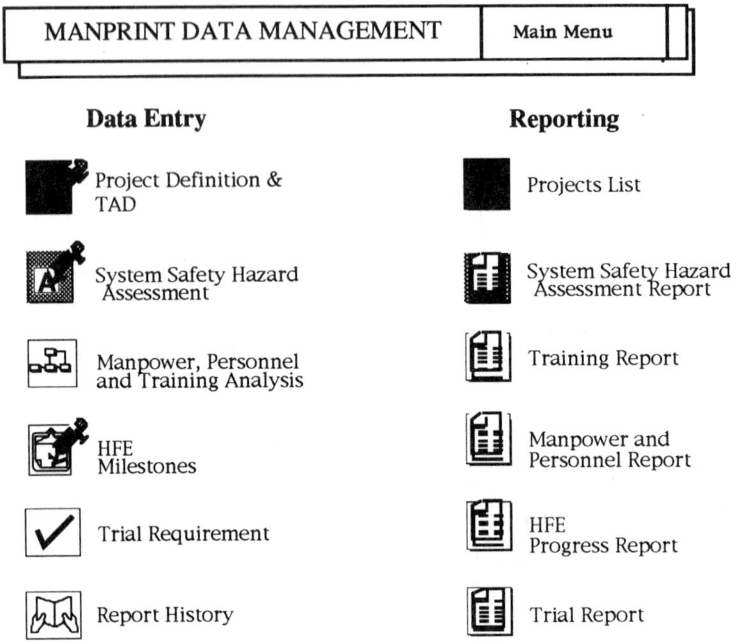

| MANPRINT DATA MANAGEMENT | Main Menu | |

Data Entry   **Reporting**

Project Definition &   Projects List
TAD

System Safety Hazard   System Safety Hazard
Assessment   Assessment Report

Manpower, Personnel   Training Report
and Training Analysis

HFE   Manpower and
Milestones   Personnel Report

Trial Requirement   HFE
                                     Progress Report

Report History   Trial Report

The safety and health hazard descriptions file has a single input/output screen in which a user may browse and edit descriptions, according to the field access limitations described above (i.e. log number and date fields may not be changed). This screen is shown in Figure 4 and may be used to demonstrate some of the program's convenient short-cut data entry features. For example, when entering a description into the hazard type classification field, the user can simply select an item from a pop-up list of choices. These choices include:-

Acoustical Energy
Biological Substances
Chemical Substances
Oxygen Deficiency
Radiated Energy
Shock
Temperature/Humidity Extremes
Physical Trauma
Vibration

Also, when the user selects items to define the hazard's probability of occurrence and classification, a hazard risk index is automatically displayed which gives the Supplier an instruction, such as "Seek Government approval for solution".

Manpower, personnel & training analysis : There are two files devoted to the manpower, personnel and training (MPT) analysis data generated by the technical assessments performed for these MANPRINT domains. In the first file each record contains a definition of each 'high driver' task which is identified in the analysis. High driver tasks are those which demand significant skill requirements on the part of operators, which impose heavy training or physical or cognitive demands, or which are associated with high error rates or training load. The importance of the task is rated according to the consequences of failure (Catastrophic, Critical, Marginal or Negligible) and the type of solution required to eliminate this problem (eliminate task, redesign equipment or change TAD). The implications of these two classifications may be used to trade-off various proposed solutions. This file also contains a specification of the numbers and types of personnel who will be required to perform the task.

Figure 4: The Hazard Description screen from Man-Q

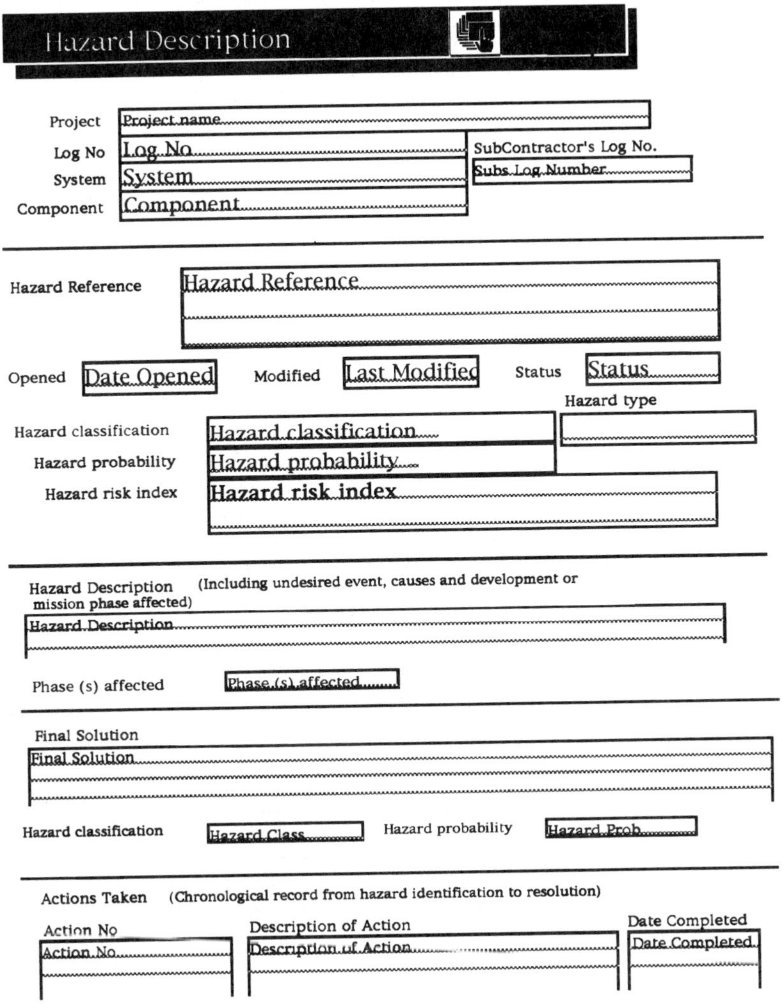

The second file includes a definition of the training required for a specific high driver task, along with the definition of the standards to be achieved and the proposed methods and conditions of training and the equipment needed. Training requirements are defined as either physical, procedural, cognitive or health and safety.

This is not recorded in the first file since not all high driver tasks will be associated with training demands and there is therefore no need to include this specification with each task definition. The records in these two files are related by the task identity field which will include a unique identifier.

There is one input/output screen associated with each data file. As usual, when both files include records for a given task, movement between the screens is effected via buttons.

HFE milestones : The human factors engineering (HFE) milestones file acts as a record of objectives for the Project's System MANPRINT Management Plan. This file will usually include a list of MMI design issues and problems. In addition, the file allows the user to enter summaries of actions taken to address specific objectives and their outcome.

There is one data entry screen for this file in which the user may browse and edit the milestone definitions, the programme of activity concerning them and the list of actions and their status.

Trial requirements : The trial requirements file is used to record trials plans. Each trial is defined in terms of its type (either MANPRINT specific or MANPRINT consequences) the type of subjects required for the assessment (TAD, other military or engineering), the facilities needed, the issues to be assessed, and the objectives and validation criteria to be met. In addition the user can record whether or not a trial report is available once the trial has been completed.

Report history : Every report generated in connection with a project is entered into the report history file. The title, reference and originator are all recorded, as is the MANPRINT domain which the report relates to. Reports can then be listed by project and by domain and a summary list generated.

Report forms

Man-Q has built-in report forms for the following record types:-

> Projects List
> Safety and Health Hazards
> Manpower Analysis (high driver tasks)
> Training Analysis (high driver tasks)
> Human Factors Engineering Milestones
> Trials Summaries

Each report form is arranged in a columnar format with row for each record and a column for each field. Not all fields are included in the reports, as this would be too unwieldy. In each case the built-in report formats summarise the main points of every entry made for a specified project. All that remains for the contractor to do is to add his covering sheets and send the completed report off to their prime contractor or MOD sponsor. In either case, the report can be sent in disk form to be added to another Man-Q data base held by the recipient.

CONCLUSION

This work has succeeded so far in demonstrating at least the feasibility of the development of such a MANPRINT data management and reporting tool. In use Man-Q is likely to bridge a vital gap in the toolset needed to address MANPRINT concerns efficiently and effectively. At the time of writing Man-Q was just commencing a phase of user assessment. The results of that work should be available for reporting by Easter 1993.

The important point about Man-Q is that it is the first 'home-grown' tool for MANPRINT in the UK. Man-Q happens to focus on an area which has not previously been addressed adequately in either the US or the UK. Consequently there was no relevant tool which could be imported for this purpose. In the context of the MANPRINT technical function there are tools available in the US which support this aspect, however we should be wary of simply adopting these wholesale as they may be based on systems of classification of the user and demographic trends which are inappropriate for the UK.

What is needed is a concerted effort to develop a suite of UK relevant MANPRINT tools. Such a set might include tools for training needs analysis based on the UK's population trends and skill base, workload assessment tools validated for the UK user population, and computer aided learning packages for the effective dissemination of UK standards and procedures for MANPRINT. Ideally this effort should be led and orchestrated by the MOD's MANPRINT office.

REFERENCES
DOD, 1989, The Manpower and Personnel Integration (MANPRINT) in the Materiel Acquisition
  Process, US Department of Defense Document N° AR602-2·

Hamilton, W. I., 1992, The Guide to MANPRINT Compliance, EPP Report N° HF/MP/922.0.

MOD, 1990, The MANPRINT Procedural Guide, DGTA & Org. MANPRINT, MOD October
  1990.

Keynote address

APPLYING ERGONOMICS IN INDUSTRY:
SOME LESSONS FROM THE MINING INDUSTRY

G.C. SIMPSON

Ergonomics Branch, TSRE British Coal Corporation
Ashby Rd. Burton on Trent
Staffs. DE15 0QD

It is somewhat surpirising given the industrial potential of ergonomics that there are still relatively few ergonomists directly employed in industry. The paper attempts to draw some general lessons for ergonomics application from the sucess achieved in recent years in the application of ergonomics in the UK mining industry.

INTRODUCTION

Even in the early 1990s, 40 years after Hwyel Murrell established the first UK industrial ergonomics department in the English Midlands there are still relatively few ergonomists directly employed in industry and even fewer departments of ergonomics in industry. This seems somewhat surprising for a discipline that offers, simultaneously, to influence health, safety and performance!

This situation clearly raises the question ...... Why has so little apparent progress been made?

The answer to such a question is bound by its very nature to be both complex and multi-faceted. Some of the possible contributory factors have been examined in previous papers by the present author (e.g. Simpson 1985, 1990), however one of the most crucial factors maybe that the very existence of (or the apparent need for) ergonomics is an implicit "threat" to other disciplines. For example if you need to invoke a "new" discipline to overcome the shortcomings of designers, managers, existing safety specialists, occupational physicians, etc. to improve safety/health/performance at work, then it can be inferred, implicitly at least, that such disciples have failed.

This is not a good starting point as a basis for collaboration.

The fact that an ergonomics function has been thriving within the British Coal Mining Industry for almost 20 years suggests that, in part at least, some success must have been achieved in overcoming industry's apparent reluctance to encompass ergonomics.

The purpose of this paper therefore is to review some aspects of the ergonomics work in UK coal mining in order to attempt to draw from that work some general principles which may be of value in a wider promotion of the utility of ergonomics in industry.

An approach has been developed which is essentially based on the need to provide a fast and efficient ergonomics service, enabling, where ever possible to leave the client self sufficient (or as nearly so as possible) in terms of the particular ergonomics issue which is being addressed. This is described in some detail below in relation to improving the ergonomic standards of the design of mobile machines used in the mining industry.

The same principles have however been applied with equal success to other ergonomic issues. These include the provision of practical advice on manual handling operations (e.g. ECSC 1990 & 1991, Simpson 1992) and on the design of auditory signals in noisy environments which also makes appropriate provision for the wearing of hearing defenders (e.g. Simpson & Coleman 1989, Coleman & Simpson 1993).

ERGONOMICS IN MINING MACHINE DESIGN

Coal mining is an industry where it appears that the very nature of the operation has been "designed" to generate ergonomic problems.

Whether you are concerned with driving roadways, cutting coal or supplying the production districts, mining equipment is, almost by definition, heavy duty. Despite this you cannot, if you are to remain remotely viable, cut bigger faces and roadways than the minimum you need - it is simply far too expensive. You immediately have, therefore, large, heavy duty and often multi-purpose equipment operating in confined space. In simple engineering terms therefore you naturally lean toward "the bigger the better" while in operational terms, "the smaller the better" is far more appealing.

When you add on the intrinsic safety requirements for use in a potentially methane rich environment, you need to accept a whole raft of design constraints which involve, among other considerations, the need to design engine compartments, power supplies etc. in flame proof enclosures, all of which tends to increase the size and design complexity of the equipment.

Frequently, these conflicting demands on size have led to the only area for compromise being the space available for the operator. This leads not only to "simple" problems of ensuring that you can get in/out, on/off easily and being remotely comfortable once you are in, but also more subtle problems in relation to sightlines and control position.

Examples of the typically cramped driving positions, restricted sightlines and occasionally bizarre control positions which can be found on mining equipment have been detailed in previous papers (e.g. Simpson 1990). In addition to the potential safety implications of poor design ergonomics, previous papers have also should that the same limitations can often have considerable production and cost implications (see, for example, Simpson 1989) and Simpson & Mason 1990).

Even a relatively cursory examination will clearly show that little consideration of ergonomic principles is normally applied to the design of mining equipment and that such a failure can have direct implications for both safety and performance standards. The question therefore arises ...

What can be done to ensure that the design process considers the operator
requirements more systematically?

IMPROVING MACHINE DESIGN ERGONOMICS

Initially there would seem a relatively straight forward solution i.e. ensure that the designers and manufacturers have the appropriate ergonomics information available to them. Unfortunately it is not quite that straight forward for four reasons:

1. Although often presented as such in text books, ergonomics data and recommendations are rarely context-free. For example there is no such thing as an intrinsically good or bad control, it only becomes good or bad dependant on the purpose and conditions of use - a tiller may be a good way to steer a sailing dinghy, but it would be a very poor way to steer a car.

2. Although a great deal of ergonomics information is now available, it is by no means complete in relation to the issues raised by the design of mining machinery.

3. Practical ergonomics is, of necessity, about compromise. It cannot be assumed that ergonomics requirements should always take preference over engineering requirements when the inevitable conflicts arise. Similarly it may not always be possible to achieve the ergonomics "ideal", as a result it is desirable that the designer is provided with a range of options for it will be better to improve the design rather than to take no action simply because the "ideal" cannot be met.

4. Designers already have a great deal of information and constraints within which they must work. This is especially true in relation to mining equipment where they have to consider additional problems such as the requirements to met intrinsic safety specifications etc. It is essential therefore, if the ergonomics information provided is to be easy to use, that only the information which is relevant to that particular design is presented.

ERGONOMICS DESIGN PRINCIPLES REPORTS

In order to provide ergonomics design recommendations which meet the above needs, the Ergonomics Branch of British Coal has developed a series of design handbooks, known as Ergonomics Principles Reports for the following range of machine families - Free Steered Vehicles(Mason & Simpson 1990a), Underground Locomotives (Mason & Simpson 1990b), Shearers (Mason & Rushworth 1991), Roadheaders (Mason & Chan 1991), Drill-Loaders (Mason & Simpson 1990c), Continuous Miners (Mason & Simpson 1990d). In addition a further Ergonomics Principles Report on Design for Maintainability (Mason et al 1986) has also been developed.

An example of the contents of these design handbooks is given below:

Preferred Working Postures	Overhead Clearances
Lateral Clearances	Seating
Lines of Sight	Access Corridor /Access Aids
Operator Reach Envelope	Operator Visual Envelope
Operator Overall Body Clearance	Design of Controls
Layout of Controls	Design of Visual Displays
Layout of Visual Displays	Labelling and Instructions
Machine Lighting	Thermal Environment
Auditory Environment	Vibratory Environment
Maintainability	Ancillary Task Aids
Machine Manuals	

In order to improve the presentation of information in the Principles Reports, the more recent versions have adopted a format of "data sheets" where each page provides the information on a particular design topic.

The Maintainability Principles Report includes a number of the areas where the required information was not readily available from the existing ergonomics literature and new information had to be developed. These issues included, for example, the forces which could be applied manually to fasteners in various configurations the effects of degree of access to fasteners on the ease of tightening or removal using a spanner. For example, in relation to fastener access, most texts and specifications state either that free (i.e. 360 degree) access should be allowed (the ideal) or that access should be adequate (without defining adequate!). Neither of these recommendations is particularly helpful to a designer. In order to overcome these problems a short series of experiments were conducted, the results of which are shown in Figure 1, which is one of the data sheets from the Maintainability Principles. It can be seen that performance is clearly limited below one flat access and that little improvement in performance is achieved beyond 3 flats access. There is therefore, in reality, little performance benefit to be gained from pursuing the "ideal" recommendation in this case (for further details see Mason et al 1985, Ferguson & Mason 1988).

These Ergonomic Design Principles Reports have now been issued by British Coal's Chief Engineer to all the UK based suppliers of primary mining equipment with the expectation that now the appropriate information has been made available to them, their new machines should show a progressive improvement in the ergonomic aspects of their design.

This is an important step, for when they were first issued it was without any overt recognition from senior positions in the industry. Although many people commented on how interesting and potentially useful they were they had, in reality, little impact.

This points to a major shortcoming in the education of ergonomists (at least in the past, in the UK). It was traditionally seen as the case that if an ergonomist wanted to improve machine/equipment design then the target for his "attack" should be the manufacturer's designers. While this may seem sensible in principle, it is, in reality, unlikely to have much impact. Designers work for commercial organisations and their breif is to design a product which will make a reasonable profit for their employers. This in its turn means that he must do the best job possible in relation to the design brief/specification given to him. If that specification does not include good ergonomics standards then he is highly unlikely to introduce them of his own volition for they will almost certainly increase cost and thereby reduce profit margin. In the "real world" the ergonomist's target if he wishes to improve machine/design standards is in fact the individual within the purchasing company who controls the specification process and who is therefore in a position to *require* ergonomic improvement.

Some improvements are now beginning to emerge, especially in relation to underground locomotives. A suite of 23 basic ergonomic design requirements for underground locomotives has been agreed between BCC and their major locomotive suppliers (see Simpson ct al 1993) and evidence of the use of the Principles Reports recommendations can be seen in new locomotive designs (see Simpson 1993 for further details).

In addition to announcing that the Corporation expected to see improvements in the ergonomic standards of new designs, the Chief Engineer also announced that the pre-purchase acceptance trails used by British Coal would now include an ergonomics appraisal in addition to the traditional mechanical and electrical engineering appraisals.

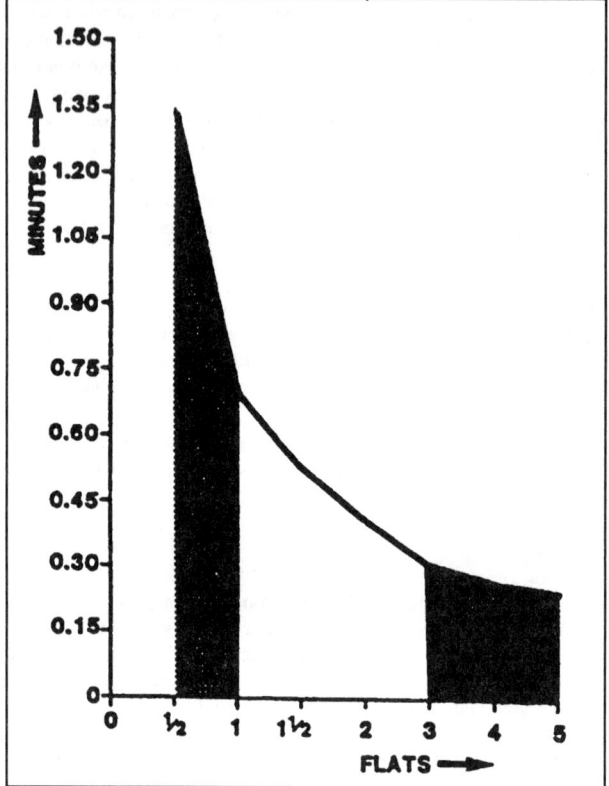

Figure 1: Spanner Access Results

ERGONOMIC DESIGN ASSESSMENTS

This new approach to improving ergonomics (as described above) although being a major step forward in getting ergonomic changes implemented, introduces new difficulties for both ergonomists and designers.

The ergonomist's report now plays a major role in whether a machine receives acceptance and it is essential therefore that the assessments must be consistent in both the issues which are considered and in the criteria which are used. Consistency in reporting format is also helpful if an organisation is to use ergonomics appraisals to compare and contrast machines. Reporting must allow a quick and accurate comparison to be made between different machines.

The same also applies to the feedback of results to the manufacturers. It is much more helpful to a designer if he is shown how his product compares with the 'opposition', albeit anonymously, than to simply list the good and bad design features as assessed by the ergonomist.

Designers have a difficult task when designing a mining machine. They are invariably faced with many design trade-offs if they are to produce a machine which will perform well and physically fit into the confined spaces underground while minimising costs.

It is important therefore that the designer has the ability to assess the ergonomics of his machine before it is submitted for evaluation. They should know in advance the detailed criteria with which the ergonomics of the total machine is going to be assessed. If all the criteria which are to be applied are available then the designer should be able to draw the same conclusions as the ergonomist. In part this is fulfilled by the Principles Reports as they provide the basis for the ergonomics criteria against which designs are evaluated. However in order to further ensure consistency of use and interpretation of criteria, two indices have been specially developed (for operability and maintainability) which together provide a total man-machine interface appraisal package.

The Operability Index: The operability of a complete machine will be dependant on the quality of the ergonomics of the following aspects of its design.

Sightlines	Workspace
Driver Protection	Access/Egress Facilities
Control and Display Location	Control Design
Display Design	Labels and Instructions
Seating	Machine Lighting
Thermal and Auditory Environment	Warning Systems

By rating the design quality of a machine against each ergonomic design feature, an overall evaluation score can be produced. However for a given type of machine, the importance of the various ergonomic features will differ. For example, sightline requirements are more critical to the safe and effective operation of a loco than are the labels and instructions.

The scoring method of the Index must therefore reflect (a) the *importance* of the relevant feature (achieved by weighting), and (b) the *quality* of the feature in the design being assessed.

The Operability Scores for all ergonomic features relevant to the safe and efficient operation of shearers are summarised in Table 1.

Table 1 Summary of operability scores for four modern shearers.
(The results have been converted to show the % of the maximum scores).

Ergonomic Feature	Machine A	Machine B	Machine C	Machine D
Sightlines	100	100	100	100
Control/Display Locn.	50	44	76	75
Control Design	59	59	64	64
Display Design	43	43	63	63
Labels & Instructions	90	88	97	97
% of Max Poss. Score	69%	67%	81%	81%

This table shows that all four shearers generally scored well. However the final scoring system can hide certain minimum requirements which are not met by balancing them against other areas of design where optimum ergonomics are achieved. It is therefore much better to look at the individual entries. Analysing the data in this way shows that all shearers had good sightlines and labelling. However the standard of control and display location varied across the machines - with A and B scoring very low and C and D being barely adequate against the ideal ergonomics requirements. Control design and particularly display designs were again only barely adequate for all the shearers.

It must be borne in mind that improved machine performances and reduced operating errors (and thereby, improved safety) are likely as the operability scores are increased.

A designer would use such data, especially when comparisons are made to other shearers, to identify where improvements are warranted. *A buyer* could use such data to identify where he required further improvements prior to purchase. Additionally he could also use such data to select the machine which would be easier for the operators to use which would also have the benefit of shorter training periods (and hence reduced training costs).

The Bretby Operability Index provides results which allow designers to identify individual features which would benefit from improved ergonomics and also to allow purchasers to compare different machines to select those with the better operability characteristics. It provides a 'fast track' means of identifying where a more detailed assessment by ergonomists would be beneficial in order to generate specific design recommendations. Finally it provides a mechanism for improving the consistency of ergonomic appraisals of machines by both ergonomists and non-specialists. For further details of the Operability Index see, for example, Mason & Simpson (1992).

The Bretby Maintainability Index: Maintenance in the mining industry, like most others, is expensive and often dangerous. Having produced the Maintainability Design Principles, it was apparent that much of our existing equipment left a great deal to be desired in terms of their 'maintenance friendliness'.

Although the Maintainability Principles Handbook gave useful information to designers, it was difficult for them to obtain an overview of the maintainability features of the design of any complete machine. In addition, as the Industry is likely to buy less new equipment, the question arose as to whether improvements to the existing fleet could be made by cost-effective retrofit solutions. Both factors raised the need to be able to assess maintainability of equipment in a quantifiable way.

The BMI was developed to satisfy these requirements (see Coleman et al 1990). Its basic elements are shown below:

> **Section A:**   **Access**
> Part 1: Hatches & Covers   Part 2: Apertures Part 3: Location
> **Section B:**   **Operations**
> Part 1: Removal & Replacement   Part 2: Slackening & Tightening
> Part 3: Carrying & Lifting   Part 4: Preparation
> Part 5: Fluid Compartment Checks Part 6: Component Checking
> Part 7: Lubrication   Part 8: Draining
> Part 9: Filling   Part 10: Cleaning
> Part 11: Adjustment   Part 12: Miscellaneous
> **Section C:**   **Additional Allowances**
> Percentage modifiers to take account of energy expenditure, posture, head room, visual demand, task requiring more than one man.
> **Section D:**   **Frequency Multiplier**
> Used to weight scores depending on whether job is done, for example, each shift, weekly, monthly etc.

The BMI score was designed to be translated into the man-hours per annum which are required to conduct each task on the maintenance schedule and, of course, the overall time requirement to perform all the operations on the maintenance schedule.

The BMI can be used iteritively during design to assess the ease of maintenance and to compare different design concepts. The overall results can equally be used in purchasing decisions by giving reasonable estimates for the total cost element in a life-cost calculation and when looking at availability times.

An example of its use as a purchasing aid can be seen from Table 2. The table shows the BMI scores, converted to man-hour scores per annum, for conducting routine oiling and greasing of three modern electric shearers. The results clearly show one shearer to be better than the others, however this shearer actually had the worst score for oiling. If a hybrid machine were to be developed using the best practices from the three machines, the results would show that this machine could have a further 60% reduction in time required for the current best machine!

Table 3 Comparison of oiling and greasing scores for three shearers.

TASK	Machine 1	Machine 2	Machine 3	IDEAL
Oiling	45	127	147	45
Greasing	322	140	20	20
TOTAL	367	268	167	65

In order to simplify the scoring procedures, especially when the BMI is used iteritively by designers, a PC version of the BMI has now been completed.

RETRO-FIT IMPROVEMENTS

Although the above services and procedures have primarily been aimed at new machines, the current state of the mining industry clearly cannot rely on improvements in ergonomics to be achieved solely through improvements in the design of new machines entering the industry. The Ergonomics Branch of British Coal has therefore developed procedures which can be used to identify a series of retrofit alterations which are both practical and effective at promoting safety for a wide range of mining machinery.

Such retrofits have been arrived at through applications of the above indices and through the application of checklists to identify the potential for unsafe situations.

For example, a maintenance schedule for a commonly used FSV was taken from the manufacturer's maintenance manual and the BMI applied to each task. Of the 32 routine maintenance tasks, the highest scoring 10 constituted 75% of the total BMI score and the top two constituted over 25% of the total. This information clearly shows those tasks where improvement would have the largest benefit to the overall maintainability of the total machine. The largest score was in fact overhauling the conditioner box. This contributed 13.6% of the total score as this relatively lengthy task was specified as having to be undertaken every 4 operating hours. The BMI was then used to identify the ergonomic features of each of the high scoring tasks which were the biggest contributors. For example, the need to check components (which accounted for nearly 21% of the total score) was the highest single contributor.

As a result of this type of analysis a small team of ergonomists and engineers were able to suggest a range of retro-fit modifications. Of these, four relatively straight forward modifications (listed in Table 3) were shown to potentially save around 35% of the total maintenance demand for this type of vehicle - a projected saving of about 170 man-hours per annum.

Table 3: Some retrofit suggestions for FSV maintainability

RETRO-FIT SUGGESTION	Man-hour saving per annum
Ducting the conditioner drain plug to the vehicle side - while still protected by the chassis	41
Reducing the numbers of joints in the exhaust pipes and flameproof components which require frequent inspection	42.5
Improving access to the exhaust flametrap fasteners	12.4
Use of remote greasing to lower pivot bearings	3.4

A current project is considering retrofit improvements to the driver's cabs of underground locomotives. Some of the changes which have been introduced so far include are given below.

The displays and gauges which were on the front of the machine above the windscreen have been moved down and angled in a similar position to the instrument panel of a car.

Moving the instruments enabled the height of the upper edge of the windscreen to be raised, thereby improving the visibility for taller drivers.

The instrument panel was not illuminated and therefore could not be seen by the driver (as he cannot use his cap-lamp whilst driving without causing reflections on the windscreen). Panel illumination has now been provided by running a fibre optic from the headlight.

The parking brake, which was a wheel placed between the seat and the bulkhead has been replaced by a simple lever in a more accessible position.

Some additional leg room has been achieved by extending the cab forwards, although this remains considerably less than would be considered ideal.

Further retrofit improvements are currently being costed.

The development of a standard suite of ergonomics recommendations, specifically tailored to the needs of the industry together with standard methodologies for assessment and the specification for improvement (both in new designs and retrofit) has led to an ergonomics service which is both quick and reliable. Standardisation has also made ergonomics information and procedures available to the none specialist.

SAFETY IMPROVEMENTS VIA HUMAN ERROR AUDITS

The most recent application of the approaches described above has been the development of a human error audit system designed to isolate the potential for human error and the best route for the reduction of human error potential. The procedure is based on the human error classifications of Reason (e.g. Reason 1990) and Rasmussen (e.g. Rasmussen 1987). The procedure allows for the identification of both active and latent failures and can be used as a means to target safety initiatives covering the ergonomics of design, training, attitudes to safety (both workforce and management), work organisation and the use of codes and rules.

An example of the specific errors identified in one aspect of an underground haulage system studied is given in Table 4 and a summary of the associated latent failures identified in the system is shown in Table 5.

Following the completion of the audit report for this system, three working parties consisting of mining, engineering, safety and ergonomics staff were established to identify solutions (in relation to specific errors) and new initiatives (in relation to latent failures).

Prior to the audit, the colliery concerned had an all-accident rate of **35.8/100,000 manshifts**

In the year following the audit and the subsequent actions described above the all-accident rate reduced to **8.4/100,000 manshifts**

This represents an 80% improvement in accident rate which took the colliery from the bottom of the 15 collieries in the Group safety league to the top, not only of the Group safety league but also of the British Coal Corporation safety league. The colliery had become the safest mine within British Coal.

Table 4: Summary of Specific Errors in the Loco Operation Category.

Potential Error	Error Type	Preferred Source of Action
Loco not returned for 24hr. service	Violation	Organisation/Management
Starting with parking brake on	Slip	Design
Driving loco with earth-tester warning light on	Violation	Design/Training
Drivers leaning out of the cab when travelling	Violation	Design/Training
Inadequate use of warning horns	Violation	Design
Misreading of displays	Slip	Design
Guards leaning out of the cab while travelling	Violation	Design/Training
Insufficient warning of objects/people on the track	Slip	Design
Inability to effectively use fire extinguishers	Mistake	Design
Incorrect operation of loco controls	Slip/Mistake	Design

Table 5: Common Latent Failures

Common Latent Failures (i.e. those which influenced more than 1 specific error)
Quality assurance in supplying companies
Supplies ordering
Loco design
Surface make-up of supplies
Lack of equipment
Training
Attitudes to safety
The safety inspection/reporting/action procedure.

In addition, during the same period, absenteeism (for whatever reason) had more than halved and the combination of reduced accidents and reduced time off work following an accident had not only significantly reduced costs but had also enabled "light duty" men to carry out work which would otherwise have involved the use of external contractors.

Full details of the audit procedure are given in Simpson et al (1992) and summarised in Simpson (1992).

A current project is developing training material to enable the human error audit procedure to be used routinely by colliery staff without ergonomics expertise thus enabling its potential to be fully capitalised rather than rely on the limited application possible if its use were restricted to a small team of specialist ergonomists.

CONCLUSIONS

If ergonomics is to establish a stronger base in the every-day life of industry it would seem apparent that it must more closely reflect the needs and requirements of industry as perceived by industry rather than as perceived by ergonomists. In short, there is a need for the development of *Industrially Sensitive Ergonomics*. There are four major tenets of industrially sensitive ergonomics:

1.There is no such thing as an ergonomics problem.

An industrial manager has enough problems already without someone with an unpronounceable job title telling him that he has more! Moreover, it is a managers job to solve problems not to collect them. If ergonomists are to help a manager, they need to offer him ergonomic solutions to the problems which he already has.

This is not at all difficult as there are indeed ergonomic elements in the solution of:

- health and safety problems,
- production problems,
- maintaining quality standards
- in the reduction of operating costs (reducing absenteeism, reducing operating errors, reducing maintenance downtime, reducing health and safety litigation costs etc.)
- in ensuring rapid return on investment (by for example, good standards of equipment design, quality training based on understanding human error potential etc.)

Although this distinction may seem somewhat semantic, it is in fact very significant for it changes the total emphasis of ergonomic studies from the definition of problems to the identification of solutions.

2.The time scales for ergonomic studies must reflect industry needs.

In the obvious sense this is almost a truism as anyone is risking a loss of credibility if they consistently take longer then is expected of them. However the point of importance is a little more subtle, what is being implied is that a great deal of industrial ergonomics has to be taken out of the "research time frame". It will only rarely be possible to satisfy yourself that you've "dotted all the i's and crossed all the t's" prior to offering the guidance requested. In an industrial context there has to be an element of pragmatism in the application of ergonomics.

"I don't know and the literature doesn't tell me, but give me 6 months and I'll have answer for you" has to be replaced with "I don't know and the literature doesn't tell me, but my best bet is ....". Obviously this can't always be the case, they are issues and problems where too little is known or what is known is so confused, or where the penalty of being wrong is so high that time to work reasonably thoroughly can be justified. It is an essential requirement however that the ergonomist should be aware of the need to justify and can do so. If ergonomics is to grow in industry the conclusion which every ergonomist should dread is...."more work is needed"!

3.The economics of the problems addressed cannot be ignored.

For many years ergonomists behaved (implicitly if not explicitly) as though their interest in the promotion of health and safety give them a form of "moral right" which took them beyond any concern with the economic implications of their work. This is totally untenable. The nonsense of such a position was encapsulated nicely in a conversation with a manager from the manufacturing sector who said:

"What is the point in my spending vast sums of money to create the healthiest and safest factory in the country, if the only people work in it are liquidators?"

While this is clearly an over statement it does indicate that changes, however desirable and for even the most laudable of motives, require investment. The investment potential of even the most profitable organisation is finite and there will always be competition for funds. Investment in ergonomics therefore needs to be justified and be seen to generate a return, not just in altruistic terms, but in financial terms.

4.An ergonomist cannot expect to have control over the implementation of his work.

This has two implications:

Firstly it will be exceedingly rare that ergonomics advice can be implemented in isolation of other considerations. For example while there may be highly desirable ergonomic reasons why a control should be positioned in a particular place, there may be sound engineering constraints which make such a position problematic. It is essential therefore to accept that some ergonomics criteria/recommendations, no matter how well documented, can, in some circumstances, be almost trivial.

Secondly, although ergonomics criteria and recommendations are often presented as though they are context free, they are in fact only very rarely so. For example, as stated earlier, there is no such thing as an intrinsically good or bad control - a control is only good or bad in the context of its intended use - a tiller may be a very good way to steer a dingy but it would be a very bad way to steer a car!

Both of these points and their related implications require that ergonomics data, criteria and recommendations must be presented in a way which ensures that they are readily understood and used by the non-ergonomist. Unfortunately, despite the fact that this problem was recognised many years ago it is still an all too common one. There is an ironic element to this issue ..... ergonomists have been leaders in the pursuit of "user friendly" equipment, yet it is clear that designers and purchasers are in clear need of some "user-friendly" ergonomics!

The potential contribution for ergonomics in industry remains considerable. In Europe in particular, this has been emphasised by the specific inclusion of the need to consider ergonomics in almost all of the recent EC Directives on health and safety at work. If however ergonomists are to capitalise on this potential then they must re-consider their interaction with industry. Industry can manage to "limp along" without ergonomics as it has done for many years. However, as ergonomics is by its very nature an applied discipline, it is doubtful whether it can continue to exist in any meaningful way for much longer without a closer, more routine involvement in the day-to-day operations in industry.

References.

Coleman, G. J., Simpson, G. C. 1993 Effective auditory warning signals: The Design Window Approach. Progress in Coal, Steel and Related Social Research. no. 14 Luxembourg: Commission of the European Communities.

Coleman, G. J., Mason, S., Rushworth, A, M., Simpson, G. C., Sims, M. T., 1990 Ergonomics guidelines for the design of mining machinery to facilitate maintenance. Burton-upon-Trent: British Coal Corporation Technical Dept. Report (Final Report on CEC contract no. 7249/12/011).

European Coal & Steel Community 1990, Guidelines for Manual Handling in the Coal Industry. (Luxembourg: Community Ergonomics Action Report no. 14 Series 3).

European Coal & Steel Community 1991, Guidelines for Manual Handling in the Steel Industry. (Luxembourg: Community Ergonomics Action Report no. 16 Series 3).

Ferguson, C. A., Mason, S., 1988 Design strategies for maximising human force capability when using spanners. Int. Jnl. Industrial Ergonomics 2 251-258.

Mason, S., Chan, W, L., 1991 Ergonomic design handbook for Roadheaders Burton-upon-Trent: British Coal Corporation TSRE Report TM/91/01.

Mason, S., Ferguson C. A., Pethick A. J., 1986 Ergonomic principles in designing for maintainability. ECSC Community Ergonomics Action Report no.8 Series 3. (Luxembourg: European Coal and Steel Community).

Mason, S. Ferguson, C.A., Golding, D., Morris L. A., 1985 Access and force requirements on spanner usage. In, Proceedings of IEA Congress Ergonomics International 1985 eds. I.D. Brown, R. Goldsmith, K. Coombes, M. Sinclair M (London: Taylor & Francis).

Mason, S., Rushworth A. M., 1991 Ergonomic design handbook for shearers. Burton-upon-Trent: British Coal Corporation TSRE Report TM/91/03.

Mason, S., Simpson G. C., 1990(a) Ergonomic Principles for the design of Free-Steered Vehicles. Burton-upon-Trent: British Coal Corporation Technical Dept. Report SSL/90/173.

Mason, S. Simpson, G. C., 1990(b) Ergonomic Principles for the design of underground Locomotives. Burton-upon-Trent: British Coal Corporation Tech Dept Report SSL/90/174

Mason, S., Simpson G. C., 1990(c) Ergonomic Principles for the design of combined Drilling and Loading machines. Burton-upon-Trent: British Coal Corporation Tech Dept Report SSL/90/165.

Mason, S., Simpson G. C., 1990(d) Ergonomic Principles for the design of Continuous Miners. Burton-upon-Trent: British Coal Corporation Tech Dept Report SSL/90/166.

Mason, S., Simpson, G. C., 1992 Ergonomic aspects in the design of integrated face control and monitoring systems. Eastwood: British Coal Corporation (Final Report on CEC contract no. 7249-11/055).

Rasmussen, J., 1987 The definition of human error and a taxonomy for technical system design. In, New Technology and Human Error. eds. J. Rasmussen, K. Duncan, J. Leplat J. (Chichester: John Wiley & Sons).

Reason, J. T., 1990 Human Error. (Cambridge: Cambridge University Press).

Simpson, G. C., 1985 Some requirements for improving the industrial utility of ergonomics. In, Proceedings of the IEA Congress Ergonomics International 1985 eds. I.D. Brown, R. Goldsmith, K. Coombes, M. Sinclair (London: Taylor & Francis).

Simpson, G. C., 1989 Is ergonomics cost-effective? In, Health Safety and Ergonomics. eds. A.S. Nicholson, J. E. Ridd J. E. (London: Butterworths) pp 154-170.

Simpson, G. C., 1990 Costs and benefits in occupational ergonomics. Ergonomics 33 261-268.

Simpson, G. C., 1992(a), The ECSC Ergonomics Programme's approach to evolving a corporate manual handling policy. Progress in Coal, Steel and Related Social Research. no. 13 Luxembourg: Commission of the European Communities.

Simpson, G. C., 1992 Human error and accidents: lessons from recent disasters. In: Proceedings of Inst. Mining Engineers Symposium Safety, Hygiene and Health in Mining. (Doncaster: Institution of Mining Engineers).

Simpson, G. C., 1993 Applying ergonomics in the mining workplace. In, Proceedings of MineSafe 93. (Perth: Western Australia Chamber of Mines & Energy).

Simpson, G. C., Coleman, G. J. 1988 The development of a procedure to ensure effective warning signals. The Mining Engineer May 511-514.

Simpson, G. C., Mason, S. 1990 Economic analysis in ergonomics. In Evaluation of Human Work eds. J.R. Wilson, E.N. Corlett EN (London: Taylor & Francis) pp 798-816.

Simpson, G. C., Mason, S., Rushworth, A. M., Talbot, C. F., 1992 Risk perception and hazard awareness as factors in safe and efficient working. Eastwood: British Coal Corporation (Final Report on CEC Contract no. 7249-12/067).

AUTHOR INDEX

Adams, L.A. 365
Allison, G. 410, 475
Anderson, M. 457
Ashby, M.C. 403

Baber, C. 152, 178, 463
Baker, C.C. 321
Balbo, S. 97
Bapat, A.R. 352
Barber, P. 52
Beagley, N.I. 142
Bekker, M.M. 79, 103
Biggin, K. 45
Birch, K. 303, 308
Blair-Ford, C.G. 64
Bloomfield 392
Bontoft, M. 148
Bremner, A. 422
Bridger, R.S. 365
Brown, W. 221
Brymer, E. 445
Buck, J. 392
Burnett, G. 397

Carthey, J. 234
Clarke, A.M. 410, 475, 478
Cobb, S.G. 64
Coleman, N. 272
Corlett, E.N. 370
Coulthard, S.E. 315
Coutaz, J. 97
Crawshaw, C.M. 253

Densmore, H. 58
Devine, M. 40
Duggan, A. 5, 358

Eckart, R. 346
Edwards, R.J. 28
Edworthy, J. 164

Fairclough, S.H. 403, 469
Fowler, N. 327

Gale, A.G. 64
Garbutt, G. 321
Gelderblom, G.J. 422
Genaidy, A. 346
Gibson, H. 202
Glendon, A.I. 228
Godin, M. 387
Goom, M. 34
Gosling, P. 52

Graham, C.M. 321
Graham, R. 463
Granda, T. 221
Gregg, V.H. 28, 370
Guangyan, L. 334
Guo, L. 346
Gyi, D.E. 291

Haisman, M.F. 5
Hamilton, W.I. 481
Haslam, R.A. 142
Haslegrave, C.M. 64, 297, 334
Hellier, E. 164
Herman, L. 215
Hewson, T.J. 475
Hirst, S. 469
Hollywell, P.D. 171
Hone, K. 153
Hoyes, T.W. 178
Hoonehout, H.C.M. 85

Jackson, J.A. 184
Jenkins, P.B. 208
Jenkins, V. 123
Johnson, G.I. 429, 435
Jordan, P.W. 416, 451

Kalick, J. 40
Kanis, H. 91, 117
Kerr, K.C. 416
Kirby, K. 45
Kirton, E. 315
Kirwan, B. 41, 266
Knowles, D.J. 17
Kolli, R. 135

Lamb, I. 247
Lee, E. 58
Lim, K Y. 69, 123
Long, J.B. 69, 123

MacLeod, I.S. 45
McFadyen, I.M. 303, 308
McGrath, E. 315
McLeod, R.W. 2
Malafi, T. 40
Marinissen, A.H. 117
Masih, A. 340
Megaw, E.D. 202, 457
Mills, S. 374
Modak, J.P. 352
Morris, N. 247
Mossel, W.P. 129
Nijhuis, H.B. 284

O'Hara, J.   221

Page, M.   190
Parkes, A.M.   397, 403
Parsons, K.C.   142, 196
Patton, J.F.   358
Plocher, T.   392
Pomfrett, S.M.   478
Porter, M.   190, 291

Rainford, J.   190
Rea, K.   234
Reed, J.   266
Reid, A.F.   2
Reilly, T.   303, 308, 315, 327, 441,
    445
Robertson, S.A.   381
Romans, J.   45

Scott, D.   260
Shek, Y.   52
Simpson, G.C.   490
Sinnerton, S.   303, 308
Skinner, S.J.   22
Smith, A.C.   291
Stanton, N.   159

Stoner, J.W.   292
Symington, L.E.   40

Temple, J.   441
Thelwell, P.J.   266
Thody, M.   28
Toms, M.   340
Townend, M.S.   445

Van Vianen, E.P.G.   429
Vermeeren, A.P.O.S.   79, 103,
    135
Vingelis, P.J.   221
Voute, C.C.C.   117

Walters, H.   481
Wayman, K.M.   11
Welch, D.   221, 240
Whalen, T.   58
Whistance, R.S.   365
Wilkinson, N.J.   297
Williams, D.I.   253
Wilson, T.   111, 387

Young, J.   253

Zwaga, H.J.G.   278

SUBJECT INDEX

Acceleration 381
Accident 171, 184, 190, 208, 240
Acoustic 164
ACR (Advanced Control Room) 221
AECB 215
Age 215
Aircrew 45
Air Traffic Control 284
Alarm 159, 164, 202, 266
Alarm rate 178
Alcohol 469
Amusements 184
Angle 321
A NOVA 11, 17, 52, 247
Anthropometry 142, 184, 291, 334
Anti-tank G.W. 17
Arm rest 291
Army 34
Arousal 28
Assessment 416
ASR 152, 463
ATC 178, 284
ATM 435
Audio 159, 164, 397, 429
Automated Teller Machine
   (ATM) 435
Automobile 392

Battlefield 22
Bicycle 352
Bill design 253
Biodegradable 40
Biostereometric 365
Blood alcohol 469
BNFL 215
Body fat 303
Body movement 346
Body weight 303
Braking 381
Breast cancer 64
BS 7179 457
Burns 196

CAD 184
Car 190, 397
Carphone 403
CCU 159
CEA 215
Chair 370
Chair-o-plane 184
Checklist 103, 457, 475
Chemical process 234

Chernobyl 215
Children 184, 190
Clock-radio 117
Coal 490
Coding 2, 278
Colour 2, 22, 278
Comfort 22
Command system 2
Communication 202, 284, 475, 478
Computer model 184
Consumer 253
Contrast 260
Controls 117, 374
Control-display 11
Control law 17
Control room 221, 240
Cooper Harper 397, 403
Coronary care 159
CRAMP 148
CRT 278

Data base 142, 148, 416
Data link 284
Demography 34
Dependency 171
Design 34, 64, 117, 123, 129, 202,
   240, 370, 422
Designer 129
Design process 79
Desk top publishing 111
DEW 22
Digital Map 52
Directive 90/270/EEC 457
Disaster 202
Discomfort 28, 291, 297
Display 2, 22, 85, 117, 266, 272,
   278, 397, 457
Dithering 381
Domestic energy 253
Domestic task 123, 196
Domestic workload 123
Driver 381, 387, 392, 397, 403, 469
Driving 381, 387, 392, 397, 403, 469
Dynamic lifting 303, 308

ECG 469
Effort 358
EH101 45
Elbow 291, 346
Electricity 253
Electricity industry 228
EMG 291

Empiracle method   103
EN29241   457
Energy   352, 358
Equipment   34
Error   17, 45, 52, 69, 91, 152, 171,
   228, 234, 266, 374, 435, 463
Error correction   153
Error rate   28, 215
European standard   196
Expert   416
Eye protection   22

Failure   171, 215
Fatigue   5, 291, 445
Feature   58, 451
Feedback   69, 135, 234, 253, 429,
   435
Fetch 'alarm'   266
Field of view   22
Field study   184
Finger   291
Fishing Vessel   374
Fitness   5
Flywheel motor   352
Focus group   451
Food   40
Fuel   253

Gas   253
Gas cooker   196
Gender   5
Glare   374
Goggles   22
Graphics package   111
Guided weapon   17
Guideline   85, 221, 410

Hand tremor   17
Hardware   111
Harness   190, 445
HCI   2, 69, 79, 97, 111, 123, 148,
   152, 221, 260, 340, 435
HDF   171
Health   34, 481
Health and Safety   272, 457
Heart rate   284, 308
Helicopter   45
Homeostasis   178
HRA   171
HRV   284
HSE   184

Human dependent failure   171
Human error   215, 228, 490

IAEA   215
IBC   410, 478
Icon   260
Impaired driver   469
Identikit   58
IDS   392
In-car   429
Incident   228, 416, 451
Industry   346, 490
Infant carrier   190
Inharmonicity   164
Injury   5, 190, 196, 208, 240, 291,
   297, 321, 327, 441
Inspection   64, 221
Interactive system   97
Interface design   79
Interview   79, 135, 202, 234, 451
Iowa   392
ISO   91
Isokinetic   321 327
Isometric lifting   303, 308

Joystick   11
Junction   381

Knowledge base   228
KTA   215

Lathe   352
Life cycle   234
Lighting   117, 374
Low back   346, 370
Lumbar   346, 370
LUSI   475

Machine   490
Machine behaviour   123
Maintenance   215, 234
Mammograph   64
MANPRINT   34, 40, 481
MAN-Q   481
Manual handling   208, 303, 308
Manufacture   297
Map   52
Measurement   91
Memory   247, 422, 463
Menu   142
Methane   490
Methodology   22, 234

Meter reading   253
MFTA   457
Micro-SAINT   45, 152
Military   5, 11, 17, 22, 28, 34, 40,
    45, 52, 481
Mimic display   266
Mining   490
Minkowski   58
Mission analysis   45
Misuse   190
MMI   416
Monitor   159, 164
Motion cues   392
Motorway   403
Mugshot   58
Multi-function   416
Multi-linear control law   17
Multi media   410
Multi viewer   64
Muscle strength   327, 441
Musculoskeletal   5, 291, 297, 321,
    327, 334

NASA-RTLX   397
NASA-TLX   52, 403
Navigation   397
Neck   346
NII   215
NIOSH   5
Noise   374, 463
Nuclear   178, 215, 234, 240
Nuclear power plant   221

OECD   215
Oil platform   202
OPERA   272
Operational sequence diagram   45, 435
OSD   45, 435
Oxygen cost   358

Packaging   40
Pain   196, 297
Palmar Sweat Index   340
Paper cup   40
Patient care   159
Performance   152, 422, 451
Personnel   34
Photofit   58
Photogrammetry   365
Physical training   5
PIPER-ALPHA   202
Plan Position Indicator (PPI)   2

Posture   64, 297, 334, 346, 365, 441
Power   240
PRA   171
PrEN563   196
Pre information   397
Pregnancy   303
Priority model   208
Processing capacity   247
Process control   266
Procurement   123
Product assessment   .129
Product evaluation   429
Prototype   79, 142, 429
PSI   340
Public inquiry   202

Quality assurance   481
QUEST   475
Questionnaire   129, 266, 297, 403,
    416, 451

RACE   410, 475, 478
Radiologist   64
Radiotelephony   284
Random variation   91
Rapid prototyping   85
Rations   40
Real time   247
Reaction time   11, 28, 52
Reclined seat   28
Redrunning   381
Relax alarm   291
Repetitive strain injury   291
Reproductive cycle   308
Restraint   184, 190
RHT   178
Risk   171, 178, 327, 457
Road running   358
Route guidance   397
Royal Mail   148
Royal Navy   208
RSI   291
Running   441

Safety   171, 190, 196, 202, 215, 221,
    234, 240, 387, 481, 490
Sailing   445
Scan   260
Schema   171
Screen   260, 272
Seating   28, 64, 184, 190, 370, 374
Semi-structured interview   79

Sewing machine 334
Ships 208
Shoulder 291, 346
Signal 381
Simulation 315, 392, 445
Sitting 28, 346
Ski 315
Skill 228, 315, 445
Sleepiness 28
Slips 208
Software 85, 103, 221, 272, 435
Spatial frequency 260, 435
Specification 69
Speech 463
Speech recognition 152
Sport 315, 327, 441, 445
SSCI 2
Standing 346, 365, 370
Stereophotogrammetry 365
Steven's power law 164
Strain 441
Strategy 135
Strength 5, 321, 327
Stress 5, 340, 346, 463
Structured notation 69
Subjective 334
Surface 196
Surface ship 2
Symbol 2, 52

Talocrural 321
Target detection 22
Task analysis 45, 97, 148, 240, 451
Task model 97
Task network 152
Taxonomy 97, 171
Telecommunications 475
Telephone 416, 422
Temperature 196
Three dimensional 365
Three Mile Island 215
Timeline analysis 45
Torque 321
Tracking task 11, 17
Traffic 381, 429
Training 5, 11, 17, 34, 272, 315, 481
Trapped driver 387
Treadmill 358
TV 429

UIPM 416
Urgency mapping 164

Usability 2, 117, 221, 416, 429, 435,
  475, 478
User 34, 40, 103, 123, 135, 148,
  184, 190, 422, 463
User interface 85, 97, 148, 451
US Navy 40
USNRC 215

VDT 457
VDU 52, 69, 159, 260, 266, 272,
  278, 457
Vehicle 28, 142, 381, 387, 403
Video conferencing 410, 478
Video recording 381
Video telephony 410
Vibration 374
Vigilance 28
Vision 22, 374
Visual 64, 392
Visual alarm 159
Visual display 11
VO_2 315, 358
Voice 284

Walking 358
Walkthrough method 103
Warning 164
Weapon system 11
Weaving 297
Wheelhouse 374
Windows 135, 142, 481
Witness 58
Woodturning 352
Word processor 135
Workload 45, 52, 123, 284, 358,
  397, 403
Workplace 340, 370
Workspace 374
Workstation 64, 272, 340
Wrist 291, 346
WYSIWYG 111

Yerkes-Dodson 28